Continued Fractions
Analytic Theory
and Applications

GIAN-CARLO ROTA, *Editor*
ENCYCLOPEDIA OF MATHEMATICS AND ITS APPLICATIONS

GIAN-CARLO ROTA, *Editor*
ENCYCLOPEDIA OF MATHEMATICS AND ITS APPLICATIONS

Other volumes in preparation

ENCYCLOPEDIA OF MATHEMATICS
and Its Applications

GIAN-CARLO ROTA, Editor
Department of Mathematics
Massachusetts Institute of Technology
Cambridge, Massachusetts

Editorial Board

GIAN-CARLO ROTA, *Editor*

ENCYCLOPEDIA OF MATHEMATICS AND ITS APPLICATIONS

Volume 11

Section: Analysis
Felix E. Browder, *Section Editor*

Continued Fractions
Analytic Theory
and Applications

William B. Jones and W. J. Thron

Department of Mathematics
University of Colorado, Boulder, Colorado

Foreword by

Felix E. Browder

University of Chicago

Introduction by

Peter Henrici

Eidgenössische Technische Hochschule, Zurich

1980

CAMBRIDGE UNIVERSITY PRESS

CAMBRIDGE UNIVERSITY PRESS
Cambridge, New York, Melbourne, Madrid, Cape Town, Singapore, São Paulo, Delhi

Cambridge University Press
The Edinburgh Building, Cambridge CB2 8RU, UK

Published in the United States of America by Cambridge University Press, New York

www.cambridge.org
Information on this title: www.cambridge.org/9780521302319

First published 1980 by Addison Wesley
First published by Cambridge University Press 1984
This digitally printed version 2008

A catalogue record for this publication is available from the British Library

Library of Congress Cataloguing in Publication data

Jones, William B   1931-
  Continued fractions.

  (Encyclopedia of mathematics and its applications; v. 11)
  Bibliography: p.
  Includes indexes.
  1.   Fractions, Continued.   I.   Thron, Wolfgang J.,
joint author.   II.   Title.   III.   Series.
QA295.J64   519.5   80-24255

ISBN 978-0-521-30231-9 hardback
ISBN 978-0-521-10152-3 paperback

To
Martha Jones
and
Ann Thron

Contents

ix

Editor's Statement

A large body of mathematics consists of facts that can be presented and described much like any other natural phenomenon. These facts, at times explicitly brought out as theorems, at other times concealed within a proof, make up most of the applications of mathematics, and are the most likely to survive changes of style and of interest.

This ENCYCLOPEDIA will attempt to present the factual body of all mathematics. Clarity of exposition, accessibility to the non-specialist, and a thorough bibliography are required of each author. Volumes will appear in no particular order, but will be organized into sections, each one comprising a recognizable branch of present-day mathematics. Numbers of volumes and sections will be reconsidered as times and needs change.

It is hoped that this enterprise will make mathematics more widely used where it is needed, and more accessible in fields in which it can be applied but where it has not yet penetrated because of insufficient information.

GIAN-CARLO ROTA

Foreword

Any commentary upon the present volume by Professors Jones and Thron on the analytic theory of continued fractions must begin with the remarkable fact that it is the first systematic treatment of the theory of continued fractions in book form for over two decades. As such, it supplants and updates to a large degree the well-known treatise by Oskar Perron which in its various editions dominated the exposition of this theory for over 50 years, a veritable monument of Germanic *Gründlichkeit*. The fact that Perron's book of 1957 (and its coadjutors by Wall in 1948 and Khovanskii in 1956) has had to wait so long for a successor (and indeed that Perron's book seems never to have been translated into English) raises some significant questions about the role of continued fractions as a focus of contemporary mathematical activity.

The study of finite continued fractions, i.e., expressions of the form

$$\cfrac{a_1}{b_1 + \cfrac{a_2}{b_2 + \cfrac{a_3}{b_3 + \cfrac{\cdots}{\cdots + \cfrac{a_n}{b_n}}}}}$$

(written more economically as

$$\frac{a_1}{b_1} + \frac{a_2}{b_2} + \cdots + \frac{a_n}{b_n}\Bigg)$$

began in its explicit form in the latter decades of the 16th century with a paper by Bombelli written when the concepts and notations of algebra were first being laid down in Italy and France. Such expressions play a natural role in connection with the iterated application of the Euclidean algorithm and some mathematical historians have claimed to have found similar usages in Hindu or even Greek mathematics. Infinite continued fractions were first considered by Lord Brouncker, first president of the Royal Society.

The earliest continued fractions had whole numbers as their entries and were applied to the rational approximation of various algebraic numbers and of π. The use of continued fractions as an important tool in number theory began with 17th century results of Schwenter, Huygens, and Wallis and came to maturity with the work of Euler in 1737 and the subsequent use of continued fractions as a number theoretic tool by Lagrange, Legendre, Gauss, Galois, and their successors. Continued fraction expansions involving functions of a complex variable rather than simply numbers were introduced by Euler and became an important tool in the approximation of special classes of analytic functions in the work of Euler, Lambert, and Lagrange. A particularly influential direction of study was the expansions in continued fractions introduced by Gauss in 1813 for ratios of hypergeometric functions, one of the earliest contexts in which orthogonal polynomials made their appearance.

The divergence in aim of the number theoretic and analytical applications of the formalism of continued fractions brought about a central bifurcation in the development of the theory in the 19th century. One of the two principal branches which was generated, the analytic theory of continued fractions, is the mathematical discipline which is exposed in detail in the present volume. Its central concern is the expansion and convergence theory of continued fractions whose terms are linear functions of a complex variable z. The importance of the study lies in the fact that the finite approximants which are generated are rational functions of z which approximate the function $f(z)$ being expanded in the sense of Hermite interpolation among rational functions whose degrees satisfy fixed bounds.

The central position of the study of continued fractions in the edifice of classical 19th century analysis is attested to by the long list of major analysts who treated problems in this area, including names like Laplace, Legendre, Jacobi, Eisenstein, Heine, Laguerre, Riemann, Stieltjes, Tchebycheff, Frobenius, and Poincaré. The conceptual consequences of these investigations were far-reaching, particularly in the work of Stieltjes where they led to such far-reaching developments as the study of the moment problem (with all its eventual impact on the development of twentieth-century functional analysis), the definition and study of the Stieltjes integral, the beginnings of the systematic study of convergence of sequences of holomorphic functions (the Stieltjes-Vitali theorem) and the application of the early machinery of spectral theory for self-adjoint operators in Hilbert space to the moment problem by Hilbert and his school. Asymptotic expansions seem to have found an early origin in the work of Poincaré and Stieltjes on the use of continued fraction expansions in connection with divergent series.

The particularly fruitful character of some of these lines of development of the theory of continued fractions in the context of the flowering of late nineteenth century classical analysis makes it somewhat ironic how the later development of the theory became so sharply separated from most of the major trends of twentieth-century mathematics. Number theorists continued to use continued fractions and to be interested in their properties. For analysts, on the other hand, even in the most classical areas, interest in continued fractions became a relative rarity. With few exceptions, one would find it difficult to think of a textbook on analytic functions of a complex variable among those in common use today, which gives any special attention or emphasis to this circle of techniques and problems. Like many other areas of earlier emphasis in mathematical development, the theory of continued fractions became a specialist area in which a limited circle of analysts continued to work with skill and energy to solve the technical problems of the field, extend concepts, and find new techniques.

What has caused an interesting and important reversal of such trends in this particular field has been a development of a type which we should find increasingly less surprising: the rapid and almost explosive involvement of techniques from this domain in the application of mathematics in the physical sciences. Beginning in the area of statistical mechanics and solid state physics (but with a rapid extension to other areas of theoretical physics), techniques elaborated by Frobenius and Padé in the late nineteenth century to obtain expansions of analytic functions in the complex plane in terms of rational functions given as approximants of continued fractions, have become a major computational tool under the general name of *Padé approximants*. The large-scale nature of the resulting computational enterprise has given a renewed impetus to the whole field of study, and in particular has made the present volume very timely. Professors Jones and Thron have given a succinct but complete development of the principal results and techniques of the field of analytical theory of continued fractions on a completely up-to-date basis, surveying not only the latest developments of the relatively complicated convergence theory of continued fractions but the domain of numerical applications as well. By so doing, they have earned the gratitude of analysts as well as applied mathematicians and numerical analysts. More significantly, perhaps, they have brought the material embodied in the relatively inaccessible periodical literature into a form where it is accessible to non-specialists in other fields of the mathematical sciences and to students.

Let me conclude these comments with some brief remarks on a more general theme. In his introduction, Professor Henrici (who himself has made a very significant contribution to bringing the technique of continued fractions into a more prominent position in numerical analysis) has put forward a view of the relative decline of the theory of continued fractions as an object of mathematical attention as being a consequence of the trend that he detects in the history of 19th and 20th century mathematics to de-emphasize *algorithmic* mathematics in favor of what Henrici calls *dialectical* mathematics. The use of the term *dialectical* in this context, though unusual is roughly equivalent to terms like *conceptual, systematic,* or *discursive*. (His *dialectic* is the Platonic one, not dialectic in the Aristotelian, Hegelian, or Marxist usage.) Even aside, however, from the issue of linguistic distinctions, Henrici's diagnosis deserves a thorough philosophical analysis. If one believes that conceptual emphasis (and in particular emphasis on logical rigor) in modern mathematics began with Gauss, it certainly did not lead in the latter's case to a disinterest in computation, algorithms, or even special problems. The rigorization of nineteenth century analysis took place at the same time and in the hands of the same group of classical analysts who carried through the development of the detailed analysis of special functions. Nor does an interest in algebra evidently preclude an involvement in the computational side of the analytic theory of semigroups, as the case of Frobenius amply shows.

The tendency toward a logical scholasticism in some circles of the contemporary mathematical world to which Henrici is addressing his analysis has a far more complex relation to mathematical history than these terms of opposition between *algorithmic* and *dialectical* allow. What seems certain is that the new phase of intensified involvement of mathematics in the sciences, natural and social, based in part on the role of electronic computers, promises a new revival of both algorithmic and conceptual mathematics based on the increasingly mathematical character of all spheres of human knowledge and practice in the world of the present and the future.

FELIX E. BROWDER
General Editor, Section on Analysis

Introduction

Mathematics derives much of its vitality from the fact that it has several faces, each face having its own sharply distinguished features. One face, which might be called the *dialectic* face of mathematics, is the face of a scholar, or even of a philosopher. It is the face which tells us whether theorems are true or false, and whether mathematical objects with specified properties do or do not *exist*. Dialectic mathematics is an intellectual game, played according to rules about which there is a high degree of consensus, and where progress can be measured sharply in terms of the generality of a result that has been achieved.

There is another, entirely different face of mathematics, which I like to call its *algorithmic* face. This is the face of an engineer. The algorithmic mathematician tells us how to *construct* the beautiful things of whose existence we are assured by the dialectic mathematician. The rules of the game of algorithmic mathematics, and in particular the significance attached to results in algorithmic mathematics, depend on the equipment that is available to carry out the required constructions.

Dialectic mathematics has experienced a continuous growth at least since the time of C. F. Gauss. Algorithmic mathematics, on the other hand, has stagnated from Euler's time until very recently, because no really new computing equipment came into existence; even the manually operated desk calculator did not significantly increase the speed of computation. It has been brought to light by H. H. Goldstine that Gauss in essence invented the fast-Fourier-transform algorithm. Nobody seems to have cared at the time because, whether fast or not, it was not really possible to meet the computational demands of the discrete Fourier transform except on an utterly trivial level.

All this has changed, of course, with the invention of the electronic computer. Accordingly, we have experienced a surge in the use of algorithmic mathematics, especially in engineering and in applied science, which is without parallel in the history of mathematics.

The theory of *continued fractions* may serve as an example of a branch of algorithmic mathematics which has bloomed in the time of Euler, but whose growth since then has been rather modest compared to typical areas of dialectic mathematics such as group theory or topology. To explore and to make use of the algorithmic potential of a continued fraction has become possible, on a large scale, only since, thanks to von Neumann, we have learned how to compute.

Just what is the algorithmic essence of a continued fraction? Let us recall the principle of *iteration*, one of the pillars of algorithmic mathematics. The word "iteration" stems from the latin root "iterare" which in the agriculturally oriented society of the old Romans meant "to plow once again." In mathematical iteration, what is always plowed once over is a given mathematical operation or function. Thus if f is a given function, we obtain an iteration sequence associated with f by choosing x_0 and generating the sequence $\{x_n\}$ by forming $x_{n+1}=f(x_n)$. If $\{a_n\}$ is a given sequence, we obtain a new sequence $\{s_n\}$ by letting $s_0=0$ and always adding a new a_n by forming $s_n=s_{n-1}+a_n$. That is, we iterate the translations $t_n(z):=z+a_n$. If we are lucky, the limit $s:=\lim s_n$ exists, and we have arrived at the concept of an infinite series. Infinite products are obtained similarly by iterating the rotations $t_n(z):=a_nz$. Nothing of interest is obtained by iterating the inversion $t(z):=1/z$. If, however, we iterate Moebius transformations of the form

$$t_n(z):=\frac{a_n}{b_n+z},$$

where $\{a_n\}$ and $\{b_n\}$ are given sequences—that is, if we let translations, inversions, and rotations alternate in the same fixed order—we are led to consider limits as $n\to\infty$ of expressions of the form

$$\cfrac{a_1}{b_1+\cfrac{a_2}{b_2+\cfrac{a_3}{b_3+\cfrac{\ddots}{{}+\cfrac{a_n}{b_n}}}}}$$

or, written in the space–saving notation adopted in this book,

$$\frac{a_1}{b_1}+\frac{a_2}{b_2}+\frac{a_3}{b_3}+\cdots+\frac{a_n}{b_n}. \tag{1}$$

A limit of this kind is called a continued fraction, and it is with such limits that this book is concerned.

It is evident from their very genesis that continued fractions are intrinsically more interesting objects than, say, infinite series or products. This becomes clear already when we consider the very elementary problem of evaluating the partial fractions (1) of a continued fraction. While there is not much to be said about the evaluation of the partial sums of an infinite series, several essentially different algorithms exist for the evaluation of the fractions (1) (see Section 2.1.4 of this book), each having its own advantages and disadvantages.

Another fascinating topic is the *convergence theory* of continued fractions. It should not be ignored that a considerable body of convergence theory also exists for infinite series. However, once we get beyond the elementary convergence tests which are always used, this theory quickly comes to rest in the recesses of academic irrelevance. Not so for continued fractions; here the convergence theory is much richer, and also much more difficult, as is already clear from the fact that continued fractions are not linear in their elements a_n and b_n; a totally new fraction is obtained when these numbers are all multiplied by one and the same constant. I am happy to report that the convergence theory of continued fractions is dealt with in an exemplary fashion in this volume, including some recent and important results and methods due to the authors.

The theory of infinite series becomes particularly relevant for the purposes of analysis when the terms of the series are allowed to depend on a parameter in certain standardized ways, as is the case, for instance, for power series or for Fourier series. Similarly, the theory of continued fractions derives much of its interest for the purposes of algorithmic or computational mathematics from situations where its elements depend on a parameter. Again, because of their richer structure, many possibilities exist to standardize this dependence in a meaningful way. This book, in addition to the classical C-fractions, also presents the more modern theories of g-fractions and of T-fractions, to name but a few.

Already the examples just quoted show that the theory of continued fractions is full of new and exciting developments. This is true of almost all parts of continued-fraction theory. We mention some additional examples.

1. One of the main reasons why continued fractions are so useful in computation is that they often provide representations for transcendental functions that are much more generally valid than the classical representation by, say, the power series. Thus, for instance, while the power series at $z=0$ of a meromorphic function represents that function only up to the nearest pole, continued-fraction representations exist for certain meromorphic functions (see Section 6.1) which represent that function everywhere in the complex plane except at the poles. Beautiful as these results are, they have, at the present, more the character of isolated gems than of concrete manifestations of an underlying general theory. It certainly would be desirable to see at least the beginnings of such a theory.

2. A famous application of continued fractions occurs in control theory. There it is often necessary to decide whether a given polynomial with real coefficients is *stable*, i.e., whether all of its zeros have negative real parts. This question can be answered, in a finite number of steps and without computing the zeros, as follows. For concreteness we consider the polynomial of degree 6,

$$p(x) = a_0 + a_1 x + a_2 x^2 + \cdots + a_6 x^6.$$

With the coefficients of p we form the rational function

$$r(x) = \frac{a_1 x + a_3 x^3 + a_5 x^5}{a_0 + a_2 x^2 + a_4 x^4 + a_6 x^6},$$

called its stability test function. By a standard algorithm the stability test function may be represented as a continued fraction,

$$r(x) = \frac{1}{b_1 x} + \frac{1}{b_2 x} + \cdots + \frac{1}{b_6 x},$$

and it may be shown that the given polynomial is stable if and only if all $b_i > 0$ in this representation. Modern control theory increasingly calls for the investigation of the stability of polynomials of several variables. It is to be hoped that algorithms similar in simplicity to the above will be invented to decide such multivariate stability questions.

3. Frequently in applied mathematics results are obtained in the form of asymptotic series, say in the form

$$f(x) \sim \frac{c_0}{x} + \frac{c_1}{x^2} + \frac{c_2}{x^3} + \cdots, \qquad x \to \infty,$$

where the series diverges for all x. If it is desired to evaluate $f(x)$ from that series, then this evaluation is not possible to arbitrary accuracy for any given value of x. An empirical approach which often works, however, is as follows: one converts the asymptotic series into a continued fraction of the form

$$\frac{a_0}{x} + \frac{a_1}{x} + \cdots + \frac{a_n}{x} + \cdots;$$

algorithms for performing this conversion are discussed in Section 7.1 of this book. One then finds that the continued fraction converges for all $x \neq 0$, and that it actually represents the function f one is looking for. In a strict mathematical sense, the validity of this technique has been established only for a very limited class of asymptotic series (namely for those series where $(-1)^n c_n$ is the nth moment of a certain mass distribution), but the method is much more widely used, especially by physicists. It certainly would be desirable to have more theoretical insight into these matters.

Of the many outstanding problems of current computational interest in continued fraction theory I have mentioned only a few. The authors of the present book are among the foremost exponents of their field, and the profundity of their knowledge and experience appears on almost every page of the book. It is my fervent wish that its appearance will bring with it a resurgence of interest in continued fraction theory, and will help to bring its outstanding problems closer to solution.

PETER HENRICI

Preface

An up-to-date exposition of the analytic theory of continued fractions has been long overdue. To remedy this is the intent of the present book. It deals with continued fractions in the complex domain, and places emphasis on applications and computational methods. All analytic functions have various expansions into continued fractions. Among those functions which have fairly simple expansions are many of the special functions of mathematical physics. Other applications deal with analytic continuation, location of zeros and singular points, stable polynomials, acceleration of convergence, summation of divergent series, asymptotic expansions, moment problems and birth-death processes.

The present volume is intended for mathematicians (pure and applied), theoretical physicists, chemists, and engineers. It is accessible to anyone familiar with the rudiments of complex analysis. We hope that it will be of interest to specialists in the theory of functions, approximation theory, and numerical analysis. Some of the material presented here has been developed for seminars given at the University of Colorado over a number of years. It also has been used in a seminar at the University of Trondheim.

The three most recent books on the analytic theory of continued fractions are those by Wall [1948], Perron [1957a] and Khovanskii [1963; the original Russian edition was published in 1956]. More recently Henrici [1977] has included an excellent chapter on continued fractions in the second volume of his treatise on *Applied and Computational Complex Analysis*. We owe much to the books of Perron and Wall, but since these books were written, many advances have been made in the subject. We have tried to incorporate the most significant of these in this volume. In addition we have stressed computation more than these two authors did and have directed our presentation more toward readers interested mainly in applications. Henrici did not intend to give an exhaustive account of the analytic theory of continued fractions. It should therefore not come as a surprise that our treatment of many topics is more detailed and comprehensive than his.

We present a systematic development which is, to a large extent, self-contained, even though for the sake of brevity, proofs of a number of theorems have been omitted. Proofs are included if they help to illuminate the meaning of a theorem or if they provide examples of general methods. For those theorems which are given without proof, bibliographical references are provided. Historical remarks and references are given throughout the text. The book ends with an extensive bibliography. In it we have placed particular emphasis on recent articles and papers concerned with applications. Many examples (some numerical) are distributed throughout the book. They are meant to illustrate methods as well as theory.

Two recent developments from outside have had a strong influence on the direction of research in continued fractions and thus on the selection of and emphasis on topics for this book. They are:

1. The discovery of Padé tables as an important tool in applications in the physical sciences (about 70 or 80 years after they had been introduced by Frobenius and Padé).
2. The advent of high-speed digital computers.

That there is indeed a great interest in Padé tables is shown by the recent publication of a number of books on the subject [Baker and Gammel, 1970; Baker, 1975; Gilewicz, 1978] and a bibliography [Brezinski, 1977] with more than 1000 references, as well as the fact that there have been five international conferences dedicated mainly to Padé tables and continued fractions. In their proceedings [Graves-Morris, 1973; Jones and Thron, 1974c; Cabannes, 1976; Saff and Varga, 1977; Wuytack, 1979] one finds applications to physics, chemistry and engineering in addition to strictly mathematical results.

We therefore considered it important to include a brief introduction to Padé tables and bring out the connection between Padé approximants and continued fractions. The main connection is that the entries of the Padé table can be realized as approximants of suitably chosen continued fractions. Questions of convergence of sequences of Padé approximants can thus in numerous instances be answered by means of known results on the convergence of continued fractions.

A consequence of the second development (computers) was a large increase in the potential for practical use of continued fractions. A great step forward in implementing this potential was taken by Rutishauser [1954a, b, c] when he introduced the quotient-difference (qd) algorithm for developing power series into continued fractions. The qd algorithm and other similar algorithms are treated in this book. The epsilon algorithm for computations involving Padé approximants was discovered by Wynn [1956].

To render continued fractions more useful in computation, it was desirable to know more about their speed of convergence as well as their numerical stability. The convergence theory which had been developed for purely theoretical purposes by Leighton, Wall, Scott and Thron, among others, proved to provide a good foundation for attacking these problems. This book contains both the convergence theory and its application to truncation-error analysis and computational stability. Another approach to truncation-error analysis was initiated by Henrici and Pfluger [1966]. It is concerned with best inclusion regions for Stieltjes fractions. This work has since been extended to other types of continued-fraction expansions and is treated in Chapter 8.

Among other areas of advance which are given space in this book are: moment problems and associated asymptotic expansions, birth and death processes, and the theory of three-term recurrence relations, where we have incorporated recent results of Gautschi [1969b] and as yet unpublished results of Henrici in our treatment (See Appendix B).

The chapter on convergence theory is by far the longest in the book. This is the case even though we have exercised a great deal of restraint in selecting its contents. Many older theorems as well as results treated adequately elsewhere (such as positive definite continued fractions, which are studied in detail by Wall [1948]) have been omitted. Unfortunately no simple proofs are known as yet for some of the most important convergence criteria, such as the general parabola theorem (Theorem 4.40). In addition much space is devoted to value-region results. These are of importance for convergence proofs, in the derivation of truncation-error bounds, and in the analysis of computational stability. They may also be of interest in other contexts, since it is rare to have value information for an infinite process.

Due to space and time limitations, the authors have found it necessary either to omit or to severely curtail the treatment of certain topics. Some of these are well covered in available books. The close connections between orthogonal polynomials, Gaussian quadratures and continued fractions have been discussed briefly in Sections 1.1.3 and 7.2.2, although perhaps a more extensive development would have been preferable (see, for example, Chihara [1978]). We have already mentioned the treatment of positive definite continued fractions taken up in [Wall, 1948]. The Ramanujan identities, a further treatment of limit periodic continued fractions and a number of other subjects can be found in [Perron, 1957a, b]. See also [Andrews, 1979] for a recent expository article on the identities of Ramanujan. Additional applications of continued fractions in problems of approximation theory can be found in [Khovanskii, 1963]. Although we have considered continued fractions whose elements lie in a normed field, we have not dealt with more general algebraic structures as has been done, for example by Wynn [1960, 1963, 1964], Fair [1972], Hayden [1974] and Roach [1974]. We have also omitted the Thiele continued fractions (see, for example, [Nörlund, 1924], [Wuytack, 1973] and [Claessens, 1976]).

We gratefully acknowledge the assistance we have received from many people in the preparation of this book. In particular, we have appreciated the excellent work of Janice Wilson, Susan LeCraft, Burt Rashbaum, and Martha Troetschel in typing the manuscript. We are grateful to Anne C. Jones for able assistance in computer programming and to Martha H. Jones for her patient care and critical eye in proofreading the typescript and proofs.

Richard Askey, Walter Gautschi, Peter Henrici, Arne Magnus, and Haakon Waadeland were kind enough to read parts or all of the manuscript and to make critical comments and suggestions. We value their help greatly.

Some of the work on the book was done while one of us was at the University of Kent and while both of us, though at different times, were at the University of Trondheim. We appreciate the stimulating environment that the Mathematics Institutes of these universities provided.

Finally, we are grateful to Gian-Carlo Rota for inviting us to contribute a volume on continued fractions to the *Encyclopedia of Mathematics and Its Applications* and to the staff of the Advanced Book Program of Addison-Wesley Publishing Company for their expert handling of the problems associated with getting this volume into print. We would like to single out our Publisher, Lore Henlein, for her persistent efforts to hold us to the deadlines that we ourselves had set.

WILLIAM B. JONES
W. J. THRON

SYMBOLS

1 Sets

$a \in A$	a is an element of the set A; a belongs to A
$A \subseteq B$	A is a subset of B
$A \subset B$	A is a proper subset of B
$A \cap B$	Intersection of sets A and B
$A \cup B$	Union of sets A and B
$c(A)$	Closure of the set A
$\text{Int}(A)$	Interior of the set A
∂A	Boundary of the set A
$F(A)$	$[F(a): a \in A]$: the set of all $F(a)$ such that $a \in A$ where F is a function defined on A
$\varnothing$	Null set
$B \sim A$	Complement of set A with respect to set B

2 Complex numbers

$\mathbb{C}$	Set of all complex numbers: finite complex plane
$\hat{\mathbb{C}}$	$\mathbb{C} \cup [\infty]$: the extended complex plane
$\mathbb{R}$	Set of all real numbers
$\text{Re}(z)$	Real part of z
$\text{Im}(z)$	Imaginary part of z
$\bar{z}$	Complex conjugate of z
$|z|$	Modulus or absolute value of z
$\arg z$	Argument of z
Domain	Open connected subset of $\mathbb{C}$ or $\hat{\mathbb{C}}$
Region	Domain together with all, part or none of its boundary
$N_z(d)$	$[w : |w - z| < d]$
Neighborhood of z_0	Open set containing z_0

3 Miscellaneous

$g_1 {\circ} g_2$	$g_1 {\circ} g_2(z) = g_1(g_2(z))$: the composition of functions g_1 and g_2
$[\![x]\!]$	Integral part of x for $x \in \mathbb{R}$ ($[\![x]\!] \leqslant x$)
$\text{Frac}(x)$	Fractional part of x for $x \in \mathbb{R}$ $[[\![x]\!] + \text{Frac}(x) = x]$
$[\![h]\!]$	Integral part of the rational function h

$f(z) \equiv g(z)$ $f(z) = g(z)$ for all z in the domain of definition of f and g

$f(z) \not\equiv g(z)$ $f(z) \neq g(z)$ for some z in the domain of definition of f and g

$f_n \sim g_n$ $\lim_{n \to \infty}(f_n/g_n) = 1$

$K \approx K^*$ K is equivalent to K^* for continued fractions K and K^*

$n \pmod m$ n modulo m

l.f.t.   Non-singular linear fractional transformation

f.L.s.   Formal Laurent series

f.p.s   Formal power series

iff   If and only if

∎   End of proof

Continued Fractions
Analytic Theory
and Applications

CHAPTER 1

Introduction

1.1 History

1.1.1 *Beginnings*

Even though the Greeks knew about the Euclidean algorithm, there is no evidence that they used it to form continued fractions.

The first known use of continued fractions is the approximate expression for $\sqrt{13}$

$$3 + \frac{4}{6} + \frac{4}{6}$$

given by R. Bombelli (ca. 1526–1573) in 1572. This is a special case of the formula

$$\sqrt{a^2+b} = a + \frac{b}{2a} + \frac{b}{2a} + \cdots. \qquad (1.1.1)$$

A second special case of (1.1.1) was given by P. Cataldi (1548–1626) in 1613. He had

$$\sqrt{18} = 4 \& 2 \over \displaystyle 8 \& 2 \over \displaystyle 8 \& 2 \over 8,$$

which he abbreviated as

$$4 \& \frac{2}{8.} \& \frac{2}{8.} \& \frac{2}{8}.$$

Cataldi also discussed the formula (1.1.1).

ENCYCLOPEDIA OF MATHEMATICS and Its Applications, Gian-Carlo Rota (ed.). Vol. 11: William B. Jones and W. J. Thron, Continued Fractions.   ISBN 0-201-13510-8

D. Schwenter in 1625 and C. Huygens (1629–1695) in a posthumous publication considered the approximants of finite regular continued fractions as a means of expressing large fractions approximately in terms of fractions involving smaller numbers. Thus Schwenter (but in very awkward notation) had

$$\frac{177}{233} = \frac{1}{1} + \frac{1}{3} + \frac{1}{6} + \frac{1}{4} + \frac{1}{2},$$

and Huygens found (in a problem concerned with the construction of cogwheels)

$$\frac{77708431}{2640858} = 29 + \frac{1}{2} + \frac{1}{2} + \frac{1}{1} + \frac{1}{5} + \frac{1}{1} + \frac{1}{4} + \cdots.$$

He was aware of the fact that the approximants are alternately greater and smaller than the number and that they provide a best rational approximation.

The first infinite continued-fraction expansion is due to Lord W. Brouncker (1620–1686), who was the first president of the Royal Society of London. Around 1659 he gave

$$\frac{4}{\pi} = 1 + \mathop{\mathbf{K}}_{n=1}^{\infty} \left(\frac{(2n-1)^2}{2} \right) \tag{1.1.2}$$

without proof. He probably derived it from the infinite-product formula for $\pi/2$ due to J. Wallis (1616–1703).

It was L. Euler (1707–1783), beginning in 1737, who gave a systematic development of continued fractions. In his work it became clear that continued fractions can be employed both in number theory and in analysis. In this book we shall be concerned almost exclusively with the analytic theory of continued fractions. Thus it may be useful to give here a very brief account of some of the major contributors to and some of the significant results in the number-theoretic part.

1.1.2   Number-Theoretic Results

Regular continued-fraction expansions (see also Section 2.1.2 below) of real irrational numbers $x > 0$ are of the form

$$b_0(x) + \frac{1}{b_1(x)} + \frac{1}{b_2(x)} + \cdots.$$

Here the $b_n(x)$ are defined by $b_n(x) = [\![x_n]\!]$, $n \geqslant 0$, where $x_0 = x$ and $x_n = 1/\mathrm{Frac}(x_{n-1})$, $n \geqslant 1$, in which $[\![x]\!]$ denotes the integral part and $\mathrm{Frac}(x)$

denotes the fractional part of x. It follows that all $b_n(x)$ are positive
integers.

Most of the number-theoretic applications rely on regular continued-
fraction expansions and their approximations to x. A regular continued
fraction $b_0(x) + \mathbf{K}(1/b_n(x))$ always converges to x. Thus there is no
convergence theory to worry about. It is the degree of approximation
which is provided by the nth approximant $p_n(x)/q_n(x)$ that is most
important.

As we already mentioned, the first examples of regular continued
fractions were given by Schwenter and Huygens. In 1685 Wallis computed
the first 35 $b_n(x)$ for $x = \pi$. All three authors appear to have been aware of
the fact that the approximants $p_n(x)/q_n(x)$ provide a best rational approximation to x in the sense that

$$|bx - a| \geqslant |q_n(x) x - p_n(x)|, \qquad n \geqslant 1, \tag{1.1.3}$$

provided a and b are integers relatively prime to each other and $0 < b <$
$q_n(x)$.

J. L. Lagrange (1736–1813) contributed many results to the theory of
regular continued fractions. He showed that quadratic irrational numbers
are exactly the numbers that have periodic expansions (from some n on).
The inequality

$$\left| x - \frac{p_n(x)}{q_n(x)} \right| \leqslant \frac{1}{[p_n(x)]^2 b_{n+1}(x)}, \qquad n \geqslant 1, \tag{1.1.4}$$

is also due to him, as is a solution of the Pell equation

$$u^2 - Dv^2 = 1, \qquad D \text{ a positive integer.} \tag{1.1.5}$$

The solutions are pairs $\langle p_n(\sqrt{D}), q_n(\sqrt{D}) \rangle$ for certain values of n.
A. Legendre (1752–1833) gave a complete solution of the problem. Partial
solutions had already been given by Euler. The equation is of interest in
part because it can be used to solve problems in additive number theory
such as the result:

Every prime number of the form $4n + 1$ is the sum of two squares.

This result was conjectured by P. Fermat (1601–1665) and first proved by
Euler. A proof based on continued fractions was given by C. F. Gauss
(1777–1855).

E. Galois (1811–1832), in his first published paper, investigated certain
periodic regular continued fractions. He determined the value of the dual
periodic regular continued fractions (see Section 3.3 below).

The first to prove that there exist transcendental (non-algebraic) numbers was J. Liouville (1809–1882). In 1851 he observed that algebraic numbers cannot be approximated too closely by rationals. He proved that if ξ is the solution of an irreducible polynomial equation, with integer coefficients, of degree n, then there exists a constant $0 < c < 1$ such that for all integers p and q

$$\left| \frac{p}{q} - \xi \right| > \frac{c}{q^n}, \qquad n \geqslant 1. \tag{1.1.6}$$

Using this result he was able to exhibit an infinite number of transcendental numbers. Among these are those x whose regular continued-fraction expansions satisfy the inequality

$$b_{n_h+1}(x) > \left[p_{n_h}(x) \right]^{n_h} \tag{1.1.7}$$

for some sequence $\{n_h\}$ of integers. That these numbers must be transcendental follows from Lagrange's estimate (1.1.4), which leads to

$$\left| \frac{p_{n_h}(x)}{q_{n_h}(x)} - x \right| < \frac{1}{\left[p_{n_h}(x) \right]^{n_h+2}},$$

which would contradict (1.1.6) if x were algebraic.

A later result, due to Hurwitz (1859–1919) [1891], is that

$$\left| x - \frac{p}{q} \right| < \frac{1}{q^2 \sqrt{5}} \tag{1.1.8}$$

always has an infinite number of rational solutions p/q. E. Borel (1871–1956) [1903] gave a simple proof of this by observing that among any three consecutive approximants of the regular continued-fraction expansion of x there is at least one which satisfies (1.1.8). Hurwitz also showed that $\sqrt{5}$ is the smallest value for which this result is true for all x.

A measure-theoretic flavor was added to the theory by Borel [1909] and F. Bernstein (1878–1956) [1912], who proved that for almost all x, $0 < x < 1$, the sequence $\{b_n(x)\}$ is unbounded. A. Khintchine (1894–1959) made further contributions in this direction (he called it the metric theory of continued fractions). We quote two of his results.

1. For almost all x

$$\limsup_{n \to \infty} \sqrt[n]{b_1(x) b_2(x) \cdots b_n(x)} < e^{e\sqrt{2\log 2}}.$$

[Khintchine, 1924].

2. There exists a constant γ, independent of x, such that for almost all x

$$\lim_{n \to \infty} \sqrt[n]{q_n(x)} = \gamma$$

[Khintchine, 1936].

1.1.3 *Analytic Theory*

Euler made important contributions to the analytic theory. He gave continued-fraction expansions (always without convergence considerations) of integrals and power series, including divergent ones. He also showed how Brouncker's expression for $4/\pi$ could be derived from either Wallis's product formula or the Gregory-Leibniz alternating series for $\pi/4$. Another of Euler's contributions was a solution of the Riccati differential equation in terms of continued fractions.

J. H. Lambert (1728–1777) expanded $\log(1+x)$, $\arctan x$ and $\tan x$ in continued fractions in 1768. His work is particularly noteworthy because it contains an adequate discussion of the convergence of the continued fraction to the function in question. Lagrange found expansions for $(1+x)^M$ and

$$\int_0^x \frac{dt}{1+t^n} .$$

In a paper published only in 1813 (well after his death), Euler found an expansion for

$$\log\left(\frac{1+x}{1-x}\right).$$

Since Euler, Lambert and Lagrange at different times were all members of the Berlin Academy, one wonders whether they ever discussed their work on continued fractions.

A method for obtaining approximate solutions of polynomial equations with numerical coefficients, using regular continued-fraction expansions, was worked out by Lagrange in 1769 and 1770 [Lagrange, 1867].

Besides applying continued fractions to number theory, Gauss [1813, 1814] employed them in analysis. In the study of hypergeometric series he generalized the earlier work of Euler, Lambert and Lagrange by giving continued-fraction expansions for ratios

$$\frac{F(a, b; c; z)}{F(a, b+1; c+1; z)}$$

of hypergeometric functions (see also Section 6.1.1). In a second paper on mechanical quadratures, that is, on the approximate evaluation of integrals, he considered

$$\int_{-1}^{+1} f(t)\,dt = \sum_{k=1}^{n} \gamma_n\big(x_k^{(n)}\big) f\big(x_k^{(n)}\big) + \text{error}. \qquad (1.1.9)$$

He showed that $\gamma_n(x)$ and $x_1^{(n)}, \ldots, x_n^{(n)}$ can be chosen, independent of f, so that equality holds in (1.1.9) for all polynomials $f(x)$ of degree not exceeding $2n-1$. To obtain this result he made use of

$$\int_{-1}^{+1} \frac{dt}{z+t} = \log \frac{z+1}{z-1}$$

$$= \frac{2}{z} - \frac{1/3}{z} - \frac{4/(3\cdot5)}{z} - \frac{9/(5\cdot7)}{z} - \cdots, \qquad (1.1.10)$$

which he had derived in his previous paper. It turns out that the function $\gamma_n(x)$ can be expressed in terms of the numerator $P_n(x)$ and the denominator $Q_n(x)$ of the nth approximant of the continued fraction (1.1.10). The numbers $x_1^{(n)}, \ldots, x_n^{(n)}$ are the zeros of the polynomial $Q_n(-x)$.

The sequence $\{Q_n(x)\}$ satisfies

$$\int_{-1}^{+1} Q_n(t) Q_m(t)\,dt = 0, \qquad m \neq n, \qquad (1.1.11)$$

that is, it is a sequence of orthogonal polynomials with respect to the weight function 1 and the interval $-1 \leqslant t \leqslant 1$. As was first observed by C. G. Jacobi (1804–1851) [1827], the $Q_n(x)$ are exactly the polynomials obtained by Legendre in 1784–1789 in connection with his investigations concerning the attraction of spheroids and the shape of planets. Jacobi [1826] had previously devoted a paper to Gauss's quadratures, deriving the result without using continued fractions. He relied instead on the formula (1.1.11). (According to Bell [1940] the name "orthogonal" was introduced only in 1833–1835 by R. Murphy.)

The nineteenth century proved to be a golden age for the analytic theory of continued fractions. Study of special functions as well as actual computational results (for example in quadratures) were still in the foreground, and it is here that continued-fraction techniques could be of use. Apparently many mathematicians were familiar with continued fractions, and a large number used them in their research and/or helped to develop the analytic theory.

Besides the expansions already mentioned, new continued-fraction expansions for special functions were found by Laplace, Legendre, Jacobi, Eisenstein, Schlömilch and Laguerre. Heine in 1846, 1847 worked on hypergeometric functions. The question of convergence of the continued fractions for the ratios of hypergeometric functions, which had been left open by Gauss, attracted the attention of Riemann and was satisfactorily disposed of by Thomé [1867].

Investigations into the problem of expanding arbitrary power series into continued fractions were begun by Stern [1832] and Heilermann [1846] and continued by G. Frobenius (1849–1917) [1881] and Stieltjes among others. They studied in particular regular C-fractions and associated continued fractions. Towards the end of the century Frobenius [1881] and H. Padé (1863–1953) [1892] developed an even more general scheme for expanding a formal power series $P(z)$ into rational functions. The resulting double-entry table is known as the Padé table of $P(z)$.

Even though he was active mainly in the twentieth century, this is probably the place to mention S. Ramanujan (1887–1920) "whose mastery of continued fractions was on the formal side at any rate, beyond that of any mathematician in the world" (G. H. Hardy in the Introduction to Ramanujan's *Collected Papers* [1927]). Ramanujan gave no proof of his formulas. The merit of having put them on a solid foundation belongs to G. N. Watson [1929, 1939, and elsewhere], Preece [1929, 1930] and Perron [1952, 1953, 1958a,b].

A problem that proved to be especially fruitful in stimulating research in continued fractions throughout the nineteenth century and into the twentieth was Gauss's mechanical quadrature. Four interrelated questions grew out of this problem. We shall state these in terms of Stieltjes integrals that were introduced by Stieltjes as a tool in the study of these problems. For this purpose we let $\psi(t)$ denote a (fixed) bounded, non-decreasing function. The four questions are then as follows:

1. To determine functions $\gamma_n(x)$ and constants $x_1^{(n)}, \ldots, x_n^{(n)}$ so that

$$\int_{-\infty}^{\infty} f(t)\, d\psi(t) = \sum_{k=1}^{n} \gamma_n\big(x_k^{(n)}\big) f\big(x_k^{(n)}\big) + \text{error},$$

with error $= 0$ if $f(t)$ is a polynomial of degree up to $2n-1$.

2. To express

$$\int_{-\infty}^{\infty} \frac{d\psi(t)}{z+t}$$

as a continued fraction and to explore its region of convergence.

3. To find a sequence $\{Q_n(x)\}$ of polynomials which is orthogonal with respect to the weight distributions $d\psi(t)$.

4. To expand "arbitrary" functions in terms of a sequence $\{Q_n(x)\}$ of orthogonal functions as

$$f(x) = \sum_{n=0}^{\infty} c_n Q_n(x)$$

and to study the convergence.

Contributions to one or more of these topics were made by many of the best analysts of the nineteenth century. Not all of them used continued fractions. Of those who did, Tchebycheff and Stieltjes were the most successful, but there are also important investigations by Christoffel, Rouché and Markoff, among others.

P. Tchebycheff (1821–1894) used continued fractions in more than twenty of his papers. The first of these was in 1854, the last in the year of his death. He was quite successful in obtaining results on all the problems mentioned above. Since Tchebycheff considered it unimportant to read the current mathematical literature, he was probably unaware of the fact that T. Stieltjes (1856–1894), beginning in 1884 and partly inspired by a paper of Tchebycheff [1858], was solving many of the problems that Tchebycheff was working on. It is ironic that one of Tchebycheff's maxims was that effort devoted to the study of the work of others detracted from the originality of one's own work. When both men died within a month of each other in 1894, Stieltjes had outdistanced Tchebycheff considerably, having (among other results) obtained continued-fraction expansions

$$\frac{a_1}{z} + \frac{a_2}{1} + \frac{a_3}{z} + \frac{a_4}{1} + \cdots, \qquad a_n > 0, \quad n \geqslant 1,$$

and full knowledge of their convergence behavior for integrals

$$\int_0^\infty \frac{d\psi(t)}{z+t}.$$

Stieltjes had been in poor health since 1890 and achieved these results by a last determined effort. His interest in these problems came not only from the quadrature problem but also from the problem of "summing" certain divergent series. By one of the coincidences so frequent in the history of mathematics, both Stieltjes (in his thesis [1886]) and H. Poincaré (1854–1912) [1866] made important contributions to this subject in the same year. Both were in Paris at the time, but they evidently did not know of each other's work. That the theory of asymptotic series, which they both studied, could make use of continued fractions had already been suggested by E. Laguerre (1834–1886) in 1879 and was known to C. Hermite (1822–1901). For asymptotic series Stieltjes used the term "semi-convergent," which had been in use at that time with a slightly more narrow meaning. Hermite was Stieltjes's protector and friend. They corresponded regularly from 1882 to 1894, and Hermite was one of the examiners on Stieltjes's thesis. The others were Darboux and Tisserand.

The theory of moments, proposed and established by Stieltjes, also answered some questions about asymptotic expansions. By determining a function $\psi(t)$ which was connected to a given sequence $\{c_n\}$ by

$$c_n = \int_0^\infty (-t)^n \, d\psi(t),$$

he was able not only to solve the moment problem but also to provide a function (in terms of a continued fraction) for which the series

$$\sum_{k=0}^\infty c_k z^{-k}$$

is an asymptotic expansion at ∞ (see Chapter 9).

Both F. Klein (1849–1925) and D. Hilbert (1862–1943) took an interest in the work of Stieltjes. Hilbert had actually met Stieltjes when he visited Paris in 1886, and sent him reprints of his publications. Hilbert's own interests overlapped those of Stieltjes, since expansion of functions in terms of systems of orthogonal functions plays an important role in the theory of integral equations.

E. B. Van Vleck (1863–1943) wrote his thesis under Klein at Göttingen in 1893 on the topic "Zur Kettenbruchentwicklung hyperelliptischer und ähnlicher Intergrale." Van Vleck continued to work on continued fractions for some time. Among his contributions are some of the basic convergence criteria [1901a, b, 1904]. Considerably later, after Van Vleck had become chairman of the mathematics department at the University of Wisconsin, H. S. Wall (1902–1971) became his student and wrote a Ph.D. thesis in 1927 "On the Padé approximants associated with the continued fraction and series of Stieltjes." Wall in turn interested W. Leighton in the subject. Between them they became the founders of an American school of continued fractions including W. T. Scott, W. J. Thron, M. Wetzel, E. Frank, R. E. Lane, E. P. Merkes, T. L. Hayden, W. B. Jones and A. Magnus among others.

Hilbert's students who wrote theses on continued fractions were O. Blumenthal (1876–1944) in 1898 and J. Grommer in 1914. Two other students of his, G. Hamel (1877–1954) and E. Hellinger (1883–1950), also made contributions to continued fractions.

Stieltjes's theory was extended from $0 \le t < \infty$ to $\infty < t < \infty$ by H. Hamburger (1889–1956) in a series of papers [1920, 1921]. Hamburger had studied both at Göttingen and at München (where he received his doctorate) and thus was familiar with the work on continued fractions that was done at those two centers.

Continued fractions arising in connection with the moment problem were studied in the 1920s and 1930s by J. Shohat (1886–1944). He came out of the St. Petersburg school of Tchebycheff and Markoff. Later some of his Ph.D. students at the University of Pennsylvania also worked in this area.

It was in the nineteenth century that careful investigations into the convergence behavior of infinite processes began. The first acceptable definition of convergence for a continued fraction is due to Seidel [1846]. Stern [1832] had earlier suggested that continued fractions oscillating between finite bounds should be considered to be convergent. Later [1848] he adopted Seidel's formulation. Seidel and Stern then proceeded to develop convergence and divergence criteria for continued fractions with real elements.

For continued fractions with complex elements the result of Worpitzky [1865]

$$\mathbf{K}(a_n/1) \text{ converges if } |a_n| \leqslant \tfrac{1}{4},\ n \geqslant 1,$$

appears to have been the first. Worpitzky's theorem was published in the annual program of the Friedrichs Gymnasium und Realschule in Berlin, and thus it is not surprising that it did not attract attention. His theorem was rediscovered by Pringsheim [1899] and Van Vleck [1901b]. It was only in 1905 that Worpitzky's article was brought to Van Vleck's attention [1905]. Apparently this article was Worpitzky's dissertation. It also contains a proof of the convergence of the Gauss continued fractions, which predates Thomé's result by two years.

The next important contributions were made by A. Pringsheim (1850–1941) and Van Vleck. In 1898 Pringsheim showed that

$$\mathbf{K}(a_n/b_n) \text{ converges if } |b_n| \geqslant |a_n| + 1, \qquad n \geqslant 1.$$

From this one can deduce the Worpitzky criterion as well as

$$\mathbf{K}(1/b_n) \text{ converges if } |b_n| \geqslant 2,\ n \geqslant 1.$$

A slightly weaker result, namely,

$$\mathbf{K}(1/b_n) \text{ converges if } |b_n| \geqslant 2 + \varepsilon,\ \varepsilon > 0,\ n \geqslant 1,$$

had been given already [1889] by S. Pincherle (1853–1936), an extremely prolific mathematician who made numerous other contributions to continued fraction theory. Among these is a result which relates the solutions of three-term recurrence relations to the convergence of a related continued fraction (see Section 5.3).

Van Vleck [1901a] proved that

$$\mathbf{K}(1/b_n) \text{ converges if } |\arg b_n| < \pi/2 - \varepsilon,\ \varepsilon > 0,\ n \geqslant 1,\ \text{and } \Sigma |b_n| = \infty.$$

Further additions to convergence theory, in particular the limit-periodic continued fractions, were made by Pringsheim in München, his student O. Perron (1880–1973), who also became a professor in München, and O. Szasz (1884–1952). Szasz spent a year in München before moving on to Frankfurt (where he became a colleague of Hellinger). Later he came to Cincinnati. Perron's substantial original contributions to the subject are outweighed by his scholarly work on continued fractions. The three editions of his book "Die Lehre von den Kettenbrüchen" [1913; 1929; 1954, 1957a] not only kept interest in the subject alive but also provided a model of expository writing.

Other historical references, mainly to more recent developments, will be found in the next section as well as in introductory sections of various chapters.

1.2 Overview of Contents

Several important infinite processes in analysis can be defined by means of the composition

$$S_n(w) = s_0 \circ s_1 \circ \cdots \circ s_n(w), \qquad n = 0, 1, 2, \ldots, \tag{1.2.1}$$

of linear fractional transformations (l.f.t.'s)

$$s_n(w) = \frac{a_n + c_n w}{b_n + d_n w}, \qquad n = 0, 1, 2, \cdots. \tag{1.2.2}$$

For example, we obtain

$$S_n(0) = \sum_{k=0}^{n} a_k$$

as the nth partial sum of an infinite series if in (1.2.2) we choose $c_k = b_k = 1$ and $d_k = 0$. Similarly we obtain the nth partial product

$$S_n(1) = \prod_{k=0}^{n} c_k$$

if we let $a_k = b_k = 0$ and $d_k = 1$. To get the nth approximant

$$S_n(0) = b_0^* + \frac{a_1}{b_1} + \frac{a_2}{b_2} + \cdots + \frac{a_n}{b_n}$$

of an infinite continued fraction, we take $c_k = 0$, $d_k = 1$ for $1 \leqslant k \leqslant n$ and $a_0 = b_0^*$, $c_0 = b_0 = 1$, $d_0 = 0$.

There are other ways of defining a continued fraction. Pringsheim and Perron, for example, preferred using the recurrence relations

$$\left. \begin{array}{l} A_n = b_n A_{n-1} + a_n A_{n-2} \\ B_n = b_n B_{n-1} + a_n B_{n-2} \end{array} \right\} \quad n > 1, \tag{1.2.3}$$

$$A_0 = b_0, \quad A_{-1} = 1, \quad B_0 = 1, \quad B_{-1} = 0.$$

With this approach the nth approximant is given as A_n/B_n. In our treatment we have emphasized the role of linear fractional transformations. This comes out strongly in (1) the derivation of basic properties such as equivalence transformations, contractions and extensions (Chapter 2), (2) the development of convergence theory (Chapters 3 and 4), (3) truncation error analysis (Chapter 8) and (4) the discussion of numerical stability in evaluating continued fractions (Chapter 10).

In the analytic theory it is convenient to distinguish the study of continued fractions whose elements a_n, b_n are complex constants from the study of those continued fractions whose elements are functions of one (or sometimes more than one) complex variable z. One of the most important topics in the first area is the study of the convergence behavior of such continued fractions.

For the second type of continued fraction there are at least two central topics:

1. The formal expansion of analytic functions $f(z)$ by means of continued fractions of the form

$$b_0(z) + \mathbf{K}\left(\frac{a_n(z)}{b_n(z)}\right). \tag{1.2.4}$$

2. The question for what values of z, if any, the continued fraction (1.2.4) converges to $f(z)$.

Since the difficulty of computing (1.2.4), as well as its convergence behavior, may vary depending on the restrictions imposed on the functions $a_n(z)$, $b_n(z)$, one finds it convenient to study a number of different types of continued-fraction expansions, such as C-fractions, regular C-fractions, associated continued fractions, J-fractions and general T-fractions among others (see Appendix A for the classification of many types of continued fractions).

Convergence theory is developed primarily in Chapters 3 and 4. Chapter 3 deals with periodic continued fractions. The convergence behavior of such continued fractions can be completely characterized (Theorem 3.1). However, due to their highly special character, periodic continued fractions are useful mainly to illustrate various kinds of convergence behavior and to serve as a basis for the derivation of necessary conditions for convergence. In 1828 and 1829 Galois proved a result for dual periodic continued fractions. In Theorems 3.4 and 3.5 we present, for the first time in a text, a generalization of Galois's result to arbitrary continued fractions and to sequences generated by arbitrary sequences $\{t_n\}$ of l.f.t.'s.

Most convergence criteria for continued fractions are of the convergence-region type (Chapter 4). This means that one tries to determine regions in the complex plane with the property that if the elements a_n, b_n

are contained in these regions, then the continued fraction $K(a_n/b_n)$ will converge. Sometimes an additional condition has to be imposed on the elements before one can conclude that the continued fraction will converge. It is of interest to note that in the cases of infinite series and products, no non-trivial convergence-region results exist, so that the convergence theory for continued fractions differs markedly from those for other infinite processes.

Convergence regions for continued fractions originated in the works of Worpitzky [1865], Pringsheim [1899] and Van Vleck [1901a,b]. A great deal of convergence theory has subsequently been developed. Among the more recent results, at least for continued fractions of the type $K(a_n/1)$, two types of convergence regions stand out: simple parabolic regions and twin convergence regions. The first parabolic region, with axis along the real axis, was found by Scott and Wall [1940a]. Paydon and Wall [1942] and Leighton and Thron [1942] added parabolic regions with axes along rays $te^{i\alpha}$, $|\alpha| < \pi$. Extensions, generalizations and new methods of proof have been given by Wall and Wetzel [1944a, b], Thron [1943, 1958, 1963] and Jones and Thron [1968]. In treating the important class of limit-periodic continued fractions $K(a_n/1)$ (with $\lim a_n = a$) we have introduced a new circular-convergence-neighborhood result (Theorem 4.45) which is derived from the parabola theorems. Twin convergence regions are pairs of regions $\langle E_1, E_2 \rangle$ in the complex plane such that $a_{2n-1} \in E_1$ and $a_{2n} \in E_2$, $n \geq 1$, insures the convergence of $K(a_n/1)$. Leighton and Wall [1936] proved that

$$E_1 = \left[w : |w| > \tfrac{25}{4} \right], \qquad E_2 = \left[w : |w| < \tfrac{1}{4} \right]$$

is a pair of twin convergence regions. This result has been greatly improved and extended, in particular by Lange and Thron [1960] and Lange [1966]. A number of other twin-convergence-region results are also given in Section 4.4.

The single most important application of these results is to continued fractions with variable elements. These applications are given in Section 4.5. Among the newer results we call particular attention to the cardioid theorem (Theorem 4.57) for regular continued fractions $K(a_n z/1)$, which follows from the parabola theorems. One of its most important corollaries is Theorem 4.58 for S-fractions, which was originally proved by Stieltjes [1894].

In close relationship with convergence theory is truncation-error analysis. By this is meant the error produced when a convergent continued fraction is replaced by one of its approximants. Sharp estimates of truncation error are of great importance in computational problems. It is therefore not surprising that most of the truncation-error results known today have been derived since the advent of electronic digital computers. Our treatment of truncation error is found in Chapter 8.

The Taylor series $\Sigma c_n z^n$ which represents a function $f(z)$ holomorphic in some neighborhood of $z=0$ is determined by the property that its nth partial sum

$$f_n(z) \sum_{k=0}^{n} c_k z^k \qquad (1.2.5)$$

satisfies the condition

$$f_n^{(j)}(0) = f^{(j)}(0), \qquad j = 0, 1, 2, \ldots, n. \qquad (1.2.6)$$

The equations (1.2.6) define a certain type of Hermite interpolation problem. The same conditions are used to determine continued-fraction expansions of functions as well as Padé approximants. For example the regular C-fraction expansion of $f(z)$, when it exists, has the property that for each n, the nth approximant

$$r_n(z) = b_0 + \frac{a_1 z}{1} + \frac{a_2 z}{1} + \cdots + \frac{a_n z}{1} \qquad (1.2.7)$$

satisfies the Hermite interpolation condition (1.2.6).

We note in passing the independence, in the sense described below, of the coefficients a_k in (1.2.7). In fact, to obtain the rational-function approximant which interpolates to $f(z)$ at $z=0$ in the derivatives of order $j = 0, 1, \ldots, n+1$, it suffices to attach one additional term to the finite continued fraction (1.2.7); the coefficients $a_1, \ldots, a_n$ remain unchanged. This property is one of the reasons why continued fractions provide a natural tool for rational approximation.

In the terminology which will be used here, the regular C-fraction $b_0 + \mathrm{K}(a_n z/1)$ is said to correspond to the function $f(z)$ if its nth approximants $r_n(z)$ satisfy (1.2.6) for all $n \geqslant 0$. A general theory of correspondence is developed in Section 5.1. Special types of corresponding continued-fraction expansions are discussed in Section 5.1 and in Chapter 7. A brief introduction to Padé approximants and their relation to continued fractions is given in Section 5.5.

A number of methods exist for obtaining continued fractions corresponding to a given function. One method, based on systems of three-term recurrence relations, originated with Gauss [1813] in his investigations of hypergeometric functions. The particular treatment of this approach described in Sections 5.1 and 5.2 is due to Jones and Thron [1979]. Many examples of continued-fraction representations of functions obtained in this way are given in Section 6.1. These include many of the special functions of mathematical physics. Another procedure for obtaining continued-fraction expansions, due to Gautschi [1967], is described in Section

6.1. His method involves minimal solutions of the three-term recurrence relations and is based on Pincherle's theorem [1894]. We give a generalization of Pincherle's theorem (Theorem 5.7) which applies to convergence not only in the usual metric of the complex plane, but also in the normed field L of formal Laurent series. Also included is a generalization of a closely related theorem of Auric [1907]. Examples of applications of Gautschi's method are given in Sections 5.3 and 6.2. Further results on minimal solutions of three-term recurrence relations are given in Appendix B.

Algorithms for obtaining continued-fraction expansions of formal power series are described in Chapter 7. Perhaps the most important of these is the quotient-difference (qd) algorithm introduced by Rutishauser [1954a] (Section 7.1.2). It provides an efficient computational procedure for calculating the regular C-fraction expansion from the coefficients of the given power series. The algorithm can also be used to compute zeros and poles of analytic functions and eigenvalues of matrices. Algorithms analogous to the qd algorithm have been developed by Bauer [1959, 1960, 1965] for g-fractions (Section 7.1.3), by McCabe [1975] and McCabe and Murphy [1976] for general T-fractions (Section 7.3.2), and by Claessens [1976] and Wuytack [1973] for Thiele fractions (not given here). Section 7.1.3 ends with a presentation of recent work of Arms and Edrei [1970] on continued fractions and Padé approximants associated with totally positive sequences and Polya frequency series (Theorem 7.12). An algorithm for obtaining J-fractions and associated continued-fraction expansions is due to Gragg (Section 7.2) although the basic ideas apparently go back to Tchebycheff [1858] and can be found in [Wall, 1948, Chapter 11]. In Section 7.2.2 we also discuss briefly the strong connection that exists between certain kinds of continued fractions and orthogonal polynomials. Stable polynomials are discussed in Section 7.4.

Moment problems and asymptotic expansions are dealt with in Chapter 9. We take up three moment problems that are closely related to continued fractions: the classical moment problems of Stieltjes [1894] and Hamburger [1920, 1921] and the strong Stieltjes moment problem recently introduced by Jones, Thron and Waadeland [1980]. Necessary and sufficient conditions are given for the existence of a solution as well as for its uniqueness in all three cases. For each problem we also discuss the associated integral representation, asymptotic expansions and continued fraction. The principal application of moment theory given here is to summability of certain divergent power series. Examples include the incomplete gamma function $\Gamma(a, z)$, the complementary error function $\mathrm{erfc}(z)$ and the gamma function $\Gamma(z)$.

Chapter 11 is concerned with an interesting new application of continued fractions to the study of birth-death processes. This work was begun by Murphy and O'Donohoe [1975] and has been extended by Jones and

Magnus [1977]. A continued-fraction representation is given for the expression

$$P_r(s) = L\{p_r(s)\},$$

where $p_r(t)$ denotes the probability that a population has size r at time t and where L is the Laplace transform. A procedure for computing $p_r(t)$, which employs the qd algorithm, is described, and a numerical example is given.

Finally in Chapter 12 we give some recent results on T-fractions and general T-fractions (mainly due to H. Waadeland) as well as brief accounts of theorems on the location of singular points of analytic functions and on univalency of functions represented by continued fractions.

CHAPTER 2

Elementary Properties of Continued Fractions

2.1 Preliminaries

2.1.1 Basic Definitions and Theorems

A *continued fraction* is an ordered pair $\langle\langle\{a_n\},\{b_n\}\rangle,\{f_n\}\rangle$, where $a_1, a_2,\ldots$ and $b_0, b_1, b_2,\ldots$ are complex numbers with all $a_n\neq 0$ and where $\{f_n\}$ is a sequence in the extended complex plane defined as follows:

$$f_n = S_n(0), \qquad n = 0, 1, 2, \ldots, \tag{2.1.1a}$$

where

$$S_0(w) = s_0(w); \qquad S_n(w) = S_{n-1}(s_n(w)), \quad n = 1, 2, 3, \ldots, \tag{2.1.1b}$$

and

$$s_0(w) = b_0 + w; \qquad s_n(w) = \frac{a_n}{b_n + w}, \quad n = 1, 2, 3, \ldots. \tag{2.1.1c}$$

The *continued-fraction algorithm* is the function $\mathbf{K}$ which assigns to each ordered pair $\langle\{a_n\},\{b_n\}\rangle$ the sequence $\{f_n\}$ defined by (2.1.1). The numbers a_n and b_n are called the nth *partial numerator* and *denominator* of the continued fraction, respectively. Sometimes they are simply called the *elements*; f_n is called the nth *approximant*. If $\{a_n\}$ and $\{b_n\}$ are infinite sequences, then $\langle\langle\{a_n\},\{b_n\}\rangle,\{f_n\}\rangle$ is called an *infinite* (or *non-terminating*) *continued fraction*. It is called a *finite* (or *terminating*) *continued fraction* if

ENCYCLOPEDIA OF MATHEMATICS and Its Applications, Gian-Carlo Rota (ed.).
Vol. 11: William B. Jones and W. J. Thron, Continued Fractions.   ISBN 0-201-13510-8

$\{a_n\}$ and $\{b_n\}$ have only a finite number of terms $a_1, a_2, \ldots, a_m$ and $b_0, b_1, \ldots, b_m$. Hereafter a continued fraction will be assumed to be infinite unless otherwise stated. The motivation for defining a continued fraction by means of the equations (2.1.1) is that (non-singular) linear fractional transformations (l.f.t.'s) play a basic role in the subsequent development of the subject.

It can be seen that the nth approximant is given by

$$f_n = b_0 + \cfrac{a_1}{b_1 + \cfrac{a_2}{b_2 + \cfrac{\ddots}{\quad + \cfrac{a_n}{b_n}}}} \qquad . \tag{2.1.2}$$

Thus a continued fraction $\langle\langle\{a_n\}, \{b_n\}\rangle, \{f_n\}\rangle$ is often denoted by the symbol

$$b_0 + \cfrac{a_1}{b_1 + \cfrac{a_2}{b_2 + \cfrac{a_3}{b_3 + \ddots}}} \qquad . \tag{2.1.3}$$

For convenience we shall generally denote a continued fraction $\langle\langle\{a_n\}, \{b_n\}\rangle, \{f_n\}\rangle$ by one of the symbols

$$b_0 + \frac{a_1}{b_1} + \frac{a_2}{b_2} + \frac{a_3}{b_3} + \cdots, \tag{2.1.4a}$$

$$b_0 + \mathop{\mathrm{K}}(a_n/b_n) \quad \text{or} \quad b_0 + \mathop{\mathrm{K}}_{n=1}^{\infty}(a_n/b_n). \tag{2.1.4b}$$

Similarly, the nth approximant $f_n = S_n(0)$ may be denoted by

$$f_n = b_0 + \frac{a_1}{b_1} + \frac{a_2}{b_2} + \cdots + \frac{a_n}{b_n}$$

or by

$$f_n = b_0 + \mathop{\mathrm{K}}_{j=1}^{n}(a_j/b_j),$$

instead of by (2.1.2). A continued fraction

$$b_0 + \frac{(-a_1)}{b_1} + \frac{(-a_2)}{b_2} + \frac{(-a_3)}{b_3} + \cdots$$

will sometimes be written as

$$b_0 - \frac{a_1}{b_1} - \frac{a_2}{b_2} - \frac{a_3}{b_3} - \cdots.$$

In addition to the symbols (2.1.3) and (2.1.4), one also finds in the literature the notation

$$b_0 + \frac{a_1|}{|b_1} + \frac{a_2|}{|b_2} + \frac{a_3|}{|b_3} + \cdots \qquad (2.1.5)$$

which was introduced by Pringsheim in 1898 (see [Cajori, 1929, p. 56]).

Determining which notation is to be preferred proved to be somewhat of a problem. The explicit writing out of the continued fraction as in (2.1.3) appeared to be too space consuming, and thus the use of an abbreviation is desirable. Over the years a number of symbols have been suggested and employed by various authors. Of these essentially two have survived. The one due to Pringsheim (2.1.5) may have the advantage of being more easily distinguished from the notation for a series than the abbreviation (2.1.4a) which we have adopted. The latter occurs in England as early as 1820 (see [Herschel, 1820] and [Cajori, 1929, p. 53–56]) and appears to be at present the most frequently used notation.

Quite often we use the symbol for continued fractions in (2.1.4b). It is analogous to Σa_n for series and to Πa_n for products. Whether the K is a Greek or a Latin capital letter does not matter. The choice of K probably comes from the German word *Kettenbruch*. The notation (2.1.4b) has the advantage of brevity. Even more important is that it brings out that the continued-fraction algorithm is a function K that maps a pair of sequences $\langle \{a_n\}, \{b_n\} \rangle$ to a sequence of approximants $\{f_n\}$.

A continued fraction is said to *converge* if its sequence of approximants $\{f_n\}$ converges to a point in the extended complex plane. When convergent, the *value* of the continued fraction is $\lim f_n$. By analogy with the theory of infinite series and products, the symbols (2.1.4) will sometimes be used to denote both the continued fraction and its value when that exists.

Corresponding to each continued fraction $b_0 + K(a_n/b_n)$, there are sequences of complex numbers $\{A_n\}, \{B_n\}$ defined by the system of second-order linear *difference equations*

$$A_{-1}=1, \quad A_0=b_0, \quad B_{-1}=0, \quad B_0=1, \tag{2.1.6a}$$

$$A_n=b_n A_{n-1}+a_n A_{n-2}, \qquad n=1,2,3,\ldots, \tag{2.1.6b}$$

$$B_n=b_n B_{n-1}+a_n B_{n-2}, \qquad n=1,2,3,\ldots. \tag{2.1.6c}$$

The numbers A_n, B_n are called the nth *numerator* and *denominator*, respectively. The significance of these numbers lies in part in

THEOREM 2.1. *If A_n, B_n and f_n denote the nth numerator, denominator and approximant of a continued fraction $b_0+\mathrm{K}(a_n/b_n)$, respectively, and if $\{S_n\}$ is the sequence of l.f.t.'s (2.1.1b), then*

$$S_n(w)=\frac{A_n+A_{n-1}w}{B_n+B_{n-1}w}, \qquad A_n B_{n-1}-A_{n-1}B_n\neq 0, \qquad n=0,1,2,\ldots, \tag{2.1.7}$$

$$f_n=S_n(0)=\frac{A_n}{B_n}, \qquad n=0,1,2,\ldots, \tag{2.1.8}$$

and

$$A_n B_{n-1}-B_n A_{n-1}=(-1)^{n-1}\prod_{k=1}^{n} a_k, \qquad n=1,2,3,\ldots. \tag{2.1.9}$$

Equations (2.1.9) will be called the *determinant formulas*. Note that A_n and B_n are not uniquely determined by (2.1.7) and (2.1.8) but that they are so determined by the difference equations (2.1.6).

THEOREM 2.2. *Let $\{A_n\},\{B_n\}$ be sequences of complex numbers such that*

$$A_{-1}=1, \quad A_0=b_0, \quad B_{-1}=0, \quad B_0=1, \tag{2.1.10a}$$

and

$$A_n B_{n-1}-A_{n-1}B_n\neq 0, \qquad n=0,1,2,\ldots. \tag{2.1.10b}$$

Then there exists a uniquely determined continued fraction $b_0+\mathrm{K}(a_n/b_n)$ with nth numerator A_n and denominator B_n for all n. Moreover,

$$b_0=A_0, \qquad a_1=A_1-A_0 B_1, \qquad b_1=B_1, \tag{2.1.11a}$$

$$a_n=\frac{A_{n-1}B_n-A_n B_{n-1}}{A_{n-1}B_{n-2}-A_{n-2}B_{n-1}}, \qquad b_n=\frac{A_n B_{n-2}-A_{n-2}B_n}{A_{n-1}B_{n-2}-A_{n-2}B_{n-1}},$$

$$n=2,3,4,\ldots. \tag{2.1.11b}$$

Occasionally, it will be useful to define continued fractions over a general field $\mathbb{F}$ (at the start of this section we took $\mathbb{F}=\mathbb{C}$). As in the case of

C, we adjoin to $\mathbb{F}$ an additional element, called infinity, and denoted by ∞. The set $\mathbb{F} \cup [\infty]$ will be denoted by $\hat{\mathbb{F}}$ and will be called the *extended field*. Arithmetic operations involving ∞ are defined as follows: For all $a, b \in \mathbb{F}$ with $a \neq 0$,

$$a \cdot \infty = \infty, \quad \frac{a}{\infty} = 0, \quad \frac{a}{0} = \infty, \text{ and } b + \infty = \infty. \tag{2.1.12}$$

A field $\mathbb{F}$ will be called a *normed field* if, for each element x in $\mathbb{F}$, there is defined a real number designated by $\|x\|$ with the following properties: For each $x, y \in \mathbb{F}$,

$$\|x\| \geqslant 0, \tag{2.1.13a}$$

$$\|x\| = 0 \quad \text{iff} \quad x = 0, \tag{2.1.13b}$$

$$\|xy\| \leqslant \|x\| \cdot \|y\| \qquad (\text{sometimes } \|xy\| = \|x\| \cdot \|y\| \text{ is required}), \tag{2.1.13c}$$

$$\|x+y\| \leqslant \|x\| + \|y\|. \tag{2.1.13d}$$

The number $\|x\|$ is called the *norm* of x. In $\mathbb{C}$ the norm is defined by

$$\|z\| = |z| \qquad \text{for} \quad z \in \mathbb{C}.$$

A sequence $\{x_n\}$ in $\hat{\mathbb{F}}$ is said to *converge to an element* $x \in \mathbb{F}$ if, for all n sufficiently large, $x_n \in \mathbb{F}$ and

$$\lim_{n \to \infty} \|x_n - x\| = 0.$$

A sequence $\{x_n\}$ in $\mathbb{F}$ is said to converge to ∞ if, for all n sufficiently large, $1/x_n \in \mathbb{F}$ and

$$\lim_{n \to \infty} \left\| \frac{1}{x_n} \right\| = 0.$$

If a sequence $\{x_n\}$ in $\hat{\mathbb{F}}$ converges to $x \in \hat{\mathbb{F}}$, this is designated by writing

$$\lim_{n \to \infty} x_n = x.$$

The following rules for limits can be verified: If $\{x_n\}$, $\{y_n\}$, $\{u_n\}$ are sequences in $\mathbb{F}$ converging to elements in $\mathbb{F}$ and if $\lim_{n \to \infty} u_n \neq 0$, then

$$\lim_{n \to \infty} (x_n + y_n) = \lim_{n \to \infty} x_n + \lim_{n \to \infty} y_n, \tag{2.1.14a}$$

$$\lim_{n \to \infty} (x_n \cdot y_n) = \left(\lim_{n \to \infty} x_n \right) \left(\lim_{n \to \infty} y_n \right), \tag{2.1.14b}$$

$$\lim_{n \to \infty} \frac{1}{u_n} = \frac{1}{\lim_{n \to \infty} u_n}. \tag{2.1.14c}$$

A *continued fraction over a normed field* F can then be defined in a manner completely analogous to the definition of a continued fraction given at the start of this section. We note that the elements a_n and b_n belong to F and the approximants f_n belong to the extended field $\hat{F}$. Theorems 2.1 and 2.2, as well as many other properties of continued fractions, also hold for continued fractions over a normed field $\hat{F}$. Hereafter when we speak of a continued fraction over a normed field F, other than C, we shall always specify the field F. If no field is specified, it is understood that we are in the field of complex numbers. Our main application of continued fractions over general normed fields comes in Chapter 5 when we consider the field $\mathcal{L}$ of formal Laurent series.

Before proving Theorems 2.1 and 2.2, we first consider some elementary examples of continued fractions and some methods for computing approximants of continued fractions.

2.1.2   *Regular Continued Fractions*

The *regular continued-fraction expansion*

$$b_0 + \frac{1}{b_1} + \frac{1}{b_2} + \frac{1}{b_3} + \cdots \tag{2.1.15a}$$

of a positive real number x is defined by

$$b_n = [\![x_n]\!], \qquad n = 0, 1, 2, \ldots, \tag{2.1.15b}$$

where

$$x_0 = x; \qquad x_n = \frac{1}{\text{Frac}(x_{n-1})}, \qquad n = 1, 2, 3, \ldots. \tag{2.1.15c}$$

Here $[\![x_n]\!]$ denotes the integral part of x_n, and $\text{Frac}(x_n)$ denotes the fractional part of x_n. If $\text{Frac}(x_{n-1}) = 0$, the regular continued fraction expansion terminates with b_{n-1}. The following is a summary of the convergence properties of these continued fractions (for proofs, see Perron [1954, pp. 23–35]), where B_n and f_n denote the nth denominator and approximant, respectively, of (2.1.15a):

(A) $f_{2n-2} < f_{2n} < x < f_{2n+1} < f_{2n-1}$, $n \geqslant 1$.
(B) $x = \lim_{n\to\infty} f_n$.
(C) Every positive irrational number x has an infinite regular continued-fraction expansion which converges to it. Every infinite regular continued

fraction (2.1.15a), where all b_n are positive integers, converges to an irrational number x of which it is the regular continued-fraction expansion.

(D) A positive rational number has a finite regular continued-fraction expansion and conversely.

(E) If $|r/s - x| < |f_n - x|$, r and s being integers, $r > 0$, $n > 1$, then $s > B_n$.

The following are regular continued-fraction expansions of some special numbers:

$$\pi = 3 + \frac{1}{7} + \frac{1}{15} + \frac{1}{1} + \frac{1}{292} + \frac{1}{1} + \frac{1}{1} + \frac{1}{1}$$

$$+ \frac{1}{2} + \frac{1}{1} + \frac{1}{3} + \frac{1}{1} + \frac{1}{14} + \cdots, \qquad (2.1.16)$$

$$e = 2 + \frac{1}{1} + \frac{1}{2} + \frac{1}{1} + \frac{1}{1} + \frac{1}{4} + \frac{1}{1} + \frac{1}{1} + \frac{1}{6} + \frac{1}{1} + \frac{1}{1} + \frac{1}{8} + \cdots, \qquad (2.1.17)$$

$$\frac{\sqrt{5} - 1}{2} = \frac{1}{1} + \frac{1}{1} + \frac{1}{1} + \cdots \qquad \text{(golden ratio)}. \qquad (2.1.18)$$

For π a general formula for the b_n is not known. However, Lehmer has computed the b_n for π to b_{100} (see Perron [1954, pp. 33–35]). For e there is a regular pattern in the b_n that can be seen in (2.1.17). For the golden ratio $(\sqrt{5} - 1)/2$ the b_n are all equal to 1. The remarkable speed of convergence of the continued fraction for π, (2.1.16), is illustrated by Table 2.1.

The approximation of e is illustrated by Table 2.2.

Table 2.1. Approximation of π by its Regular Continued-Fraction Expansion (Here $\bar{\pi} = 3.141592654$ is the approximate value of π rounded in the 9th decimal place and $\bar{f}_n$ is the nth approximant computed with floating-decimal arithmetic using 10 decimal digits.)

n	A_n	B_n	$\bar{f}_n$	$\bar{\pi} - \bar{f}_n$
0	3	1	3.0	0.141592654
1	22	7	3.142857143	−0.001264489
2	333	106	3.141509434	0.000083220
3	355	113	3.141592920	−0.000000266
4	103,993	33,102	3.141592654	0.000000000

Table 2.2. Approximation of e by its Regular Continued-Fraction Expansion $\bar{e} = 2.718281828$ is the approximate value of e rounded in the 9th decimal place. $\bar{f}_n$ is the nth approximant computed with 10-decimal-digit floating arithmetic.

n	A_n	B_n	$\bar{f}_n$	$\bar{e} - \bar{f}_n$
0	2	1	2.0	0.718281828
1	3	1	3.0	−0.281718172
2	8	3	2.666666667	0.051615161
3	11	4	2.750000000	−0.031718172
4	19	7	2.714285714	0.003996114
5	87	32	2.718750000	−0.000468172
6	106	39	2.717948718	0.000333110
7	193	71	2.718309859	−0.000028031
8	1264	465	2.718279570	0.000002258
9	1457	536	2.718283582	−0.000001754
10	2721	1001	2.718281718	0.000000110

2.1.3 Other Continued-Fraction Expansions

Although regular continued fractions have interesting properties, they do not always provide a practical method for determining a number, since the application of the algorithm (2.1.15) requires an *a priori* knowlege of x. Our interest here is mainly in continued-fraction expansions of analytic functions that can be used to represent the functions and compute their values. The following are examples of such expansions:

$$\arctan z = \frac{z}{1} + \frac{z^2}{3} + \frac{4z^2}{5} + \frac{9z^2}{7} + \cdots, \tag{2.1.19}$$

$$\log(1+z) = \frac{z}{1} + \frac{z}{2} + \frac{z}{3} + \frac{4z}{4} + \frac{4z}{5} + \frac{9z}{6} + \frac{9z}{7} + \cdots, \tag{2.1.20}$$

$$e^z = \frac{1}{1} - \frac{z}{1} + \frac{z}{2} - \frac{z}{3} + \frac{z}{2} - \frac{z}{5} + \frac{z}{2} - \cdots. \tag{2.1.21}$$

These and other expansions will be developed in Chapters 5, 6, 7 and 9. For now it will suffice to make the following remarks: (2.1.19) converges throughout the z-plane cut along the imaginary axis from i to $+i\infty$ and from $-i$ to $-i\infty$. (2.1.20) converges throughout the z-plane cut along the negative real axis from -1 to $-\infty$. (2.1.21) converges in the entire finite complex plane. In each of these examples the general pattern of the elements of the continued fraction is apparent. Our interest is in methods for obtaining such expansions and for proving that they converge to the given functions. By setting $z = 1$ in (2.1.19) we obtain an expansion for π

$$\pi = \frac{4}{1} + \frac{1^2}{3} + \frac{2^2}{5} + \frac{3^2}{7} + \frac{4^2}{9} + \cdots, \qquad (2.1.22)$$

for which the elements can be expressed in closed form. It can be seen from Table 2.3 that the convergence of (2.1.22) is not as fast as that for the regular continued fraction (2.1.16).

Table 2.3. Approximation of π by the Continued Fraction (2.1.22) ($\bar{\pi} = 3.141592654$ is the approximate value of π rounded in the 9th decimal place. $\bar{f}_n$ is the nth approximant computed with 10-decimal-digit floating arithmetic.)

n	A_n	B_n	$\bar{f}_n$	$\bar{\pi} - \bar{f}_n$
0	0	1	0	3.141592654
1	4	1	4.0	-0.858407346
2	12	4	3.0	0.141592653
3	76	24	3.166666667	-0.025074013
4	640	204	3.137254902	0.004337752
5	6976	2220	3.142342342	-0.000749688
6	92736	29520	3.141463413	0.000129241

On the other hand, setting $z = 1$ in (2.1.21), we obtain the expansion

$$e = \frac{1}{1} - \frac{1}{1} + \frac{1}{2} - \frac{1}{3} + \frac{1}{2} - \frac{1}{5} + \frac{1}{2} - \cdots, \qquad (2.1.23)$$

which converges at about the same speed as the regular continued-fraction expansion (2.1.17). This is illustrated by Table 2.4.

Table 2.4. Approximation of e by the Continued Fraction (2.1.23) ($\bar{e} = 2.718281828$ is the approximate value of e rounded in the 9th decimal place. $\bar{f}_n$ is the nth approximant computed with 10-decimal-digit floating arithmetic.)

n	A_n	B_n	$\bar{f}_n$	$\bar{e} - \bar{f}_n$
0	0	1	0.0	-2.718281828
1	1	1	1.0	1.718281828
2	1	0	∞	$-\infty$
3	3	1	3	-0.281718172
4	8	3	2.666666667	0.051615161
5	19	7	2.714285715	0.003996113
6	87	32	2.718750000	-0.000468172
7	193	71	2.718309860	-0.000028032
8	1264	465	2.718279569	0.000002259
9	2721	1001	2.718281719	0.000000109
10	23225	8544	2.718281834	-0.000000006

2.1.4   *Algorithms for Computing Approximants*

A number of algorithms are available for computing the nth approximant f_n of a continued fraction $b_0 + \mathrm{K}(a_n/b_n)$. The *forward recurrence algorithm* (FR algorithm) consists of an application of the difference equations (2.1.6) to obtain $f_n = A_n/B_n$. In this process $4n+1$ operations of multiplication or division are required to obtain f_n. However, one also obtains $A_1, A_2, \ldots, A_n$ and $B_1, B_2, \ldots, B_n$, and hence the approximants $f_1, f_2, \ldots, f_{n-1}$ can be computed with only $n-1$ additional divisions.

The *backward recurrence algorithm* (BR algorithm) consists of the following: Set

$$G^{(n)}_{n+1} = 0, \tag{2.1.24a}$$

and compute successively from "tail to head"

$$G^{(n)}_k = \frac{a_k}{b_k + G^{(n)}_{k+1}}, \qquad k = n, n-1, \ldots, 1, \tag{2.1.24b}$$

to obtain $f_n = b_0 + G^{(n)}_1$. Only n operations of multiplication or division are required to compute f_n. But in this case one does not obtain the other approximants $f_1, f_2, \ldots, f_{n-1}$.

Another algorithm for computing f_n is based on the series

$$f_n = \frac{A_n}{B_n} = \sum_{k=1}^{n} \left(\frac{A_k}{B_k} - \frac{A_{k-1}}{B_{k-1}} \right) = \sum_{k=1}^{n} \left[\frac{(-1)^{k+1} \prod\limits_{j=1}^{k} a_j}{B_k B_{k-1}} \right], \tag{2.1.25}$$

which follows immediately from the determinant formula (2.1.9) and is valid if $B_k \neq 0$, $k = 1, 2, \ldots, n$.

In Chapter 10 it is shown that in many circumstances the BR algorithm is numerically stable in the sense that the rounding error produced in the computation of f_n either is bounded or else grows very slowly as n increases. This suggests that the BR algorithm may be particularly suitable for computational purposes. What makes the BR algorithm less attractive than other (forward) algorithms is the uncertainty as to the proper value of n.

2.2   Sequences Generated by Linear Fractional Transformations

Let $\{t_n\}$ be a sequence of non-singular linear fractional transformations (l.f.t.'s)

$$t_n(w) = \frac{a_n + c_n w}{b_n + d_n w}, \qquad a_n d_n - b_n c_n \neq 0, \qquad n = 0, 1, 2, \ldots . \tag{2.2.1}$$

An l.f.t. is sometimes called a Moebius transformation. The sequence $\{t_n\}$ is said to *generate* the sequence $\{T_n\}$ if

$$T_0(w) = t_0(w); \qquad T_n(w) = T_{n-1}(t_n(w)), \quad n = 1, 2, 3, \ldots. \qquad (2.2.2)$$

We shall assume here knowledge of elementary properties of linear fractional transformations, their composition and their mapping properties. Good references for this subject are Caratheodory [1932, Chapter I; 1950, pp. 21–45], Ford [1929, Chapter 1], Hille [1959, pp. 46–48], and Lehner [1966, pp. 1–10]. Corresponding to a sequence of l.f.t.'s (2.2.1) are sequences of complex numbers $\{A_n\}$, $\{B_n\}$, $\{C_n\}$, $\{D_n\}$ defined by the systems of second-order linear *difference equations*

$$A_0 = a_0, \quad B_0 = b_0, \quad C_0 = c_0, \quad D_0 = d_0, \qquad (2.2.3a)$$

$$A_n = a_n C_{n-1} + b_n A_{n-1}, \qquad n = 1, 2, 3, \ldots, \qquad (2.2.3b)$$

$$B_n = a_n D_{n-1} + b_n B_{n-1}, \qquad n = 1, 2, 3, \ldots, \qquad (2.2.3c)$$

$$C_n = c_n C_{n-1} + d_n A_{n-1}, \qquad n = 1, 2, 3, \ldots, \qquad (2.2.3d)$$

$$D_n = c_n D_{n-1} + d_n B_{n-1}, \qquad n = 1, 2, 3, \ldots. \qquad (2.2.3e)$$

THEOREM 2.3. *Let $\{t_n\}$ be a given sequence of l.f.t.'s (2.2.1), and let $\{A_n\}$, $\{B_n\}$, $\{C_n\}$, $\{D_n\}$ be sequences of complex numbers defined by the difference equations (2.2.3). If $\{T_n\}$ is the sequence of l.f.t.'s generated by $\{t_n\}$, then*

$$T_n(w) = \frac{A_n + C_n w}{B_n + D_n w}, \qquad n = 0, 1, 2, \ldots, \qquad (2.2.4a)$$

and

$$A_n D_n - B_n C_n \neq 0, \qquad n = 0, 1, 2, \ldots. \qquad (2.2.4b)$$

Moreover

$$B_n C_n - A_n D_n = (b_n c_n - a_n d_n)(B_{n-1}C_{n-1} - A_{n-1}D_{n-1}), \qquad n = 1, 2, \ldots, \qquad (2.2.5)$$

$$A_n D_n - B_n C_n = (-1)^n \prod_{k=0}^{n} (a_k d_k - b_k c_k), \qquad n = 0, 1, 2, \ldots, \qquad (2.2.6)$$

and

$$A_{n-1}B_n - A_n B_{n-1} = a_n(A_{n-1}D_{n-1} - B_{n-1}C_{n-1}), \qquad n = 1, 2, \ldots, \qquad (2.2.7a)$$

$$A_n D_{n-1} - B_n C_{n-1} = b_n(A_{n-1}D_{n-1} - B_{n-1}C_{n-1}), \qquad n = 1, 2, \ldots, \qquad (2.2.7b)$$

$$A_{n-1}D_n - B_{n-1}C_n = c_n(A_{n-1}D_{n-1} - B_{n-1}C_{n-1}), \qquad n = 1, 2, \ldots, \qquad (2.2.7c)$$

$$C_n D_{n-1} - C_{n-1}D_n = d_n(A_{n-1}D_{n-1} - B_{n-1}C_{n-1}), \qquad n = 1, 2, \ldots. \qquad (2.2.7d)$$

Proof. The formulas (2.2.5) and (2.2.7) are obtained from (2.2.3) by straightforward computation. The expression (2.2.4a) is obtained by induction on n using (2.2.2) and (2.2.3). Since the composition of non-singular linear fractional transformations is non-singular, (2.2.4b) follows. Finally, (2.2.6) is obtained by induction from (2.2.5). ∎

Proof of Theorem 2.1.   In Theorem 2.3 suppose that

$$b_0 = 1, \quad c_0 = 1, \quad d_0 = 0,$$
$$a_n \neq 0, \quad c_n = 0, \quad d_n = 1, \qquad n = 1, 2, 3, \ldots,$$

so that

$$t_0(w) = a_0 + w; \qquad t_n(w) = \frac{a_n}{b_n + w}, \qquad n = 1, 2, 3, \ldots.$$

To conform with standard continued-fraction notation, we shall replace the above a_0 by b_0. Thus t_n has the same form as s_n in (2.1.1c), and (2.2.3d, e) reduce to

$$C_n = A_{n-1}, \quad D_n = B_{n-1}, \qquad n = 1, 2, 3, \ldots.$$

Inserting these into (2.2.3b, c) gives

$$A_n = a_n A_{n-2} + b_n A_{n-1}, \qquad n = 1, 2, 3, \ldots,$$
$$B_n = a_n B_{n-2} + b_n B_{n-1}, \qquad n = 1, 2, 3, \ldots.$$

Hence $\{A_n\}$ and $\{B_n\}$ satisfy the difference equations (2.1.6). Theorem 2.3 implies that

$$S_n(w) = \frac{A_n + A_{n-1}w}{B_n + B_{n-1}w}, \qquad n = 0, 1, 2, \ldots,$$

and

$$A_n B_{n-1} - A_{n-1} B_n \neq 0, \qquad n = 1, 2, 3, \ldots.$$

This proves (2.1.7) and (2.1.8). (2.1.9) follows from (2.2.6). ∎

THEOREM 2.4. *Corresponding to a given sequence of l.f.t.'s*

$$T_n(w) = \frac{A_n + C_n w}{B_n + D_n w}, \quad A_n D_n - B_n C_n \neq 0, \qquad n = 0, 1, 2, \ldots, \qquad (2.2.8)$$

there exists a uniquely determined sequence

$$t_n(w) = \frac{a_n + c_n w}{b_n + d_n w}, \quad a_n d_n - b_n c_n \neq 0, \qquad n = 0, 1, 2, \ldots, \qquad (2.2.9)$$

which generates $\{T_n\}$. Moreover,

$$t_0(w) = \frac{A_0 + C_0 w}{B_0 + D_0 w}, \qquad (2.2.10a)$$

and

$$t_n(w) = \frac{(A_{n-1}B_n - A_n B_{n-1}) + (A_{n-1}D_n - B_{n-1}C_n)w}{(A_n D_{n-1} - B_n C_{n-1}) + (C_n D_{n-1} - C_{n-1}D_n)w},$$

$$n = 1, 2, \ldots . \quad (2.2.10b)$$

Proof. From (2.2.2) we have $t_n(w) = T_{n-1}^{-1}(T_n(w))$ for $n > 1$. Thus t_n is uniquely determined and can be computed to be given by (2.2.10a, b). Note however that the coefficients a_n, b_n, c_n, d_n in t_n are not unique and that the choice used in (2.2.10a, b) is only one of the possible ones. ∎

Proof of Theorem 2.2. In Theorem 2.4 we set

$$B_0 = 1, \qquad C_0 = 1, \qquad D_0 = 0$$

and

$$C_n = A_{n-1}, \qquad D_n = B_{n-1}, \qquad n = 1, 2, 3, \ldots,$$

so that

$$A_{-1} = 1 \quad \text{and} \quad B_{-1} = 0.$$

Then from (2.2.10) the sequence $\{t_n\}$ which generates $\{T_n\}$ is given by

$$t_0(w) = A_0 + w \qquad (2.2.11a)$$

and

$$t_n(w) = \frac{\dfrac{A_{n-1}B_n - A_n B_{n-1}}{A_{n-1}B_{n-2} - A_{n-2}B_{n-1}}}{\dfrac{A_n B_{n-2} - A_{n-2} B_n}{A_{n-1}B_{n-2} - A_{n-2}B_{n-1}} + w}, \qquad n = 1, 2, 3, \ldots . \quad (2.2.11b)$$

Thus defining the a_n and b_n by (2.1.11), we see that the continued fraction $b_0 + \mathbf{K}(a_n/b_n)$ has nth approximant $f_n = A_n/B_n$. That A_n and B_n are the nth numerator and denominator of the continued fraction, respectively, is shown by verifying by direct substitution that the A_n and B_n satisfy the equations (2.1.6). ∎

In our definition of a continued fraction $b_0 + K(a_n/b_n)$ the nth approximant was defined by $f_n = S_n(0)$, where $\{S_n\}$ is the sequence of l.f.t.'s generated by the sequence

$$s_0(w) = b_0 + w; \qquad s_n(w) = \frac{a_n}{b_n + w}, \qquad n = 1, 2, 3, \ldots, \qquad (2.2.12)$$

with $a_n \neq 0$ for all n. Thus a sequence of l.f.t.'s of the form (2.2.12) is called a *continued-fraction-generating* (c.f.g.) sequence. A characterization of sequences generated by c.f.g. sequences is given in the following:

THEOREM 2.5. *A sequence of l.f.t.'s (T_n) is generated by a c.f.g. sequence iff*

$$T_0(w) = A_0 + w \qquad (2.2.13a)$$

and

$$T_n(\infty) = T_{n-1}(0), \qquad n = 1, 2, 3, \ldots. \qquad (2.2.13b)$$

Proof. Suppose $\{T_n\}$ is generated by a c.f.g. sequence $\{s_n\}$ (2.2.12). Then clearly (2.2.13) holds, since $T_n(w) = T_{n-1}(s_n(w))$ and

$$s_n(\infty) = 0, \qquad n = 1, 2, 3, \ldots.$$

Conversely, suppose that (2.2.13) holds, where

$$T_n(w) = \frac{A_n + C_n w}{B_n + D_n w}, \qquad A_n D_n - B_n C_n \neq 0, \qquad n = 0, 1, 2, \ldots.$$

Then

$$\frac{C_n}{D_n} = T_n(\infty) = T_{n-1}(0) = \frac{A_{n-1}}{B_{n-1}}, \qquad n = 1, 2, 3, \ldots,$$

so that

$$B_{n-1} C_n - A_{n-1} D_n = 0, \qquad n = 1, 2, 3, \ldots.$$

It follows from this and from (2.2.10b) that

$$t_n(w) = \frac{\dfrac{A_{n-1} B_n - A_n B_{n-1}}{C_n D_{n-1} - C_{n-1} D_n}}{\dfrac{A_n D_{n-1} - B_n C_{n-1}}{C_n D_{n-1} - C_{n-1} D_n} + w}, \qquad n = 1, 2, 3, \ldots,$$

where $\{t_n\}$ is the sequence of l.f.t.'s that generates $\{T_n\}$. Since each t_n is non-singular, it follows that $C_n D_{n-1} - C_{n-1} D_n \neq 0$ for all $n \geq 1$. Thus $\{t_n\}$ has the form of a c.f.g. sequence. ∎

2.3 Equivalence Transformations

2.3.1 Equivalent Continued Fractions

Continued fractions $b_0 + \mathbf{K}(a_n/b_n)$ and $b_0^* + \mathbf{K}(a_n^*/b_n^*)$ with nth approximants f_n and f_n^*, respectively, are said to be *equivalent* if

$$f_n = f_n^*, \qquad n = 0, 1, 2, \ldots. \tag{2.3.1}$$

The resulting equivalence relation for continued fractions has no analogue for infinite series or products. The equivalence of two continued fractions is denoted by writing

$$b_0 + \mathbf{K}(a_n/b_n) \approx b_0^* + \mathbf{K}(a_n^*/b_n^*). \tag{2.3.2}$$

The following theorem characterizes equivalence of continued fractions:

THEOREM 2.6. *Continued fractions $b_0 + \mathbf{K}(a_n/b_n)$ and $b_0^* + \mathbf{K}(a_n^*/b_n^*)$ are equivalent iff there exists a sequence of non-zero constants $\{r_n\}$ with $r_0 = 1$ such that*

$$a_n^* = r_n r_{n-1} a_n, \qquad n = 1, 2, 3, \ldots, \tag{2.3.3a}$$

$$b_n^* = r_n b_n, \qquad n = 0, 1, 2, \ldots. \tag{2.3.3b}$$

Proof. Let

$$s_0(w) = b_0 + w, \quad s_0^*(w) = b_0^* + w \tag{2.3.4a}$$

$$s_n(w) = \frac{a_n}{b_n + w}, \quad s_n^*(w) = \frac{a_n^*}{b_n^* + w}, \qquad n = 1, 2, 3, \ldots, \tag{2.3.4b}$$

and let $\{S_n\}$ and $\{S_n^*\}$ denote the sequences generated by the c.f.g. sequences $\{s_n\}$ and $\{s_n^*\}$ respectively [see (2.2.2)]. Then

$$s_n^{-1}(w) = -b_n + \frac{a_n}{w}, \qquad n = 1, 2, 3, \ldots, \tag{2.3.5}$$

$$S_n^{-1}(w) = s_n^{-1} \circ S_{n-1}^{-1}(w), \qquad n = 1, 2, 3, \ldots, \tag{2.3.6}$$

$$s_n(w) = S_{n-1}^{-1} \circ S_n(w), \qquad n = 1, 2, 3, \ldots, \tag{2.3.7}$$

and similar formulas hold for the corresponding "star" sequences.

Suppose that the two continued fractions are equivalent. Our proof of (2.3.3) is by induction. If f_n and f_n^* denote the nth approximants of $b_0 + \mathrm{K}(a_n/b_n)$ and $b_0^* + \mathrm{K}(a_n^*/b_n^*)$, respectively, then

$$S_n(0) = f_n = f_n^* = S_n^*(0), \qquad n = 0, 1, 2, \ldots . \qquad (2.3.8)$$

Thus

$$b_0 = S_0(0) = S_0^*(0) = b_0^*,$$

so that $S_0(w) = S_0^*(w)$ and hence

$$(S_0^*)^{-1}(w) = r_0 S_0^{-1}(w).$$

Since $S_1^*(0) = S_1(0)$, it is easy to see that $a_1^*/b_1^* = a_1/b_1$ and hence that there exists an $r_1 \neq 0$ such that $a_1^* = r_0 r_1 a_1$ and $b_1^* = r_1 b_1$. Therefore $(S_1^*)^{-1}(w) = (s_1^*)^{-1} \circ (S_0^*)^{-1}(w) = -b_1^* + a_1^*/r_0 S_0^{-1}(w) = r_1 s_1^{-1} \circ S_0^{-1}(w) = r_1 S_1^{-1}(w)$. Now assume that, for some positive integer m, there exist non-zero constants $r_1, r_2, \ldots, r_m$ such that

$$a_k^* = r_k r_{k-1} a_k, \quad b_k^* = r_k b_k, \qquad k = 1, 2, \ldots, m \qquad (2.3.9)$$

and

$$(S_k^*)^{-1}(w) = r_k S_k^{-1}(w), \qquad k = 1, 2, \ldots, m. \qquad (2.3.10)$$

Then

$$\frac{a_{m+1}^*}{b_{m+1}^*} = s_{m+1}^*(0) = (S_m^*)^{-1} \circ S_{m+1}^*(0), \qquad \text{by (2.3.7)}$$

$$= r_m S_m^{-1} \circ S_{m+1}(0), \qquad \text{by (2.3.10) and (2.3.8)}$$

$$= r_m s_{m+1}(0) = \frac{r_m a_{m+1}}{b_{m+1}}, \qquad \text{by (2.3.7)}.$$

Hence there exists a non-zero constant r_{m+1} such that

$$a_{m+1}^* = r_{m+1} r_m a_{m+1}, \qquad b_{m+1}^* = r_{m+1} b_{m+1}. \qquad (2.3.11)$$

From (2.3.11), (2.3.5) and (2.3.6) it is easily shown that

$$(S_{m+1}^*)^{-1}(w) = r_{m+1} S_{m+1}^{-1}(w).$$

This proves (2.3.3).

Next suppose that there exists a sequence of non-zero constants $\{r_n\}$ satisfying (2.3.3). In order to prove (2.3.8) it will suffice to verify the relations

$$s_n^*(r_n w) = r_{n-1} s_n(w), \qquad n = 1, 2, \dots, \qquad (2.3.12)$$

and

$$S_n^*(r_n w) = S_n(w), \qquad n = 0, 1, 2, \dots . \qquad (2.3.13)$$

The equations (2.3.12) follow immediately from (2.3.3) and (2.3.4). A simple induction using (2.3.12) proves (2.3.13). ∎

The mapping of $b_0 + \mathrm{K}(a_n/b_n)$ into $b_0^* + \mathrm{K}(a_n^*/b_n^*)$ defined by (2.3.3) is called an *equivalence transformation*. In subsequent chapters it is shown that certain properties of continued fractions are invariant under equivalence transformations. An equivalence transformation can be expressed by writing

$$b_0 + \frac{a_1}{b_1} + \frac{a_2}{b_2} + \frac{a_3}{b_3} + \cdots \approx b_0 + \frac{r_1 a_1}{r_1 b_1} + \frac{r_2 r_1 a_2}{r_2 b_2} + \frac{r_3 r_2 a_2}{r_3 b_3} + \cdots \quad (2.3.14)$$

for infinite continued fractions and

$$b_0 + \frac{a_1}{b_1} + \frac{a_2}{b_2} + \cdots + \frac{a_n}{b_n} \approx b_0 + \frac{r_1 a_1}{r_1 b_1} + \frac{r_2 r_1 a_2}{r_2 b_2} + \cdots + \frac{r_n r_{n-1} a_n}{r_n b_n} \quad (2.3.15)$$

for finite continued fractions. Before looking at examples of equivalence transformations, we consider briefly the question of what sequences can form the approximants of continued fractions.

For an arbitrary sequence $\{f_n\}$ in the extended complex plane there exists a sequence of l.f.t.'s (T_n) with the property

$$T_n(0) = f_n, \qquad n = 0, 1, 2, \dots .$$

In fact such a sequence will be generated by any sequence of l.f.t.'s $\{t_n\}$ in which t_n has the form

$$t_n(w) = \begin{cases} \dfrac{T_{n-1}^{-1}(f_n) + c_n w}{1 + d_n w} & \text{if } T_{n-1}^{-1}(f_n) \neq \infty, \\[2mm] \dfrac{a_n}{w} + c_n & \text{if } T_{n-1}^{-1}(f_n) = \infty, \end{cases}$$

where the coefficients a_n, c_n, d_n are arbitrary except for the requirement that t_n be non-singular. In the above we have set $T_{-1}(w) = w$. This situation cannot always be attained by means of c.f.g. sequences, as is shown by the following:

THEOREM 2.7. *A sequence* $\{f_n\}$ *in the extended complex plane can be the sequence of approximants of a continued fraction* $b_0 + K(a_n/b_n)$ *iff*

$$f_0 \neq \infty; \quad f_n \neq f_{n-1}, \quad n = 1, 2, 3, \ldots . \tag{2.3.16}$$

Proof. Suppose (2.3.16) holds. Let $A_{-1} = 1$, $B_{-1} = 0$. For $n \geq 0$, let $A_n = f_n$ and $B_n = 1$ if f_n is finite, and let $A_n = 1$ and $B_n = 0$ if $f_n = \infty$. Applying Theorem 2.2 we obtain a continued fraction with sequence of approximants $\{f_n\}$. Conversely, suppose there exists a continued fraction $b_0 + K(a_n/b_n)$ with sequence of approximants $\{f_n\}$. Let $\{S_n\}$ be the sequence of l.f.t.'s generated by the c.f.g. sequence $\{s_n\}$ (see (2.1.1)). Then

$$f_n = S_n(0) = S_{n-1}(s_n(0)) \neq S_{n-1}(s_n(\infty)) = S_{n-1}(0) = f_{n-1},$$

since $s_n(0) \neq s_n(\infty)$ and hence $S_{n-1}(s_n(0)) \neq S_{n-1}(s_n(\infty))$, since s_n and S_{n-1} are non-singular. Since $f_0 = b_0$ it must be in $\mathbb{C}$. ∎

A consequence of Theorem 2.7 is that it is possible to have a continued fraction which converges at any speed (fast or slow) or which diverges in any manner.

THEOREM 2.8. *A sequence* $\{f_n\}$ *in the extended complex plane can be the sequence of approximants of a continued fraction of the form* $b_0 + K(a_n/1)$ *iff*

$$f_0 \neq \infty; \quad f_n \neq f_{n-1}, \quad n = 1, 2, 3, \ldots, \tag{2.3.17}$$

and

$$f_{n+1} \neq f_{n-1}, \quad n = 1, 2, 3, \ldots . \tag{2.3.18}$$

Proof. Suppose (2.3.17) and (2.3.18) hold. Set $A_{-1} = 1$, $B_{-1} = 0$. For $n \geq 0$, let $A_n = f_n$ and $B_n = 1$ if f_n is finite, and let $A_n = 1$ and $B_n = 0$ if $f_n = \infty$. Applying Theorem 2.2, we obtain a continued fraction $b_0^* + K(a_n^*/b_n^*)$ with sequence of approximants $\{f_n\}$. Moreover, (2.1.11b) and (2.3.18) imply that $b_n^* \neq 0$ for $n \geq 1$. An application of Theorem 2.6, with $r_n = 1/b_n$ for $n \geq 1$, gives an equivalent continued fraction of the form $b_0 + K(a_n/1)$.

Conversely, suppose there exists a continued fraction of the form $b_0 + K(a_n/1)$ with sequence of approximants $\{f_n\}$. Then by Theorem 2.7, (2.3.17) holds. For $n \geq 1$, $f_{n+1} = S_{n+1}(0) = S_{n-1}(s_n \circ s_{n+1}(0))$ and $f_{n-1} = S_{n-1}(0) = S_{n-1}(s_n(\infty))$, and hence $f_{n+1} \neq f_{n-1}$ follows from the fact that $\infty \neq s_{n+1}(0) = a_{n+1}/1$. ∎

If $b_n \neq 0$ for all $n \geq 1$, then

$$b_0 + K(a_n/b_n) \approx b_0 + K(a_n^*/1), \tag{2.3.19a}$$

where

$$a_1^* = \frac{a_1}{b_1}; \qquad a_n^* = \frac{a_n}{b_n b_{n-1}}, \qquad n=2,3,\ldots . \qquad (2.3.19b)$$

This can also be expressed by writing

$$b_0 + \frac{a_1}{b_1} + \frac{a_2}{b_2} + \frac{a_3}{b_3} + \cdots$$

$$\approx b_0 + \frac{a_1/b_1}{1} + \frac{a_2/b_1 b_2}{1} + \frac{a_3/b_2 b_3}{1} + \cdots + \frac{a_n/b_{n-1} b_n}{1} + \cdots . \qquad (2.3.20)$$

As examples, the continued-fraction expansions (2.1.22) for π and (2.1.23) for e can be written in the equivalent forms

$$\pi = \frac{4}{1} + \frac{1^2/3}{1} + \frac{2^2/3\times 5}{1} + \frac{3^2/5\times 7}{1} + \cdots \qquad (2.3.21)$$

and

$$e = \frac{1}{1} + \frac{-1}{1} + \frac{1/2}{1} + \frac{-1/2\times 3}{1}$$

$$+ \frac{1/2\times 3}{1} + \frac{-1/2\times 5}{1} + \frac{(1/2\times 5)}{1} + \cdots \qquad (2.3.22)$$

respectively.

By means of equivalence transformation it is always possible to express a continued fraction in the (equivalent) form $b_0 + \mathrm{K}(1/b_n^*)$. In fact

$$b_0 + \mathrm{K}(a_n/b_n) \approx b_0 + \mathrm{K}(1/b_n^*) \qquad (2.3.23a)$$

where

$$b_n^* = b_n \prod_{k=1}^n a_k^{(-1)^{n+k-1}}, \qquad n=1,2,3,\ldots . \qquad (2.3.23b)$$

This can also be expressed by writing

$$b_0 + \frac{a_1}{b_1} + \frac{a_2}{b_2} + \frac{a_3}{b_3} + \cdots$$

$$\approx b_0 + \frac{1}{\frac{1}{a_1}b_1} + \frac{1}{\frac{a_1}{a_2}b_2} + \frac{1}{\frac{a_2}{a_1 a_3}b_3} + \frac{1}{\frac{a_1 a_3}{a_2 a_4}b_4} + \cdots . \qquad (2.3.24)$$

Thus, for example, the expansions for π and e in (2.3.21) and (2.3.22) can be written in the equivalent forms

$$\pi = \frac{1}{\frac{1}{4}} + \frac{1}{3\frac{4}{1}} + \frac{1}{5\frac{1}{4.2^2}} + \frac{1}{7\frac{4\times 2^2}{3^2}} + \frac{1}{9\frac{3^2}{4\times 2^2\times 4^2}} + \cdots \qquad (2.3.25)$$

and

$$e = \frac{1}{1} + \frac{1}{-1} + \frac{1}{-2} + \frac{1}{3} + \frac{1}{2} + \frac{1}{-5} + \frac{1}{-2} + \frac{1}{7} + \frac{1}{2} + \cdots, \quad (2.3.26)$$

respectively.

Examples of continued fractions that do not satisfy the condition (2.3.18) in Theorem 2.8 are

$$\frac{1}{1} + \frac{1}{0} + \frac{1}{0} + \frac{1}{0} + \cdots$$

and

$$\frac{1}{0} + \frac{1}{0} + \frac{1}{0} + \cdots.$$

In the first case

$$f_{2m} = 0, \quad f_{2m+1} = 1, \quad m = 0, 1, 2, \ldots,$$

and in the second case

$$f_{2m} = 0, \quad f_{2m+1} = \infty, \quad m = 0, 1, 2, \ldots.$$

2.3.1 *Euler's* [1748] *Connection between Continued Fractions and Infinite Series*

Let $\{c_n\}$ be a sequence of complex numbers with $c_n \neq 0$, $n = 1, 2, 3, \ldots$, and let

$$f_n = \sum_{k=0}^{n} c_k, \quad n = 0, 1, 2, \ldots. \quad (2.3.27)$$

Since $f_n \neq f_{n-1}$, $n \geqslant 1$, it follows from Theorem 2.7 that there exists a continued fraction $b_0 + \mathrm{K}(a_n/b_n)$ with nth approximant f_n for all n. In fact, if we define sequences $\{A_n\}$ and $\{B_n\}$ by

$$A_{-1} = 1, \quad B_{-1} = 0,$$
$$A_n = f_n, \quad B_n = 1, \quad n = 0, 1, 2, \ldots,$$

it follows that

$$\frac{A_{n-1}B_n - A_n B_{n-1}}{A_{n-1}B_{n-2} - A_{n-2}B_{n-1}} = \frac{f_{n-1} - f_n}{f_{n-1} - f_{n-2}} = -\frac{c_n}{c_{n-1}}, \quad n = 2, 3, 4, \ldots,$$

$$\frac{A_n B_{n-2} - A_{n-2}B_n}{A_{n-1}B_{n-2} - A_{n-2}B_{n-1}} = \frac{f_n - f_{n-2}}{f_{n-1} - f_{n-2}} = \frac{c_n + c_{n-1}}{c_{n-1}}, \quad n = 2, 3, 4, \ldots.$$

Thus by Theorem 2.2 [equations (2.1.11)] the elements of $b_0 + K(a_n/b_n)$ can be taken to be

$$b_0 = f_0 = c_0, \qquad a_1 = f_1 = c_0 + c_1, \qquad b_1 = 1,$$

$$a_n = -\frac{c_n}{c_{n-1}}, \qquad b_n = \frac{c_{n-1} + c_n}{c_{n-1}}, \qquad n = 2,3,4,\ldots.$$

Now defining the sequence $\{\rho_n\}$ by

$$\rho_0 = c_0, \qquad \rho_1 = c_1, \qquad \rho_n = \frac{c_n}{c_{n-1}}, \qquad n = 2,3,4,\ldots,$$

we obtain

$$c_0 = \rho_0, \qquad c_n = \rho_1 \rho_2 \cdots \rho_n, \qquad n = 1,2,3,\ldots.$$

It follows from the preceding discussion that

$$\rho_0 + \sum_{k=1}^{n} \rho_1 \rho_2 \cdots \rho_k = \rho_0 + \frac{\rho_1}{1} \; \underset{-}{} \; \frac{\rho_2}{1+\rho_2} \; \underset{-}{} \; \frac{\rho_3}{1+\rho_3} \; \underset{-}{} \cdots \underset{-}{} \; \frac{\rho_n}{1+\rho_n},$$

$$n = 0,1,2,\ldots, \quad (2.3.28)$$

provided $\rho_k \neq 0$, $k = 1,2,3,\ldots$. As an illustration, the partial sums of a power series can be expressed by

$$\sum_{k=0}^{n} c_k z^k = c_0 + \frac{c_1 z}{1} \; \underset{-}{} \; \frac{\dfrac{c_2}{c_1} z}{1 + \dfrac{c_2}{c_1} z} \; \underset{-}{} \; \frac{\dfrac{c_3}{c_2} z}{1 + \dfrac{c_3}{c_2} z} \; \underset{-}{} \cdots \underset{-}{} \; \frac{\dfrac{c_n}{c_{n-1}} z}{1 + \dfrac{c_n}{c_{n-1}} z},$$

$$n = 0,1,2,\ldots, \quad (2.3.29)$$

provided the $c_k \neq 0$.

Conversely, suppose that $b_0 + K(a_n/b_n)$ is a continued fraction with nth approximant $f_n \neq \infty$ for $n = 0,1,2,\ldots$. Then there exists an infinite series $\sum_{k=0}^{\infty} c_k$ satisfying (2.3.27). In fact, by (2.1.25) it suffices to set

$$c_0 = f_0; \qquad c_k = \frac{(-1)^{k-1} \prod_{j=1}^{k} a_j}{B_k B_{k-1}}, \qquad k = 1,2,3,\ldots,$$

where the B_k are defined by the difference equations (2.1.6). We note in passing that the connection between infinite series and continued fractions described above is only of limited interest, since in this situation both have exactly the same convergence (or divergence) behavior. Other means of connecting continued fractions and infinite series (described in Chapters 5, 6, 7 and 9) sometimes permit the convergence of the continued fraction to be more favorable than that of the corresponding series.

2.4   Contractions and Extensions

A continued fraction $b_0^* + \mathrm{K}(a_n^*/b_n^*)$ with nth approximant f_n^* is said to be a *contraction* of a continued fraction $b_0 + \mathrm{K}(a_n/b_n)$ with nth approximant f_n if $\{f_n^*\}$ is a subsequence of $\{f_n\}$. If a continued fraction is a contraction of another continued fraction, then the latter is called an *extension* of the former. Necessary and sufficient conditions for a subsequence of the approximants $\{f_n\}$ of a continued fraction to be the sequence of approximants of another continued fraction are given in Theorem 2.7. In this section we describe methods for constructing contractions and extensions.

2.4.1   *Contraction of Continued Fractions*

Let

$$f_k^* = f_{n_k}, \qquad k = 0, 1, 2, \ldots, \tag{2.4.1}$$

where $\{n_k\}$ denotes a subsequence of the positive integers and where f_n is the nth approximant of $b_0 + \mathrm{K}(a_n/b_n)$. Let

$$s_0(w) = b_0 + w; \qquad s_n(w) = \frac{a_n}{b_n + w}, \qquad n = 1, 2, 3, \ldots, \tag{2.4.2}$$

and let $\{S_n\}$ denote the sequence generated by the c.f.g. sequence $\{s_n\}$. We seek a contraction of $b_0 + \mathrm{K}(a_n/b_n)$ with kth approximant f_k^* such that

$$f_k^* = f_{n_k} = S_{n_k}(0) = s_0 \circ s_1 \circ \cdots \circ s_{n_k}(0), \qquad k = 0, 1, 2, \ldots. \tag{2.4.3}$$

Thus it is natural to consider the l.f.t.'s

$$t_0(w) = s_0 \circ s_1 \circ \cdots \circ s_{n_0}(w), \tag{2.4.4a}$$

$$t_k(w) = s_{n_{k-1}+1} \circ s_{n_{k-1}+2} \circ \cdots \circ s_{n_k}(w), \qquad k = 1, 2, \ldots, \tag{2.4.4b}$$

since the generated sequence $\{T_n\}$ has the property

$$T_k(0) = S_{n_k}(0) = f_k^*, \qquad k = 0, 1, 2, \ldots. \tag{2.4.5}$$

If $\{t_k\}$ happened to be a c.f.g. sequence, our task would be complete; however, this usually will not be the case. For our purposes it suffices to construct a c.f.g. sequence $\{s_k^*\}$ in such a manner that its generated sequence $\{S_k^*\}$ satisfies

$$S_k^*(0) = T_k(0), \qquad k = 0, 1, 2, \ldots . \tag{2.4.6}$$

For if

$$s_0^*(w) = b_0^* + w; \qquad s_k^*(w) = \frac{a_k^*}{b_k^* + w}, \qquad k = 1, 2, 3, \ldots, \tag{2.4.7}$$

the continued fraction $b_0^* + \mathbf{K}(a_k^*/b_k^*)$ has kth approximant f_k^* and hence is a desired contraction.

We consider l.f.t.'s of the form

$$s_k^*(w) = v_{k-1}^{-1} \circ t_k \circ v_k(w), \qquad k = 0, 1, 2, \ldots, \tag{2.4.8}$$

where $\{v_k\}$ is a sequence of l.f.t.'s with $v_{-1}(w) = w$. It follows that $\{s_k^*\}$ will be a c.f.g. sequence of the form (2.4.7) iff

$$t_0(v_0(w)) = b_0^* + w \tag{2.4.9a}$$

and

$$t_k(v_k(\infty)) = v_{k-1}(0), \qquad k = 1, 2, 3, \ldots . \tag{2.4.9b}$$

Thus the s_k^* are obtained by altering the t_k in such a manner that, in the composition $S_k^* = s_0^* \circ s_1^* \circ \cdots \circ s_k^*$, the v_m^{-1} and v_m, $m = 0, 1, \ldots, k-1$, are "canceled" and

$$S_k^*(w) = T_k(v_k(w)), \qquad k = 0, 1, 2 \ldots . \tag{2.4.10}$$

It follows that (2.4.6) will hold iff

$$v_k(0) = 0, \qquad k = 0, 1, 2, \ldots . \tag{2.4.11}$$

It will be possible to determine l.f.t.'s v_k satisfying both (2.4.9) and (2.4.11) iff

$$t_0(0) \neq \infty; \qquad t_k(0) \neq 0, \quad k = 1, 2, 3, \ldots . \tag{2.4.12}$$

Now we show how the s_k^* can be constructed. Let

$$t_k(w) = \frac{\alpha_k + \gamma_k w}{\beta_k + \delta_k w}, \qquad \alpha_k \delta_k - \beta_k \gamma_k \neq 0, \qquad k = 0, 1, 2, \ldots, \tag{2.4.13a}$$

where

$$\beta_0 \neq 0; \qquad \alpha_k \neq 0, \quad k = 1, 2, 3, \ldots . \tag{2.4.13b}$$

The conditions (2.4.13b) are equivalent to (2.4.12). Let

$$v_k(w) = \frac{\mu_k w}{\lambda_k + \nu_k w}, \quad \lambda_k \mu_k \neq 0, \quad k = 0, 1, 2, \ldots, \quad (2.4.14)$$

so that (2.4.11) is satisfied. Moreover, (2.4.9) holds if

$$v_0(w) = \frac{\beta_0^2 w}{(\beta_0 \gamma_0 - \alpha_0 \delta_0) - \beta_0 \delta_0 w} \quad (2.4.15a)$$

and

$$v_k(w) = \frac{\alpha_k w}{\lambda_k - \gamma_k w}, \quad \alpha_k \lambda_k \neq 0, \quad k = 1, 2, 3, \ldots. \quad (2.4.15b)$$

Substituting into (2.4.8), we obtain

$$s_0^*(w) = \frac{\alpha_0}{\beta_0} + w,$$

$$s_1^*(w) = \frac{\dfrac{\lambda_1 \alpha_1 (\beta_0 \gamma_0 - \alpha_0 \delta_0)}{\beta_0^2 (\alpha_1 \delta_1 - \beta_1 \gamma_1)}}{\dfrac{\lambda_1 (\beta_0 \beta_1 + \alpha_1 \delta_0)}{\beta_0 (\alpha_1 \delta_1 - \beta_1 \gamma_1)} + w},$$

$$s_k^*(w) = \frac{\dfrac{\alpha_k \lambda_k \lambda_{k-1}}{\alpha_{k-1} (\alpha_k \delta_k - \beta_k \gamma_k)}}{\dfrac{\lambda_k (\alpha_{k-1} \beta_k + \alpha_k \gamma_{k-1})}{\alpha_{k-1} (\alpha_k \delta_k - \beta_k \gamma_k)} + w}, \quad k = 2, 3, 4, \ldots.$$

By taking $\lambda_k = \alpha_k \delta_k - \beta_k \gamma_k$, $k = 1, 2, 3, \ldots$, we arrive at the following

THEOREM 2.9. *Let $\{t_n\}$ be a sequence of l.f.t.'s*

$$t_k(w) = \frac{\alpha_k + \gamma_k w}{\beta_k + \delta_k w}, \quad \alpha_k \delta_k - \beta_k \gamma_k \neq 0, \quad k = 0, 1, 2, \ldots, \quad (2.4.16)$$

such that

$$\beta_0 \neq 0; \quad \alpha_k \neq 0, \quad k = 1, 2, 3, \ldots. \quad (2.4.17)$$

Let $\{s_k^*\}$ be the c.f.g. sequence

$$s_0^*(w) = b_0^* + w; \quad s_k^*(w) = \frac{a_k^*}{b_k^* + w}, \quad k = 1, 2, 3, \ldots, \quad (2.4.18)$$

where

$$b_0^* = \frac{a_0}{\beta_0}, \qquad \alpha_1^* = \frac{\alpha_1(\beta_0\gamma_0 - \alpha_0\delta_0)}{\beta_0^2}, \qquad b_1^* = \frac{\beta_0\beta_1 + \alpha_1\delta_0}{\beta_0}, \qquad (2.4.19a)$$

$$a_k^* = \frac{\alpha_k(\alpha_{k-1}\delta_{k-1} - \beta_{k-1}\gamma_{k-1})}{\alpha_{k-1}}, \qquad b_k^* = \frac{\alpha_{k-1}\beta_k + \alpha_k\gamma_{k-1}}{\alpha_{k-1}},$$

$$k = 2, 3, 4, \ldots. \quad (2.4.19b)$$

If $\{S_k^*\}$ is the sequence generated by $\{s_k^*\}$, then

$$S_k^*(0) = T_k(0), \qquad k = 0, 1, 2, \ldots. \qquad (2.4.20)$$

2.4.2 Even Part of a Continued Fraction

Let $b_0^* + \mathrm{K}(a_n^*/b_n^*)$ and $b_0 + \mathrm{K}(a_n/b_n)$ denote continued fractions with nth approximants f_n^* and f_n, respectively. If

$$f_n^* = f_{2n}, \qquad n = 0, 1, 2, \ldots,$$

then $b_0^* + \mathrm{K}(a_n^*/b_n^*)$ is called the *even part* of $b_0 + \mathrm{K}(a_n/b_n)$. It follows from Theorem 2.7 that $\{f_{2n}\}$ can be the sequence of approximants of a continued fraction iff

$$f_{2n} \neq f_{2n-2}, \qquad n = 1, 2, 3, \ldots. \qquad (2.4.21)$$

By (2.1.11b) of Theorem 2.2, the condition (2.4.21) holds iff $b_{2n} \neq 0$ for all $n \geqslant 1$. Now let $n_k = 2k$ and $t_k(w) = s_{2k-1} \circ s_{2k}(w)$. Then

$$t_0(w) = b_0 + w,$$

$$t_k(w) = \frac{a_{2k-1}b_{2k} + a_{2k-1}w}{(a_{2k} + b_{2k-1}b_{2k}) + b_{2k-1}w}, \qquad k = 1, 2, 3, \ldots,$$

where $s_0(w) = b_0 + w$, $s_k(w) = a_k/(b_k + w)$ for $k \geqslant 1$. Applying Theorem 2.9 with $\alpha_0 = b_0$, $\gamma_0 = \beta_0 = 1$, $\delta_0 = 0$ and, for $k \geqslant 1$, $\alpha_k = a_{2k-1}b_{2k}$, $\gamma_k = a_{2k-1}$, $\beta_k = a_{2k} + b_{2k-1}b_{2k}$ and $\delta_k = b_{2k-1}$, we obtain the even part $b_0^* + \mathrm{K}(a_n^*/b_n^*)$ of $b_0 + \mathrm{K}(a_n/b_n)$, where

$$b_0^* = b_0, \qquad a_1^* = a_1 b_1, \qquad b_1^* = a_2 + b_1 b_2,$$

$$a_k^* = \frac{-a_{2k-2}a_{2k-1}b_{2k}}{b_{2k-2}}, \qquad k = 2, 3, 4, \ldots,$$

$$b_k^* = \frac{a_{2k-1}b_{2k} + a_{2k}b_{2k-2} + b_{2k-2}b_{2k-1}b_{2k}}{b_{2k-2}}, \qquad k = 2, 3, 4, \ldots.$$

Using this together with an equivalence transformation (Theorem 2.6 with $r_0 = r_1 = 1$ and $r_k = b_{2k-2}$ for $k \geqslant 2$), we obtain

THEOREM 2.10. *A continued fraction* $b_0 + \mathrm{K}(a_n/b_n)$ *has an even part iff*

$$b_{2k} \neq 0, \qquad k = 1, 2, 3, \ldots . \tag{2.4.22}$$

If (2.4.22) holds, then the elements of the even part $b_0^* + \mathrm{K}(a_n^*/b_n^*)$ *are given (up to equivalence transformation) by*

$$b_0^* = b_0, \quad a_1^* = a_1 b_2, \quad b_1^* = a_2 + b_1 b_2, \quad a_2^* = -a_2 a_3 b_4, \tag{2.4.23a}$$
$$a_k^* = -a_{2k-2} a_{2k-1} b_{2k-4} b_{2k}, \qquad\qquad k = 3, 4, \ldots, \tag{2.4.23b}$$
$$b_k^* = a_{2k-1} b_{2k} + b_{2k-2}(a_{2k} + b_{2k-1} b_{2k}), \quad k = 2, 3, 4, \ldots . \tag{2.4.23c}$$

For convenience we note that the even part of a continued fraction $b_0 + \mathrm{K}(a_n/b_n)$ is given by

$$b_0 + \cfrac{a_1 b_2}{a_2 + b_1 b_2} - \cfrac{a_2 a_3 b_4}{a_3 b_4 + b_2(a_4 + b_3 b_4)}$$
$$- \cfrac{a_4 a_5 b_2 b_6}{a_5 b_6 + b_4(a_6 + b_5 b_6)} - \cfrac{a_6 a_7 b_4 b_8}{a_7 b_8 + b_6(a_8 + b_7 b_8)} - \cdots . \tag{2.4.24}$$

As an illustration, the even part of the expression (2.1.21) for e^z exists and is given by

$$e^z = \cfrac{1}{1-z} + \cfrac{3z^2}{2(1 \cdot 3) + 2z} + \cfrac{(1 \cdot 5)z^2}{2(3 \cdot 5) + 2z} + \cfrac{(3 \cdot 7)z^2}{2(5 \cdot 7) + 2z} + \cdots . \tag{2.4.25}$$

2.4.3 *Odd Part of a Continued Fraction*

Let $b_0^* + \mathrm{K}(a_n^*/b_n^*)$ and $b_0 + \mathrm{K}(a_n/b_n)$ be continued fractions with nth approximants f_n^* and f_n, respectively. If

$$f_n^* = f_{2n+1}, \qquad n = 0, 1, 2, \ldots,$$

then $b_0^* + \mathrm{K}(a_n^*/b_n^*)$ is called the *odd part* of $b_0 + \mathrm{K}(a_n/b_n)$. It follows from Theorem 2.7 that $\{f_{2n+1}\}$ can be the sequence of approximants of a continued fraction iff

$$f_{2n+1} \neq f_{2n-1}, \qquad n = 1, 2, 3, \ldots . \tag{2.4.26}$$

By (2.1.11b) of Theorem 2.2, the condition (2.4.26) holds iff $b_{2n+1} \neq 0$ for all $n \geqslant 1$. An argument analogous to that used in the proof of Theorem 2.10

now yields:

THEOREM 2.11. *A continued fraction* $b_0 + \mathbf{K}(a_n/b_n)$ *has an odd part iff*

$$b_{2k+1} \neq 0, \qquad k = 0, 1, 2, \dots . \tag{2.4.27}$$

If (2.4.27) holds, then the elements of the odd part $b_0^* + \mathbf{K}(a_n^*/b_n^*)$ *are given (up to equivalence transformation) by*

$$b_0^* = \frac{a_1 + b_0 b_1}{b_1} \tag{2.4.28a}$$

$$a_1^* = -\frac{a_1 a_2 b_3}{b_1}, \qquad b_1^* = a_2 b_3 + b_1(a_3 + b_2 b_3), \tag{2.4.28b}$$

$$a_k^* = -a_{2k-1} a_{2k} b_{2k+1} b_{2k-3}, \qquad k = 2, 3, 4, \dots, \tag{2.4.28c}$$

$$b_k^* = a_{2k} b_{2k+1} + b_{2k-1}(a_{2k+1} + b_{2k} b_{2k+1}), \qquad k = 2, 3, 4, \dots . \tag{2.4.28d}$$

For convenience we note that the odd part of $b_0 + \mathbf{K}(a_n/b_n)$ is given by

$$\frac{a_1 + b_0 b_1}{b_1} \;-\; \frac{a_1 a_2 b_3 / b_1}{a_2 b_3 + b_1(a_3 + b_2 b_3)}$$

$$-\; \frac{a_3 a_4 b_1 b_5}{a_4 b_5 + b_3(a_5 + b_4 b_5)} \;-\; \frac{a_5 a_6 b_3 b_7}{a_6 b_7 + b_5(a_7 + b_6 b_7)} \;-\; \cdots . \tag{2.4.29}$$

As an illustration, the odd part of the expansion (2.1.21) of e^z exists and is given by

$$e^z = 1 + \frac{2z}{2-z} + \frac{2z^2}{2 \cdot 3} + \frac{4z^2}{2 \cdot 5} + \frac{4z^2}{2 \cdot 7} + \frac{4z^2}{2 \cdot 9} + \cdots,$$

$$\approx 1 + \frac{2z}{2-z} + \frac{z^2}{3} + \frac{z^2}{5} + \frac{z^2}{7} + \frac{z^2}{9} + \cdots . \tag{2.4.30}$$

2.4.4. *Extension of a Continued Fraction*

The problem we shall consider here is described in the following

THEOREM 2.12. *Let* $b_0 + \mathbf{K}(a_n/b_n)$ *be a continued fraction with* n *th approximant* $f_n = A_n/B_n$. *Let* k *be a positive integer, and let* g *be a point in the extended complex plane such that*

$$f_{k-1} \neq g \neq f_k . \tag{2.4.31}$$

Then there exists a continued fraction $b_0^* + K(a_n^*/b_n^*)$ whose nth approximants f_n^* satisfy

$$f_n^* = \begin{cases} f_n, & 0 \leqslant n \leqslant k-1, \\ g, & n=k, \\ f_{n-1}, & n \geqslant k+1. \end{cases} \qquad (2.4.32)$$

The elements of this continued fraction are given by

$$a_n^* = \begin{cases} a_n, & 1 \leqslant n \leqslant k \\ \rho, & n=k+1 \\ -\dfrac{a_{k+1}}{\rho}, & n=k+2 \\ a_{n-1}, & n \geqslant k+3, \end{cases}$$

$$b_n^* = \begin{cases} b_n, & 0 \leqslant n \leqslant k-1, \\ b_k - \rho, & n=k, \\ 1, & n=k+1 \\ b_{k+1} + \dfrac{a_{k+1}}{\rho}, & n=k+2, \\ b_{n-1}, & n \geqslant k+3, \end{cases} \qquad (2.4.33)$$

where

$$\rho = \frac{A_k - B_k g}{A_{k-1} - B_{k-1} g}.$$

Proof. It is clear from Theorem 2.7 that (2.4.31) is necessary and sufficient for the existence of a continued fraction with nth approximants f_n^* satisfying (2.4.32). To construct such a continued fraction we consider l.f.t.'s

$$s_0(w) = b_0 + w; \qquad s_n(w) = \frac{a_n}{b_n + w}, \quad n=1,2,3,\dots, \qquad (2.4.34a)$$

$$s_0^*(w) = b_0^* + w; \qquad s_n^*(w) = \frac{a_n^*}{b_n^* + w}, \quad n=1,2,3,\dots, \qquad (2.4.34b)$$

and sequences $\{S_n\}$ and $\{S_n^*\}$ generated by $\{s_n\}$ and $\{s_n^*\}$, respectively. When the elements a_n^*, b_n^* have been properly determined, we will have $f_n^* = S_n^*(0)$ satisfying (2.4.32).

First we define

$$s_n^*(w) = s_n(w), \quad n=0,1,\dots,k-1, \qquad (2.4.35)$$

and hence the first part of (2.4.32) is satisfied. Next we set

$$s_k^*(w) = s_k \circ v(w), \qquad (2.4.36)$$

where v is a l.f.t. Since $s_k^*(w)$ is to have the form (2.4.34), $v(w)$ must satisfy $v(\infty) = \infty$. To insure that $f_k^* = s_k^*(0) = g$, $v(w)$ must also satisfy

$$v(0) = S_k^{-1}(g) = \frac{A_k - B_k g}{A_{k-1} - B_{k-1} g} = -\rho.$$

Hence it suffices to set

$$v(w) = mw - \rho, \qquad \text{where} \quad \rho = -S_k^{-1}(g),$$

with m an arbitrary non-zero complex number. That ρ cannot be 0 or ∞ follows from (2.4.31). The motivation for writing s_k^* in the form (2.4.36) is that it is convenient for the subsequent restoration of the sequence $\{S_n\}$.

Now we wish to define

$$s_{k+1}^*(w) = u(w) = \frac{\alpha}{\beta + w}, \qquad \alpha \neq 0,$$

in such a manner that

$$f_{k+1}^* = S_{k+1}^*(0) = S_k \circ v \circ u(0) = S_k(0) = f_k.$$

Hence we must have $v \circ u(0) = 0$, or

$$\frac{\alpha}{\beta} = u(0) = v^{-1}(0) = \frac{\rho}{m}.$$

We take $\alpha = \rho$ and $\beta = m = 1$ and obtain

$$s_{k+1}^*(w) = u(w) = \frac{\rho}{1+w}.$$

To restore the sequence $\{S_n\}$ it suffices to set

$$s_{k+2}^*(w) = u^{-1} \circ v^{-1} \circ s_{k+1}(w) = \frac{-a_{k+1}/\rho}{b_{k+1} + (a_{k+1}/\rho) + w},$$

and

$$s_n^*(w) = s_{n-1}(w), \qquad n = k+3, k+4, \ldots . \qquad \blacksquare$$

Theorem 2.12 shows how a continued fraction can be modified to insert a single approximant into its sequence of approximants. This procedure can be used to insert countably many additional approximants.

CHAPTER 3.

Periodic Continued Fractions

3.1 Introduction

The systematic derivation of results on continued fractions starts in the nineteenth century. This is true for general convergence theory (Chapter 4) as well as for the theory of periodic continued fractions. A continued fraction $K(a_n/b_n)$ is said to be *periodic with period k* if its elements satisfy the conditions

$$a_{rk+m} = a_m, \quad b_{rk+m} = b_m, \quad m = 1, 2, \ldots, k, \quad r = 0, 1, 2, \ldots. \quad (3.1.1)$$

The first to have a complete statement of the convergence behavior of periodic continued fractions appears to have been Stolz [1886]. Earlier work was done by D. Bernoulli [1775] for period 1. This was rediscovered by Clausen [1828]. Kahl [1852] considered period 2; further progress was made by Günther [1872]; Thiele [1879] pointed out the divergence phenomenon named after him. Proofs of the complete theorem were modified and/or improved by Landsberg [1892], Pringsheim [1900], Perron [1905], Lane [1945] and Schwerdtfeger [1946]. Our proof (Section 3.2, Theorem 3.1) uses the approach of the last two authors.

The other main result dealt with in this chapter is the convergence relationship between a periodic continued fraction and its dual continued fraction (Section 3.3, Theorem 3.4). Galois [1828/9] first proved the theorem for regular continued fractions; hence it bears his name. Landsberg [1892] generalized to $K(a_n/b_n)$ with a_n and b_n arbitrary rational numbers. The theorem in full generality was proved by Pringsheim [1900]. The case where both continued fractions converge was rediscovered by Bankier and Leighton [1942]. Merkes and Scott [1960] rediscovered the general theorem and generalized to mixed periodic continued fractions.

ENCYCLOPEDIA OF MATHEMATICS and Its Applications, Gian-Carlo Rota (ed.). Vol. 11: William B. Jones and W. J. Thron, Continued Fractions.   ISBN 0-201-13510-8

3.2 Convergence of Periodic Continued Fractions

THEOREM 3.1. *Let* $K(a_n/b_n)$ *be a periodic continued fractio with period* k, *and let*

$$s_n(w) = \frac{a_n}{b_n + w}, \quad n = 1, 2, 3, \ldots, \tag{3.2.1a}$$

$$S_1(w) = s_1(w); \quad S_n(w) = S_{n-1}(s_n(w)), \, n = 2, 3, 4, \ldots, \tag{3.2.1b}$$

$$P(w) = S_k(w). \tag{3.2.1c}$$

Let w_1 *and* w_2 *be the (not necessarily distinct) fixed points of the l.f.t.* P *$[P(w) \not\equiv w]$, and let* B_n *denote the* nth *denominator of* $K(a_n/b_n)$. *Then* $K(a_n/b_n)$ *converges to* w_1 *if either of the following conditions is satisfied:*

$$w_1 = w_2, \tag{3.2.2}$$

$$\text{or} \quad w_1 \neq w_2, \quad \left| \frac{B_{k-1}w_2 + B_k}{B_{k-1}w_1 + B_k} \right| < 1$$

$$\text{and} \quad S_m(0) \neq w_2, \quad m = 1, 2, \ldots, k. \tag{3.2.3}$$

In all other cases the continued fraction diverges.

Before proving Theorem 3.1 we consider the special case with period $k = 1$. The continued fraction has the form

$$\frac{a}{b} + \frac{a}{b} + \frac{a}{b} + \cdots, \quad a \neq 0. \tag{3.2.4}$$

If $b = 0$ the approximants of (3.2.4) are alternatel∵ ∞ and 0. If $b \neq 0$ then (3.2.4) is equivalent to

$$\frac{a/b}{1} + \frac{a/b^2}{1} + \frac{a/b^2}{1} + \frac{a/b^2}{1} + \cdots.$$

Hence for the case $k = 1$ it suffices to study continued fractions of the form

$$\frac{a}{1} + \frac{a}{1} + \frac{a}{1} + \cdots, \quad a \neq 0. \tag{3.2.5}$$

For this case $P(w) = a/(1 + w)$ has fixed points

$$w_1 = \frac{-1 + \sqrt{1 + 4a}}{2}, \quad w_2 = \frac{-1 - \sqrt{1 + 4a}}{2}.$$

Thus $w_1 = w_2$ iff $a = -\frac{1}{4}$, in which case (3.2.5) converges to the value $w_1 = -\frac{1}{2}$. On the other hand, if $a \neq -\frac{1}{4}$, we let $a = -\frac{1}{4} + c$, where $c = |c|e^{i\gamma}$

$\neq 0$, $-\pi < \gamma \leqslant \pi$ and $\sqrt{1+4a} = \sqrt{4c} = 2\sqrt{|c|}\ e^{i\gamma/2}$. In this case the condition (3.2.3) of Theorem 3.1 will be satisfied iff

$$a = S_1(0) \neq w_2 = \frac{-1 - \sqrt{1+4a}}{2} \tag{3.2.6}$$

and

$$\left| \frac{B_0 w_2 + B_1}{B_0 w_1 + B_1} \right| = \left| \frac{w_2 + 1}{w_1 + 1} \right| < 1. \tag{3.2.7}$$

It is easily shown that (3.2.6) is implied by the fact that $a \neq 0$; (3.2.7) will hold iff

$$\mathrm{Re}(\sqrt{1+4a}) = 2\sqrt{|c|}\ \cos\frac{\gamma}{2} > 0.$$

Thus we obtain

THEOREM 3.2. *The periodic continued fraction* $\mathbf{K}(a/1)$ *converges for all non-zero complex numbers* a *unless* $a = -\frac{1}{4} + c$, *where* c *is real and negative. For* $a = -\frac{1}{4} + c$, $c = |c| e^{i\gamma}$, $-\pi < \gamma < \pi$ *the value of* $\mathbf{K}(a/1)$ *is*

$$w_1 = \frac{-1 + 2\sqrt{|c|}\ e^{i\gamma/2}}{2}. \tag{3.2.8}$$

When $a = 1$ in Theorem 3.2 we obtain the golden ratio

$$\frac{1}{1} + \frac{1}{1} + \frac{1}{1} + \cdots = \frac{\sqrt{5} - 1}{2} \doteq 0.61803398,$$

[see (2.1.5)]. From the difference equations (2.1.6) it follows that the nth numerator A_n and denominator B_n of $\mathbf{K}(1/1)$ satisfy

$$A_{-1} = 1, \quad A_0 = 0, \quad B_{-1} = 0, \quad B_0 = 1,$$
$$A_n = A_{n-1} + A_{n-2}, \quad B_n = B_{n-1} + B_{n-2}, \qquad n = 1, 2, 3, \ldots.$$

Here $A_{n+1} = B_n$ is the nth term in the well-known sequence of Fibonacci numbers $1, 2, 3, 5, 8, 13, 21, 34, 55, \ldots$.

For use in proving Theorem 3.1 we summarize some results on the classification of l.f.t.'s. An l.f.t.

$$P(w) = \frac{A + Cw}{B + Dw}, \qquad AD - BC \neq 0, \tag{3.2.9}$$

may have exactly one or exactly two fixed points, or else every point in the extended plane is a fixed point, in which case P is the identity transformation.

The fixed points of P satisfy the equation $P(w)=w$, or equivalently,

$$Dw^2+(B-C)w-A=0. \tag{3.2.10}$$

Infinity is a fixed point iff $D=0$. Every point is a fixed point of P iff

$$A=D=0 \quad \text{and} \quad C=B\neq0, \tag{3.2.11}$$

in which case

$$(C-B)^2+4AD=0. \tag{3.2.12}$$

If (3.2.12) holds but (3.2.11) does not hold, then there is exactly one fixed point w_1, which may be either finite or infinite. In that case P is said to be *parabolic*. If w_1 is finite, then $D\neq0$ and P can be written as

$$P(w)=\mu^{-1}\circ\varphi\circ\mu(w), \tag{3.2.13a}$$

where

$$\mu(w)=\frac{1}{w-w_1}, \quad \mu^{-1}(w)=w_1+\frac{1}{w}, \quad w_1=\frac{C-B}{2D}, \tag{3.2.13b}$$

and

$$\varphi(w)=w+l, \quad l=\frac{2D}{B+C}\neq0,\infty. \tag{3.2.13c}$$

We note that (3.2.13) can also be expressed as

$$\frac{1}{P(w)-w_1}=\frac{2D}{B+C}+\frac{1}{w-w_1}. \tag{3.2.14}$$

That $l\neq0$ follows from the assumption $D\neq0$; also $l\neq\infty$, since $B+C=0$ together with (3.2.12) imply that $AD-BC=0$, a contradiction of (3.2.9). If, on the other hand, $w_1=\infty$, then $D=0$, and hence by (3.2.12) and (3.2.9), $B=C\neq0$ and $A\neq0$. Thus P reduces to

$$P(w)=w+\frac{A}{B}, \quad A\neq0, \quad B\neq0. \tag{3.2.15}$$

The results of (3.2.13) and (3.2.15) may be combined into one statement as follows:

In the parabolic case

$$P(w)=\mu^{-1}\circ\varphi\circ\mu(w), \tag{3.2.16a}$$

where

$$\mu(w) = \begin{cases} \dfrac{1}{w-w_1} & \text{if} \quad w_1 \neq \infty, \\[2mm] w & \text{if} \quad w_1 = \infty, \end{cases} \qquad (3.2.16b)$$

and

$$\varphi(w) = w + l. \qquad (3.2.16c)$$

Here

$$l = \begin{cases} \dfrac{2D}{B+C} & \text{if} \quad w_1 \neq \infty, \\[2mm] \dfrac{A}{B} & \text{if} \quad w_1 = \infty. \end{cases} \qquad (3.2.16d)$$

In both cases we have

$$\mu(w_1) = \infty, \qquad l \neq 0, \infty. \qquad (3.2.16e)$$

If (3.2.12) does not hold, then P has exactly two fixed points w_1 and w_2. If they are both finite, then P can be written in the form

$$P(w) = \lambda^{-1} \circ \tau \circ \lambda(w), \qquad (3.2.17a)$$

where

$$\lambda(w) = \frac{w-w_1}{w-w_2}, \qquad \lambda^{-1}(w) = \frac{w_2 w - w_1}{w-1}, \qquad (3.2.17b)$$

and

$$\tau(w) = qw, \qquad q = \frac{C-w_1 D}{C-w_2 D} = \frac{B+w_2 D}{B+w_1 D} \neq 0, \infty. \qquad (3.2.17c)$$

Here the subscripts for w_1 and w_2 may be so chosen that $|q| < 1$. To obtain the second expression for q we note that

$$P(w) - w_i = P(w) - P(w_i) = \frac{(AD - BC)(w_i - w)}{(B+Dw)(B+Dw_i)}, \qquad i = 1, 2,$$

and hence

$$\lambda \circ P(w) = \frac{P(w) - w_1}{P(w) - w_2} = \frac{B+Dw_2}{B+Dw_1} \cdot \frac{w-w_1}{w-w_2}.$$

The first expression for q in (3.2.17c) can be obtained from the second by an application of the identity

$$(B+Dw_i)(C-Dw_i)=BC-D\left[Dw_i^2+(B-C)w_i\right]=BC-AD. \quad (3.2.18)$$

The last equality in (3.2.18) follows from (3.2.10). Now $w_j \neq -B/D, j=1,2$, for otherwise (3.2.18) would imply that $BC-AD=0$. It follows that $q \neq 0, \infty$. If $w_1 \neq w_2$ and one of the fixed points is ∞, we let w^* denote the finite fixed point. Then $D=0$, $B \neq C$, $BC \neq 0$. If in addition $|C/B| < 1$, we shall write

$$P(w)=\lambda^{-1} \circ \tau \circ \lambda(w), \quad (3.2.19a)$$

where

$$\lambda(w)=w-w^*, \quad \lambda^{-1}(w)=w+w^*, \quad w^*=\frac{A}{B-C}, \quad (3.2.19b)$$

$$\tau(w)=qw, \quad q=\frac{C}{B} \neq 0. \quad (3.2.19c)$$

If $|q| < 1$, we take $w_1=w^*$ and $w_2=\infty$. If $|C/B|>1$, we take $w_1=\infty$ and $w_2=w^*$ and set $\lambda(w)=1/(w-w_2)$. In this case

$$P(w)=\lambda^{-1} \circ \tau \circ \lambda(w), \quad (3.2.20a)$$

where

$$\lambda(w)=\frac{1}{w-w_2}, \quad w_2=\frac{A}{B-C}, \quad (3.2.20b)$$

and

$$\tau(w)=qw, \quad q=\frac{B}{C} \neq 0. \quad (3.2.20c)$$

If $w_1 \neq w_2$ it is convenient to distinguish three cases:

Case I. w_1 and w_2 are both finite (the subscripts have been chosen in such a way that $|C-w_1 D| < |C-w_2 D|$).
Case II. $D=0$, $|C/B| < 1$, $w_1=A/(B-C)$, $w_2=\infty$.
Case III. $D=0$, $|C/B|>1$, $w_1=\infty$, $w_2=A/(B-C)$.

The relations (3.2.17), (3.2.19) and (3.2.20) can now be combined into a common statement:

$$P(w)=\lambda^{-1} \circ \tau \circ \lambda(w), \quad \tau(w)=qw, \quad (3.2.21a)$$

where

$$
\lambda(w) = \begin{cases} \dfrac{w-w_1}{w-w_2} & \text{in case I,} \\ w-w_1 & \text{in case II,} \\ \dfrac{1}{w-w_2} & \text{in case III,} \end{cases} \tag{3.2.21b}
$$

and

$$
q = \begin{cases} \dfrac{C-w_1 D}{C-w_2 D} = \dfrac{B+w_2 D}{B+w_1 D} & \text{in case I,} \\ \dfrac{C}{B} & \text{in case II,} \\ \dfrac{B}{C} & \text{in case III.} \end{cases} \tag{3.2.21c}
$$

In all three cases

$$
\lambda(w_1)=0, \quad \lambda(w_2)=\infty, \quad |q|\le 1. \tag{3.2.21d}
$$

Depending on whether $|q|=1$, $q>0$, or q satisfies neither of these conditions, the transformation P is called *elliptic*, *hyperbolic*, or *loxodromic*. If $|q|<1$ then w_1 is called the *attractive* and w_2 the *repulsive* fixed point. If $|q|=1$, then both fixed points are called *indifferent*.

Using the chain rule of differentiation, one can easily show that if $w_1 \neq w_2$ and both fixed points are finite, then

$$
|P'(w_1)|=|q|\le 1, \quad |P'(w_2)|=|1/q|\ge 1.
$$

If either fixed point is infinite, then

$$
P'(w)=\frac{C}{B},
$$

and hence the result is valid in this case also.

If $w_1=w_2$, one finds in all cases that

$$
P'(w_1)=P'(w_2)=1.
$$

Since the indifferent fixed points of elliptic transformations satisfy

$$
|P(w_i)|=1,
$$

it makes sense to refer to the single fixed point of parabolic transformations also as indifferent. With this agreement w_1 will always be attractive or indifferent, while w_2 will be repulsive or indifferent.

Theorem 3.1 is a special case of the following more general

THEOREM 3.3. *Let* $\{t_n\}$ *be a periodic sequence of l.f.t.'s with period* k; *that is*,

$$t_{rk+m} = t_m, \quad m = 1, 2, \ldots, k; \quad r = 0, 1, 2, \ldots . \tag{3.2.22}$$

Let $\{T_n\}$ *denote the sequence generated by* $\{t_n\}$, *let* $P = T_k$, *and if* P *is not the identity transformation, let* w_1 *and* w_2 *denote the (not necessarily distinct) fixed points of* P.

(A) *If* $w_1 = w_2$ (P *parabolic*), *then* $\{T_n(w)\}$ *converges to* w_1 *for all* w.

(B) *If* $w_1 \neq w_2$ *and* P *is not elliptic, let* w_1 *denote the attractive fixed point of* P. *Then* $\{T_n(w)\}$ *converges to* w_1 *for all* $w \notin \Delta = [T_m^{-1}(w_2): m = 1, 2, \ldots, k]$. *If* Δ *contains more than one point, then* $\{T_n(w)\}$ *diverges by oscillation for all* $w \in \Delta$. *If* Δ *contains only one point, then* $\Delta = [w_2]$, $t_m(w_2) = w_2$ *for all* $m = 1, 2, \ldots, k$, *and* $\{T_n(w_2)\}$ *converges to* w_2.

(C) *If* $w_1 \neq w_2$ *and* P *is elliptic, then* $\{T_n(w)\}$ *diverges for all* $w \notin [w_1, w_2]$. *For* $j = 1, 2$, $\{T_n(w_j)\}$ *converges iff* $t_m(w_j) = w_j$ *for all* $m = 1, 2, \ldots, k$. *In this case*

$$T_n(w_j) = w_j, \quad n = 1, 2, \ldots .$$

(D) *If* P *is the identity transformation then* $\{T_n(w)\}$ *converges iff* $t_m(w) = w$ *for all* $m = 1, 2, \ldots, k$.

Proof. (A): Suppose $w_1 = w_2$. Then by (3.2.16)

$$T_{rk+m}(w) = P^r \circ T_m(w) = \mu^{-1} \circ \varphi^r \circ \mu(T_m(w))$$

$$= \mu^{-1}(\mu(T_m(w)) + rl),$$

$$m = 1, 2, \ldots, k, \quad r = 0, 1, 2, \ldots .$$

Hence

$$\lim_{n \to \infty} T_n(w) = \lim_{r \to \infty} T_{rk+m}(w) = \mu^{-1}(\infty) = w_1.$$

(B): Suppose $w_1 \neq w_2$ and P is not elliptic. Then by (3.2.21)

$$T_{rk+m}(w) = P^r \circ T_m(w) = \lambda^{-1} \circ \tau^r \circ \lambda \circ T_m(w)$$

$$= \lambda^{-1}(q^r \lambda(T_m(w))). \tag{3.2.23}$$

Thus if $T_m(w) \neq w_2$ for all $m = 1, 2, \ldots, k$, then

$$\lim_{r \to \infty} T_{rk+m}(w) = \lambda^{-1}(0) = w_1, \qquad m = 1, 2, \ldots, k,$$

and

$$\lim_{n \to \infty} T_n(w) = w_1 \qquad \text{for all} \quad w \notin \Delta.$$

Now suppose Δ contains at least two points w' and w'' and write

$$w' = T_{m'}^{-1}(w_2), \qquad 1 \leqslant m' \leqslant k,$$
$$w'' = T_{m''}^{-1}(w_2), \qquad 1 \leqslant m'' \leqslant k.$$

Clearly, since $w' \neq w''$, it follows that $m' \neq m''$. From (3.2.23) one then deduces

$$T_{rk+m'}(w') = \lambda^{-1}\big(q^r \lambda(T_{m'}(w')) = \lambda^{-1}(q^r \lambda(w_2))$$
$$= \lambda^{-1}(\infty) = w_2, \qquad r = 0, 1, 2, \ldots,$$

and

$$\lim_{r \to \infty} T_{rk+m''}(w') = \lim_{r \to \infty} \lambda^{-1}\big(q^r \lambda(T_{m''}(w'))\big)$$
$$= \lambda^{-1}(0) = w_1.$$

Hence $\{T_n(w')\}$ diverges by oscillation.

If Δ contains only one point, then

$$T_1^{-1}(w_2) = T_2^{-1}(w_2) = \cdots = T_k^{-1}(w_2) = P^{-1}(w_2) = w_2.$$

It follows that $\Delta = [w_2]$ and that $t_m(w_2) = w_2$ for all $m = 1, 2, \ldots, k$. Hence by (3.2.1b)

$$\lim_{n \to \infty} T_n(w_2) = w_2.$$

(C): Suppose $w_1 \neq w_2$ and P is elliptic. Then $q = e^{i\theta}$, where $\theta \neq 0 \pmod{2\pi}$. Let $w \notin [w_1, w_2]$. Then $\lambda(w) \notin [0, \infty]$, and (3.2.21) yields

$$T_{rk}(w) = \lambda^{-1}\big(e^{ir\theta} \lambda(w)\big), \qquad r = 1, 2, 3, \ldots,$$

and it is clear that $\{T_{rk}(w)\}_{r=1}^{\infty}$ diverges by oscillation. Hence $\{T_n(w)\}$ diverges.

If $t_m(w_1) = w_1$, $m = 1, 2, \ldots, k$, then $T_n(w_1) = w_1$ for all $n = 1, 2, \ldots$ and hence $\{T_n(w_1)\}$ converges to w_1. A similar statement holds if w_1 is replaced by w_2.

Now suppose that $t_m(w_1) \neq w_1$, for some m with $1 \leqslant m \leqslant k$. Then $T_m(w_1) \neq w_1$ for some m with $1 \leqslant m \leqslant k$ and

$$T_{rk+m}(w_1) = \lambda^{-1}(e^{ir\theta}\lambda(T_m(w_1))).$$

Hence $\{T_{rk+m}(w_1)\}_{r=0}^{\infty}$ diverges by oscillation unless $\lambda(T_m(w_1)) \in [0, \infty]$. Since $T_m(w_1) \neq w_1$, $\lambda(T_m(w_1)) \neq 0$. Moreover $\lambda(T_m(w_1)) = \infty$ iff $T_m(w_1) = w_2$. In that case

$$T_{rk+m}(w_1) = \lambda^{-1}(\infty) = w_2, \qquad r = 0, 1, 2, \ldots,$$

and

$$T_{rk}(w_1) = P^r(w_1) = w_1, \qquad r = 1, 2, 3, \ldots.$$

Hence $\{T_n(w_1)\}$ diverges. A similar argument holds if w_1 and w_2 are interchanged. This proves (C).

(D): Suppose $P(w) = w$. Then

$$T_{rk+m}(w) = T_m(w), \qquad m = 1, 2, \ldots, k, \quad r = 0, 1, 2, \ldots,$$

so that

$$\lim_{r \to \infty} T_{rk+m}(w) = T_m(w), \qquad m = 1, 2, \ldots, k.$$

Thus $\{T_n(w)\}$ is convergent iff

$$T_1(w) = T_2(w) = \cdots = T_k(w) \qquad (w \text{ fixed}).$$

This holds iff $t_m(w) = w$, $m = 1, 2, \ldots, k$.   ∎

Proof of Theorem 3.1. It follows from (2.1.7) in Theorem 2.1 that

$$P(w) = \frac{A_k + A_{k-1}w}{B_k + B_{k-1}w}, \qquad A_k B_{k-1} - A_{k-1} B_k \neq 0.$$

The proof is an application of Theorem 3.3. If P is parabolic ($w_1 = w_2$), the continued fraction converges to w_1. If P is not elliptic and not the identity transformation, and if $w_1 \neq w_2$ and both are finite, then the parameter q of (3.2.21c) is given by

$$q = \frac{B_k + B_{k-1}w_2}{B_k + B_{k-1}w_1}.$$

Hence if w_1 is the attractive fixed point of P, then $|q| < 1$ and so the continued fraction converges to w_1 if $S_m(0) \neq w_2$ for all $m = 1, 2, \ldots, k$. If $w_1 \neq w_2 = \infty$, then $\{S_n(0)\}$ can converge only if $S_m(0) \neq w_2$ for all $m = 1, 2, \ldots, k$. This cannot happen, since $S_{k-1}(0) = A_{k-1}/B_{k-1} = \infty$. If $w_1 = \infty$, then $\lim_{n \to \infty} S_n(0) = w_1 = \infty$. Thus $\{S_n(0)\}$ converges to ∞. In each of the other cases for which Theorem 3.3 might assert convergence of $\{S_n(0)\}$, it would be required that $S_m(0)=0$ for all $m=1,2,\ldots,k$. But $s_m(0)=a_m\neq0$ for all $m=1,2,\ldots,k$.   ■

3.3 Dual Periodic Continued Fractions

THEOREM 3.4 (Galois generalized). *Let* $\mathrm{K}(a_n/b_n)$ *be a periodic continued fraction of period* k, *let* A_n *and* B_n *denote the* n *th numerator and denominator, respectively, and let* P *denote the l.f.t.*

$$P(w)=\frac{A_k+A_{k-1}w}{B_k+B_{k-1}w}. \tag{3.3.1}$$

Then the continued fractions

$$\mathrm{K}(a_n/b_n)=\frac{a_1}{b_1}+\frac{a_2}{b_2}+\cdots+\frac{a_k}{b_k}+\frac{a_1}{b_1}+\frac{a_2}{b_2}+\cdots+\frac{a_k}{b_k}+\frac{a_1}{b_1}+\cdots \tag{3.3.2}$$

and

$$b_k+\frac{a_k}{b_{k-1}}+\frac{a_{k-1}}{b_{k-2}}+\cdots+\frac{a_2}{b_1}+\frac{a_1}{b_k}+\frac{a_k}{b_{k-1}}+\cdots+\frac{a_2}{b_1}+\frac{a_1}{b_k}+\cdots \tag{3.3.3}$$

converge and diverge together unless all of the following conditions are satisfied:

(i) *P has exactly two distinct fixed points* w_1 *and* w_2, *and* P *is not elliptic,*
(ii)

$$|B_k-w_2B_{k-1}|<|B_k-w_1B_{k-1}| \tag{3.3.4}$$

(i.e., w_1 *is the attractive fixed point of* P*), and*
(iii) *either*
(a) $A_m/B_m=w_1$ *for some but not all* $m=1,2,\ldots,k$, *and* $A_m/B_m\neq w_2$ *for all* $m=1,2,\ldots,k$, *in which case (3.3.2) converges to* w_1 *and (3.3.3) diverges, or*
(b) $A_m/B_m=w_2$ *for some but not all* $m=1,2,\ldots,k$ *and* $A_m/B_m\neq w_1$ *for all* $m=1,2,\ldots,k$, *in which case (3.3.2) diverges and (3.3.3) converges to* $-w_2$.

If both continued fractions converge, then (3.3.2) converges to w_1, *(the attractive or indifferent fixed point of* P*) and (3.3.3) converges to* $-w_2$.

For reasons pointed out below, (3.3.3) is called the *dual continued fraction* of the periodic continued fraction (3.3.2). Our proof of Theorem 3.4 is based on the more general result for l.f.t.'s given by Theorem 3.5. For that purpose we give the following definitions. Let $\{t_n\}$ be a periodic sequence of l.f.t.'s with period k, so that

$$t_{rk+m}=t_m, \qquad m=1,2,\ldots,k, \quad r=0,1,2,\ldots. \tag{3.3.5}$$

Corresponding to each t_n we define a l.f.t., called the *dual* of t_n with respect to the periodic sequence $\{t_n\}$ and denoted by t_n^D, as follows:

$$t_{rk+m}^D = t_m^D = t_{k-m+1}^{-1}, \qquad m=1,2,\ldots,k, \quad r=0,1,2,\ldots. \qquad (3.3.6)$$

It is easily seen that $(t_n^D)^D = t_n$, $n=1,2,3,\ldots$. Clearly $\{t_n^D\}$ is periodic, with period k, and it is called the *dual sequence* of $\{t_n\}$. Next let $\{T_n\}$ and $\{T_n^D\}$ denote the sequences generated by $\{t_n\}$ and $\{t_n^D\}$, respectively, and let $P = T_k$ and $P^D = T_k^D$. Then it can be seen that $P^D = P^{-1}$ and that

$$T_{rk+m} = P^r \circ T_m, \qquad m=1,2,\ldots,k, \quad r=0,1,2,\ldots, \qquad (3.3.7)$$

and

$$T_{rk+m}^D = P^{-r} \circ T_m^D = P^{-r} \circ t_k^{-1} \circ t_{k-1}^{-1} \circ \cdots \circ t_{k-m+1}^{-1},$$

$$m=1,2,\ldots,k, \quad r=0,1,2,\ldots. \qquad (3.3.8)$$

THEOREM 3.5. *Let $\{t_n\}$ be a periodic sequence of l.f.t's with period k, let $\{t_n^D\}$ denote its dual sequence, and let $\{T_n\}$ and $\{T_n^D\}$ denote the sequences generated by $\{t_n\}$ and $\{t_n^D\}$, respectively. Then the two sequences $\{T_n\}$ and $\{T_n^D\}$ converge and diverge together unless all of the following conditions are satisfied:*

(i) *P has exactly two fixed points w_1 and w_2 ($P = t_1 \circ \cdots \circ t_k$),*
(ii) *w_1 is the attractive fixed point of P (P is not elliptic), and*
(iii) *either*

(a) *$T_m(w) = w_1$ for some but not all $m=1,2,\ldots,k$ and $T_m(w) \neq w_2$ for all $m=1,2,\ldots,k$, in which case $\{T_m(w)\}$ converges to w_1 and $\{T_n^D(w)\}$ diverges, or*

(b) *$T_m(w) = w_2$ for some but not all $m=1,2,\ldots,k$ and $T_m(w) \neq w_1$ for all $m=1,2,\ldots,k$, in which case $\{T_m(w)\}$ diverges and $\{T_n^D(w)\}$ converges to w_2.*

If both $\{T_n(w)\}$ and $\{T_n^D(w)\}$ converge, then one of the following occurs:

(a) *P is parabolic and $\lim T_n(w) = \lim T_n^D(w) = w_1$, the fixed point of P.*
(b) *P is hyperbolic or loxodromic, and $\lim T_n(w) = w_1$ (the attractive fixed point of P) while $\lim T_n^D(w) = w_2$ (the repulsive fixed point of P).*
(c) *$T_m(w) = w$ for all $m=1,2,\ldots,k$, and $\lim T_n(w) = \lim T_n^D(w) = w$.*

Proof of Theorem 3.5. Since $\{T_n\}$ and $\{T_n^D\}$ are generated by periodic sequences of l.f.t.'s, their convergence behavior is governed by Theorem 3.3. Since P and P^{-1} have the same fixed points, one obtains from the remarks on the classification of l.f.t.'s following (3.2.21) that if w_1 is an attractive, repulsive or indifferent fixed point of P, then w_2 is a repulsive, attractive or indifferent fixed point, respectively, of P^{-1}. Further P and P^{-1} have the same classification—elliptic, hyperbolic, loxodromic, parabolic or the identity. In the following we describe the consequences of Theorem 3.3 for each of the possible classifications of P.

If P is the *identity*, then $\{T_n\}$ and $\{T_n^D\}$ converge and diverge together for all w; they converge iff $t_m(w) = w$ for all $m = 1, 2, \ldots, k$, and when convergent, $\lim T_n(w) = \lim T_n^D(w) = w$.

If P is *parabolic*, then $\{T_n(w)\}$ and $\{T_n^D(w)\}$ both converge for all w to the fixed point of P.

If P is *elliptic*, then for all $w \notin [w_1, w_2]$ (the fixed points of P) $\{T_n(w)\}$ and $\{T_n^D(w)\}$ both diverge. For $j = 1$ or 2, $\{T_n(w_j)\}$ converges iff $t_m(w_j) = w_j$ for all $m = 1, 2, \ldots, k$; under these conditions $T_n(w_j) = w_j$. Similarly $t_m(w_j) = w_j$ for all $m = 1, 2, \ldots, k$ iff $t_m^D(w_j) = w_j$ for all $m - 1, 2, \ldots, k$. Thus $\{T_n(w_j)\}$ and $\{T_n^D(w_j)\}$ converge and diverge together, and when convergent their common limit is w_j.

If P is either *hyperbolic or loxodromic* (so that P has exactly two fixed points w_1 and w_2, and P is not elliptic), then we let

$$\Delta = \left\{ T_m^{-1}(w_2) : m = 1, 2, \ldots, k \right\}$$

and

$$\Delta^D = \left[(T_m^D)^{-1}(w_1) : m = 1, 2, \ldots, k \right]$$

where w_1 is the attractive fixed point of P and w_2 is the attractive fixed point of $P^D = P^{-1}$. Then for all $w \notin \Delta \cup \Delta^D$, $\{T_n(w)\}$ converges to w_1 and $\{T_n^D(w)\}$ converges to w_2. It is easily seen that

$$T_{kr+m}(w) = P^r \circ T_m(w) = \lambda^{-1}\left[q^r \lambda(T_m(w)) \right],$$
$$m = 1, 2, \ldots, k, \quad r = 0, 1, 2, \ldots, \tag{3.3.9}$$

where P is expressed in terms of λ and q as in (3.2.21). Now suppose that $T_m(w) \neq w_2$ for all $m = 1, 2, \ldots, k$. It follows from (3.2.21) that $\lambda(T_m(w)) \neq \infty$ for $m = 1, 2, \ldots, k$, and hence by (3.3.9) that $\lim T_n(w) = \lambda^{-1}(0) = w_1$. On the other hand, if $T_m(w) = w_1$ for some but not all $m = 1, 2, \ldots, k$, then $\{T_n^D(w)\}$ diverges by oscillation, since Δ^D contains more than one point and since it is readily seen that

$$\Delta^D = \left\{ T_m^{-1}(w_1) : m = 1, 2, \ldots, k \right\}. \tag{3.3.10}$$

Note that by (3.2.21) we can write

$$T_{kr+m}^D(w) = P^{-r-1} \circ T_{k-m}(w) = \lambda^{-1}\left[(1/q)^{r+1} \lambda(T_{k-m})(w) \right],$$
$$m = 1, 2, \ldots, k, \quad r = 0, 1, 2, \ldots, \tag{3.3.11}$$

where $|q| < 1$ and $T_0(w) = w$. The condition $T_m(w) \neq w_2$ for $m = 1, 2, \ldots, k$ implies the condition $T_{k-m}(w) \neq w_2$ for $m = 1, 2, \ldots, k$. It follows that $\lambda(T_{k-m}(w)) \neq \infty$ for $m = 1, 2, \ldots, k$ and hence, by (3.3.11), $\lim T_n^D(w) = \lambda^{-1}(\infty) = w_2$. ∎

Proof of Theorem 3.4. Let

$$t_n(w) = \frac{a_n}{b_n + w}, \qquad n = 1, 2, 3, \ldots \tag{3.3.12}$$

where $a_{rk+m}=a_m$ and $b_{rk+m}=b_m$, $m=1,2,\ldots,k$ and $r=0,1,2,\ldots$. Then the dual sequence $\{T_n^D\}$ is given by

$$t_{kr+m}^D(w)=t_m^D(w)=t_{k-m+1}^{-1}(w)=-b_{k-m+1}+\frac{a_{k-m+1}}{w},$$

$$m=1,2,\ldots,k,\quad r=0,1,2,\ldots. \tag{3.3.13}$$

The convergence of the sequence $\{T_n(w)\}$ and $\{T_n^D(w)\}$ generated by $\{t_n\}$ and $\{t_n^D\}$, respectively, is governed by Theorem 3.5. We note that $T_n(0)$ is the nth approximant of (3.3.2). We shall also consider a periodic c.f.g. sequence $\{s_n\}$ of period k defined as follows:

$$s_{rk+m}(w)=s_m(w)=v_{m-1}^{-1}\circ t_m^D\circ v_m(w),$$

$$m=1,2,\ldots,k,\quad r=0,1,2,\ldots, \tag{3.3.14a}$$

$$v_{rk+m}(w)=v_m(w)=w-b_{k-m},$$

$$m=1,2,\ldots,k,\quad r=0,1,2,\ldots \quad (b_0=b_k). \tag{3.3.14b}$$

It is easily seen that

$$s_m(w)=\frac{a_{k-m+1}}{-b_{k-m}+w},\qquad m=1,2,\ldots,k,$$

since $b_0=b_k$. Thus if $\{S_n\}$ denotes the sequence generated by $\{s_n\}$, we see that $S_n(0)$ is the nth approximant of the continued fraction

$$\frac{a_k}{-b_{k-1}}+\frac{a_{k-1}}{-b_{k-2}}+\cdots+\frac{a_2}{-b_1}+\frac{a_1}{-b_k}+\frac{a_k}{-b_{k-1}}+\cdots$$

$$+\frac{a_2}{-b_1}+\frac{a_1}{-b_k}+\cdots,$$

which is equivalent to the continued fraction

$$\frac{-a_k}{b_{k-1}}+\frac{a_{k-1}}{b_{k-2}}+\cdots+\frac{a_2}{b_1}+\frac{a_1}{b_k}+\frac{a_k}{b_{k-1}}+\cdots+\frac{a_2}{b_1}+\frac{a_1}{b_k}+\cdots.$$

It follows that $b_k-S_n(0)$ is the nth approximant of the continued fraction (3.3.3). It follows from (3.3.14) that

$$T_{rk+m}^D(0)=-b_k+S_{rk+m}(b_{k-m}),\qquad m=1,2,\ldots,k,\quad r=0,1,2,\ldots$$

$$=-b_k+S_{rk+m-2}(0),\qquad rk+m=3,4,5,\ldots.$$

Thus $\{T_n^D(0)\}$ and the continued fraction (3.3.3) converge and diverge together, and the limit of (3.3.3), when convergent, is equal to $-\lim T_n^D(0)$. The remainder of the proof consists in applying Theorem 3.5 to the sequences $\{T_n(0)\}$ and $\{T_n^D(0)\}$. We leave this straightforward argument to the reader, but point out that the condition $T_m(0)=0$ for $m=1,2,\ldots,k$ cannot be satisfied, since $T_1(0)=t_1(0)=a_1\neq0$. ∎

CHAPTER 4

Convergence of Continued Fractions

4.1 Introduction

There exists such a wealth of convergence criteria for continued fractions that a selection has been unavoidable. We present the results that have been the most useful for applications, as well as the theorems that appear to us as having the most promise. That the potential of this latter group has not been fully realized we believe is due to the fact that it is relatively unknown. Many of these results are of recent vintage.

Among the best known and most often used convergence criteria are those of Worpitzky [1865, Corollary 4.36(B)], Pringsheim [1899, Theorem 4.35, Corollary 4.36] and Van Vleck [1901a, Theorem 4.29]. Not quite so well known, but surely of great importance, are the parabola theorems, of which we give a simple version in Theorem 4.40, and stronger versions in Theorems 4.42 and 4.43. Of the relatively unknown results we would like to emphasize Theorem 4.33, which is an extremely general criterion for continued fractions $K(1/b_n)$, and Theorem 4.49, which is a powerful twin-convergence-region theorem for continued fractions $K(a_n/1)$. Not enough attention has been paid in the past to results which insure convergence of the even or odd part of a continued fraction. Such results have been considered as only a first step in proving convergence of the whole continued fraction. We feel that these criteria should be considered in their own right and give examples in Theorem 4.31 and Theorem 4.50. More results of this kind can be found in [Jones and Thron, 1970].

While the theorems of Worpitzky, Pringsheim and Van Vleck have stood the test of time, there are others which have fallen by the wayside or are only of historical interest and have been omitted. Among these are the

ENCYCLOPEDIA OF MATHEMATICS and Its Applications, Gian-Carlo Rota (ed.). Vol. 11: William B. Jones and W. J. Thron, Continued Fractions. ISBN 0-201-13510-8

elaborate and difficult-to-prove results on limit-periodic continued fractions. For our purposes we have been able to replace these by Theorem 4.45, a convergence-neighborhood result, which can be derived from the parabola theorem (Theorem 4.40). Perhaps even more important than the statements of convergence criteria is the description of methods of proof contained in this chapter. The methods that have been most useful and those that appear most promising in establishing new general or *ad hoc* criteria have been emphasized. Most of the early proofs of the classical theorems and those exploiting the "fundamental inequalities" involve largely special manipulations of the difference equations (2.1.6) for the nth numerator A_n and denominator B_n and thus do not constitute general approaches. Nevertheless a few examples of these are also included (e.g., Theorem 4.19).

All of the known general methods are based on value-region considerations; hence value regions and corresponding element regions are treated at length in Section 4.2. The interplay between element regions and value regions leads to convergence region criteria, that is, results of the form: *If* $\langle a_n, b_n \rangle \in \Omega_n \subseteq \mathbb{C} \times \mathbb{C}$ *for all* $n = 1, 2, 3, \ldots$, *then the continued fraction* $\mathbf{K}(a_n/b_n)$ *converges.* In addition, the relationship between element regions and value regions provides one with knowledge of the location of the approximants of the continued fraction $\mathbf{K}(a_n/b_n)$ whose elements $\langle a_n, b_n \rangle$ lie in some convergence regions Ω_n. Both of these phenomena (i.e., the convergence regions and the information about the location of the approximants) are not to be found for most common infinite processes, such as series and products. One general method is not dealt with here: the theory of positive definite continued fractions. The reasons for this omission are that an adequate presentation would require a great deal of space and that a rather detailed discussion of the theory can be found in [Wall, 1948]. In this connection we note that truncation-error bounds for positive definite continued fractions are treated in Chapter 8.

Three general methods for developing convergence criteria are described in this chapter. The first method involves knowledge of the location of the continued fraction approximants. It then uses this information together with the Stieltjes-Vitali theorem (Theorem 4.30) to extend convergence already known for a small region to a larger region. This method is illustrated in the proof of Theorem 4.29.

In the second method we start with the assumption that there exist sequences of element regions $\{\Omega_n\}$ and value regions $\{V_n\}$ such that

$$s_n(V_n) \subseteq V_{n-1} \quad \text{for all} \ \langle a_n, b_n \rangle \in \Omega_n, \quad n = 1, 2, 3, \ldots. \quad (4.1.1)$$

Here a_n and b_n denote the elements of the continued fraction $\mathbf{K}(a_n/b_n)$, and s_n is the l.f.t. $s_n(w) = a_n/(b_n + w)$. From this one deduces (Theorem 4.2 in Section 4.2) that all approximants f_{n+m} $(m = 0, 1, 2, \ldots)$ lie in nested regions $K_n = S_n(V_n)$, where $\{S_n\}$ is the sequence of l.f.t. generated by $\{s_n\}$. If, in particular, the K_n are circular disks, one can determine the radius R_n of K_n either by using the fact that the circumference of K_n is $2\pi R_n$ so that

$$2\pi R_n = \int_{\partial V_n} |S_n'(w)| \, dw, \qquad n = 1, 2, 3, \ldots, \qquad (4.1.2)$$

or by using the geometry of inversions. The first approach [using (4.1.2)] has the advantage that it could be used even if the K_n are not circular disks but more general regions. It is employed in the proof of Theorem 4.46. The other approach, using inversion of circles, is illustrated in the proof of Theorem 4.40. Once an explicit formula for R_n is obtained, one considers the ratio R_n/R_{n-1}, which turns out to be a function of the various parameters defining the V_n, the elements a_n, b_n of the continued fraction $K(a_n/b_n)$ and the ratios B_n/B_{n-1}, where B_n denotes the nth denominator of $K(a_n/b_n)$. For continued fractions $K(a_n/1)$ one can show (Corollary 4.7) that

$$\frac{B_n}{B_{n-1}} = 1 + \frac{a_n}{1} + \frac{a_{n-1}}{1} + \cdots + \frac{a_2}{1}, \qquad (4.1.3)$$

so that these quantities can be estimated by applying value-theoretic considerations. One then shows that $R_n \to 0$ and hence the continued fraction converges. Although this method requires a great deal of close analysis, it does yield useful truncation-error bounds, since one obtains explicit estimates of R_n. This is further exploited in Chapter 8.

The third method, which is illustrated in the proof of Theorem 4.37, also uses element regions, value regions and the relations (4.1.1) to arrive at nested circular disks K_n which contain the approximants f_{n+m} ($m = 0, 1, 2, \ldots$) of the continued fraction $K(a_n/b_n)$. However, instead of obtaining an explicit estimate of R_n, one argues that either $\lim R_n = 0$ (the limit-point case, in which the continued fraction converges to a finite value), or else $\lim R_n = R > 0$ (the limit-circle case, in which case by an indirect argument it can be shown that the continued fraction still converges to a finite value). This method has proved to be very powerful and has led to many new results (see, for example, [Jones and Thron, 1968, 1970] and [Jones and Snell, 1972]). One disadvantage of the method is that it gives no information about the speed of convergence of the continued fraction.

In choosing examples for these methods, we have selected theorems which are significant but not so involved that the flavor of the procedure is lost under the weight of the detail.

Section 4.4 is devoted to a study of continued fractions $K(a_n/b_n)$ such that, for each n, the elements a_n, b_n are constants. Although it is by far the largest section of this chapter, its main importance lies in preparing the way for the last section, in which continued fractions with variable elements $a_n(z), b_n(z)$ are considered. In particular we are interested in the cases where $a_n(z)$ and $b_n(z)$ are holomorphic functions of one (or several) complex variables z for z in some domain D.

Whenever information is available about the values taken on by the approximants, as it is in most of the situations considered here, one need not be much concerned about uniform convergence in Section 4.4. By adapting Montel's generalization of the Stieltjes-Vitali theorem to our purposes, we obtain in Theorem 4.54 in Section 4.5 as much uniformity as is needed, for functions of one or several complex variables.

Section 4.5 is the culmination of this chapter. In it we give the basic convergence theorems for the most important forms of continued-fraction expansions, including the regular C-fractions, J-fractions, and general T-fractions. Stieltjes's theorem for $K(a_n z/1)$ with all $a_n > 0$ is obtained as a corollary of the cardioid theorem, which in turn is derived from the parabola theorem. We present a fairly strong result for J-fractions as well as a number of convergence criteria for general T-fractions. For other special situations which one may encounter, the reader should be able to combine theorems in Section 4.4 with Theorem 4.54 to derive convergence results for the particular continued-fraction expansion of interest.

Convergence behavior of general sequences of linear fractional transformations has been studied systematically by Piranian and Thron [1957], Erdös and Piranian [1959], De Pree and Thron [1962], Thron [1963], Mandell and Magnus [1970] and Gill [1973].

Finally we mention the following special features of the present chapter: In Section 4.3 we give two necessary conditions for convergence regions in terms of properties of their associated value regions. In Theorems 4.52 and 4.53 we give two rather curious results due to Farinha [1954] and Wall [1956] relating boundedness of a subsequence of the elements with the convergence behavior of the odd and even part of continued fractions $K(a_n/1)$.

4.2 Element Regions, Value Regions, and Sequences of Nested Circular Disks

Throughout this chapter we shall consider continued fractions of the form $K(a_n/b_n)$. Since we are concerned here with matters of convergence, there is clearly no loss of generality in having the initial element $b_0 = 0$. Thus, in accordance with (2.1.1), we have $s_0(w) = w$ and

$$s_n(w) = \frac{a_n}{b_n + w}, \qquad n = 1, 2, 3, \ldots, \qquad (4.2.1a)$$

and

$$S_1(w) = s_1(w); \qquad S_n(w) = S_{n-1}(s_n(w)), \qquad n = 2, 3, \ldots . \qquad (4.2.1b)$$

Hence

$$S_n(w) = s_1 \circ s_2 \circ \cdots \circ s_n(w), \qquad n = 1, 2, 3, \ldots .$$

The term *region* will be used loosely to mean any subset of $\hat{\mathbb{C}}$ or $\mathbb{C} \times \mathbb{C}$, respectively. For the more important applications the sets are usually regions in the strict sense, i.e. open connected sets plus part, all or none of their boundaries. A number of the basic theorems are, however, valid for the more general sets. This is made clear as we proceed.

Most of the important convergence criteria for continued fractions are of the convergence-region type. This leads us to consider element and value regions. More specifically we will be concerned with sequences $\{V_n\}$ and $\{\Omega_n\}$, where $\varnothing \neq V_n \subseteq \hat{\mathbb{C}}$, $n = 0, 1, 2, \ldots$, and $\varnothing \neq \Omega_n \subseteq \mathbb{C} \times \mathbb{C}$, $n = 1, 2, 3, \ldots$, for which the following conditions are satisfied:

$$a_n / b_n \in V_{n-1} \quad \text{for } \langle a_n, b_n \rangle \in \Omega_n, \qquad n = 1, 2, 3, \ldots, \qquad (4.2.2a)$$

$$s_n(V_n) \subseteq V_{n-1} \quad \text{for } \langle a_n, b_n \rangle \in \Omega_n, \qquad n = 1, 2, 3, \ldots . \qquad (4.2.2b)$$

Note that

$$0 \in V_n, \qquad n = 1, 2, 3, \ldots, \qquad (4.2.3)$$

together with (4.2.2b) implies (but is not equivalent to) (4.2.2a).

We shall call $\{\Omega_n\}$ a *sequence of element regions* and $\{V_n\}$ a *sequence of value regions* if the two sequences are connected by (4.2.2). If the two sequences $\{\Omega_n\}$ and $\{V_n\}$ are connected by (4.2.2), this will be expressed by saying that $\{V_n\}$ is a sequence of value regions *corresponding* to or *induced* by $\{\Omega_n\}$, or that $\{\Omega_n\}$ is a sequence of element regions *belonging* to the sequence of value regions $\{V_n\}$. If $\{V_n\}$ is a sequence of value regions corresponding to a sequence of elements regions $\{\Omega_n\}$, and if for every sequence $\{V_n'\}$ of value regions corresponding to $\{\Omega_n\}$ we have $V_n \subseteq V_n'$, $n = 0, 1, 2, \ldots$, then $\{V_n\}$ will be called the *best* sequence of value regions corresponding to the sequence of element regions $\{\Omega_n\}$. Similarly we can speak of a *best* sequence of element regions belonging to a sequence of value regions.

In the following theorem it is shown that, for any given sequence of element regions, a best sequence of value regions exists and is unique.

THEOREM 4.1. *Let* $\{\Omega_n\}$ *be a sequence of element regions, and let* $\{V_n\}$ *be a sequence of value regions corresponding to it. Then*:

(A) *For* $k = 1, 2, 3, \ldots$,

$$s_k \circ s_{k+1} \circ \cdots \circ s_m(0) \in V_{k-1} \qquad (4.2.4)$$

for all m *such that* $1 \leqslant k \leqslant m$, *provided*

$$\langle a_\nu, b_\nu \rangle \in \Omega_\nu, \qquad for \quad k \leqslant \nu \leqslant m.$$

(B) *If, for* $k = 1, 2, 3, \ldots$, *one defines*

$$W_{k-1} = \left[s_k \circ s_{k+1} \circ \cdots \circ s_m(0) : \langle a_\nu, b_\nu \rangle \in \Omega_\nu, 1 \leqslant k \leqslant \nu \leqslant m \right], \quad (4.2.5)$$

then $\{W_n\}$ *is a sequence of value regions corresponding to* $\{\Omega_n\}$. *In* (4.2.5) m *is allowed to vary over all* $m \geqslant k$.

(C) *For* $n = 0, 1, 2, \ldots$ *we have* $W_n \subseteq V_n$, *and hence* $\{W_n\}$ *is the uniquely determined best sequence of value regions corresponding to* $\{\Omega_n\}$.

Proof. (A): The condition (4.2.2a) insures that $s_k(0) = a_k / b_k \in V_{k-1}$; (4.2.2b) and induction then establish (4.2.4) for all permissible k and m. (B): It follows directly from the definition of $\{W_n\}$ that the sequence satisfies (4.2.2) and hence is a sequence of value regions corresponding to $\{\Omega_n\}$. (C): The inclusion relation $W_n \subseteq V_n$ is a restatement of (4.2.4). Since $\{V_n\}$ is an arbitrary sequence of value regions, it follows that $\{W_n\}$ is the uniquely determined best sequence of value regions for $\{\Omega_n\}$. ∎

If $\{\Omega_n\}$ is any sequence of non-empty regions in $\mathbb{C} \times \mathbb{C}$, then regions W_n as defined in (4.2.5) always exist. Some or all of the W_n may be equal to $\hat{\mathbb{C}}$.

At this point it is convenient to introduce

$$\mathfrak{R} = [V : V \subseteq \hat{\mathbb{C}} \text{ such that } V \text{ is closed and the boundary of } V \text{ is a circle or straight line}].$$

$\mathfrak{R}$ then is the family of all closed circular disks, half planes and complements of circular disks. The following result plays a key role in many of our convergence proofs.

THEOREM 4.2. *Let* $\{\Omega_n\}$ *be a sequence of element regions, let* $\{V_n\}$ *be a corresponding sequence of value regions, and define*

$$K_n = S_n(V_n), \qquad n = 1, 2, 3, \ldots. \qquad (4.2.6)$$

Then:

(A)

$$K_{n+1} \subseteq K_n \subseteq V_0, \qquad n = 1, 2, 3, \ldots \qquad (4.2.7)$$

If, in addition, $V_n \in \Re$ for all $n \geqslant n_0$ and if K_{n_0} is a closed circular disk, then:

(B) *All K_n, $n \geqslant n_0$, are closed circular disks. If R_n is the radius and C_n the center of K_n, then*

$$\lim R_n = R \text{ and } \lim C_n = C \text{ both exist.}$$

(C) *If $R > 0$, then the closed circular disk K of radius R and center C is called the limit circle, and*

$$K = \bigcap \; [K_n : n \geqslant n_0].$$

(D) *If $R = 0$ then C is called the limit point and*

$$[C] = \bigcap \; [K_n : n \geqslant n_0].$$

Proof. (A): Using (4.2.1) and (4.2.2) one has, for $n = 1, 2, 3, \ldots,$

$$K_{n+1} = S_{n+1}(V_{n+1}) = S_n(s_{n+1}(V_{n+1})) \subseteq S_n(V_n) = K_n \subseteq V_0.$$

(B): Since K_n is the image under the l.f.t. S_n of V_n, it follows from $V_n \in \Re$ that $K_n \in \Re$. Since $K_n \subseteq K_{n_0}$ for all $n \geqslant n_0$ and since the only elements of $\Re$ which are subsets of a circular disk are themselves circular disks, it follows that all K_n, $n \geqslant n_0$, are closed circular disks. From (4.2.7) it now follows that

$$|C_{n+1} - C_n| + R_{n+1} \leqslant R_n, \qquad n \geqslant n_0. \qquad (4.2.8)$$

From this it is clear that the sequence $\{R_n\}$ is a monotone decreasing sequence of positive numbers and hence $\lim R_n = R$ exists and is positive or zero. Successive application of (4.2.8) yields $|C_{n+m} - C_n| \leqslant R_n - R_{n+m}$. Since $R_{n+m} \geqslant R$, we then have

$$|C_{n+m} - C_n| \leqslant R_n - R; \qquad (4.2.9)$$

that is, $\{C_n\}$ is a Cauchy sequence in $\mathbb{C}$ and hence converges to a limit in $\mathbb{C}$.

Now let $P \in K$. Then $|P - C| \leqslant R$ and hence, for $n \geqslant n_0$,

$$|P - C_n| \leqslant |P - C| + |C - C_n| \leqslant R + R_n - R = R_n. \qquad (4.2.10)$$

Here the inequality $|C - C_n| \leqslant R_n - R$ is obtained from (4.2.9) by letting $m \to \infty$. From (4.2.10) the intersection property of K (or $[C]$ if $R = 0$) follows immediately, proving (C) and (D). ∎

Depending upon whether $R > 0$ or $R = 0$ in Theorem 4.2, one speaks of the *limit-circle case* or of the *limit-point case* for the sequence $\{K_n\}$.

It would appear natural that, for a given sequence of element regions $\{\Omega_n\}$, one would determine the best corresponding value regions $\{W_n\}$ as defined in Theorem 4.1. However, this process is in general extremely difficult to carry out. Instead it has proved more feasible to compute $\{\Omega_n\}$ for a given sequence $\{V_n\}$. We describe here two approaches for doing this. The first goes back to Lane [1945] and is illustrated by the following:

THEOREM 4.3. *Let* $\{V_n\}$ *be a sequence of circular disks*

$$V_n = [\, w : |w - \Gamma_n| \leqslant \rho_n \,], \qquad |\Gamma_n| < \rho_n, \qquad n = 0, 1, 2, \ldots . \quad (4.2.11)$$

Let

$$\Omega_n = \Big[\, \langle a_n, b_n \rangle : \big| a_n(\bar{b}_n + \bar{\Gamma}_n) - \Gamma_{n-1}(|b_n + \Gamma_n|^2 - \rho_n^2) \big| + |a_n|\rho_n$$

$$\leqslant \rho_{n-1}(|b_n + \Gamma_n|^2 - \rho_n^2) \,\Big], \qquad n = 1, 2, 3, \ldots . \quad (4.2.12)$$

Then $\{V_n\}$ *is a sequence of value regions corresponding to the sequence of element regions* $\{\Omega_n\}$.

An application of this result in proving convergence of continued fractions is given by Theorem 4.39.

Proof. First we observe that Ω_n is not empty. This follows from the second inequality in (4.2.11) by taking $a_n = 0$ in (4.2.12). In order for the inequality in (4.2.12) to hold for a pair of complex numbers $\langle a_n, b_n \rangle$ it is necessary that $|b_n + \Gamma_n| \geqslant \rho_n$; if $a_n \neq 0$ it is necessary that

$$|b_n + \Gamma_n| > \rho_n. \qquad (4.2.13)$$

The region $(b_n + V_n)$ is the closed circular disk with center at $(b_n + \Gamma_n)$ and radius ρ_n. The condition (4.2.13) insures that $0 \notin (b_n + V_n)$, so that $1/(b_n + V_n)$ is itself a circular disk. Let $\gamma_n = \arg(b_n + \Gamma_n)$, and consider the two points

$$c_n = b_n + \Gamma_n - \rho_n e^{i\gamma_n}, \qquad d_n = b_n + \Gamma_n + \rho_n e^{i\gamma_n}$$

which lie on opposite sides of a diameter of the circle $\partial(b_n + V_n)$. Then $1/c_n$ and $1/d_n$ are on opposite sides of a diameter of the circle $\partial(1/(b_n + V_n))$. This can be seen by observing that the ray $\arg w = \gamma_n$ is orthogonal to the circle $\partial(b_n + V_n)$, so that the ray $\arg w = -\gamma_n$ must be orthogonal to the circle $\partial(1/(b_n + V_n))$. For the radius h_n and center k_n of the circle $\partial(1/(b_n + V_n))$ we thus obtain

$$h_n = \frac{1}{2}\left|\frac{1}{c_n} - \frac{1}{d_n}\right|, \qquad k_n = \frac{1}{2}\left(\frac{1}{c_n} + \frac{1}{d_n}\right).$$

For the circular disk $a_n/(b_n + V_n)$, with $a_n \neq 0$, one arrives at radius p_n and center q_n given by

$$p_n = |a_n| h_n, \qquad q_n = a_n k_n.$$

Substituting the values of c_n and d_n, one obtains

$$p_n = \frac{|a_n| \rho_n}{|b_n + \Gamma_n|^2 - \rho_n^2}, \qquad q_n = \frac{a_n(\bar{b}_n + \bar{\Gamma}_n)}{|b_n + \Gamma_n|^2 - \rho_n^2}.$$

The condition for $a_n/(b_n + V_n) = s_n(V_n)$ to be contained in V_{n-1} becomes

$$|\Gamma_{n-1} - q_n| + p_n \leqslant \rho_{n-1},$$

which is easily seen to be equivalent to the defining inequality for Ω_n in (4.2.12). ∎

We note that in Theorem 4.3 and in other situations later it is convenient to permit elements in Ω_n of the form $\langle 0, b_n \rangle$ in which the first member $a_n = 0$. However, such a pair of numbers cannot be elements of a continued fraction $K(a_n/b_n)$, since (as stated in the definition of continued fraction (Section 2.1)) the elements $a_n \neq 0$.

A proof similar to the one given in Theorem 4.3, which we shall not give here, but which can be found for example in [Jones and Thron, 1968, p. 1040] establishes the following result.

THEOREM 4.4. *Let $\{V_n\}$ be a sequence of half planes*

$$V_n = \left[w : \mathrm{Re}(w e^{-i\psi_n}) \geqslant -p_n \right], \quad p_n > 0, \qquad n = 0, 1, 2, \ldots. \quad (4.2.14)$$

Let

$$\Omega_n = \left[\langle a_n, b_n \rangle : |a_n| - \mathrm{Re}(a_n e^{-i(\psi_n + \psi_{n-1})}) \right.$$
$$\left. \leqslant 2 p_{n-1}\left(\mathrm{Re}(b_n e^{-i\psi_n}) - p_n \right) \right], \qquad n = 1, 2, 3, \ldots. \quad (4.2.15)$$

Then $\{V_n\}$ is a sequence of value regions corresponding to the sequence of element regions $\{\Omega_n\}$.

This result is used in the multiple parabola theorem (Theorem 4.43). Special cases of Theorem 4.4 are employed in the parabola theorem (Theorem 4.42) and in the uniform parabola theorem (Theorem 4.40).

Note that the region V_n defined by (4.2.14) is a closed half plane containing 0 in its interior. The boundary of this half plane is orthogonal to the ray $\arg w = \psi_n$. If b_n is held fixed, then the inequality in (4.2.15) insures that a_n lies in a region bounded by a parabola (see discussion in Corollary 4.16).

We now consider two results which depend on the existence of an idempotent l.f.t. $v(w)$ such that

$$s_n(w) = v \circ s_n^{-1} \circ v(w) \qquad \text{for all } \langle a_n, b_n \rangle \in \Omega, \qquad (4.2.16)$$

where $s_n(w) = a_n/(b_n + w)$ and v is independent of the individual elements of Ω_n. A transformation v is said to be *idempotent* if $(v \circ v)(w) = v(v(w)) = w$. There are only three types of idempotent l.f.t.'s. They are

$$v(w) = w, \qquad v(w) = \frac{\beta}{w}, \qquad v(w) = \frac{w + \beta}{\gamma w - 1}. \qquad (4.2.17)$$

By direct computation one verifies easily that for

$$\Omega_n = [\langle a_n, 1 \rangle] \quad \text{and} \quad v(w) = -1 - w \qquad (4.2.18)$$

and for

$$\Omega_n^* = [\langle 1, b_n \rangle] \quad \text{and} \quad v(w) = \frac{-1}{w}, \qquad (4.2.19)$$

the formula (4.2.16) is valid. The ideas discussed here originated with Thron [1944b] and were further developed by Jones and Thron [1970].

THEOREM 4.5. *Let*

$$S_n(w) = s_1 \circ s_2 \circ \cdots \circ s_n(w) = \frac{A_n + A_{n-1}w}{B_n + B_{n-1}w}, \qquad n = 1, 2, 3, \ldots, \qquad (4.2.20)$$

and let (4.2.16) hold for all $s_m(w) = a_m/(b_m + w)$, $1 \leqslant m \leqslant n$, with the same v. Then

$$\frac{B_n}{B_{n-1}} = -S_n^{-1}(\infty) = -v \circ s_n \circ s_{n-1} \circ \cdots \circ s_1 \circ v(\infty).$$

Proof. Since $S_n^{-1} = s_n^{-1} \circ s_{n-1}^{-1} \circ \cdots \circ s_1^{-1}$, the result follows directly from (4.2.16). ∎

THEOREM 4.6. *Let $\{V_n\}$ be a best sequence of value regions, with*

$$V_{2n+1} = V_1, \quad V_{2n} = V_2, \qquad n = 0, 1, 2, \ldots, \qquad (4.2.21)$$

corresponding to a sequence of element regions $\{\Omega_n\}$. Further assume that

(4.2.16) *holds for all* Ω_n *and the same* $v(w)$. *Then*

$$v(0) \in V_1 \cup V_2 \quad or \quad V_1 \cap v(V_2) = \varnothing. \tag{4.2.22}$$

Proof. Recalling that the set-theoretic complement of V is denoted by $\hat{C} \sim V$, we see that, for any l.f.t. $t(w)$,

$$t(\hat{C} \sim V) = \hat{C} \sim t(V). \tag{4.2.23}$$

Further $s_n(V_n) \subseteq V_{n-1}$ implies $\hat{C} \sim V_{n-1} \subseteq s_n(\hat{C} \sim V_n)$ and hence

$$s_n^{-1}(\hat{C} \sim V_{n-1}) \subseteq \hat{C} \sim V_n. \tag{4.2.24}$$

Now define

$$W_{2n+1} = W_1 = v(\hat{C} \sim V_2) = v(\hat{C} \sim V_{2n})$$

and

$$W_{2n} = W_2 = v(\hat{C} \sim V_1) = v(\hat{C} \sim V_{2n+1}).$$

We shall show that either $\{W_n\}$ is a sequence of value regions for $\{\Omega_n\}$ or that $v(0) \in V_1 \cup V_2$. The following is easily verified by use of (4.2.24):

$$s_{2n+1}(W_{2n+1}) = s_{2n+1} \circ v(\hat{C} \sim V_{2n}) = v \circ s_{2n+1}^{-1}(\hat{C} \sim V_{2n})$$
$$\subseteq v(\hat{C} \sim V_{2n+1}) = v(\hat{C} \sim V_1) = W_{2n}.$$

An analogous argument yields $s_{2n}(W_{2n}) \subseteq W_{2n+1}$. Note that we have made essential use in these arguments of the fact that the V_n (and hence the W_n) alternate between V_1 and V_2 (W_1 and W_2). The sequence $\{W_n\}$ then will be a sequence of value regions for $\{\Omega_n\}$ iff the condition $(a_n/b_n) \in W_{n-1}$ is satisfied for all $\langle a_n, b_n \rangle \in \Omega_n$ and all n. This condition is equivalent to $s_n(0) \in W_{n-1}$ or $v \circ s_n^{-1} \circ v(0) \in v(\hat{C} \sim V_n)$. This is the same as $s_n^{-1} \circ v(0) \in \hat{C} \sim V_n$ or $v(0) \in s_n(\hat{C} \sim V_n)$ or, finally $v(0) \notin s_n(V_n)$. Thus if $\{W_n\}$ is not a sequence of value regions, then for some $n \geqslant 1$ and some $\langle a_n, b_n \rangle \in \Omega_n$, $s_n(0) \notin W_{n-1}$; that is, $v(0) \in s_n(V_n) \subseteq V_{n-1}$. In this case the first condition in (4.2.22) holds.

Now, suppose that $\{W_n\}$ is a sequence of value regions corresponding to $\{\Omega_n\}$. It is clear from the definition of value regions that if $\{W_n\}$ and $\{V_n\}$ are both sequences of value regions for the same sequence of element regions, then $\{W_n \cap V_n\}$ is also a sequence of value regions for $\{\Omega_n\}$. Since $\{V_n\}$ was assumed to be a best sequence of value regions for $\{\Omega_n\}$, it follows that $V_n \subseteq W_n$ or, equivalently,

$$v(V_2) \cap V_1 = \varnothing \quad and \quad v(V_1) \cap V_2 = \varnothing. \tag{4.2.25}$$

Since v is idempotent, the two conditions in (4.2.25) are identical. Hence the second condition in (4.2.22) holds. ∎

Somewhat more can be said about element regions and value regions in the two special cases where either all $b_n = 1$ or all $a_n = 1$. Then the element regions can be considered as subsets of $\mathbb{C}$ instead of $\mathbb{C} \times \mathbb{C}$ and shall be denoted by E_n for continued fractions $K(a_n/1)$ (i.e., $a_n \in E_n$ and $b_n = 1$) and by G_n for continued fractions $K(1/b_n)$ (i.e., $b_n \in G_n$ and $a_n = 1$).

Using the relations (4.2.18) and (4.2.19), we have the following corollaries to Theorems 4.5 and 4.6.

COROLLARY 4.7. *For* $K(a_n/1)$,

$$\frac{B_n}{B_{n-1}} = 1 + \frac{a_n}{1} + \frac{a_{n-1}}{1} + \cdots + \frac{a_2}{1}, \qquad n = 2, 3, \ldots, \qquad (4.2.26)$$

where B_n *is the n th denominator of* $K(a_n/1)$.

COROLLARY 4.8. *For* $K(1/b_n)$,

$$\frac{B_n}{B_{n-1}} = b_n + \frac{1}{b_{n-1}} + \cdots + \frac{1}{b_1}, \qquad n = 2, 3, \ldots, \qquad (4.2.27)$$

where B_n *is the n th denominator of* $K(1/b_n)$.

COROLLARY 4.9. *Let* $\{V_n\}$ *be a best sequence of value regions corresponding to a sequence of element regions* $\{E_n\}$ *for continued fractions* $K(a_n/1)$. *Further assume that*

$$V_{2n+1} = V_1, \quad V_{2n} = V_2, \qquad n = 0, 1, 2, \ldots .$$

Then

$$-1 \in V_1 \cup V_2 \quad \text{or} \quad V_1 \cap (-1 - V_2) = \varnothing. \qquad (4.2.28)$$

COROLLARY 4.10. *Let* $\{V_n\}$ *be a best sequence of value regions corresponding to a sequence of element regions* $\{G_n\}$ *for continued fractions* $K(1/b_n)$. *Further assume that*

$$V_{2n+1} = V_1, \quad V_{2n} = V_2, \qquad n = 0, 1, 2, \ldots .$$

Then

$$\infty \in V_1 \cup V_2 \quad \text{or} \quad V_1 \cap (-1/V_2) = \varnothing. \qquad (4.2.29)$$

We now return to the determination of element regions for a given sequence of value regions. The proof of the result given below can be found in [Jones and Thron, 1970, p. 110].

THEOREM 4.11. *Let* $\{V_n\}$ *and* $\{E_n\}$ *be sequences of regions (in* $\hat{\mathbb{C}}$ *and* $\mathbb{C}$, *respectively) defined as follows: for* $n = 0, 1, 2, \ldots$,

$$V_{2n+1} = V_1 = [w : |w - \Gamma_1| \geqslant \rho_1], \qquad |1 + \Gamma_1| < \rho_1, \qquad (4.2.30a)$$

$$V_{2n} = V_2 = [w : |w - \Gamma_2| \leqslant \rho_2], \qquad |\Gamma_2| < \rho_2 \neq |1 + \Gamma_2|, \quad (4.2.30b)$$

and for $n = 1, 2, 3, \ldots$,

$$E_{2n-1} = E_1 = \left[w : \left| w\left(1 + \bar{\Gamma}_1\right) + \Gamma_2\left(\rho_1^2 - |1 + \Gamma_1|^2\right) \right| + \rho_1 |w| \right.$$

$$\left. \leqslant \rho_2\left(\rho_1^2 - |1 + \Gamma_1|^2\right) \right] \qquad (4.2.31a)$$

and

$$E_{2n} = E_2 = \left[w : \left| w\left(1 + \bar{\Gamma}_2\right) + \Gamma_1\left(\rho_2^2 - |1 + \Gamma_2|^2\right) \right| - \rho_2 |w| \right.$$

$$\left. \geqslant \rho_1\left(|1 + \Gamma_2|^2 - \rho_2^2\right) \right] \qquad (4.2.31b)$$

if

$$\rho_2 < |1 + \Gamma_2|,$$

whereas

$$E_{2n} = E_2 = \left[w : \rho_2 |w| - \left| w\left(1 + \bar{\Gamma}_2\right) + \Gamma_1\left(\rho_2^2 - |1 + \Gamma_2|^2\right) \right| \right.$$

$$\left. \geqslant \rho_1\left(\rho_2^2 - |1 + \Gamma_2|^2\right) \right] \qquad (4.2.31c)$$

if

$$|1 + \Gamma_2| < \rho_2.$$

Then $\{V_n\}$ *is a sequence of value regions corresponding to the sequence of element regions* $\{E_n\}$ *for continued fractions* $\mathbf{K}(a_n/1)$.

Reid in his thesis [1978] has given more transparent formulations for the regions E_1 and E_2 in Theorem 4.11. For E_1 he obtained

$$E_1 = \left[w : w = re^{i\theta}, 0 \leqslant r \leqslant h_1(\theta) \right],$$

where

$$h_1(\theta) = \rho_1 \rho_2 - d\cos(\theta - \beta)$$

$$- \sqrt{\left[\rho_1 \rho_2 - d\cos(\theta - \beta)\right]^2 - \left(\rho_1^2 - |1 + \Gamma_1|^2\right)\left(\rho_2^2 - |\Gamma_2|^2\right)}$$

and

$$(1+\Gamma_1)(-\Gamma_2)=d\,e^{i\beta}.$$

Thus the region $E_1 \cup [0]$ is always star-shaped with respect to $w=0$. If

$$\rho_2 < |1+\Gamma_2|,$$

then

$$E_2 = E_{2a} \cup E_{2b} \cup E_{2c},$$

where

$$E_{2a} = \left[w : w=re^{i\theta}, r>0, |\theta-\gamma| \leqslant \alpha \right],$$
$$E_{2b} = \left[w : w=re^{i\theta}, 0<r\leqslant h_2(\theta), \alpha<\theta-\gamma<2\pi-\alpha \right],$$
$$E_{2c} = \left[w : w=re^{i\theta}, r\geqslant h_3(\theta), \alpha<\theta-\gamma<2\pi-\alpha \right],$$
$$ge^{i\gamma} = (1+\Gamma_2)(-\Gamma_1),$$

$$\alpha = \cos^{-1} \left[\frac{\rho_1\rho_2 - \sqrt{(|\Gamma_1|^2 - \rho_1^2)(|1+\Gamma_2|^2 - \rho_2^2)}}{g} \right],$$

$$h_2(\theta) = \rho_1\rho_2 - g\cos(\theta-\gamma)$$
$$- \sqrt{[\rho_1\rho_2 - g\cos(\theta-\gamma)]^2 - (|\Gamma_1|^2 - \rho_1^2)(|1+\Gamma_2|^2 - \rho_2^2)} ,$$
$$h_3(\theta) = \rho_1\rho_2 - g\cos(\theta-\gamma)$$
$$+ \sqrt{[\rho_1\rho_2 - g\cos(\theta-\gamma)]^2 - (|\Gamma_1|^2 - \rho_1^2)(|1+\Gamma_2|^2 - \rho_2^2)} .$$

If

$$|1+\Gamma_2| < \rho_2,$$

then one obtains

$$E_2 = \left[w : w=re^{i\theta}, r\geqslant h_3(\theta), 0\leqslant\theta\leqslant 2\pi \right].$$

Another result of Reid [1978] shows that if $\Gamma_1, \Gamma_2, \rho_1, \rho_2$ are suitably restricted, then E_1 and E_2 can be obtained as intersections of circular regions expressed in terms of v, where $v^2 = w$.

COROLLARY 4.12. *Let* $0 \leqslant k \leqslant k' < 1$ *be given. Let*

$$|\Gamma_1 + 1| = k|\Gamma_1|, \qquad k|\Gamma_2 + 1| = |\Gamma_2|,$$

or equivalently

$$\left| \Gamma_1 + \frac{1}{1-k^2} \right| = \frac{k}{1-k^2}, \qquad \left| \Gamma_2 + \frac{k^2}{1-k^2} \right| = \frac{k}{1-k^2}.$$

Further let

$$\rho_1 = k'|\Gamma_1|, \qquad \rho_2 = k'|1+\Gamma_2|.$$

Then

$$|1+\Gamma_1| < \rho_1 < |\Gamma_1| \quad and \quad |\Gamma_2| < \rho_2 < |1+\Gamma_2|$$

are satisfied. Finally let

$$V_{2n+1} = V_1 = [\, w : |w - \Gamma_1| \geqslant \rho_1 \,], \tag{4.2.32a}$$

$$V_{2n} = V_2 = [\, w : |w - \Gamma_2| \leqslant \rho_2 \,], \tag{4.2.32b}$$

$$E_{2n+1} = E_1 = \Big[\, w : w = v^2, \, |v - i\Gamma^\dagger|$$

$$\leqslant \sqrt{\rho_1\rho_2}\,, \, |v + i\Gamma^\dagger| \leqslant \sqrt{\rho_1\rho_2}\,\Big], \tag{4.2.33a}$$

$$E_{2n} = E_2 = \Big[\, w : w = v^2, \, |v - i\Gamma^*|$$

$$\leqslant \sqrt{\rho_1\rho_2}\,, \, |v + i\Gamma^*| \leqslant \sqrt{\rho_1\rho_2}\,\Big], \tag{4.2.33b}$$

where

$$(\Gamma^\dagger)^2 = (1+\Gamma_1)(-\Gamma_2) \quad and \quad (\Gamma^*)^2 = (-\Gamma_1)(1+\Gamma_2).$$

Then $\{V_n\}$ is a sequence of value regions corresponding to the sequence of element regions $\{E_n\}$ for continued fractions $K(a_n/1)$.

An application of Corollary 4.12 is given in the proof of the twin-convergence-region results (Theorem 4.46 and Theorem 4.49).

If

$$|\Gamma| < \rho < |\Gamma + 1|,$$

then setting $\Gamma = (1+\Gamma_1) = -\Gamma_2$ and $\rho_1 = \rho_2 = \rho$ is a special case of Corollary 4.12 which was known to Thron [1943c]. In this case (4.2.32c, d) reduce to

$$E_1 = [\, w : w = v^2, \, |v - i\Gamma| \leqslant \rho, \, |v + i\Gamma| \leqslant \rho \,], \tag{4.2.34a}$$

$$E_2 = [\, w : w = v^2, \, |v - i(1+\Gamma)| \geqslant \rho, \, |v + i(1+\Gamma)| \geqslant \rho \,], \tag{4.2.34b}$$

respectively.

The following corollaries of Theorem 4.3 are also of interest.

COROLLARY 4.13. *For continued fractions* $K(a_n/1)$ *let*

$$V_n = \left[w : |w| < \tfrac{1}{2} \right], \qquad n = 0, 1, 2, \ldots, \tag{4.2.35}$$

$$E_n = \left[w : |w| < \tfrac{1}{4} \right], \qquad n = 1, 2, 3, \ldots. \tag{4.2.36}$$

Then $\{V_n\}$ *is a sequence of value regions corresponding to the sequence of element regions* $\{E_n\}$.

The preceding result is used in the Worpitzky criterion for convergence of continued fractions $K(a_n/1)$ [Corollary 4.36(B)].

COROLLARY 4.14. *For continued fractions* $K(1/b_n)$ *let*

$$V_n = \left[w : |w - c| < \sqrt{1 + c^2} \right], \qquad c \text{ real}, \quad n = 0, 1, 2, \ldots, \tag{4.2.37}$$

$$G_n = \left[w : |w + 2c| \geqslant 2\sqrt{1 + c^2} \right], \qquad n = 1, 2, 3, \ldots. \tag{4.2.38}$$

Then $\{V_n\}$ *is a sequence of value regions corresponding to the sequence of element regions* $\{G_n\}$ *for continued fractions* $K(1/b_n)$.

Although the preceding is indeed a corollary of Theorem 4.3, the computation is extremely tedious. The result was originally obtained in a different manner (see [Thron, 1944a]) and would have been difficult to deduce from Theorem 4.3. The following, however, can be obtained without difficulty.

COROLLARY 4.15. *For continued fractions* $K(1/b_n)$ *let*

$$V_n = \left[w : |w| < \rho_n \right], \qquad n = 0, 1, 2, \ldots, \tag{4.2.39}$$

$$G_n = \left[w : |w| > \rho_n + \frac{1}{\rho_{n-1}} \right], \qquad n = 1, 2, 3, \ldots. \tag{4.2.40}$$

Then $\{V_n\}$ *is a sequence of value regions corresponding to the sequence of element regions* $\{G_n\}$.

The following is a corollary of Theorem 4.4:

COROLLARY 4.16. *For continued fractions* $K(a_n/1)$ *let* $-\pi/2 < \alpha < \pi/2$, *and let* V_n *be the half plane*

$$V_n = V(\alpha) = \left[w : \operatorname{Re}(we^{-i\alpha}) > -\tfrac{1}{2}\cos\alpha \right], \qquad n = 0, 1, 2, \ldots, \tag{4.2.41}$$

$$E_n = E(\alpha) = \left[w : |w| - \operatorname{Re}(we^{-i2\alpha}) < \tfrac{1}{2}\cos^2\alpha \right], \qquad n = 1, 2, 3, \ldots. \tag{4.2.42}$$

Then $\{V_n\}$ is a sequence of value regions corresponding to the sequence of element regions $\{E_n\}$. The region $E(\alpha)$ is bounded by the parabola which passes through $w = -\frac{1}{4}$ and has its focus at the origin, its vertex at $-\frac{1}{4}\cos^2\alpha\, e^{i2\alpha}$, and its axis along the ray $\arg w = 2\alpha$ (see Figure 4.2.1).

The last result is used in the uniform parabola theorem (Theorem 4.40).

Another approach to the problem of finding element regions for given value regions will now be described. It was first used by Leighton and Thron [1942].

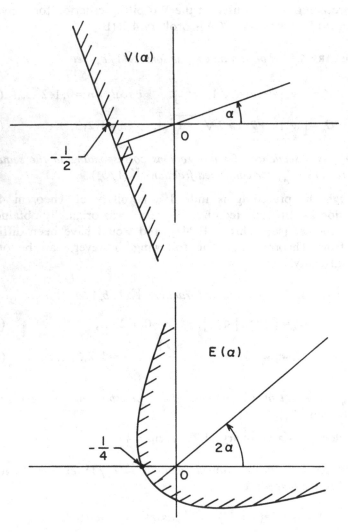

Figure 4.2.1. (a) Half-plane value region $V(\alpha) = [w : \mathrm{Re}(we^{-i\alpha}) > -\frac{1}{2}\cos\alpha]$. (b) Parabolic element region $E(\alpha) = [w : |w| - \mathrm{Re}(we^{-i2\alpha}) < \frac{1}{2}\cos^2\alpha]$.

THEOREM 4.17. *Let $\{V_n\}$ be a given sequence of regions in $\hat{\mathbb{C}}$ such that if $-1 \in V_n$, then $\infty \in V_{n-1}$, and if $\infty \in V_n$, then $0 \in V_{n-1}$. Define*

$$E_n^* = \bigcap \left[(1+w)V_{n-1} : w \in V_n \sim [-1, \infty] \right] \cap V_{n-1}, \qquad n = 1, 2, 3, \dots .$$

(4.2.43)

If $E_n^* \neq \varnothing$ for all $n = 1, 2, 3, \dots$, then $\{V_n\}$ is a sequence of value regions corresponding to the sequence of element regions $\{E_n^*\}$ for continued fractions $K(a_n/1)$. Moreover if $\{E_n'\}$ is any sequence of element regions for the sequence of value regions $\{V_n\}$, then

$$E_n' \subseteq E_n^*, \qquad n = 1, 2, 3, \dots , \tag{4.2.44}$$

so that $\{E_n^*\}$ is the best sequence of element regions corresponding to the sequence of value regions $\{V_n\}$.

Proof. Unless all $E_n^* \neq \varnothing$ we do not have a sequence of element regions. It follows from the definition of E_n^* that $E_n^* \subseteq V_{n-1}$. Thus (4.2.2a) is satisfied. According to (4.2.43), if $a_n \in E_n^*, a_n \neq 0$, then for every $w \in V_n \sim [-1, \infty]$ there exists $u_w \in V_{n-1}$ such that

$$a_n = u_w(1 + w).$$

Thus for every $w \in V_n \sim [-1, \infty]$,

$$s_n(w) = \frac{a_n}{1+w} = \frac{u_w(1+w)}{1+w} = u_w \in V_{n-1}.$$

For $w \in [-1, \infty]$ our assumption takes care of $s_n(w) \in V_{n-1}$. It follows that $s_n(V_n) \subseteq V_{n-1}$, so that (4.2.2b) holds. This establishes that $\{V_n\}$ is a sequence of value regions corresponding to the sequence of element regions $\{E_n^*\}$.

Now suppose that $\{E_n'\}$ is a sequence of element regions for $\{V_n\}$. Then by (4.2.2a) $E_n' \subseteq V_{n-1}$ for all $n = 1, 2, 3, \dots$ and, by (4.2.2b),

$$\frac{a_n}{1+w} \in V_{n-1}$$

for every $a_n \in E_n'$, $w \in V_n$ and $n = 1, 2, 3, \dots$. Hence

$$E_n' \subseteq \bigcap \left[(1+w)V_{n-1} : w \in V_n \sim [-1, \infty] \right] \cap V_{n-1}, \qquad n = 1, 2, 3, \dots . \quad \blacksquare$$

Theorem 4.17 gives a necessary and sufficient condition for a sequence of regions $\{V_n\}$ to be a sequence of value regions for continued fractions $K(a_n/1)$ corresponding to a sequence of element regions $\{E_n\}$. The condition is: $E_n^* \neq \varnothing$ for all $n = 1, 2, 3, \dots$, where E_n^* are defined in terms of the sequence $\{V_n\}$ by (4.2.43).

By an argument similar to that used in the proof of Theorem 4.17, one can obtain the following result for continued fractions $K(1/b_n)$.

THEOREM 4.18. *Let $\{V_n\}$ be a given sequence of regions in $\hat{C}$. Define*

$$G_n^* = \bigcap \left[\frac{1}{V_{n-1}} - w : w \in V_n \sim [\infty] \right] \cap \frac{1}{V_{n-1}}, \qquad n = 1, 2, 3, \ldots .$$

If $G_n^* \neq \varnothing$ for all $n = 1, 2, 3, \ldots$, then $\{V_n\}$ is a sequence of value regions corresponding to the sequence of element regions $\{G_n^*\}$ for continued fractions $K(1/b_n)$. Moreover, if $\{G_n'\}$ is any sequence of element regions for the sequence of value regions $\{V_n\}$, then

$$G_n' \subseteq G_n^*, \qquad n = 1, 2, 3, \ldots,$$

so that $\{G_n^*\}$ is the best sequence of element regions belonging to the sequence of value regions $\{V_n\}$.

4.3 Necessary Conditions for Convergence

4.3.1 *Stern-Stolz Theorem*

A sequence $\{\Omega_n\}$ of element regions will be called a *sequence of convergence regions for* $K(a_n/b_n)$ if

$$\langle a_n, b_n \rangle \in \Omega_n, \qquad n = 1, 2, 3, \ldots \qquad (4.3.1)$$

insures the convergence of $K(a_n/b_n)$. Similarly we call a sequence $\{E_n\}$ of element regions a sequence of convergence regions for $K(a_n/1)$ if

$$a_n \in E_n, \qquad n = 1, 2, 3, \ldots \qquad (4.3.2)$$

is sufficient for the convergence of $K(a_n/1)$. Analogously one defines a sequence $\{G_n\}$ of convergence regions for $K(1/b_n)$ if

$$b_n \in G_n, \qquad n = 1, 2, 3, \ldots \qquad (4.3.3)$$

implies convergence of $K(1/b_n)$. If $\Omega_n = \Omega_1$ for all $n = 1, 2, 3, \ldots$ (or $E_n = E_1$ or $G_n = G_1$), one calls Ω_1 (or E_1 or G_1) a *simple convergence region*. If $\Omega_{2n+1} = \Omega_1$ and $\Omega_{2n} = \Omega_2$ (or similarly for E_n or G_n) one speaks of *twin convergence regions*.

We begin with what is probably the oldest necessary condition for convergence of continued fractions. It can be found in the textbooks of Stern [1860] and Stolz [1886].

THEOREM 4.19 (Stern-Stolz). *A continued fraction* $K(1/b_n)$ *diverges if* $\Sigma|b_n|<\infty$.

Proof. Using the difference equations (2.1.6) for the nth numerator A_n and denominator B_n of $K(1/b_n)$, one obtains by induction

$$|A_n|\leqslant(1+|b_1|)(1+|b_2|)\cdots(1+|b_n|),\qquad n=1,2,3,\ldots,$$

and

$$|B_n|\leqslant(1+|b_1|)(1+|b_2|)\cdots(1+|b_n|),\qquad n=1,2,3,\ldots.$$

Hence $\{|A_n|\}$ and $\{|B_n|\}$ are bounded, since the products on the right in the above inequalities converge to finite values, and by hypothesis $\Sigma|b_n|<\infty$. It follows that the two series $\Sigma b_n A_{n-1}$ and $\Sigma b_n B_{n-1}$ converge absolutely. Using the difference equations again, one has $A_n-A_{n-2}=b_n A_{n-1}$ and hence

$$A_{2n}=\sum_{m=1}^{n}b_{2m}A_{2m-1},\qquad A_{2n+1}=A_1+\sum_{m=1}^{n}b_{2m+1}A_{2m}.$$

Analogous results can be obtained for the B_n. Thus

$$\lim_{n\to\infty}A_{2n+1}=A^{(1)},\quad \lim_{n\to\infty}B_{2n+1}=B^{(1)},\quad \lim_{n\to\infty}A_{2n}=A^{(2)},\quad \lim_{n\to\infty}B_{2n}=B^{(2)}$$

all exist. It follows from the determinant formula (2.1.9) that

$$A^{(1)}B^{(2)}-A^{(2)}B^{(1)}=1,$$

and hence at most one of $B^{(1)}$, $B^{(2)}$ can vanish. Hence the two limits

$$\lim_{n\to\infty}\frac{A_{2n+1}}{B_{2n+1}}=\frac{A^{(1)}}{B^{(1)}},\qquad \lim_{n\to\infty}\frac{A_{2n}}{B_{2n}}=\frac{A^{(2)}}{B^{(2)}}$$

are distinct (one may be infinite), and the continued fraction $K(1/b_n)$ diverges by oscillation. ∎

As an immediate consequence of the above result we have

COROLLARY 4.20. *The continued fraction* $K(a_n/b_n)$ *diverges if* $\Sigma|b_n^*|<\infty$, *where*

$$b_{2n-1}^*=b_{2n-1}\frac{a_2 a_4\cdots a_{2n-2}}{a_1 a_3\cdots a_{2n-1}},\qquad n=1,2,3,\ldots,\qquad(4.3.4a)$$

$$b_{2n}^*=b_{2n}\frac{a_1 a_3\cdots a_{2n-1}}{a_2 a_4\cdots a_{2n}},\qquad n=1,2,3,\ldots.\qquad(4.3.4b)$$

Proof. $K(a_n/b_n)$ is equivalent to $K(1/b_n^*)$ [see (2.3.23)]. ∎

4.3.2 Necessary Conditions for Best Value Regions and Convergence Regions

If a continued fraction of the form $K(a_n/1)$ is equivalent to $K(1/b_n^*)$, then $b_n^* b_{n-1}^* = 1/a_n$, so that $\Sigma|b_n^*| < \infty$ only if $a_n \to \infty$. It is also true that no simple convergence region for $K(a_n/1)$ can be unbounded and that there probably do not exist any maximal simple convergence regions for $K(a_n/1)$. In view of these results we introduce the following: A subset E of $\mathbb{C}$ is called a simple *conditional* convergence region for continued fractions $K(a_n/1)$ if

$$a_n \in E, \quad n = 1, 2, 3, \ldots, \quad \text{and} \quad \sum|b_n^*| = \infty$$

are sufficient for the convergence of $K(a_n/1)$. Here b_n^* is defined by (4.3.4) with $b_n = 1$.

THEOREM 4.21. *Every bounded subset B of a simple conditional convergence region E for $K(a_n/1)$ is a simple convergence region.*

Proof. If all $a_n \in B$, then $\Sigma|b_n^*| = \infty$. ∎

We have excluded the possibility that $a_n = 0$ for any n in Theorem 4.21. However, we do admit $b_n = 0$ for some n in $K(1/b_n)$. If all $b_{2n+1} = 0$ in $K(1/b_n)$, then

$$\left.\begin{array}{l} B_{2n+1} = 0, \quad B_{2n} = 1 \\[2mm] A_{2n+1} = 1, \quad A_{2n} = \displaystyle\sum_{m=1}^{n} b_{2m} \end{array}\right\} \quad n = 1, 2, 3, \ldots.$$

Thus the continued fraction $K(1/b_n)$ diverges by oscillation or converges to ∞.

In the next theorem (due to Jones and Thron [1970, p. 100]) we continue the investigation begun in Theorem 4.6 and Corollary 4.9 which attempts to characterize twin convergence regions in terms of their value regions. We know of no analogous result for continued fractions $K(1/b_n)$.

THEOREM 4.22. *Let V_1, V_2 be the best twin value regions corresponding to conditional twin convergence regions E_1, E_2 for continued fractions $K(a_n/1)$. Then*

$$-1 \notin V_1 \cap V_2. \tag{4.3.5}$$

Before proving the theorem, we establish the following

LEMMA 4.23. *If*

$$s_n(w) = \frac{a_n}{1+w}, \quad a_n \neq 0, \quad n = 1, 2, \ldots, m,$$

and

$$-1 = s_1 \circ s_2 \circ \cdots \circ s_m(0), \tag{4.3.6}$$

then

$$-1 = s_m \circ s_{m-1} \circ \cdots \circ s_1(0). \tag{4.3.7}$$

Proof. We recall from (4.2.16) and (4.2.18) that if $v(w) = -1 - w$, then

$$s_n(w) = v \circ s_n^{-1} \circ v(w) \tag{4.3.8}$$

and v is an idempotent l.f.t. The hypothesis (4.3.6) can be written as $v(0) = s_1 \circ s_2 \circ \cdots \circ s_m(0)$ and hence $s_m^{-1} \circ s_{m-1}^{-1} \circ \cdots \circ s_1^{-1} \circ v(0) = 0$. In view of (4.3.8) we obtain $v \circ s_m \circ v \circ \cdots \circ v \circ s_1 \circ v \circ v(0) = v \circ s_m \circ \cdots \circ s_1(0) = 0$, from which (4.3.7) follows immediately. ∎

Proof of Theorem 4.22. Assuming $-1 \in V_1 \cap V_2$, we shall prove the existence of at least one divergent periodic continued fraction $\mathbf{K}(a_n/1)$ with elements satisfying $a_n \in E_{n \pmod 2}$, $a_n \neq 0$. This is sufficient for the proof of the theorem, since a periodic continued fraction $\mathbf{K}(a_n/1)$ always satisfies $\Sigma |b_n^*| = \infty$, where the b_n^* are defined by (4.3.4) with $b_n = 1$. Thus the result holds for conditional convergence regions.

It follows from the assumption $-1 \in V_1 \cap V_2$ and Theorem 4.1 [equation (4.2.5)] that there exist integers n and k and complex numbers $a_1, \ldots a_n$ and $d_2, \ldots d_k$ such that

$$-1 = \frac{a_1}{1 +} \ \cdots \ \frac{a_n}{+ \ 1}, \tag{4.3.9}$$

$$-1 = \frac{d_2}{1 +} \ \cdots \ \frac{d_k}{+ \ 1}, \tag{4.3.10}$$

where

$$a_\nu, d_\nu \in \begin{cases} E_1 & \text{if } \nu \text{ is odd,} \\ E_2 & \text{if } \nu \text{ is even.} \end{cases}$$

It will suffice to consider three cases:

(a) If n is even, we consider the two periodic continued fractions of period $n + 2$

$$\frac{a_1}{1 +} \ \cdots \ \frac{a_n}{+ \ 1 +} \ \frac{x_1}{1 +} \ \frac{x_2}{1 +} \ \frac{a_1}{1 +} \ \cdots, \tag{4.3.11}$$

$$\frac{x_2}{1 +} \ \frac{x_1}{1 +} \ \frac{a_n}{1 +} \ \cdots \ \frac{a_1}{+ \ 1 +} \ \frac{x_2}{1 +} \ \cdots, \tag{4.3.12}$$

where x_1 and x_2 denote arbitrary non-zero elements of E_1 and E_2, respectively.

(b) If k is odd, then applying Lemma 4.23 to (4.3.10), we have a situation equivalent to (4.3.9) with n even. Hence case (b) reduces to case (a).

(c) If n is odd and k is even, and if $y_1 \in E_1$ and $y_2 \in E_2$ ($y_i \neq 0$), then

$$-1 = \frac{a_1}{1\ +} \ \cdots \ \frac{a_n}{+\ 1\ +} \frac{y_2}{1\ +} \frac{y_1}{1\ +} \frac{d_2}{1\ +} \ \cdots \ \frac{d_k}{+\ 1}$$

involves an even number of elements and hence could be used to construct periodic continued fractions of the same type considered in case (a).

We shall show that in case (a) either (4.3.11) or (4.3.12) diverges for some $x_1 \in E_1$, $x_2 \in E_2$ ($x_i \neq 0$).

The continued fractions (4.3.11) and (4.3.12) are so constructed that the approximants of (4.3.11) of order $n + m(n+2)$ are all -1 and that the approximants of (4.3.12) of order $m(n+2)$ are all zero. This follows from (4.3.9) and (4.3.7) of Lemma 4.23. Since by hypothesis E_1 and E_2 are conditional twin convergence regions, the continued fractions (4.3.11) and (4.3.12) must converge to -1 and 0, respectively. By the generalized Galois theorem (Theorem 3.4) it follows that $w_1 = w_2 = -1$ are the solutions of the equation

$$S_{n+2}(w) = w,$$

where

$$S_{n+2}(w) = \frac{a_1}{1\ +} \ \cdots \ \frac{a_n}{+\ 1\ +} \frac{x_1}{1\ +} \frac{x_2}{1 + w}.$$

By (2.1.7)

$$S_{n+2}(w) = \frac{A_{n+2} + A_{n+1}w}{B_{n+2} + B_{n+1}w},$$

where A_m and B_m denote the mth numerator and denominator of (4.3.11), respectively. Thus the quadratic equation

$$B_{n+1}w^2 + (B_{n+2} - A_{n+1})w - A_{n+2} = 0$$

has the double root $w_1 = w_2 = -1$, and so we must have

$$\frac{A_{n+2}}{B_{n+1}} = -1 \quad \text{and} \quad \frac{B_{n+2} - A_{n+1}}{B_{n+1}} = 2. \tag{4.3.13}$$

From the difference equations (2.1.6) we have $A_{n+2} = A_{n+1} + x_2 A_n$ and

$B_{n+2} = B_{n+1} + x_2 B_n$. Hence (4.3.13) becomes

$$x_2 A_n = -(A_{n+1} + B_{n+1}) \quad \text{and} \quad x_2 B_n = (A_{n+1} + B_{n+1}). \quad (4.3.14)$$

These equations can hold for more than one x_2 iff $A_n = B_n = 0$, which we know is impossible [see (2.1.7)]. Hence E_2 must contain only one point. From (4.3.14) it follows that $x_2(A_n + B_n) = 0$ and hence $A_n + B_n = 0$ and $A_n B_n \neq 0$. Again using (4.3.14) and the difference equations (2.1.6), we obtain

$$x_2 = \frac{A_{n+1} + B_{n+1}}{B_n} = \frac{(A_n + x_1 A_{n-1}) + (B_n + x_1 B_{n-1})}{B_n}$$

$$= x_1 \frac{A_{n-1} + B_{n-1}}{B_n}.$$

It follows that there is but one choice of x_1, and hence E_1 contains only one point. If E_1 and E_2 both contain only one point, then the continued fraction (4.3.11) is of period 2. Thus its approximants of order $(m+1)(n+2)$ are all 0, and so it diverges. This contradicts the hypothesis that E_1, E_2 are twin convergence regions, and so the assumption $-1 \in V_1 \cap V_2$ is false. ∎

For simple convergence regions we can deduce from Theorem 4.22 and Corollary 4.9 the following necessary condition.

THEOREM 4.24. *A region E is a simple conditional convergence region for continued fractions $K(a_n/1)$ only if its best value region V satisfies the condition*

$$V \cap (-1 - V) = \varnothing. \quad (4.3.15)$$

Proof. From Corollary 4.9 we have that either $-1 \in V$ or $V \cap (-1-V) = \varnothing$. In Theorem 4.23 we showed that $-1 \notin V$. ∎

The following very recent result of Roach [1977] is also concerned with characterizing convergence regions in terms of their value regions. He has an analogous result for $K(1/b_n)$ which will not be given here.

THEOREM 4.25. *Let E be an element region for continued fractions $K(a_n/1)$, and let V be the corresponding best value region. If -1 is not a limit point of $V + V$, then a necessary and sufficient condition for E to be a simple convergence region for continued fractions $K(a_n/1)$ is that V is bounded.*

We know that $-1 \in V$ implies that E is not a convergence region (see Theorem 4.22). Thus $-1 \notin V$ is a necessary condition for E to be a convergence region. The condition $-1 \in V$ is easily shown to be equivalent to $-1 \in V + V$. This explains why one wants $-1 \notin V + V$. Requiring that

-1 is not a limit point of $V+V$ is, however, a much stronger condition, and it is pointed out by Roach [1977] that this condition is not satisfied by many of the well-known value regions for simple convergence regions.

Proof of Theorem 4.25. The crux of the proof is the following interesting identity: For $n \geqslant 1, m \geqslant 1$,

$$S_{n+m}(0) - S_n(0) = \frac{(-1)^{n-1} \prod_{\nu=1}^{n} a_\nu}{B_{n-1}^2 [1 + s_n \circ \cdots \circ s_2(0) + s_{n+1} \circ \cdots \circ s_{n+m}(0)]},$$
(4.3.16)

where s_n and S_n are defined by (2.1.1) with $b_n = 1$, and B_{n-1} is the $(n-1)$th denominator of $K(a_n/1)$. It is proved by observing that

$$D = (1 + s_n \circ \cdots \circ s_2(0) + s_{n+1} \circ \cdots \circ s_{n+m}(0))$$

$$= \frac{B_n}{B_{n-1}} + \frac{a_n}{x} - 1, \quad \text{where} \quad x = s_n \circ \cdots \circ s_{n+m}(0). \quad (4.3.17)$$

To establish (4.3.17), Corollary 4.7 is employed. Using (4.3.17) and the difference equations (2.1.6), we arrive at

$$D = \frac{xB_n - xB_{n-1} + a_n B_{n-1}}{xB_{n-1}} = \frac{a_n(xB_{n-2} + B_{n-1})}{xB_{n-1}}.$$

Thus

$$\frac{(-1)^{n-1} \prod_{\nu=1}^{n} a_\nu}{B_{n-1}^2 D} = \frac{(-1)^{n-1} xB_{n-1} \prod_{\nu=1}^{n} a_\nu}{a_n(xB_{n-2} + B_{n-1})B_{n-1}^2} = \frac{(-1)^{n-1} x \prod_{\nu=1}^{n-1} a_\nu}{B_{n-1}(xB_{n-2} + B_{n-1})}$$

$$= \frac{A_{n-1} + xA_{n-2}}{B_{n-1} + xB_{n-2}} - \frac{A_{n-1}}{B_{n-1}} = S_{n+m}(0) - S_{n-1}(0), \quad (4.3.18)$$

which proves (4.3.16). To prove the next to the last equality in (4.3.18) we use the determinant formulas (2.1.9). The last equality in (4.3.18) follows from (2.1.7).

By hypothesis $V + V$ is bounded away from -1. Suppose that V is bounded. Then there exist positive constants m and M such that

$$m < |1 + v + v'| < M \quad \text{for all} \quad v, v' \in V.$$

It also follows from the boundedness of V that if $\{S_n(0)\}$ is a sequence of approximants of a continued fraction $\mathbf{K}(a_n/1)$ with elements in E, then there exists a convergent subsequence $\{S_{n_p}(0)\}$. Let $\{k_p\}$ be a sequence of natural numbers such that $n_p \leqslant k_p < n_{p+1}$. Then from (4.3.16)

$$|S_{n_{p+1}}(0) - S_{n_p}(0)| = \frac{\left| \prod_{\nu=1}^{n_p} a_\nu \right|}{|B_{n_p-1}|^2 \cdot |1 + v_{n_p} + v_{n_p, n_{p+1}}|},$$

and

$$|S_{k_p}(0) - S_{n_p}(0)| = \frac{\left| \prod_{\nu=1}^{n_p} a_\nu \right|}{|B_{n_p-1}|^2 \cdot |1 + v_{n_p} + v_{n_p, k_p}|}.$$

Here the numbers v_{n_p}, $v_{n_p, n_{p+1}}$ and v_{n_p, k_p} belong to V and are defined by (4.3.16); as indicated by the subscripts, they depend upon n_p, (n_p, n_{p+1}) and (n_p, k_p), respectively. Hence

$$\frac{|S_{n_p}(0) - S_{k_p}(0)|}{|S_{n_p}(0) - S_{n_{p+1}}(0)|} = \left(\frac{|1 + v_{n_p} + v_{n_p, k_p}|}{|1 + v_{n_p} + v_{n_p, n_{p+1}}|} \right)^{-1} \leqslant \frac{M}{m} = K.$$

It follows that any two convergent subsequences of $\{S_n(0)\}$ converge to the same limit and hence $\{S_n(0)\}$ converges. Thus E is a convergence region.

Now assume that V is unbounded and that E is a convergence region. It will suffice to show that -1 is a limit point of $V + V$. As we observed earlier, E is bounded (see remarks following Corollary 4.20). Since V is unbounded, there exists a sequence

$$v_1 = \frac{a_1^{(1)}}{1}, \quad v_2 = \frac{a_1^{(2)}}{1} + \frac{a_2^{(2)}}{1}, \quad v_3 = \frac{a_1^{(3)}}{1} + \frac{a_2^{(3)}}{1} + \frac{a_3^{(3)}}{1}, \dots,$$
$$a_k^{(m)} \in E,$$

such that the sequence of absolute values of these finite continued fractions is unbounded. Since the $a_k^{(m)}$ are bounded, it follows that the numbers

$$u_2 = \frac{a_2^{(2)}}{1}, \quad u_3 = \frac{a_2^{(3)}}{1} + \frac{a_3^{(3)}}{1}, \quad u_3 = \frac{a_2^{(4)}}{1} + \frac{a_3^{(4)}}{1} + \frac{a_4^{(4)}}{1}, \dots$$

tend to -1. Let $a \in E$, $a \neq 0$. The $\lim_{n \to \infty} a/(1 + v_n) = 0$. Hence

$$\lim_{n \to \infty} \left(\frac{a}{1 + v_n} + u_{n-1} \right) = -1.$$

Since all of the numbers $a/(1+v_n)$ and u_{n-1} are in V, the proof is complete.   ∎

One of the important applications of our results on periodic continued fractions is that they provide examples of divergent continued fractions. In the following we give two necessary conditions for convergence which are obtained in this manner.

THEOREM 4.26. *Let E be a simple convergence region for continued fractions $K(a_n/1)$. Then E cannot contain points on the negative real axis less than $-\frac{1}{4}$.*

Proof. The periodic continued fraction $K(a/1)$ with a real and $a < -\frac{1}{4}$ is known to diverge (see Theorem 3.2).   ∎

For periodic continued fractions of period 2,

$$\frac{c^2}{1} + \frac{d^2}{1} + \frac{c^2}{1} + \frac{d^2}{1} + \cdots,$$

the transformation

$$P(w) = \frac{c^2}{1} + \frac{d^2}{1+w}$$

is parabolic if $d = c \pm i$ or $d = -c \pm i$ (see remarks following Theorem 3.2). Hence one can find complex numbers η of arbitrarily small absolute value so that for $d' = d + \eta$ the transformation

$$P(w) = \frac{c^2}{1} + \frac{(d')^2}{1+w}$$

is elliptic and hence the continued fraction

$$\frac{c^2}{1} + \frac{(d')^2}{1} + \frac{c^2}{1} + \frac{(d')^2}{1} + \cdots$$

diverges. This establishes the following:

THEOREM 4.27. *Let E_1 and E_2 be twin convergence regions for continued fractions $K(a_n/1)$. If $a = c^2 \in E_1$, then no neighborhood of either $(c+i)^2$ or $(c-i)^2$ can be contained in E_2.*

4.4 Sufficient Conditions for Convergence: Constant Elements

4.4.1 Classical Results and Generalizations

We begin with a classical result due to Seidel [1846] and Stern [1848].

THEOREM 4.28 (Seidel-Stern). *Let* $K(1/b_n)$ *be a continued fraction with positive elements* b_n.

(A) *If* f_n *denotes the n th approximant, then*

$$f_{2n-1} < f_{2n+1} < f_{2n+2} < f_{2n}, \qquad n=1,2,3,\ldots, \qquad (4.4.1)$$

so that the even and odd parts of $K(1/b_n)$ *both converge to finite values.*

(B) *If, in addition,* $\Sigma b_n = \infty$, *then the continued fraction converges to a finite value f and*

$$|f-f_n| < |f_n - f_{n-1}|, \qquad n=2,3,4,\ldots. \qquad (4.4.2)$$

Proof. (A): From the difference equations (2.1.6) we have

$$B_n = b_n B_{n-1} + B_{n-2} > B_{n-2}, \qquad n=1,2,3,\ldots, \qquad (4.4.3)$$

where B_n denotes the nth denominator of $K(1/b_n)$. It follows that $\{B_{2n-1}\}$ and $\{B_{2n}\}$ are increasing sequences of positive numbers; hence each sequence converges either to a finite value or to ∞. Moreover, from (4.4.3),

$$0 < \frac{1}{B_n B_{n-1}} < \frac{1}{B_{n-1} B_{n-2}}, \qquad n=2,3,4,\ldots. \qquad (4.4.4.)$$

In (2.1.22) it was shown that

$$f_n = \sum_{k=1}^{n} \frac{(-1)^{k+1}}{B_k B_{k-1}}, \qquad n=1,2,3,\ldots. \qquad (4.4.5)$$

The assertion (4.4.1) follows from (4.4.4) and (4.4.5).

(B): Again using the difference equations (2.1.6), we obtain $B_{2n} \geqslant 1$ and $B_{2n-1} \geqslant b_1$ and hence, by induction,

$$B_{2n-1} \geqslant b_1 + b_3 + \cdots + b_{2n-1}, \qquad n=1,2,3,\ldots, \qquad (4.4.6a)$$

$$B_{2n} \geqslant 1 + b_1(b_2 + b_4 + \cdots + b_{2n}), \qquad n=1,2,3,\ldots,. \qquad (4.4.6b)$$

Since by hypothesis $\Sigma b_n = \infty$, it follows that at least one of the sequences

$\{B_{2n-1}\}$ or $\{B_{2n}\}$ tends to ∞ and hence

$$\lim_{n\to\infty} B_n B_{n-1} = \infty.$$

Thus the Leibniz test implies that the alternating series $\Sigma(-1)^{k+1}/B_k B_{k-1}$ converges and so, by (4.4.5), $\{f_n\}$ converges to a finite value. Since $f_{2n-1} < f < f_{2n} < f_{2n-2}$, (4.4.2) follows. ∎

As a special case of Theorem 4.28 one obtains the result quoted in Section 2.1 that a regular continued fraction is convergent. The inequalities (4.4.1) and (4.4.2), satisfied by the approximants, are a special case of the more general situation

$$|f - f_n| \leqslant c|f_n - f_{n-1}|, \qquad n = 2, 3, 4, \ldots, \qquad (4.4.7)$$

which will be investigated in more detail in Chapter 8 on truncation-error bounds.

Theorem 4.28 can be substantially improved by replacing the condition $b_n > 0$ with the condition

$$-\frac{\pi}{2} + \varepsilon < \arg b_n < \frac{\pi}{2} - \varepsilon, \qquad \varepsilon > 0.$$

Since $\arg 0$ is not defined, this condition is understood to imply that $b_n \neq 0$. The result obtained by Van Vleck [1901a] is the following:

THEOREM 4.29 (Van Vleck). *Let the elements of the continued fraction* $K(1/b_n)$ *satisfy*

$$-\frac{\pi}{2} + \varepsilon < \arg b_n < \frac{\pi}{2} - \varepsilon, \qquad n = 1, 2, 3, \ldots,$$

where ε is an arbitrarily small positive number. Then:

(A) *The n th approximant f_n is finite and satisfies*

$$-\frac{\pi}{2} + \varepsilon < \arg f_n < \frac{\pi}{2} - \varepsilon, \qquad n = 1, 2, 3, \ldots.$$

(B) *The even and odd parts of the continued fraction converge to finite values.*

(C) *The continued fraction converges iff, in addition,*

$$\sum_{n=1}^{\infty} |b_n| = \infty.$$

(D) *When convergent, the value f of the continued fraction is finite and satisfies*

$$|\arg f| < \frac{\pi}{2}.$$

Besides Van Vleck's original proof, a number of others have been given for this result. We shall use a proof of Theorem 4.29 based on the Stieltjes-Vitali theorem. For convenience we state the theorem here in the form in which it will be applied. For proofs the reader is referred to [Hille, 1962, pp. 248–251], [Thron, 1953, p. 142], or [Wall, 1948, pp. 104–109].

THEOREM 4.30 (Stieltjes-Vitali). *Let D be a domain. Let $\{F_n(z)\}$ be a sequence of functions such that, for $n = 1, 2, 3, \ldots$,*

$F_n(z)$ *is holomorphic for $z \in D$,*
$F_n(z) \neq a$, $F_n(z) \neq b$ *for all $z \in D$, where a and b are distinct complex numbers.*

Further, let Δ be an infinite set with at least one limit point in D. Finally assume that $\{F_n(z)\}$ converges to a finite value for all $z \in \Delta$. Then $\{F_n(z)\}$ converges uniformly in each compact subset of D, to a function holomorphic in D.

There are two aspects to Theorem 4.30. First it is a "convergence continuation" theorem, extending convergence from Δ to D. It is this part of it that we shall use here. In the following section we shall make use of its second aspect, namely that it insures uniform convergence.

Proof of Theorem 4.29.   Let

$$W = \left[w : w \in \mathbb{C}, |\arg w| < \frac{\pi}{2} - \delta \right], \qquad 0 < \delta < \frac{\pi}{2}.$$

Then

$$\frac{1}{W} = W \quad \text{and} \quad b + W \subset W$$

provided $|\arg b| < \pi/2 - \delta$. Hence the nth approximants $f_n(z)$ of the continued fraction $\mathbf{K}(1/b_n(z))$, where

$$b_n(z) = |b_n| e^{i\beta_n z}, \qquad \beta_n = \arg b_n, \qquad n = 1, 2, 3, \ldots,$$

satisfy $f_n(z) \in W$ and therefore are holomorphic functions of z, provided

$$|\arg b_n(z)| < \frac{\pi}{2} - \delta \qquad n = 1, 2, 3, \ldots. \tag{4.4.8}$$

Let $\varepsilon > 0$ be the number given in the theorem, so that $|\beta_n| < \pi/2 - \varepsilon$. Set δ in (4.4.8) equal to $\varepsilon/2$. Then (4.4.8) is satisfied for all z such that $|\text{Re}(z)| < 1 + \varepsilon/(\pi - 2\varepsilon)$. Let D be defined by

$$D = \left[z : |\text{Re}(z)| < 1 + \frac{\varepsilon}{\pi - 2\varepsilon}, |\text{Im}(z)| < 1 \right].$$

For $z \in D$ the approximants $f_n(z)$ are holomorphic functions of z and assume no value in the left half of the complex plane. Thus in applying the Stieltjes-Vitali theorem (Theorem 4.30) we can choose, for example, $a = -1$ and $b = -2$. Let Δ be the subset of D defined by

$$\Delta = [z : \text{Re}(z) = 0, |\text{Im}(z)| < 1].$$

Then for $z \in \Delta$ the $b_n(z)$ are all positive. Hence by Theorem 4.28 the even and odd parts of $\overset{\text{K}}{}(1/b_n(z))$ converge to finite values for all $z \in \Delta$. If $\Sigma|b_n| = \infty$, then $\Sigma|b_n(z)| = \infty$ so that, by the second part of Theorem 4.28, $\overset{\text{K}}{}(1/b_n(z))$ itself converges to a finite value for all $z \in \Delta$. If $\Sigma|b_n|$ converges, then the continued fraction diverges by Theorem 4.19. An application of the Stieltjes-Vitali theorem (Theorem 4.30) then yields the convergence of the even and odd parts or of the whole continued fraction for all z in any compact subset of D. The singleton [1] is such a compact subset, and $b_n(1) = b_n$.   ∎

The following theorem is given in [Perron, 1957a, p. 66]. To the best of our knowledge Perron was the first to formulate it, although it is easily provable by methods developed by Van Vleck [1901a, b] and Jensen [1909]. It is of interest to us because it is one of the first results in which sufficient conditions for convergence of the odd part of a continued fraction are given.

THEOREM 4.31. *Let f_n denote the nth approximant of a continued fraction* $\overset{\text{K}}{}(1/b_n)$. *Then $\{f_{2n-1}\}$ converges if, for some $\varepsilon > 0$,*

$$\text{Re}(b_{2n}) \geq 0, \qquad |\arg b_{2n-1}| \leq \frac{\pi}{2} - \varepsilon, \qquad n = 1, 2, 3, \ldots. \qquad (4.4.9)$$

It follows from Theorem 4.29 and Theorem 4.19 that if $|\arg b_n| < \pi/2 - \varepsilon$ for all n, then $\Sigma|b_n| = \infty$ is necessary and sufficient for the convergence of $\overset{\text{K}}{}(1/b_n)$. Building on earlier results of Hamburger [1921] and Mall [1939], Scott and Wall [1947] were able to prove the following result. Again we omit the proof; it can be found in [Perron, 1957a, pp. 69–73] or [Wall, 1948, pp. 122–134].

THEOREM 4.32. *If in the continued fraction* $\overset{\text{K}}{}(1/b_n)$ *the elements satisfy* (4.4.9), *then a necessary and sufficient condition for* $\overset{\text{K}}{}(1/b_n)$ *to converge is that at least one of the following requirements is fulfilled*:

(A) $\displaystyle\sum_{n=0}^{\infty} |b_{2n+1}| = \infty,$

(B) $\displaystyle\sum_{n=1}^{\infty} |b_{2n+1}(b_2+b_4+\cdots+b_{2n})^2| = \infty,$

(C) $\displaystyle\lim_{n\to\infty} |b_2+b_4+\cdots+b_{2n}| = \infty.$

A very general twin-convergence-region theorem of a different type for continued fractions $K(1/b_n)$ was proved by Thron [1944a, 1949]. That result is the following

THEOREM 4.33. *Let* $\alpha(\theta)$ *be a real-valued continuous function of period* 2π *which satisfies the two conditions*

$$|\alpha(\theta)| < \frac{\pi}{2} - \varepsilon_1, \tag{4.4.10a}$$

$$\frac{|\alpha(\theta)-\alpha(\varphi)|}{|\theta-\varphi|} < 1-\varepsilon_2, \qquad \theta \neq \varphi, \tag{4.4.10b}$$

for some constants $\varepsilon_1 > 0$, $\varepsilon_2 > 0$. *Let*

$$g(\theta) = b_0 \exp\left(\int_{\pi/2}^{\theta} \tan\alpha(t)\,dt\right), \qquad b_0 > 0. \tag{4.4.11}$$

Then the continued fraction $K(1/b_n)$ *converges if*

$$|b_{2n}| \geq g(\arg b_{2n}), \qquad n=1,2,3,\ldots, \tag{4.4.12a}$$

and

$$|b_{2n-1}| \geq \frac{4}{g(\pi-\arg b_{2n-1})}, \qquad n=1,2,3,\ldots. \tag{4.4.12b}$$

The definition of $g(\theta)$ insures that the element regions for b_{2n} and b_{2n-1} both have convex complements. The theorem was proved using the Stieltjes-Vitali theorem (Theorem 4.30) with

$$b_{2n}(z) = b_{2n}\exp\left(\int_{\pi/2}^{\arg b_{2n}} \{\tan[z\alpha(t)] - \tan\alpha(t)\}\,dt\right) \tag{4.4.13}$$

and $b_{2n-1}(z)$ defined similarly. Thron [1944a] also discussed and gave results for the case where 0 is a boundary point for one, but not both of the element regions G_1, G_2. A corollary of Theorem 4.33 as well as of Theorem 4.37 (which we shall prove) is the following

COROLLARY 4.34. *The continued fraction* $K(1/b_n)$ *converges if for all* $n \geqslant 1$

$$|b_n + 2c| \geqslant 2\sqrt{1+c^2} \, , \tag{4.4.14}$$

where c *is a real number.*

We turn now to a classical result of Pringsheim [1899].

THEOREM 4.35 (Pringsheim). *The continued fraction* $K(a_n/b_n)$ *converges to a finite value if*

$$|b_n| \geqslant |a_n| + 1, \qquad n = 1, 2, 3, \ldots . \tag{4.4.15}$$

If f_n *denotes its* n^{th} *approximant, then*

$$|f_n| < 1, \qquad n = 1, 2, 3, \ldots . \tag{4.4.16}$$

Proof. We first prove (4.4.16) by induction. Let $s_n(w) = a_n/(b_n + w)$. Then by (4.4.15)

$$|s_n(0)| = \frac{|a_n|}{|b_n|} \leqslant \frac{|a_n|}{|a_n| + 1} < 1, \qquad n = 1, 2, 3, \ldots .$$

Assume that $|s_{n+1} \circ s_{n+2} \circ \cdots \circ s_{n+m}(0)| < 1$. Then

$$|s_n \circ s_{n+1} \circ \cdots \circ s_{n+m}(0)| = \left| \frac{a_n}{b_n + s_{n+1} \circ \cdots \circ s_{n+m}(0)} \right|$$

$$< \frac{|a_n|}{|b_n| - 1} \leqslant 1, \qquad \text{by (4.4.15)} \, .$$

Thus $|f_{n+m}| = |s_1 \circ s_2 \circ \cdots \circ s_{n+m}(0)| < 1$, as asserted by (4.4.16). Now from the difference equations (2.1.6) and (4.4.15) one obtains

$$|B_n| \geqslant |b_n||B_{n-1}| - |a_n||B_{n-2}|$$

$$\geqslant |b_n||B_{n-1}| - (|b_n| - 1)|B_{n-2}|$$

and hence

$$|B_n| - |B_{n-1}| \geqslant (|b_n| - 1)(|B_{n-1}| - |B_{n-2}|).$$

It follows that

$$|B_n| - |B_{n-1}| \geqslant \prod_{k=1}^{n} (|b_k| - 1) \geqslant \prod_{k=1}^{n} |a_k|, \qquad n = 1, 2, 3, \ldots . \tag{4.4.17}$$

From (2.1.22) one concludes that $K(a_n/b_n)$ converges to a finite value iff the series

$$\sum_{n=1}^{\infty} \frac{(-1)^{n-1} \prod_{k=1}^{n} a_k}{B_{n-1} B_n}$$

converges to a finite value. Since, by (4.4.17),

$$\frac{\left| \prod_{k=1}^{n} a_k \right|}{|B_{n-1} B_n|} < \frac{|B_n| - |B_{n-1}|}{|B_{n-1} B_n|} = \frac{1}{|B_{n-1}|} - \frac{1}{|B_n|},$$

we have

$$\sum_{n=1}^{m} \frac{\left| \prod_{k=1}^{n} a_k \right|}{|B_{n-1} B_n|} = \frac{1}{|B_0|} - \frac{1}{|B_m|}, \qquad m = 1, 2, 3, \dots .$$

Since $\{|B_m|\}$ is montonic increasing [see (4.4.17)], it follows that (4.4.18) converges absolutely and hence $K(a_n/b_n)$ converges to a finite value. ∎

More can be proved about the limit f of the convergent continued fraction $K(a_n/b_n)$. If

(a) $|b_n| = |a_n| + 1$, $n = 1, 2, 3, \dots,$
(b) $\arg(a_{n+1}/b_n b_{n+1}) = \pi$, $n = 1, 2, 3, \dots,$ and
(c) $\sum_{n=1}^{\infty} \prod_{k=1}^{n} |a_k| = \infty$

all hold, then $|f| = 1$. A proof can be found in [Perron, 1957a, p. 59].

Since the condition (4.4.15) insures that $s_n(U) \subset U$, where $U = [w : |w| \le 1]$, it follows from Theorem 4.2 that the disks $K_n = S_n(U)$ [where S_n is defined by (4.2.1b)] are nested and that there exists either a limit circle K or a limit point C. Few cases are known in which the limit-circle case actually occurs. It is therefore of interest to note that if (a) and (b) hold and the series in (c) converges to a finite value, then the limit-circle case occurs; that is,

$$K = \bigcap [K_n : n \ge n_0],$$

where K is the circular disk defined Theorem 4.2(C). A proof of this can be found in [Thron, 1963].

By using equivalence transformations and/or restricting a_n to 1 or b_n to 1, a number of other convergence criteria can be deduced from Theorem 4.35. We shall now state a few of these.

COROLLARY 4.36.

(A) $K(a_n/b_n)$ *converges to a finite value if*

$$\left| \frac{a_1}{b_1} \right| \leqslant \frac{p_1 - 1}{p_1}, \tag{4.4.19a}$$

$$\frac{|a_n|}{|b_{n-1}b_n|} \leqslant \frac{p_n - 1}{p_{n-1}p_n}, \qquad n = 2,3,4,\ldots, \tag{4.4.19b}$$

where all p_n are real and $p_n > 1$.
  (B) (Worpitzky) $K(a_n/1)$ *converges to a finite value if*

$$|a_n| \leqslant \tfrac{1}{4}, \qquad n = 1,2,3\ldots. \tag{4.4.20}$$

(C) $K(1/b_n)$ *converges to a finite value if*

$$\frac{1}{|b_{2n-1}|} + \frac{1}{|b_{2n}|} \leqslant 1, \qquad n = 1,2,3,\ldots. \tag{4.4.21}$$

Proof. (A) is obtained by an application of Theorem 4.35 to the equivalent continued fraction

$$\frac{p_1 a_1}{p_1 b_1} + \frac{p_1 p_2 a_2}{p_2 b_2} + \frac{p_2 p_3 a_3}{p_3 b_3} + \cdots$$

and then setting $\rho_n = p_n/b_n$ for $n \geqslant 1$.
  (B) follows from (A) if one sets $b_n = 1$ and $p_n = 2$ for all $n \geqslant 1$. (The first condition becomes $|a_1| < \tfrac{1}{2}$.)
  Statement (C) can be established from (A) by setting

$$a_n = 1 \text{ and } p_{2n-1} = p_{2n} = |b_{2n}|, \qquad n = 1,2,3,\ldots. \qquad \blacksquare$$

Statement (B) was first proved by Worpitzky [1865] and is the earliest known convergence-region criterion for continued fractions with complex elements.
  One of the general approaches to convergence theory was initiated by Thron [1963] and further developed by Hillam and Thron [1965], Jones and Thron [1968, 1970] and Jones and Snell [1972]. The method is well illustrated by the result of Hillam and Thron [1965] which we now state and prove.

THEOREM 4.37. *Let D be the circular region defined by*

$$D = [w : |w - c| \leqslant r], \qquad \text{where} \quad |c| < r.$$

Let the continued fraction $K(a_n/b_n)$ *be such that*

$$s_n(D) \subseteq D, \qquad n = 1, 2, 3, \ldots,$$

where $s_n(w) = a_n/(b_n + w)$. *Then the continued fraction converges to a value* $f \in D$.

The proof depends on the following:

LEMMA 4.38. *Let* $\{t_n(w)\}$ *be a sequence of l.f.t.'s such that*

$$t_n(U) \subseteq U, \qquad n = 1, 2, 3, \ldots, \tag{4.4.22}$$

where $U = [w : |w| \leqslant 1]$ *and such that*

$$t_n(\infty) = k, \qquad n = 1, 2, 3, \ldots, \tag{4.4.23}$$

for some complex number k such that $|k| < 1$. *Let* $\{T_n\}$ *denote the sequence of l.f.t's generated by* $\{t_n\}$; *i.e.,* $T_n = t_1 \circ t_2 \circ \cdots \circ t_n$. *Then* $\{T_n(w)\}$ *converges to a finite value for all w in* Int(U) (*the interior of U*).

Proof. Repeated application of (4.4.22) yields

$$T_n(U) \subseteq T_{n-1}(U) \subseteq \cdots \subseteq T_1(U) \subseteq U, \qquad n = 1, 2, 3, \ldots. \tag{4.4.24}$$

Hence if one denotes by C_n and R_n the center and radius of the circular disk $T_n(U)$, respectively, arguments analogous to those used in the proof of Theorem 4.2 establish that

$$|C_{n-1} - C_n| \leqslant R_{n-1} - R_n \tag{4.4.25}$$

and that

$$\lim C_n = C \quad \text{and} \quad \lim R_n = R$$

both exist.

In view of the mapping properties (4.4.24), one can write

$$T_n(w) = C_n + R_n e^{i\omega_n} \frac{w + \overline{Q}_n}{Q_n w + 1}, \qquad n = 1, 2, 3, \ldots, \tag{4.4.26}$$

where $|Q_n| = q_n < 1$ and ω_n is a real number. This follows from the fact that the most general l.f.t. which maps U onto U is given by

$$e^{i\omega} \frac{w + \overline{Q}}{Qw + 1}, \qquad |Q| < 1, \quad \omega \text{ real.}$$

For convenience we set

$$P_n = R_n e^{i\omega_n}.$$ (4.4.27)

Condition (4.4.23) leads to

$$T_n(\infty) = T_{n-1}(t_n(\infty)) = T_{n-1}(k), \qquad n=2,3,4,\ldots,$$

which can be expressed in terms of the parameters in (4.4.26) and (4.4.27) by

$$\frac{P_n}{Q_n} = P_{n-1}\frac{k+\overline{Q}_{n-1}}{Q_{n-1}k+1} + C_{n-1} - C_n, \qquad n=2,3,4,\ldots.$$

Combining this with (4.4.25) and noting that for $|k|<1$

$$\left|\frac{k+\overline{Q}_{n-1}}{Q_{n-1}k+1}\right| < 1,$$

one arrives at

$$\left|\frac{P_n}{Q_n}\right| = \frac{R_n}{q_n} < R_{n-1} + (R_{n-1} - R_n).$$

From this one obtains

$$\frac{R_n}{R_{n-1}} < \frac{2q_n}{1+q_n} = 1 - \frac{1-q_n}{1+q_n} = 1 - \frac{\delta_n}{1+q_n} < 1 - \frac{\delta_n}{2},$$ (4.4.28)

where we have set $\delta_n = 1 - q_n$. Repeated application of (4.4.28) leads to

$$R_n < R_1 \prod_{m=2}^{n}\left(1 - \frac{\delta_m}{2}\right).$$

Now $\Sigma\delta_m$ and $\prod(1-\delta_m/2)$ converge or diverge together. If $\Sigma\delta_m = \infty$, then $\prod(1-\delta_m/2)$ diverges to zero, since its sequence of partial products is monotone decreasing. In this case $\lim R_n = 0$ and the sequence $\{T_n(w)\}$ converges to $\lim C_n = C$ for all $w \in U$.

If $\Sigma\delta_m < \infty$, then $\prod(1-\delta_m/2)$ converges, so that $\lim R_n = R \geqslant 0$. For the limit-circle case $R>0$, we therefore must have $\Sigma\delta_m < \infty$ and hence, in particular, $\delta_m \to 0$, so that $q_m \to 1$. We shall use these facts to show that the sequence $\{P_n/Q_n\}$ converges. One has

$$\frac{P_n}{Q_n} - \frac{P_{n-1}}{Q_{n-1}} = P_{n-1}\frac{k+\bar{Q}_{n-1}}{Q_{n-1}k+1} + (C_{n-1}-C_n) - \frac{P_{n-1}}{Q_{n-1}}$$

$$= -\frac{P_{n-1}}{Q_{n-1}}\frac{1-q_{n-1}^2}{1+kQ_{n-1}} + (C_{n-1}-C_n). \qquad (4.4.29)$$

Applying (4.4.29) repeatedly yields

$$\frac{P_{n+m}}{Q_{n+m}} - \frac{P_n}{Q_n} = C_n - C_{n+m} - \sum_{j=1}^m \frac{P_{n+j-1}}{Q_{n+j-1}} \cdot \frac{1-q_{n+j-1}^2}{1+kQ_{n+j-1}}.$$

Using (4.4.25), one obtains

$$\left|\frac{P_{n+m}}{Q_{n+m}} - \frac{P_n}{Q_n}\right| \le R_n - R_{n+m} + \sum_{j=1}^m \frac{R_{n+j-1}}{q_{n+j-1}} \cdot \frac{(1+q_{n+j-1})(1-q_{n+j-1})}{|1+kQ_{n+j-1}|}.$$

Since $q_n \to 1$, for sufficiently large n one has $q_{n+j-1} > 1/2$. Using this, together with $R_{n+j-1} < 1$ and $\delta_{n+j-1} = 1 - q_{n+j-1}$, one obtains, for sufficiently large n,

$$\left|\frac{P_{n+m}}{Q_{n+m}} - \frac{P_n}{Q_n}\right| \le R_n - R_{n+m} + \frac{4}{1-|k|}\sum_{j=1}^m \delta_{n+j-1}.$$

Since $\lim R_n = R$ and $\Sigma\delta_m < \infty$, it follows that the sequence $\{P_n/Q_n\}$ converges to a finite value. Using (4.4.26), one can write

$$T_n(w) = C_n + \frac{P_n}{Q_n}\left(1 + \frac{q_n^2-1}{Q_n w+1}\right).$$

Hence in the limit-circle case, the sequence $\{T_n(w)\}$ converges to a finite limit for all w such that $|w| < 1$. ∎

Proof of Theorem 4.37. Let $v(w) = rw + c$, $v^{-1}(w) = (w-c)/r$, so that $v(U) = D$. Let $\{t_n\}$ be the sequence of l.f.t. defined by

$$t_n(w) = v^{-1}\circ s_n\circ v(w), \qquad n = 1,2,3,\dots. \qquad (4.4.30)$$

It follows that $t_n(U) \subseteq U$, since by hypothesis $s_n(D) \subseteq D$. Moreover,

$$t_n(\infty) = -\frac{c}{r}.$$

Since $|-c/r|<1$, it follows from Lemma 4.38 that $\{T_n(w)\}$ converges to a finite limit for $|w|<1$. From (4.4.30) one obtains $T_n(w)=v^{-1}\circ S_n\circ v(w)$, where $\{S_n\}$ is the sequence of l.f.t.'s generated by $\{s_n\}$. If f_n denotes the nth approximant of the continued fraction $\mathbf{K}(a_n/b_n)$, one has

$$f_n=S_n(0)=v\circ T_n\circ v^{-1}(0)=v\circ T_n\left(-\frac{c}{r}\right),\qquad n=1,2,3,\dots.$$

Since $|-c/r|<1$, it follows that $\mathbf{K}(a_n/b_n)$ converges to a finite value.   ∎

Among the corollaries of Theorem 4.37 are the following: (1) Pringsheim's criterion (Theorem 4.35), since $|b_n|\geqslant|a_n|+1$ implies $s_n(U)\subseteq U$; (2) Worpitzky's criterion [Corollary 4.36(B)], since $|a_n|\leqslant\frac{1}{4}$ implies $|s_n(w)|\leqslant\frac{1}{2}$ for $|w|\leqslant\frac{1}{2}$ (see Corollary 4.13); and (3) Corollary 4.34, which can now be proved using Theorem 4.37 and Corollary 4.14.

Using Theorem 4.3 we can state the necessary and sufficient condition on $s_n(w)=a_n/(b_n+w)$ to map D into D (in Theorem 4.37) and thus arrive at

THEOREM 4.39. *The continued fraction* $\mathbf{K}(a_n/b_n)$ *converges to a finite value if*

$$|a_n(\bar{b}_n+\bar{c})-c(|b_n+c|^2-r^2)|+|a_n|r\leqslant r(|b_n+c|^2-r^2),$$
$$n=1,2,3,\dots,\qquad(4.4.31a)$$

where

$$|c|<r.\qquad(4.4.31b)$$

The value f to which the continued fraction converges satisfies $|f-c|\leqslant r$.

Theorem 4.39 was generalized to variable element and value regions by Jones and Thron [1968, Theorem 3.2]. The result is based on Theorem 4.3 with Γ_n and ρ_n variable. However, the following restriction is required:

$$|\Gamma_n|/q_n\leqslant 1-\varepsilon,\qquad\varepsilon>0,\qquad n=1,2,3,\dots.\qquad(4.4.32)$$

4.4.2 Parabolic Convergence Regions

A new chapter was opened in the theory of convergence criteria for continued fractions when Scott and Wall [1940a] proved that the parabolic region

$$P=\left[w:|w|-\operatorname{Re}(w)\leqslant\tfrac{1}{2}\right]$$

is a simple conditional convergence region for continued fractions $\mathbf{K}(a_n/1)$.

Paydon and Wall [1942] and Leighton and Thron [1942] improved this result by showing that any bounded set contained in the interior of one of the parabolic regions P_α defined by (4.4.34) (see Theorem 4.40) is a simple convergence region for continued fractions $\mathbf{K}(a_n/1)$. (Actually each proved a slightly stronger result.) That each P_α is a simple conditional convergence region was established by Thron [1943a]. Generalizations and proofs of various kinds have been given by Wall [1948], Thron [1958, 1963] and Jones and Thron [1968]. Here we shall prove that every bounded part of a P_α is a simple convergence region. In the proof we give an explicit estimate of the radii of the nested circles, which can be used to establish truncation errors.

First we make the following definitions. Let f_n denote the nth approximant of a continued fraction $\mathbf{K}(a_n/b_n)$ [or $\mathbf{K}(a_n/1)$ or $\mathbf{K}(1/b_n)$], and let $f = \lim f_n$ when that limit exists. A sequence $\{\Omega_n\}$ of convergence regions is called a *uniform sequence* of *convergence regions* if there exists a sequence of positive numbers $\{\varepsilon_n\}$ depending only on $\{\Omega_n\}$ with $\lim \varepsilon_n = 0$ such that

$$\langle a_n, b_n \rangle \in \Omega_n, \qquad n = 1,2,3,\ldots$$

implies that $\lim f_n = f$ is finite and

$$|f - f_n| \leq \varepsilon_n, \qquad n = 1,2,3,\ldots. \qquad (4.4.33)$$

When this occurs we say that $\mathbf{K}(a_n/b_n)$ *converges uniformly with respect to* $\{\Omega_n\}$. Analogous definitions apply to convergence regions for continued fractions $\mathbf{K}(a_n/1)$ and $\mathbf{K}(1/b_n)$. If $\Omega_n = \Omega_1$ (or $E_n = E_1$ or $G_n = G_1$) for $n = 1,2,3,\ldots$, we call Ω_1 (or E_1 or G_1) a *uniform simple convergence region*. If $\Omega_{2n+1} = \Omega_1$ and $\Omega_{2n} = \Omega_2$ (or similarly for E_n or G_n) one speaks of *uniform twin convergence regions*.

THEOREM 4.40 (Uniform parabola theorem). *Let*

$$P_\alpha = \left[w : |w| - \mathrm{Re}(we^{-2i\alpha}) \leq \tfrac{1}{2}\cos^2\alpha \right], \qquad -\frac{\pi}{2} < \alpha < \frac{\pi}{2}, \qquad (4.4.34)$$

$$Q_M = [w : |w| < M], \qquad M > 0, \qquad (4.4.35)$$

and

$$E = P_\alpha \cap Q_M. \qquad (4.4.36)$$

Then E is a uniform simple convergence region for continued fractions $\mathbf{K}(a_n/1)$.

A description of the parabolic boundary of P_α is given in Corollary 4.16, where P_α is denoted by $E(\alpha)$ (see Figure 4.2.1). Before proving the above theorem we shall state and prove the following:

LEMMA 4.41. *Let* $x \geqslant c > 0$ *and* $v^2 \leqslant 4u + 4$. *Then*

$$\min \operatorname{Re}\left(\frac{u + iv}{x + iy}\right) = -\frac{1}{c}.$$

Proof. We have

$$F = \operatorname{Re}\left(\frac{u + iv}{x + iy}\right) = \frac{ux + vy}{x^2 + y^2}.$$

Leaving u, v and x fixed, we first determine the value of y that minimizes F. From elementary calculus this value is known to be one of the solutions of the equation

$$vy^2 + 2uxy - x^2 v = 0.$$

That is,

$$y = \frac{x\left(-u \pm \sqrt{u^2 + v^2}\right)}{v}.$$

Thus

$$\min_{-\infty < y < \infty} F = \frac{-v^2 \sqrt{u^2 + v^2}}{2x\left(u^2 + v^2 + u\sqrt{u^2 + v^2}\right)} = \frac{-v^2}{2x\left(u + \sqrt{u^2 + v^2}\right)}.$$

Continuing to keep u and x fixed, we now allow v^2 to vary. Its range is $0 < v^2 \leqslant 4u + 4$. This leads to

$$\min_{y, v} F = \frac{-(4u + 4)}{2x[u + (u + 2)]} = -\frac{1}{x}.$$

Since $x \geqslant c > 0$, the lemma is proved.   ∎

Proof of Theorem 4.40.   In Corollary 4.16 it was shown that $a_n / (1 + V_\alpha)$ $\subseteq V_\alpha$, where

$$V_\alpha = \left[w : \operatorname{Re}(we^{-i\alpha}) \geqslant -\tfrac{1}{2}\cos\alpha\right].$$

It follows from Corollary 4.7 that if one defines

$$L_n = \frac{B_{n-1}}{B_{n-2}}, \qquad n = 2, 3, 4, \ldots,$$

then

$$L_n \in 1 + V_\alpha, \qquad n = 2, 3, 4, \dots.$$

Here B_n denotes the nth denominator of the continued fraction $\mathbf{K}(a_n/1)$. Hence we can write

$$L_n = e^{i\alpha} \frac{\cos \alpha}{2} (x_n + i y_n), \qquad x_n \geqslant 1, \quad -\infty < y_n < \infty. \qquad (4.4.37)$$

Similarly $a_n \in P_\alpha$ implies that a_n can be written as

$$a_n = e^{i2\alpha} \frac{\cos^2 \alpha}{4} (u_n + i v_n), \qquad v_n^2 \leqslant 4 u_n + 4. \qquad (4.4.38)$$

Now let C_n denote the center of the circular disk $K_n = S_n(V_\alpha)$ (see Theorem 4.2). Then C_n and ∞ are inverses of each other with respect to the circular boundary ∂K_n of K_n. It follows that $H_n = S_n^{-1}(C_n)$ and $-B_n/B_{n-1} = S_n^{-1}(\infty)$ are inverses of each other with respect to the boundary of V_α. Since the boundary of V_α is a straight line, one easily computes

$$H_n e^{-i\alpha} = \overline{\left(\frac{B_n}{B_{n-1}} \right)} e^{i\alpha} - \cos \alpha. \qquad (4.4.39a)$$

For a short description of the pertinent facts on the geometry of inversions, the reader is referred to [Henrici, 1974, pp. 316–322] and [Thron, 1953, pp. 191–194]. From (4.4.39a) one obtains

$$H_n = \frac{\overline{B}_n e^{i2\alpha} - (\cos \alpha) e^{i\alpha} \overline{B}_{n-1}}{\overline{B}_{n-1}}$$

$$= \frac{\overline{B}_n (2 \cos \alpha \, e^{i\alpha} - 1) - (\cos \alpha) e^{i\alpha} \overline{B}_{n-1}}{\overline{B}_{n-1}}$$

$$= \frac{(2 \overline{B}_n - \overline{B}_{n-1}) \beta - \overline{B}_n}{\overline{B}_{n-1}}. \qquad (4.4.39b)$$

Here we have set

$$\beta = e^{i\alpha} \cos \alpha$$

and have replaced $e^{i2\alpha}$ by $2\beta-1$, since it is true that

$$
\begin{aligned}
e^{i2\alpha} &= (2\cos^2\alpha - 1) + i(2\sin\alpha\cos\alpha) \\
&= 2\cos\alpha(\cos\alpha + i\sin\alpha) - 1 \\
&= 2e^{i\alpha}\cos\alpha - 1 = 2\beta - 1.
\end{aligned}
$$

For $C_n = S_n(H_n)$ one easily computes [using (2.1.7)]

$$
S_n(H_n) = \frac{(2A_{n-1}\bar{B}_n - A_{n-1}\bar{B}_{n-1})\beta - A_{n-1}\bar{B}_n + A_n\bar{B}_{n-1}}{(2B_{n-1}\bar{B}_n - B_{n-1}\bar{B}_{n-1})\beta - B_{n-1}\bar{B}_n + B_n\bar{B}_{n-1}}.
$$

Next observe that $S_n(-\frac{1}{2})$ is on the boundary of $K_n = S_n(V_\alpha)$, since $-\frac{1}{2}$ is on the boundary of V_α. Hence the radius R_n of K_n is given by

$$
R_n = |S_n(H_n) - S_n(-\tfrac{1}{2})|.
$$

A straightforward but lengthy computation yields

$$
S_n(H_n) - S_n(-\tfrac{1}{2})
$$

$$
= \frac{(2\beta-1)(2\bar{B}_n - \bar{B}_{n-1})(A_{n-1}B_n - A_n B_{n-1})}{(2B_n - B_{n-1})\left[(2B_{n-1}\bar{B}_n - B_{n-1}\bar{B}_{n-1})\beta - B_{n-1}\bar{B}_n + B_n\bar{B}_{n-1}\right]}.
$$

Hence

$$
R_n = \frac{\prod\limits_{m=1}^{n} |a_m|}{|B_{n-1}(2\bar{B}_n - \bar{B}_{n-1})\beta - (B_{n-1}\bar{B}_n - B_n\bar{B}_{n-1})|}.
$$

If one replaces B_n with $B_{n-1} + a_n B_{n-2}$ by means of the difference equation (2.1.6), then one obtains

$$
\frac{R_n}{R_{n-1}} = \frac{|a_n|\,|B_{n-1}(2\bar{B}_{n-1} - \bar{B}_{n-2})\beta - (B_{n-2}\bar{B}_{n-1} - B_{n-1}\bar{B}_{n-2})|}{|B_{n-1}(\bar{B}_{n-1} + 2\bar{a}_n\bar{B}_{n-2})\beta - (\bar{a}_n B_{n-1}\bar{B}_{n-2} - a_n\bar{B}_{n-1}B_{n-2})|}.
$$

By dividing both numerator and denominator through by $B_{n-1}\bar{B}_{n-2}$ and setting $L_n = B_{n-1}/B_{n-2}$, one arrives at

$$
\begin{aligned}
\frac{R_n}{R_{n-1}} &= \frac{|a_n|\,|(2\bar{L}_n - 1)\beta - (\bar{L}_n - L_n)|}{|L_n(2\bar{a}_n + \bar{L}_n)\beta - (\bar{a}_n L_n - a_n\bar{L}_n)|} \\
&= \frac{|a_n|\,|\bar{L}_n e^{i2\alpha} + L_n - (\cos\alpha)e^{i\alpha}|}{|\bar{a}_n L_n e^{i2\alpha} + a_n\bar{L}_n + |L_n|^2(\cos\alpha)e^{i\alpha}|}.
\end{aligned}
$$

Now from (4.4.37), (4.4.38) and (4.4.39) we obtain

$$\bar{L}_n e^{i2\alpha} + L_n - (\cos\alpha)e^{i\alpha} = e^{i\alpha}(\cos\alpha)(x_n - 1)$$

and

$$\bar{a}_n L_n e^{i2\alpha} + a_n \bar{L}_n + |L_n|^2(\cos\alpha)e^{i\alpha} = e^{i\alpha}\frac{\cos^3\alpha}{4}(u_n x_n + v_n y_n + x_n^2 + y_n^2),$$

so that

$$\frac{R_n}{R_{n-1}} = \frac{\frac{1}{4}\cos^2\alpha(u_n^2 + v_n^2)^{1/2}(\cos\alpha)(x_n - 1)}{\frac{1}{4}\cos^3\alpha|u_n x_n + v_n y_n + x_n^2 + y_n^2|}$$

$$= \frac{(u_n^2 + v_n^2)^{1/2}(x_n - 1)}{|x_n^2 + u_n x_n + (y_n + \frac{1}{2}v_n)^2 - \frac{1}{4}v_n^2|}.$$

Since

$$x_n^2 + u_n x_n + (y_n + \tfrac{1}{2}v_n)^2 - \tfrac{1}{4}v_n^2 \geqslant x_n^2 + u_n x_n - u_n - 1$$

$$= (x_n - 1)(x_n + 1 + u_n),$$

the absolute-value sign in the denominator can be omitted. Hence

$$\frac{R_n}{R_{n-1}} < \frac{(u_n^2 + v_n^2)^{1/2}(x_n - 1)}{(x_n - 1)(x_n + 1 + u_n)} \leqslant \frac{u_n + 2}{u_n + 2 + x_n - 1}$$

$$= \frac{1}{1 + \dfrac{x_n - 1}{u_n + 2}}. \qquad (4.4.40)$$

We know that $x_n > 1$. To obtain a useful estimate of R_n/R_{n-1}, it is desirable to investigate x_n further. The following identity will be useful:

$$1 = e^{i\alpha}e^{-i\alpha} = e^{i\alpha}(\cos\alpha - i\sin\alpha). \qquad (4.4.41)$$

Now

$$L_2 = e^{i\alpha}\frac{\cos\alpha}{2}(x_2 + iy_2) = 1 = e^{i\alpha}(\cos\alpha - i\sin\alpha),$$

and hence $d_2 = \min x_2 = x_2 = 2$. Similarly

$$L_{n+1} = 1 + \frac{a_n}{L_n}$$

leads to

$$e^{i\alpha}\frac{\cos\alpha}{2}(x_{n+1}+iy_{n+1})=1+\frac{e^{i2\alpha}\dfrac{\cos^2\alpha}{2}(u_n+iv_n)}{e^{i\alpha}\dfrac{\cos\alpha}{2}(x_n+iy_n)}$$

$$=e^{i\alpha}(\cos\alpha-i\sin\alpha)+e^{i\alpha}\frac{\cos\alpha}{2}\frac{u_n+iv_n}{x_n+iy_n}.$$

Hence

$$d_{n+1}=\min x_{n+1}=2+\min\mathrm{Re}\left(\frac{u_n+iv_n}{x_n+iy_n}\right).$$

Applying Lemma 4.41 and assuming d_n exists and is positive, one has

$$d_{n+1}=2-\frac{1}{d_n}.$$

By induction

$$d_n=\frac{n}{n-1},\qquad n=2,3,4,\ldots \qquad\qquad (4.4.42)$$

is easily obtained. We return now to (4.4.40). Since $|a_n|<M$, we have $|u_n+iv_n|<4M/\cos^2\alpha$ and hence

$$1\leqslant u_n+2\leqslant\frac{4M}{\cos^2\alpha}+2. \qquad\qquad (4.4.43)$$

Setting

$$\delta=\delta(M,\alpha)=\frac{1}{\dfrac{4M}{\cos^2\alpha}+2},$$

it follows from (4.4.40), (4.4.42) and (4.4.43) that

$$\frac{R_n}{R_{n-1}}\leqslant\frac{1}{1+\dfrac{\delta}{n-1}},\qquad n=2,3,4,\ldots$$

and hence

$$R_n=R_1\prod_{m=1}^{n-1}\frac{R_{m+1}}{R_m}\leqslant\frac{R_1}{\displaystyle\prod_{m=1}^{n-1}\left(1+\frac{\delta}{m}\right)},\qquad n=2,3,4,\ldots. \qquad (4.4.44)$$

Since $R_1 = |a_1|/\cos \alpha < M/\cos \alpha$,

$$R_n < \cfrac{M}{(\cos \alpha) \prod\limits_{m=1}^{n-1}\left(1 + \dfrac{\delta}{m}\right)} = \frac{\varepsilon_n}{2}, \qquad n = 2, 3, 4, \ldots.$$

Clearly $\{\varepsilon_n\}$ is independent of the particular sequence $\{a_n\}$ chosen from E, and $\lim_{n\to\infty}\varepsilon_n = 0$. Let f_n denote the nth approximant of $\mathrm{K}(a_n/1)$. Since $0 \in V_\alpha$, it follows that

$$f_{n+k} = S_{n+k}(0) = S_n(s_{n+1}\circ \cdots \circ s_{n+k}(0)) \in S_n(V_\alpha).$$

Hence

$$|f_{n+k} - f_{n+1}| < 2R_n < \varepsilon_n, \qquad n = 2, 3, 4, \ldots, \quad k = 1, 2, 3, \ldots. \qquad (4.4.45)$$

This proves that E is a uniform simple convergence region for $\mathrm{K}(a_n/1)$. ∎

Given $\varepsilon > 0$, there exists an $\eta > 0$ such that

$$\frac{1}{1 + \delta x} \leqslant (1 - x)^{\delta(1 - \varepsilon)} \qquad \text{for} \quad 0 < x < \eta.$$

If one chooses $n_0 > 1/\eta$, it follows from (4.4.44) that

$$R_n \leqslant R_{n_0} \prod_{m=n_0}^{n-1}\left(1 - \frac{1}{m}\right)^{\delta(1-\varepsilon)} = R_0\left(\frac{n_0 - 1}{n - 1}\right)^{\delta(1-\varepsilon)},$$

$$n = n_0 + 1, n_0 + 2, \ldots. \qquad (4.4.46)$$

The speed of convergence of the continued fraction $\mathrm{K}(a_n/1)$ considered in Theorem 4.40 can be substantially improved if the elements a_n are further restricted. The proof given here can easily be modified to yield such results. This was done by Thron [1958] and by Jones and Snell [1969] (see also Chapter 8).

Theorem 4.40 does not give the very best result that has been established for parabolic convergence regions. The best known result for simple parabolic regions is

THEOREM 4.42 (Parabola theorem). *If all elements of a continued fraction* $\mathrm{K}(a_n/1)$ *lie in a parabolic region*

$$P_\alpha = \left[w : |w| - \mathrm{Re}(we^{-2i\alpha}) \leqslant \tfrac{1}{2}\cos^2\alpha\right], \qquad -\frac{\pi}{2} < \alpha < \frac{\pi}{2}, \qquad (4.4.47)$$

then the continued fraction converges to a finite value iff at least one of the series

$$\sum_{n=1}^{\infty} \left| \frac{a_2 a_4 \cdots a_{2n}}{a_3 a_5 \cdots a_{2n+1}} \right|, \qquad \sum_{n=1}^{\infty} \left| \frac{a_3 a_5 \cdots a_{2n+1}}{a_4 a_6 \cdots a_{2n+2}} \right| \qquad (4.4.48)$$

is divergent (*see Figure* 4.2.1).

This theorem was first proved by Thron [1943a] by means of the Stieltjes-Vitali theorem (Theorem 4.30). He later gave [Thron, 1963] an elementary proof of Theorem 4.42 using methods illustrated in the proof of Theorem 4.37. Neither of these methods allows insight into the speed of convergence of the continued fraction.

Extensions to variable parabolic regions have been given by a number of people. One approach based on the concept of positive definite continued fractions was developed by Wall and Wetzel [1944a, b] building on earlier work by Hellinger [1922] and Hellinger and Wall [1943]. This work is treated in detail by Wall [1948, pp. 64–117]. Using methods discussed in the proof of Theorem 4.37, Jones and Thron [1968] obtained results overlapping those of Wall and Wetzel referred to above. The relations between the two results are discussed in [Jones and Thron, 1968, pp. 1040–1042 and 1049].

We now state a result of Jones and Thron [1968, Theorem 5.1] which is typical of the known criteria.

THEOREM 4.43 (Multiple parabola theorem). *Let the elements a_n of the continued fraction* $\mathbf{K}(a_n/1)$ *lie in the parabolic regions defined by*

$$|a_n| - \mathrm{Re}\left(a_n e^{-i(\psi_n + \psi_{n-1})} \right) \leqslant 2 p_{n-1}(\cos \psi_n - p_n), \qquad n = 1, 2, 3, \ldots,$$
$$(4.4.49)$$

where $p_n > 0$ and ψ_n is real and where

$$\left| p_n e^{i\psi_n} - \tfrac{1}{2} \right| \leqslant d < \tfrac{1}{2}, \qquad n = 0, 1, 2, 3, \ldots. \qquad (4.4.50)$$

(A) *Then the even part and the odd part of the continued fraction both converge to finite values.*

(B) *The values of all approximants are finite and lie in the half plane*

$$V_0 = \left[w : \mathrm{Re}(w e^{-i\psi_0}) \geqslant -p_0 \right]. \qquad (4.4.51)$$

(C) *The continued fraction itself converges to a finite value iff at least one of the series* (4.4.48) *is divergent.*

(D) *A sufficient condition for the convergence of the continued fraction to a finite value is that the a_n lie in the respective parabolic regions* (4.4.49) *and* *that for some* $M > 0$,

$$|a_n| < M, \qquad n = 1, 2, 3, \ldots . \qquad (4.4.52)$$

There is also a group of convergence criteria for continued fractions $K(a_n/b_n)$ related to the parabola theorems described here. A first result of this type is due to Thron [1943b]. Substantial improvements were made by Jones and Thron [1968]. We state one of their results below (Theorem 4.44). Similar results can also be obtained from the theory of positive definite continued fractions.

THEOREM 4.44. *A continued fraction* $K(a_n/b_n)$ *converges to a finite value if its elements* a_n, b_n *satisfy*

$$|a_n| - \mathrm{Re}\left(a_n e^{-i(\psi_n + \psi_{n-1})}\right) < 2p_{n-1}\left[\mathrm{Re}\left(b_n e^{-i\psi_n}\right) - p_n\right], \qquad n = 1, 2, 3, \ldots ,$$
$$(4.4.53)$$

where $p_n > 0$ *and* ψ_n *is real, and the sequence*

$$\left\{\frac{a_n}{p_n p_{n-1}}\right\} \qquad (4.4.54)$$

is bounded. The approximants of the continued fraction are all finite and lie in the half plane

$$\left[w : \mathrm{Re}\left(w e^{i(\psi_1 - \arg a_1)}\right) > 0\right] \qquad (4.4.55)$$

if $\mathrm{Re}(b_1 e^{-i\psi_n}) = p_1$ *and in the circular disk*

$$\left[w : \left|w - \frac{a_1 e^{-i\psi_1}}{2\left[\mathrm{Re}(b_1 e^{-i\psi_1}) - p_1\right]}\right| < \frac{|a_1|}{2\left[\mathrm{Re}(b_1 e^{-i\psi_1}) - p_1\right]}\right] \qquad (4.4.56)$$

if $\mathrm{Re}(b_1 e^{-i\psi_1}) > p_1$.

4.4.3 *Convergence Neighborhoods for* $K(a_n/1)$

In all three editions of his book on continued fractions Perron devotes a good deal of space to "*limitärperiodische Kettenbrüche.*" These are continued fractions $K(a_n/b_n)$ satisfying

$$\lim_{n \to \infty} a_n = a, \qquad \lim_{n \to \infty} b_n = b.$$

Perron [1957a, pp. 89–94] gives a result insuring uniform convergence of

$K(a_n/b_n)$ with respect to a region

$$\Omega = [\langle a_n, b_n \rangle : |a_n - a| \leqslant d_1(a, b), |b_n - b| \leqslant d_2(a, b)]$$

where the $d_1(a, b)$ and $d_2(a, b)$ are defined in a very involved manner. No regions exist for certain values of a and/or b. Since, in particular, $b = 0$ is excluded in Perron's result, it is sufficient to consider continued fractions of the form $K(a_n/1)$.

A circular disk E will be called a *convergence neighborhood of a point* $a \in E$ if E is a simple convergence region for continued fractions $K(a_n/1)$. If $a_n = a$ is real and $a < -\frac{1}{4}$, it follows from Theorem 3.2 that the periodic continued fraction $K(a_n/1)$ diverges. Hence no convergence neighborhood exists for those values of a. If a is not of this form, there is always at least one parabolic region P_α of the form (4.4.34) such that $a \in \text{Int}(P_\alpha)$. The largest circular disk with center at a and contained in P_α is then a convergence neighborhood of a by Theorem 4.40. Perron [1957a, p. 90] was aware of this but preferred his proof, since at the time of his writing no proof of the parabola theorem proving uniform convergence was available. In view of our Theorem 4.54, this shortcoming was not as important as Perron thought, since value behavior was available. It is not known to us whether the result in Perron is indeed better than that which can be obtained from the parabola theorem. (Perron himself thought that it was not.) This is another illustration of the difficulty in resolving whether one result is contained in another.

The question of determining the largest circular convergence neighborhood of a point is intriguing but not of great importance, since the main application of whatever theorem one can prove is likely to be to "limit periodic" continued fractions. The following simple theorem can be deduced from Theorem 4.40. In view of the results in [Thron, 1944b], it is clear that it does not provide the best possible circular convergence neighborhood for some a.

THEOREM 4.45 (Convergence neighborhood theorem).

(A) *If a is a complex number satisfying*

$$|\arg a| < \pi \quad \text{and} \quad |a| > \tfrac{1}{4}, \tag{4.4.57}$$

then

$$E = \left[w : |w - a| \leqslant \frac{1}{\sqrt{2}} [|a| + \text{Re}(a)]^{1/2} \right] \tag{4.4.58}$$

is a uniform simple convergence region for $K(a_n/1)$.

(B) *If a is a complex number satisfying*

$$|a| < \tfrac{1}{4}, \tag{4.4.59}$$

then

$$E = \left[w : |w - a| \leqslant |a + \tfrac{1}{4}| \right] \tag{4.4.60}$$

is a uniform simple convergence region for $K(a_n/1)$.

Proof. Our proof is based on the parabola theorem (Theorem 4.40); hence we begin with some facts about parabolas that will be used.

Let P denote a parabola with vertex v, focus f and distance $k/2$ between f and v. Let a be a given point on the axis of P and for each point z in C, let $D(a, z)$ denote the distance from a to z.

Property 1. *If* $D(a, f) \leqslant k/2$, *then the minimum of* $[D(a, q): q \in P]$ *is attained at the unique point* v, *the vertex of* P.

Property 2. *If* $D(a, f) > k/2$, *then the minimum of* $[D(a, q): q \in P]$ *is attained at exactly two points* q_1 *and* q_2 *on* P, *which are symmetric with respect to the axis of* P. *Moreover*,

$$D(f, q_0) = D(f, a), \qquad q_0 \in P,$$

iff

$$D(f, q_0) = \min_{q \in P} D(a, q).$$

To verify these properties it will suffice to consider a parabola P with focus f at the origin, axis along the real axis and vertex $v = -k/2$. The equation of such a parabola in polar coordinates (ρ, φ) is given by

$$\rho = \frac{k}{1 - \cos \varphi}, \qquad 0 < |\varphi| \leqslant \pi, \quad k > 0.$$

Thus a is real and $a \geqslant -k/2$ (see Figure 4.4.1).

Writing $q = \rho e^{i\varphi}$ and $D = D(a, q)$, we obtain from the law of cosines

$$D^2 = a^2 + \frac{k^2}{(1 - \cos \varphi)^2} - \frac{2ka \cos \varphi}{1 - \cos \varphi}$$

$$= \left(\frac{k}{1 - \cos \varphi} - a \right)^2 + 2ak.$$

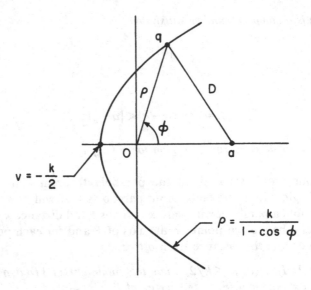

Figure 4.4.1. Parabola with focus at the origin, axis along the real axis and vertex at $v = -k/2$.

If $-k/2 \leqslant a < 0$, it follows that

$$D^2 = \left(\frac{k}{1-\cos\varphi} + |a|\right)^2 - 2|a|k$$

and hence D is a monotone decreasing function of φ as φ varies from 0 to π. Thus D attains its minimum at the vertex v (when $\varphi = \pi$) as asserted. If $a \geqslant 0$, then differentiation gives

$$2D\frac{dD}{d\varphi} = -2\left(\frac{k}{1-\cos\varphi} - a\right)\frac{k\sin\varphi}{(1-\cos\varphi)^2}.$$

It follows that if $0 \leqslant a \leqslant k/2$, then again D is a monotone decreasing function of φ as φ varies from 0 to π, and hence the minimum of D is attained at the vertex v. This establishes Property 1. If $a > k/2$, it follows from the preceding equations that the minimum of D is at $\varphi = \pm\varphi_0$ such that

$$\frac{k}{1-\cos\varphi_0} = a, \, 0 < |\varphi_0| < \pi.$$

Property 2 is an immediate consequence of this.

Proof of (A). Let a satisfy (4.4.57), let $\alpha = (\arg a)/2$, and let P_α denote the parabolic region (4.4.34) of Theorem 4.40. Writing $q = \rho e^{i\varphi}$, we obtain for the boundary ∂P_α of P_α the equation in polar form

$$\rho = \frac{\frac{1}{2}\cos^2\alpha}{1 - \cos(\varphi - 2\alpha)}.$$

Recall that the parabola ∂P_α has focus f at the origin, vertex at $v = \frac{1}{4}(\cos^2\alpha)e^{i(\pi+2\alpha)}$, and axis in the direction of the ray $\arg w = 2\alpha$ (see Figure 4.4.2). In the terminology used above we have

$$\frac{k}{2} = \frac{1}{4}\cos^2\alpha < \frac{1}{4}.$$

Therefore $|a| > \frac{1}{4} > k/2$ and so, by Property 2, $D(a, q)$ attains its minimum at two points q_1 and q_2 on ∂P_α. Let $q_1 = \rho_1 e^{i\varphi_1}$. It follows from Property 2 that

$$\rho_1 = |a|.$$

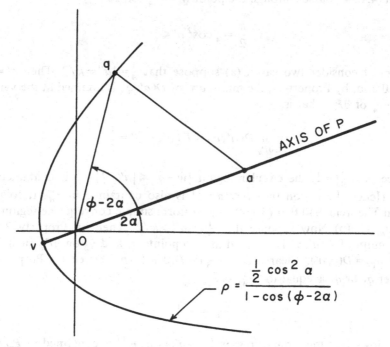

Figure 4.4.2. Parabola with focus at the origin, axis along the ray $\arg z = 2\alpha$ and vertex $v = \frac{1}{4}(\cos^2\alpha)e^{i(\pi+2\alpha)}$.

Using this with the cosine law

$$|a-q_1|^2=2\rho_1^2-2\rho_1^2\cos(\varphi_1-2\alpha),$$

one obtains (using the equation for ∂P_α)

$$|a-q_1|^2=\rho_1\cos^2\alpha=|a|\frac{\cos 2\alpha+1}{2}$$

$$=\frac{|a|}{2}[1+\cos(\arg a)]$$

$$=\tfrac{1}{2}[|a|+\operatorname{Re}(a)].$$

Since the circular region $|w-a|\leqslant|a-q_1|$ is contained in P_α, it follows from Theorem 4.40 that (4.4.58) is a uniform simple convergence region for $K(a_n/1)$. This proves (A).

   Proof of (B). Let a satisfy (4.4.59), and let a' denote the point where line L, passing through $-\tfrac{1}{4}$ and a, meets the circle $|w|=\tfrac{1}{4}$ (see Figure 4.4.3). Let $2\alpha'=\arg a'$, and recall that every parabola ∂P_α [where P_α has the form (4.4.34)] passes through the point $w=-\tfrac{1}{4}$. Since

$$\frac{k}{2}=\tfrac{1}{4}\cos^2\alpha'\leqslant\tfrac{1}{4},$$

we must consider two cases. (a) suppose that $\tfrac{1}{4}=|a'|=k/2$. Then $a'=\tfrac{1}{4}$, $\alpha'=0$ and, by Property 1, the minimum of $D(a',q)$ is attained at the vertex $v=-\tfrac{1}{4}$ of $\partial P_{\alpha'}$; that is,

$$\min_{q\in\partial P_{\alpha'}}D(a',q)=D\big(a',-\tfrac{1}{4}\big)=\tfrac{1}{2}.$$

Since $|a'+\tfrac{1}{4}|=\tfrac{1}{2}$, the circular region $|w-a'|\leqslant|a'+\tfrac{1}{4}|=\tfrac{1}{2}$ is contained in $P_{\alpha'}$. Hence the region $|w-a|\leqslant|a+\tfrac{1}{4}|$ is also contained in $P_{\alpha'}$. It follows from Theorem 4.40 that (4.4.60) is a uniform simple convergence region for $K(a_n/1)$. (b) Now suppose that $\tfrac{1}{4}=|a'|>k/2$. Then by Property 2 the minimum of $D(a',q)$ is attained at two points q_1 and q_2 on $\partial P_{\alpha'}$ such that $D(0,q_1)=D(0,a')$. Clearly $D(0,-\tfrac{1}{4})=D(0,a')$ and hence by Property 2 either q_1 or q_2 is equal to $-\tfrac{1}{4}$, say

$$q_1=-\tfrac{1}{4}.$$

It follows that the circular region $|w-a'|\leqslant|a'+\tfrac{1}{4}|$ is contained in $P_{\alpha'}$ and therefore the region $|w-a|\leqslant|a+\tfrac{1}{4}|$ is also contained in $P_{\alpha'}$. By Theorem 4.40, the set (4.4.60) is a uniform simple convergence region for $K(a_n/1)$. ∎

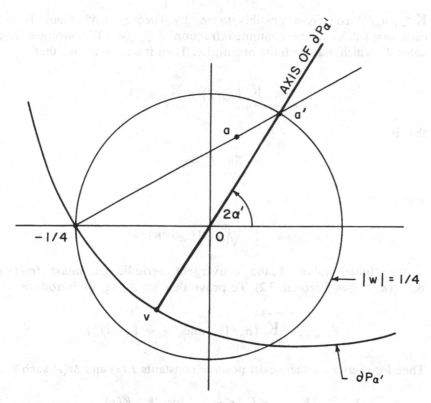

Figure 4.4.3. Parabola with focus at the origin, axis along the ray $\arg z = 2\alpha'$ and vertex v.

Remarks on Theorem 4.45.

1. For a point b on L [case (b)] on the opposite side of a' (as seen from $-\frac{1}{4}$), the circle with center b and passing through $-\frac{1}{4}$ does not lie entirely in $P_{\alpha'}$. If we had been interested in circular convergence regions containing a but with center not necessarily at a, then, of course, the region given by $|w - a'| < |a' + \frac{1}{4}|$ would have been a better convergence region than (4.4.60).

2. Scott and Wall [Wall, 1948, p. 62] showed that

$$|a_n - a| < \frac{|1 + 2a| - 2|a|}{4} \tag{4.4.61}$$

is a convergence neighborhood of a for continued fractions $K(a_n/1)$. It is easily shown that this result is contained in Theorem 4.45.

3. A continued fraction $K_{n=1}^{\infty}(a_n/1)$ is called *limit periodic* if

$$\lim_{n \to \infty} a_n = a.$$

If $|\arg(a + \frac{1}{4})| < \pi$ and $a \neq -\frac{1}{4}$, then the limit periodic continued fraction

$K_{n-1}^{\infty}(a_n/1)$ converges, possibly to ∞, by Theorem 4.45. Similarly, for each $m=1,2,3,\ldots$, the continued fraction $K_{n-m}^{\infty}(a_n/1)$ converges to a value F_m which may be finite or infinite. Then it can be shown that

$$\lim_{m\to\infty} \underset{n=m}{\overset{\infty}{K}} (a_n/1) = \underset{n=1}{\overset{\infty}{K}} (a/1),$$

that is,

$$\lim_{m\to\infty} F_m = g,$$

where

$$g = -\tfrac{1}{2} + \sqrt{|a+\tfrac{1}{4}|} \; e^{i\frac{1}{2}\arg(a+\frac{1}{4})}$$

is the (finite) value of the convergent periodic continued fraction $K_{n-1}^{\infty}(a/1)$ (see Theorem 3.2). To prove that $\lim F_m = g$, we introduce

$$F_{m,m+k} = \underset{n=m}{\overset{m+k}{K}} (a_n/1) \quad \text{and} \quad g_k \doteq \underset{n=1}{\overset{k}{K}} (a/1).$$

Then for every $\varepsilon > 0$ there exist positive constants $K(\varepsilon)$ and $M(\varepsilon)$ such that

$$|F_{m,m+k} - F_m| < \frac{\varepsilon}{3} \qquad \text{for} \quad k > K(\varepsilon)$$

and

$$|g_{k+1} - g| < \frac{\varepsilon}{3} \qquad \text{for} \quad k > M(\varepsilon).$$

The first inequality follows from Theorem 4.45 [note that $K(\varepsilon)$ is independent of m], and the second from Theorem 3.2. Let

$$k_1 = 1 + \max[\, K(\varepsilon), M(\varepsilon)\,].$$

Since $\lim_{n\to\infty} a_n = a$ and $F_{m,m+k}$ is a continuous function of $a_m,\ldots,a_{m+k}$, there exists an $N(\varepsilon) > 0$ such that

$$|F_{m,m+k_1} - g_{k_1+1}| < \frac{\varepsilon}{3} \qquad \text{for} \quad m > N(\varepsilon).$$

Hence $|F_m - g| < \varepsilon$ for $m > N(\varepsilon)$, and $\lim F_m = g$ as asserted.

If one sets

$$a_1 = r_1 x_1, \qquad a_n = r_n(1 - r_{n-1})x_n, \qquad n = 2,3,4,\ldots, \qquad (4.4.62a)$$

where

$$|x_n| \leqslant 1 \text{ and } 0 < r_n < 1, \qquad n = 1, 2, 3, \ldots, \qquad (4.4.62b)$$

the convergence of the continued fraction

$$\mathbf{K}(a_n/1) = \frac{r_1 x_1}{1} + \frac{r_2(1-r_1)x_2}{1} + \frac{r_3(1-r_2)x_3}{1} + \cdots \qquad (4.4.63)$$

follows from Corollary 4.36. This result is essentially due to Van Vleck [1901b]. Wall [1948, pp. 45–49] gives a number of variants of this theorem and discusses in detail who is responsible for which version.

In this context, as well as in the theory of positive definite continued fractions, sequences of the form

$$\{r_n(1-r_{n-1})\}, \quad 0 \leqslant r_n \leqslant 1, \qquad n = 0, 1, 2, \ldots,$$

play an important role. They are called *chain sequences*. Work on chain sequences goes back to Wall and Wetzel [1944b] and Dennis and Wall [1945] (see also [Wall, 1948, pp. 79–86], [Edrei, 1948], [Haddad, 1966] and references contained therein). The numbers r_n are called the *parameters* of the chain sequence. An important result is the existence of maximal and minimal sequences of parameters.

4.4.4 Twin Convergence Regions

Next to the parabola theorems the most important group of convergence criteria for continued fractions $\mathbf{K}(a_n/1)$ may well be the twin-convergence-region results, whose simplest cases go back to Thron [1943a] and Cowling, Leighton and Thron [1944], and which were subsequently generalized and improved by Singh and Thron [1956a], Thron [1959], Lange and Thron [1960], Lange [1966] and Jones and Thron [1970]. For the statement of some of these results it is convenient to replace a_n in $\mathbf{K}(a_n/1)$ by c_n^2 and thus consider the continued fraction $\mathbf{K}(c_n^2/1)$. The parabola theorems likewise can be stated more simply in terms of $\mathbf{K}(c_n^2/1)$. In fact, the condition $a_n \in P_\alpha$ [see (4.4.34) of Theorem 4.40] is equivalent to the condition

$$|\mathrm{Im}(c_n e^{-i\alpha})| < \frac{\cos \alpha}{2}. \qquad (4.4.64)$$

We begin with

THEOREM 4.46. *The continued fraction* $\mathbf{K}(c_n^2/1)$ *converges to a finite value provided*

$$|c_{2n-1}| \leqslant \rho \text{ and } |c_{2n} \pm i| \geqslant \rho, \qquad 0 < \rho < 1,$$

$$n = 1, 2, 3, \ldots. \qquad (4.4.65)$$

The convergence is uniform with respect to the regions (4.4.65).

Our proof of Theorem 4.46 makes use of the following two lemmas.

LEMMA 4.47. *If $a>0$ and $0 \leqslant k<1$, then*

$$H=\left|1+\frac{ae^{i\alpha}}{1+ke^{i\varphi}}\right| > \frac{|a^2+2a\cos\alpha+1-k^2|}{|1-k^2+ae^{i\alpha}|+ka},$$

where α and φ are arbitrary real numbers.

Proof. The set $[w: w=1/(1+ke^{i\varphi}),\ 0 \leqslant \varphi \leqslant 2\pi]$ is the image of the circle $[z:|z-1|=k]$ under the mapping $w=1/z$. Hence the image set is a circle and, by an argument analogous to the one used in the proof of Theorem 4.3, the circle is given by

$$\left|w-\frac{1}{1-k^2}\right|=\frac{k}{1-k^2}.$$

Thus one can write

$$\frac{1}{1+ke^{i\varphi}}=\frac{1}{1-k^2}+\frac{ke^{i\varphi'}}{1-k^2}=\frac{1+ke^{i\varphi'}}{1-k^2}, \qquad 0 \leqslant \varphi' \leqslant 2\pi,$$

for some real value of φ' which depends on φ. Therefore

$$H=\left|1+\frac{ae^{i\alpha}}{1+ke^{i\varphi}}\right|=\left|\frac{1-k^2+ae^{i\alpha}+kae^{i(\varphi'+\alpha)}}{1-k^2}\right|$$

$$>\left|\frac{|1-k^2+ae^{i\alpha}|-ka}{1-k^2}\right|=\left|\frac{|1-k^2+ae^{i\alpha}|^2-k^2a^2}{(1-k^2)(ka+|1-k^2+ae^{i\alpha}|)}\right|$$

$$=\left|\frac{(1-k^2)^2+2a(1-k^2)\cos\alpha+a^2(1-k^2)}{(1-k^2)(ka+|1-k^2+ae^{i\alpha}|)}\right|$$

$$=\frac{|a^2+2a\cos\alpha+1-k^2|}{ka+|1-k^2+ae^{i\alpha}|}. \qquad \blacksquare$$

LEMMA 4.48. *If*

$$a_{2n}=c_{2n}^2, \qquad where \quad |c_{2n}\pm i|>\rho, \quad 0<\rho<1, \tag{4.4.66}$$

and if $|z-1| \leqslant k<\rho$, then

$$\left|1+\frac{a_{2n}}{z}\right|>\frac{2\rho-\rho^2-k}{1-k}. \tag{4.4.67}$$

Proof. It will first be shown that

$$1 + \frac{a_{2n}}{z} \neq 0 \tag{4.4.68}$$

for all permissible values of a_{2n} and z. In fact, if (4.4.68) did not hold, then it would follow that

$$1 + a_{2n} = k'e^{i\varphi}, \quad \text{where } 0 < k' < k, \ 0 < \varphi < 2\pi. \tag{4.4.69}$$

However, $1 + a_{2n} \neq 0$ and

$$\min|1 + a_{2n}| = \min|1 + (i + \rho e^{i\theta})^2| = \rho(2 - \rho) > \rho > k.$$

Hence (4.4.69) is not possible, and so (4.4.68) must hold. Moreover, $|1 + a_{2n}/z|$ has a non-zero minimum value which is attained for values of a_{2n} and z on the boundaries of their respective regions. Hence, using Lemma 4.47, we obtain

$$
\begin{aligned}
\left|1 + \frac{a_{2n}}{z}\right| &\geqslant \min\left|1 + \frac{(i + \rho e^{i\theta})^2}{1 + ke^{i\varphi}}\right| \\
&\geqslant \min \frac{\left||i + \rho e^{i\theta}|^4 + 2\operatorname{Re}(i + \rho e^{i\theta})^2 + 1 - k^2\right|}{|1 - k^2 + (i + \rho e^{i\theta})^2| + k|i + \rho e^{i\theta}|^2} \\
&= \min \frac{\left|(1 + 2\rho \sin\theta + \rho^2)^2 - 2 - 4\rho \sin\theta + 2\rho^2 \cos 2\theta + 1 - k^2\right|}{|2\rho ie^{i\theta} + \rho^2 e^{i2\theta} - k^2| + k(1 + 2\rho \sin\theta + \rho^2)} \\
&= \min \frac{|\rho^4 - k^2 + 2x\rho^2|}{D + k(x - 1 + \rho^2)}, \tag{4.4.70}
\end{aligned}
$$

where $x = 2 + 2\rho \sin\theta$ and $D = |2\rho ie^{i\theta} + \rho^2 e^{i2\theta} - k^2|$. The permissible range for x is

$$2(1 - \rho) < x < 2(1 + \rho).$$

Hence

$$
\begin{aligned}
\rho^4 - k^2 + 2x\rho^2 &\geqslant \rho^4 - k^2 + 4(1 - \rho)\rho^2 = \rho^2(\rho^2 - 4\rho + 3) + \rho^2 - k^2 \\
&= \rho^2(3 - \rho)(1 - \rho) + \rho^2 - k^2 > 0.
\end{aligned}
$$

It follows that the absolute-value sign in the numerator of (4.4.70) can be deleted, and hence

$$\left|1 + \frac{a_{2n}}{z}\right| \geqslant \min \frac{\rho^4 - k^2 + 2x\rho^2}{D + k(x - 1 + \rho^2)}. \tag{4.4.71}$$

For D^2 an easy computation yields

$$D^2 = [D(x)]^2 = (\rho^2 - k^2)^2 + 2(\rho^2 - k^2)x + k^2 x^2.$$

Thus

$$D^2 - k^2(x - 1 + \rho^2)^2 = (1 - k^2)(\rho^4 - k^2 + 2\rho^2 x). \qquad (4.4.72)$$

Multiplying the right side of (4.4.71) by $D - k(x - 1 + \rho^2)$ in the numerator and denominator, and using (4.4.72), one obtains

$$\left| 1 + \frac{a_{2n}}{z} \right| \geqslant \min \frac{D - k(x - 1 + \rho^2)}{1 - k^2}.$$

Letting $m(x) = [D - k(x - 1 + \rho^2)]/(1 - k^2)$, one easily verifies that $dm(x)/dx$ is always positive and hence $m(x)$ takes on its minimum value at $x = 2(1 - \rho)$. At this value of x one obtains

$$D[2(1 - \rho)] = 1 - k^2 - (1 - \rho)^2$$

and hence

$$\min m(x) = m(2(1 - \rho)) = \frac{(1 + k)(2\rho - \rho^2 - k)}{1 - k^2} = \frac{2\rho - \rho^2 - k}{1 - k}. \qquad \blacksquare$$

Proof of Theorem 4.46.   If we let

$$V_1 = [w : |w + 1| \geqslant \rho] \quad \text{and} \quad V_2 = [w : |w| \leqslant \rho], \qquad (4.4.73)$$

it follows from Corollary 4.12 and (4.2.44) that

$$s_{2n}(V_2) = \frac{c_{2n}^2}{1 + V_2} \subseteq V_1 \qquad \text{if} \quad |c_{2n} \pm i| \geqslant \rho \qquad (4.4.74a)$$

and

$$s_{2n-1}(V_1) = \frac{c_{2n-1}^2}{1 + V_1} \subseteq V_2 \qquad \text{if} \quad |c_{2n-1}| \leqslant \rho. \qquad (4.4.74b)$$

For convenience we let $V_{2n} = V_2$ and $V_{2n-1} = V_1$. Then by Theorem 4.2, the sets $K_n = S_n(V_n)$ are nested circular disks [that $K_1 = S_1(V_1)$ is a circular disk follows from the fact that it is contained in the disk V_2]. Here $\{S_n\}$ is the sequence of l.f.t.'s generated by $\{s_n\}$ [see (4.2.1)]. To prove the theorem it thus suffices to show that the radii R_n of the disks K_n (of Theorem 4.2) tend to 0 as $n \to \infty$. In the proof of Theorem 4.40 we used the geometry of inversions to obtain an expression for R_n. Here, as a means of illustration, we shall use another method which could be used even if $S_n(V_n)$ were not a circular disk, since we first compute the circumference of $S_n(V_n)$.

Let C be a rectifiable curve, and let f be a function holomorphic in a domain containing C. If $l(f(C))$ denotes the length of $f(C)$, then

$$l(f(C)) = \int_C |f'(w)| \, |dw|.$$

Applying this to our case, we have

$$2\pi R_n = \int_{\partial V_n} |S_n'(w)| \, |dw|. \tag{4.4.75}$$

If A_n and B_n denote the nth numerator and denominator, respectively, of $K(c_n^2/1)$, then by (2.1.7) and the determinant formula (2.1.9) we have

$$|S_n'(w)| = \frac{|A_{n-1}B_n - B_{n-1}A_n|}{|B_n + B_{n-1}w|^2} = \frac{\prod\limits_{m=1}^{n} |c_m|^2}{|B_n + B_{n-1}w|^2}. \tag{4.4.76}$$

Now $\partial V_{2n-1} = \partial V_1 = [w : |w+1| = \rho]$, so that

$$\partial V_{2n-1} = [w : w = -1 + \rho e^{i\theta}, 0 \leqslant \theta \leqslant 2\pi].$$

Hence if $B_{2n-2} \neq 0$,

$$R_{2n-1} = \frac{1}{2\pi} \int_0^{2\pi} \frac{\prod\limits_{m=1}^{2n-1} |c_m|^2 \rho \, d\theta}{|B_{2n-1} + B_{2n-2}(-1 + \rho e^{i\theta})|^2}$$

$$= \frac{1}{2\pi} \int_0^{2\pi} \frac{\prod\limits_{m=1}^{2n-1} |c_m|^2 \rho \, d\theta}{|c_{2n-1}^2 B_{2n-3} + B_{2n-2}\rho e^{i\theta}|^2}$$

$$= \frac{\prod\limits_{m=1}^{2n-1} |c_m|^2}{2\pi |B_{2n-2}|^2} \int_0^{2\pi} \frac{\rho \, d\theta}{b^2 + \rho^2 + 2b\rho \cos(\theta - \beta)}, \tag{4.4.77}$$

where we have set $c_{2n-1}^2 B_{2n-3}/B_{2n-2} = b e^{i\beta}$ and previously had replaced

$B_{2n-1} - B_{2n-2}$ by $c_{2n-1}^2 B_{2n-3}$ using the difference equations (2.1.6). It follows that

$$R_{2n-1} = \frac{\rho \prod\limits_{m=1}^{2n-1} |c_m|^2}{\left| |c_{2n-1}^2 B_{2n-3}|^2 - \rho^2 |B_{2n-2}|^2 \right|}. \tag{4.4.78}$$

From (4.4.82) it can be seen that $\rho > b$. If $B_{2n-2} = 0$, then (4.4.78) follows immediately from the first equation in (4.4.77).

Next $\partial V_{2n} = \partial V_2 = [w : |w| = \rho]$, so that

$$\partial V_{2n} = \left[w : w = \rho e^{i\theta}, 0 \leqslant \theta \leqslant 2\pi \right].$$

Hence by (4.4.75) and (4.4.76)

$$R_{2n} = \frac{\prod\limits_{m=1}^{2n} |c_m|^2}{2\pi} \int_0^{2\pi} \frac{\rho \, d\theta}{|B_{2n} + \rho B_{2n-1} e^{i\theta}|^2}$$

$$= \frac{\rho \prod\limits_{m=1}^{2n} |c_m|^2}{\left| |B_{2n}|^2 - \rho^2 |B_{2n-1}|^2 \right|}. \tag{4.4.79}$$

By Corollary 4.7

$$\frac{B_{2n-1}}{B_{2n-2}} = 1 + \frac{a_{2n-1}}{1} + \cdots + \frac{a_2}{1}, \tag{4.4.80}$$

where $a_n = c_n^2$. It follows from (4.4.80) and (4.4.74) and the fact that $a_2 \in \operatorname{Int} V_1$, that $B_{2n-1}/B_{2n-2} \in \operatorname{Int}(1 + V_2)$; that is,

$$|B_{2n-1} - B_{2n-2}| < \rho |B_{2n-2}|. \tag{4.4.81}$$

Using the difference equations (2.1.6), we see that (4.4.81) is equivalent to

$$|c_{2n-1}^2 B_{2n-3}| < \rho |B_{2n-2}|. \tag{4.4.82}$$

Similarly one shows that

$$|B_{2n}| - \rho |B_{2n-1}| > 0. \tag{4.4.83}$$

One thus obtains, from (4.4.78), (4.4.79), (4.4.81), (4.4.83) and the difference equations

$$\frac{R_{2n}}{R_{2n-1}} = \frac{|c_{2n}|^2 \left(\rho^2 - \left| 1 - \dfrac{B_{2n-1}}{B_{2n-2}} \right|^2 \right)}{\left| c_{2n}^2 + \dfrac{B_{2n-1}}{B_{2n-2}} \right|^2 - \rho^2 \left| \dfrac{B_{2n-1}}{B_{2n-2}} \right|^2}. \tag{4.4.84}$$

We wish to maximize the right side of (4.4.84) with respect to c_{2n} and B_{2n-1}/B_{2n-2}, and we note that these two quantities are entirely independent of each other. In (4.4.84), B_{2n-1}/B_{2n-2} plays a key role. From (4.4.81) we have

$$\max \left| \frac{B_{2n-1}}{B_{2n-2}} - 1 \right| = k_n \leqslant \rho.$$

Since $B_1/B_0 = 1$, $k_1 = 0$. In general we have from the difference equations (2.1.6)

$$\left| \frac{B_{2n-1}}{B_{2n-2}} - 1 \right| = \frac{|c_{2n-1}|^2}{\left| 1 + \dfrac{c_{2n-2}^2}{B_{2n-3}/B_{2n-4}} \right|}.$$

Hence by Lemma 4.48 and the fact that B_{2n-3}/B_{2n-4} lies in the interior of $(1 + V_2)$, it follows that

$$\left| \frac{B_{2n-1}}{B_{2n-2}} - 1 \right| \leqslant k_n \leqslant \frac{\rho^2(1 - k_{n-1})}{2\rho - \rho^2 - k_{n-1}}.$$

Define

$$k_1^* = 0; \qquad k_n^* = \frac{\rho^2(1 - k_{n-1}^*)}{2\rho - \rho^2 - k_{n-1}^*}, \qquad n = 2, 3, \dots.$$

By induction one easily shows that

$$k_n^* = \frac{(n-1)\rho}{n - \rho} < \rho, \qquad n = 1, 2, 3, \dots.$$

Since the function

$$F(k) = \frac{\rho^2(1 - k)}{2\rho - \rho^2 - k}, \qquad 0 \leqslant k < \rho$$

is strictly increasing, it follows that

$$k_n \leqslant k_n^* = \frac{(n-1)\rho}{n-\rho} < \rho, \qquad n=1,2,3,\dots . \qquad (4.4.85)$$

We now return to the problem of determining $Q_n = \max R_{2n}/R_{2n-1}$. From the maximum-modulus principle it follows that Q_n will be obtained for values of c_{2n} and B_{2n-1}/B_{2n-2} on the boundaries of their respective regions. Hence

$$Q_n = \max_{\theta,\varphi} \frac{(\rho^2 - k_n^2)(\rho^2 + 1 + 2\rho\sin\theta)}{|(\rho e^{i\theta} + i)^2 + 1 + k_n e^{i\varphi}|^2 - \rho^2|1 + k_n e^{i\varphi}|^2}. \qquad (4.4.86)$$

A direct computation gives for the denominator of (4.4.86)

$$D_n = |(\rho e^{i\theta} + i)^2 + 1 + k_n e^{i\varphi}|^2 - \rho^2|1 + k_n e^{i\varphi}|^2$$

$$= \rho^4 + 3\rho^2 + k_n^2 - \rho^2 k_n^2 + 4\rho^3\sin\theta$$

$$+ 2k_n\left[\cos\varphi(\rho^2\cos2\theta - 2\rho\sin\theta - \rho^2)\right.$$

$$\left. + \sin\varphi(\rho^2\sin2\theta + 2\rho\cos\theta)\right]. \qquad (4.4.87)$$

Leaving θ fixed, we obtain the minimum of D_n with respect to φ by using the well-known property

$$\min_{\varphi}(A\cos\varphi + B\sin\varphi) = -(A^2 + B^2)^{1/2}.$$

In our case $A = \rho^2\cos2\theta - 2\rho\sin\theta - \rho^2$, $B = \rho^2\sin2\theta + 2\rho\cos\theta$ and hence it can be shown that

$$(A^2 + B^2)^{1/2} = 2\rho(1 + \rho\sin\theta).$$

It follows that the minimum of D_n with respect to φ is given by

$$\min_{\varphi} D_n = \rho^4 + 3\rho^2 + k_n^2(1 - \rho^2) + 4\rho^3\sin\theta - 2k_n\rho x, \qquad (4.4.88)$$

where $x = 2(1 + \rho\sin\theta)$. We note that $2(1-\rho) \leqslant x \leqslant 2(1+\rho)$. A slight re-arrangement of (4.4.88) yields

$$\min_{\varphi} D_n = (\rho^2 - k_n^2)(\rho^2 - 1) + 2\rho(\rho - k_n)x$$

and hence

$$Q_n = \max_\theta \frac{(\rho + k_n)(\rho^2 - 1 + x)}{(\rho + k_n)(\rho^2 - 1) + 2\rho x}$$

$$= \max_\theta \frac{\rho + k_n}{2\rho} \left(1 - \frac{(1 - \rho^2)(\rho - k_n)}{2\rho x - (1 - \rho^2)(\rho + k_n)} \right). \tag{4.4.89}$$

From the second expression we see that the maximum is attained for $x = 2(1 + \rho)$. Substituting this value into (4.4.89) leads to

$$Q_n = \frac{(\rho + k_n)(\rho + 1)}{(\rho + k_n)(\rho - 1) + 4\rho} = \frac{(\rho + k_n)(\rho + 1)}{(\rho + k_n)(\rho + 1) + 2(\rho - k_n)}$$

$$= \frac{1}{1 + \dfrac{2(\rho - k_n)}{(\rho + 1)(\rho + k_n)}}. \tag{4.4.90}$$

Since the function

$$G(k) = \frac{1}{1 + \dfrac{2(\rho - k)}{(\rho + 1)(\rho + k)}}, \qquad 0 \leqslant k < \rho,$$

is strictly increasing, it follows from (4.4.85) and (4.4.90) that

$$Q_n = G(k_n) < G(k_n^*) = G\left(\frac{(n-1)\rho}{n - \rho} \right)$$

$$= \frac{1}{1 + \dfrac{2(1 - \rho)}{(1 + \rho)(2n - 1 - \rho)}}, \qquad n = 1, 2, 3, \dots . \tag{4.4.91}$$

Since the disks K_n are nested, the R_n are monotone decreasing and hence, in particular, $R_{2n-1}/R_{2n-2} \leqslant 1$. It follows from this, the fact that $Q_n = \max R_{2n}/R_{2n-1}$, and (4.4.91), that

$$R_{2n} = R_1 \prod_{m=1}^{2n-1} \frac{R_{m+1}}{R_m} < R_1 \prod_{m=1}^{n} Q_m = \frac{R_1}{\displaystyle\prod_{m=1}^{n} \left(1 + \frac{2(1 - \rho)}{(1 + \rho)(2m - 1 - \rho)} \right)} .$$

$$\tag{4.4.92}$$

Thus $\lim_{n\to\infty} R_{2n}=0$, and hence the continued fraction $K(c_n^2/1)$ converges to a finite value uniformly with respect to the twin convergence regions (4.4.65).   ∎

Using the same method of proof but involving much more delicate arguments in the estimate of B_{2n}/B_{2n-1}, Lange [1966] proved

THEOREM 4.49. *The continued fraction $K(c_n^2/1)$ converges to a finite value provided that*

$$|c_{2n-1}\pm ia|\leqslant\rho, \quad |c_{2n}\pm i(1+a)|\geqslant\rho \qquad n=1,2,3,\ldots, \qquad (4.4.93)$$

where a is a complex number and a and ρ satisfy the inequality

$$|a|<\rho<|a+1|. \qquad (4.4.94)$$

The convergence is uniform with respect to the regions defined by (4.4.93).

Except for uniformity, Theorem 4.49 had previously been proved by Lange and Thron [1960], using the Stieltjes-Vitali theorem (Theorem 4.30) for non-real a. In (4.4.94) the condition $|a|<|1+a|$ is equivalent to $\mathrm{Re}(a)>-\frac{1}{2}$.

Theorem 4.49 reduces to Theorem 4.46 when $a=0$ and $0<\rho<1$. For $\rho=1$, Thron [1959] proved that $|c_{2n-1}|\leqslant1$ and $|c_{2n}\pm i|\geqslant1$ together with the condition $|c_{2n}|>\epsilon$, where ϵ can be an arbitrary positive number, is sufficient for the convergence of $K(c_n^2/1)$ to a finite value. Moreover, the convergence is uniform with respect to the given regions.

For $\rho>1$, Thron [1959] improved the result of Cowling, Leighton and Thron [1944] to the following: *The continued fraction $K(a_n/1)$ converges to a finite value provided that*

$$|a_{2n-1}|\leqslant\rho^2, \quad |a_{2n}|\geqslant2(\rho^2-\cos\arg a_{2n}), \qquad n=1,2,3,\ldots. \qquad (4.4.95)$$

It was further shown that the convergence is not uniform with respect to the region (4.4.95). All of the twin-convergence-region results stated above (except for uniformity) are subsumed under the general theorem proved by Jones and Thron [1970, p. 110] using the methods illustrated in the proof of Theorem 4.37. The result is as follows: *The continued fraction $K(a_n/1)$ converges to a finite value provided*

$$a_{2n-1}\in E_1 \text{ and } a_{2n}\in E_2, \qquad n=1,2,3,\ldots, \qquad (4.4.96)$$

where E_1 and E_2 are defined by Theorem 4.11 *and where ρ_1, ρ_2 are positive real numbers and Γ_1, Γ_2 are complex numbers satisfying*

$$|1+\Gamma_1|<\rho_1\leqslant|\Gamma_1| \text{ and } |\Gamma_2|<\rho_2\neq|1+\Gamma_2|. \qquad (4.4.97)$$

Inspired by earlier work of Wall [1956] and Copp [1950], Perron [1957b] studied the convergence behavior of the even part of the continued fraction. This in turn motivated Thron [1959, 1964] to establish the following improvement of Perron's result:

THEOREM 4.50. *The even part of the continued fraction* $K(c_n^2/1)$ *converges to a finite value if*

$$|c_{2n-1} \pm ia| \leqslant \rho, \qquad |c_{2n} \pm i(1+a)| \geqslant \rho, \qquad |c_{2n}^2| \geqslant \rho^2 - |1+a|^2 + \epsilon,$$

$$n = 1, 2, 3, \ldots . \quad (4.4.98)$$

Here a may be an arbitrary complex number, ϵ an arbitrary positive number, and ρ must satisfy

$$\rho > |a| \quad and \quad \rho \geqslant |1+a|. \quad (4.4.99)$$

Just as Theorem 4.49 was generalized by Jones and Thron [1970], so was Theorem 4.50; the exact statement of the generalization may be found in that article.

Of historical interest is the result of Leighton and Wall [1936] that $|a_{2n-1}| < \frac{1}{4}$ and $|a_{2n}| > \frac{25}{4}$ is sufficient for the convergence of $K(a_n/1)$ to a finite value. It was obtained by applying the Pringsheim criterion to a continued fraction whose approximants were the same as those of the original one but arranged in a different order. This idea was further pursued by Jordan and Leighton [1938]; they have among other results some triple- and quadruple-convergence-region criteria. Wall in [1957] once again considered convergence criteria obtained by a rearrangement of the approximants.

4.4.5 *Miscellaneous Convergence Criteria*

Another approach to convergence theory of continued fractions $K(a_n/1)$ begins with the system of inequalities

$$r_1|1+a_1| \geqslant |a_1|, \qquad r_2|1+a_1+a_2| \geqslant |a_2|, \quad (4.4.100a)$$

$$r_n|1+a_{n-1}+a_n| \geqslant r_n r_{n-2}|a_{n-1}| + |a_n|, \qquad n = 3, 4, 5, \ldots , \quad (4.4.100b)$$

where $r_n \geqslant 0$. These conditions are obtained by applying the Pringsheim criterion (Theorem 4.35) to the continued fractions equivalent to the even and odd parts of the given continued fraction $K(a_n/1)$. Hence the conditions (4.4.100), called the *fundamental inequalities* by Wall [1948, p. 40], insure the convergence of the even and odd part of $K(a_n/1)$. The first to use this approach was Leighton [1938, 1939]. Scott and Wall based their proof of the parabola theorem (Theorem 4.42 with $\alpha = 0$) on these inequalities, and Wall made extensive use of them in his book [1948]. Since then Lane and Wall [1949] have used these inequalities in studying continued fractions with absolutely convergent even and odd parts. A continued fraction with nth approximant denoted by f_n is said to *converge absolutely* if

$$\sum_{n=2}^{\infty} |f_n - f_{n-1}| < \infty. \tag{4.4.101}$$

Similar investigations involving absolute convergence and the fundamental inequalities were made by Dawson [1959, 1960, 1962, 1967] and Farinha [1954b].

Another result of some interest is the following, which is due to Hayden [1963]:

THEOREM 4.51. *The continued fraction* $\mathbf{K}(a_n/1)$, $a_1 = 1$, *converges to a finite value if, for each* $n = 2, 3, 4, \ldots$, *the following conditions are all satisfied*:

(i) $a_n \in E_n$, *where* E_n *is either a circle with center at the origin plus its interior or a circle with center at the origin plus its exterior.*

(ii) *At least one of* E_n, E_{n+1} *is a circle plus its interior.*

(iii) *There exist numbers* g_n, $0 < g_n < 1$, *and a number* r, $0 < r < 1$, *such that*

$$E_n = \begin{cases} \left[w : |w| \leqslant r(1 - g_{n-1})g_n \right] & \text{if } E_n \text{ is bounded,} \\ \left[w : |w| \geqslant (1 + g_{n-1})(2 - g_n) \right] & \text{if } E_n \text{ is unbounded.} \end{cases}$$
$$\tag{4.4.102}$$

A theorem of a similar kind but with the unbounded regions alternating with the bounded ones and the unbounded regions of the form

$$\left[w : |w + 1 - g_{2n-1}^2| \geqslant (1 - g_{2n})(1 - g_{2n-1}^2) + g_{2n-1}|w| \right], \qquad 0 < g_n < 1,$$
$$\tag{4.4.103}$$

is given by Hayden [1962]. Another variation on Theorem 4.51 can be found in [Jones and Snell, 1972, Corollary 3.4].

There are convergence criteria in addition to the ones discussed above, due to Pringsheim, Perron, Szász and von Koch, among others, which today are mainly of historical interest. Some of them can be found in the books of Perron [1957a] and Wall [1948].

We conclude this section with an interesting result of Wall [1956] which improves a weaker theorem of Farinha [1954a]:

THEOREM 4.52. *Let* $\{f_n\}$ *be the sequence of approximants of a continued fraction* $\mathbf{K}(a_n/1)$. *Assume that there exist positive numbers* M, L, n_0 *and a subsequence* $\{m_k\}$ *of the sequence of positive integers such that*

$$|f_n| < M, \qquad n = n_0, n_0 + 1, n_0 + 2, \dots , \qquad (4.4.104)$$

and

$$|a_{m_k}| < L, \qquad k = 1, 2, 3, \dots . \qquad (4.4.105)$$

Further assume that the odd (even) part of $\mathbf{K}(a_n/1)$ *converges to a finite value* v. *Then there exists a subsequence of the even (odd) part which converges to* v.

Proof. Let A_n and B_n denote the nth numerator and denominator of $\mathbf{K}(a_n/1)$, respectively. Then by the determinant formulas (2.1.9)

$$f_n - f_{n+1} = \frac{A_n B_{n+1} - A_{n+1} B_n}{B_n B_{n+1}} = \frac{(-1)^{n+1} a_1 a_2 \cdots a_{n+1}}{B_n B_{n+1}},$$

$$n = 1, 2, 3, \dots . \qquad (4.4.106)$$

An application of the difference equations (2.1.6) leads to

$$f_n - f_{n+2} = \frac{A_n B_{n+2} - A_{n+2} B_n}{B_n B_{n+2}} = \frac{(-1)^{n+1} a_1 a_2 \cdots a_{n+1}}{B_n B_{n+2}},$$

$$n = 1, 2, 3, \dots . \qquad (4.4.107)$$

Combining these expressions, one obtains

$$|a_{n+3}(f_n - f_{n+2})(f_{n+1} - f_{n+3})| = |(f_n - f_{n+1})(f_{n+2} - f_{n+3})|,$$

$$n = 1, 2, 3, \dots , \qquad (4.4.108)$$

a formula first noted by Wall [1934]. We suppose that the even part of the continued fraction converges to v, and assume that no subsequence of $\{f_{2n-1}\}$ converges to v. This leads to the conclusion that for some $c > 0$ and $n_1 > 1$,

$$|(f_n - f_{n+1})(f_{n+2} - f_{n+3})| > c \quad \text{for} \quad n = n_1 + 1, n_1 + 2, n_1 + 3, \dots . \qquad (4.4.109)$$

Suppose that the sequence $\{m_k\}$ contains an infinite sequence $\{2r_j\}$ of even integers (the other case can be handled analogously). Since $f_{2n} \to v$, there exists a positive number n_2 such that

$$|f_{2n} - f_{2n+2}| < \frac{c}{2LM} \qquad \text{for all} \quad 2n > n_2. \qquad (4.4.110)$$

Let $2r_j$ be such that $2r_j - 3$ is larger than $\max(n_0, n_1, n_2)$. Then on the one hand, by (4.4.104), (4.4.105) and (4.4.110),

$$|a_{2r_j}(f_{2r_j-3} - f_{2r_j-1})(f_{2r_j-2} - f_{2r_j})| < (L)\left(\frac{c}{2LM}\right)(2M) = c; \quad (4.4.111)$$

on the other hand by (4.4.108) and (4.4.109), we see that the left side of (4.4.111) is greater than c. This is a contradiction and hence there exists a subsequence of $\{f_{2n-1}\}$ converging to v.   ∎

Among the applications of Theorem 4.52 is the following:

THEOREM 4.53. *Let* $K(a_n/1)$ *be a continued fraction such that its odd and even parts both converge to finite values. If, in addition, there exists a positive number* L *and there exists a subsequence* $\{n_k\}$ *of the sequence of positive integers such that*

$$|a_{n_k}| < L \quad \text{for all } n_k,$$

then the continued fraction converges to a finite value.

It follows from the preceding theorem that the fundamental inequalities (4.4.100) together with the boundedness of an infinite subsequence of the elements of $K(a_n/1)$ are sufficient for the convergence of the continued fraction to a finite limit. Further implications of these ideas have been studied by Wall [1957] and Dawson [1960, 1962].

Certain other results which could be considered as convergence criteria, but where the main emphasis is on truncation error bounds for continued fractions, will be considered in Chapter 8.

4.5   Sufficient Conditions for Convergence: Variable Elements

4.5.1   *Introduction; Classification of Continued Fractions*

A number of types of continued fractions $K(a_n(z)/b_n(z))$, with elements $a_n(z)$, $b_n(z)$ which are functions of a complex variable z, are of importance in expanding holomorphic functions in terms of continued fractions. Some of the most important types are listed below. We adopt the convention that if an arbitrary function of z is added to a particular type of continued fraction, it remains of the same type.

1. *C-fractions* are continued fractions of the form

$$1 + \frac{a_1 z^{\alpha_1}}{1} + \frac{a_2 z^{\alpha_2}}{1} + \frac{a_3 z^{\alpha_3}}{1} + \cdots, \qquad a_n \neq 0, \qquad (4.5.1)$$

where all α_n are positive integers and the a_n are non-zero complex constants. If all $\alpha_n = 1$, then (4.5.1) is called a *regular C-fraction*. If, for $n > 1$, $\alpha_n = 1$ and $a_n > 0$, then (4.5.1) is called an *S-fraction*. Continued fractions of the form

$$\frac{s_0}{1} + \frac{g_1 z}{1} + \frac{(1-g_1)g_2 z}{1} + \frac{(1-g_2)g_3 z}{1} + \frac{(1-g_3)g_4 z}{1} + \cdots , \qquad (4.5.2a)$$

where

$$s_0 > 0, \qquad 0 < g_n < 1, \qquad n = 1,2,3,\ldots , \qquad (4.5.2b)$$

are called *g-fractions*.

2. *Associated continued fractions* are of the form

$$1 + \frac{k_1 z}{1 + l_1 z} - \frac{k_2 z^2}{1 + l_2 z} - \frac{k_3 z^2}{1 + l_3 z} - \frac{k_4 z^2}{1 + l_4 z} - \cdots , \qquad k_n \neq 0 \quad (4.5.3)$$

where the k_n and l_n are complex constants. We note that the even part of a regular C-fraction [see (2.4.24)]

$$\frac{a_1 z}{1} + \frac{a_2 z}{1} + \frac{a_3 z}{1} + \cdots , \qquad a_n \neq 0,$$

is the associated continued fraction

$$\frac{a_1 z}{1 + a_2 z} - \frac{a_2 a_3 z^2}{1 + (a_3 + a_4)z} - \frac{a_4 a_5 z^2}{1 + (a_5 + a_6)z} - \cdots .$$

Closely related to associated continued fractions are the following.

3. *J-fractions* are continued fractions of the form

$$\frac{1}{d_1 + z} - \frac{c_1^2}{d_2 + z} - \frac{c_2^2}{d_3 + z} - \frac{c_3^2}{d_4 + z} - \cdots , \qquad c_n^2 \neq 0, \qquad (4.5.4)$$

where the c_n^2 and d_n are complex constants. A *J-fraction* is said to be *positive definite* if there exists a sequence of positive numbers $\{g_n\}$ such that

$$|c_n^2| - \mathrm{Re}(c_n^2) \leqslant 2\delta_n \delta_{n+1}(1 - g_{n-1})g_n, \qquad n = 1,2,3,\ldots , \qquad (4.5.5a)$$

where

$$\delta_n = \mathrm{Im}(d_n) > 0 \text{ and } 0 < g_{n-1} < 1, \qquad n = 1,2,3,\ldots . \qquad (4.5.5b)$$

If all c_n and d_n are real numbers, then (4.5.4) is called a *real J-Fraction*. It is easily seen that real J-fractions are positive definite.

4. *General T-Fractions* are continued fractions of the form

$$\frac{z}{e_1+d_1z} + \frac{z}{e_2+d_2z} + \frac{z}{e_3+d_3z} + \cdots, \qquad e_n \neq 0, \qquad (4.5.6)$$

where the e_n and d_n are complex constants. If $e_n = 1$ for all n, then (4.5.6) is called a *T-fraction*. A general T-fraction (4.5.6) can also be expressed in the form given by the equivalent continued fraction

$$\frac{F_1z}{1+G_1z} + \frac{F_2z}{1+G_2z} + \frac{F_3z}{1+G_3z} + \cdots, \qquad F_n \neq 0, \qquad (4.5.6')$$

where

$$F_n = \frac{1}{e_n e_{n-1}}, \quad G_n = \frac{d_n}{e_n} \quad (e_0 = 1), \qquad n = 1,2,3,\ldots.$$

A general T-fraction (4.5.6) [or (4.5.6′)] in which all coefficients e_n, d_n, F_n, G_n, for $n \geq 1$, are positive real numbers is called a *positive T-fraction*. Closely connected to general T-fractions are the *M-fractions*

$$\frac{F_1}{1+G_1z} + \frac{F_2z}{1+G_2z} + \frac{F_3z}{1+G_3z} + \cdots, \qquad F_n, G_n \neq 0,$$

introduced by Murphy [1971] and McCabe [1975].

In this section we shall give some of the more important convergence criteria for continued fractions of the above types. This will also illustrate how other criteria can be obtained by using the results of Section 4.4. The following theorem, which is a direct application of the Stieltjes-Vitali Theorem (Theorem 4.30), is basic for this section.

THEOREM 4.54. *Let* $\mathbf{K}(a_n(z)/b_n(z))$ *be a continued fraction such that, for each* $n = 1,2,3,\ldots$, *the elements* $a_n(z)$, $b_n(z)$ *are holomorphic functions of a complex variable* z *for all* z *in a domain* D. *Further assume that there exist distinct complex numbers* α *and* β *such that for each* $n = 1,2,3,\ldots$, *the* nth *approximant* $f_n(z)$ *satisfies the conditions*

$$f_n(z) \neq \alpha, \beta, \infty \qquad \text{for all} \quad z \in D.$$

If the continued fraction converges to a finite value $f(z)$ *for each* $z \in \Delta$, *where* Δ *is an infinite set with at least one limit point in* D, *then it converges uniformly on every compact subset of* D, *and hence its limit function is holomorphic in* D.

Remarks on Theorem 4.54. (1)The theorem will frequently be used in the case where $\Delta = D$. It then serves to establish uniform convergence for a sequence whose convergence is known. (2) The conclusion of the theorem is still valid if $D = \bigcup_{j=1}^{m} D_j$, where the D_j are domains and the theorem holds for each D_j (that is, there may be α_j and β_j such that $f_n(z) \neq \alpha_j, \beta_j, \infty$ for all $z \in D_j$ and all n, even though there may be no α, β such that $f_n(z) \neq \alpha, \beta, \infty$ for all $z \in D$ and all n). (3) If the theorem is valid for a "tail" of the continued fraction, then the original continued fraction converges to a function which is meromorphic in D or identically ∞. (4) Theorem 4.54 can be generalized to functions of several complex variables [Cazacu, 1976].

A distinction will be made between a function meromorphic in D and a meromorphic function. A function $f(z)$ will be called *meromorphic in a domain D* if it is holomorphic in D except for poles. It is called a *meromorphic function* if it is meromorphic in $\mathbb{C}$.

4.5.2 Regular C-Fractions

We give now a number of convergence theorems for regular C-fractions. Theorems 4.55 and 4.56 treat the limit periodic case. The classical result of Stieltjes is given by Theorem 4.58, and the convergence of g-fractions is described in Theorem 4.60. *We adopt the convention that the regular C-fraction $\mathbf{K}(a_n z / 1)$ and all of its approximants have value zero at $z = 0$.*

THEOREM 4.55. *Let $\mathbf{K}(a_n z / 1)$ be a regular C-fraction such that*

$$\lim_{n \to \infty} a_n = 0. \tag{4.5.7}$$

(A) *For each positive number M, there exists an integer n_M such that the continued fraction*

$$\frac{a_{n_M} z}{1} + \frac{a_{n_M+1} z}{1} + \frac{a_{n_M+2} z}{1} + \cdots \tag{4.5.8}$$

converges to a holomorphic function for $|z| < M$.

(B) *The convergence of (4.5.8) is uniform on each compact subset of $|z| < M$.*

(C) *The continued fraction $\mathbf{K}(a_n z / 1)$ converges to a function $f(z)$ which is either meromorphic or identically infinity.*

Remark. In Theorem 5.14 (Section 5.4) it is shown that $\mathbf{K}(a_n z / 1)$ converges uniformly to a meromorphic function $f(z)$ on compact sets which contain no poles of $f(z)$.

Proof. Let $M>0$ be given. By (4.5.7) there exists an n_M such that $|a_n|<1/4M$ for all $n \geqslant n_M$. The elements $a_n z$ of (4.5.7) satisfy

$$|a_n z|<\tfrac{1}{4} \qquad \text{for} \quad |z|<M, \quad n \geqslant n_M.$$

Thus a simple application of Corollary 4.13 shows that the approximants of (4.5.8) are in absolute value less than $\tfrac{1}{2}$, for all z such that $|z|<M$. By Worpitzky's theorem [Corollary 4.36 (B)], the continued fraction (4.5.8) converges to a finite value for each z such that $|z|<M$. It follows from Theorem 4.54 that the convergence is uniform on compact subsets of $|z|<M$ and that the limit function is holomorphic in $|z|<M$. This proves (A) and (B). To prove (C) it suffices to note that $M>0$ is arbitrary and that the functions defined in the various circles are analytic continuations of each other.   ∎

The following theorem is due to Van Vleck [1904]:

THEOREM 4.56. *Let* $\mathrm{K}(a_n z/1)$ *be a regular C-fraction such that*

$$\lim_{n\to\infty} a_n = a \neq 0, \tag{4.5.9}$$

where a *is a complex constant, and let*

$$R_a = \left[z : \left| \arg\left(az + \tfrac{1}{4} \right) \right| < \pi \right]. \tag{4.5.10}$$

(A) *If* K *is a compact subset of* R_a, *then there exists a domain* D_K *with*

$$K \subset D_K \subset R_a \tag{4.5.11}$$

and a positive integer n_K *such that*

$$\frac{a_{n_K} z}{1} + \frac{a_{n_K+1} z}{1} + \frac{a_{n_K+2} z}{1} + \cdots \tag{4.5.12}$$

converges uniformly on every compact subset of D_K *to a function holomorphic in* D_K (*see Figure* 4.5.1).
(B) *The continued fraction* $\mathrm{K}(a_n z/1)$ *converges to a function* $f(z)$ *which is either meromorphic in* R_a *or identically infinity.*

Remark. In Theorem 5.15 it is shown that $\mathrm{K}(a_n z/1)$ converges uniformly to a meromorphic function $f(z)$ on compact subsets of R_a which contain no poles of $f(z)$.

Proof. We shall use the notation

$$N_z(d) = [w : |w-z|<d].$$

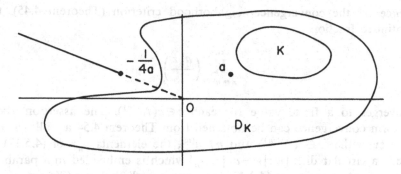

Figure 4.5.1. Domain D_K and compact set K with $K \subset D_K \subset R_a$.

Let a be a given non-zero complex constant, and let K be an arbitrary compact subset of R_a. If K is not connected, we embed it in a connected compact set K' by adding to K all line segments joining $z = 0$ to points of K. Cover K' by circular disks $N_z(d)$, $z \in K'$, where $N_z(d)$ is determined in such a way that $N_z(2d) \subset R_a$. Since K' is compact, it can be covered by a finite number of these disks, say

$$N_{z_1}(d_1), N_{z_2}(d_2), \ldots, N_{z_k}(d_k).$$

Set

$$N_{z_m}(\tfrac{3}{2}d_m) = N^{(m)}, \qquad 1 \leqslant m \leqslant k.$$

For $z \in c(N^{(m)})$, the closure of $N^{(m)}$, define

$$\rho(z) = \begin{cases} \sqrt{[|az| + \mathrm{Re}(az)]/2} & \text{if } |az| > \tfrac{1}{4}, \\ |az + \tfrac{1}{4}| & \text{if } |az| < \tfrac{1}{4}. \end{cases} \qquad (4.5.13)$$

Then $\rho(z)$ is a continuous, positive-valued function of z on the compact set $c(N^{(m)}) \subset R_a$, and hence

$$\rho_m = \min\left[\rho(z) : z \in c(N^{(m)})\right] > 0$$

exists. Now choose $n^{(m)}$ so that

$$|a_n - a| < \frac{\rho_m}{\tfrac{3}{2}d_m + |z_m|} \qquad \text{for } n > n^{(m)}. \qquad (4.5.15)$$

It follows from (4.5.14) and (4.5.15) that for each m such that $1 \leqslant m \leqslant k$,

$$|a_n z - az| < \rho(z) \qquad \text{for all } z \in c(N^{(m)}), \ n > n^{(m)}. \qquad (4.5.16)$$

Hence by the convergence-neighborhood criterion (Theorem 4.45), the continued fraction

$$\mathop{\mathbf{K}}_{n=n^{(m)}}^{\infty}\left(\frac{a_n z}{1}\right)$$

converges to a finite value for each $z \in c(N^{(m)})$. The assertion about uniform convergence can be obtained from Theorem 4.54 as follows. For all z such that $z \in c(N^{(m)})$ and $n \geqslant n^{(m)}$, the elements $a_n z$ of (4.5.17) lie inside a circular disk $[w : |w - az| < \rho_m]$ which is embedded in a parabolic region $E(\alpha)$ of the form (4.2.42) for some α such that $-\pi/2 < \alpha < \pi/2$. It follows from Corollary 4.16 that the approximants of (4.5.17) are contained in the half plane

$$V(\alpha) = \left[w : \text{Re}(we^{-i\alpha}) \geqslant -\tfrac{1}{2}\cos\alpha \right].$$

Hence no value on the negative real axis which is less than $-\tfrac{1}{2}$ is taken on by an approximant of (4.5.17) for $z \in c(N^{(m)})$. Therefore by Theorem 4.54, the continued fraction (4.5.17) converges uniformly on compact subsets of $N^{(m)}$ to a holomorphic function. ■

One reason why the parabolic convergence regions are of such importance is that they form the basis of the cardioid theorem, which is essentially due to Paydon and Wall [1942] and Leighton and Thron [1942]. We give a strong version that allows for different restrictions on different a_n. The theorem is based on Theorem 4.43.

THEOREM 4.57 (Cardioid theorem). *Let* $\mathbf{K}(a_n z/1)$ *be a regular C-fraction such that the* a_n *are non-zero complex constants satisfying*

$$|a_n| - \text{Re}(a_n) \leqslant k g_{n-1}(1 - g_n), \qquad n = 1, 2, 3 \ldots, \tag{4.5.18}$$

where k is a positive number and where $\{g_n\}$ is a sequence of real numbers such that

$$0 < \epsilon < g_n < 1 - \epsilon, \qquad n = 0, 1, 2, \ldots \tag{4.5.19}$$

where $0 < \epsilon < \tfrac{1}{2}$. Then:

(A) *The even and odd parts of* $\mathbf{K}(a_n z/1)$ *converge to holomorphic functions for all z in the cardioid domain*

$$C_k = \left[z : |z| < \frac{1 + \cos \arg z}{k} \right]. \tag{4.5.20}$$

The convergence is uniform in every compact subset of C_k.

(B) *The regular C-fraction* $K(a_n z/1)$ *converges to a function holomorphic in* C_k *iff at least one of the series*

$$\sum \left| \frac{a_1 a_3 \cdots a_{2n-1}}{a_2 a_4 \cdots a_{2n}} \right|, \quad \sum \left| \frac{a_2 a_4 \cdots a_{2n-2}}{a_1 a_3 \cdots a_{2n-1}} \right| \tag{4.5.21}$$

diverges.

(C) *A sufficient condition for the convergence of* $K(a_n z/1)$ *to a function holomorphic in* C_k *is that there exists a constant* $M > 0$ *such that*

$$|a_n| \leqslant M, \qquad n = 1, 2, 3, \ldots . \tag{4.5.22}$$

Proof. We set $z = re^{i\theta}$, $\psi_n = \frac{1}{2}\theta$, $g_n = p_n/\cos\frac{1}{2}\theta$. Then

$$|a_n z| - \operatorname{Re}(a_n z e^{-i\theta}) \leqslant rk g_{n-1}(1 - g_n)$$

$$= \frac{rk}{\cos^2\frac{1}{2}\theta} p_{n-1}\left(\cos\frac{1}{2}\theta - p_n\right). \tag{4.5.23}$$

The right side of (4.5.23) is less than or equal to $2p_{n-1}(\cos\psi_n - p_n)$ iff $rk/\cos^2\frac{1}{2}\theta < 2$, that is, iff $z \in C_k$. It follows from Theorem 4.43(A) that the even and odd parts of $K(a_n z/1)$ converge to finite values for all $z \in C_k$, provided there exists a constant d, $0 < d < \frac{1}{2}$, such that

$$\left| p_n e^{i\psi_n} - \tfrac{1}{2} \right| \leqslant d, \qquad n = 0, 1, 2, \ldots . \tag{4.5.24}$$

Since $\psi_n = \frac{1}{2}\theta$ and $|\theta| < \pi$, the conditions (4.5.19) will insure the existence of a $d(\epsilon, \theta)$ for which $0 < d(\epsilon, \theta) < \frac{1}{2}$ and

$$\left| p_n e^{i\psi_n} - \tfrac{1}{2} \right| \leqslant d(\epsilon, \theta), \qquad n = 0, 1, 2, \ldots .$$

Let $f_n(z)$ denote the nth approximant of $K(a_n z/1)$. It follows from Theorem 4.43(B) that for each $z \in C_k$, there exists a half plane $V_0(z)$ [defined by (4.4.51)] such that

$$f_n(z) \in V_0(z) \text{ and } f_n(z) \neq \infty, \qquad n = 1, 2, 3, \ldots .$$

Moreover, it is easily shown that, for all $n \geqslant 1$ and all $z \in C_k$, $f_n(z)$ does not lie on the negative real axis to the left of $-g_0$. An application of Theorem 4.54 then yields the uniform convergence of the even and odd parts of $K(a_n z/1)$ to holomorphic functions on all compact subsets of C_k. This proves part (A). The proofs of parts (B) and (C) follow readily from parts (C) and (D), respectively, of Theorem 4.43. ∎

Perhaps the most important application of Theorem 4.57 is the following classical theorem of Stieltjes [1894]:

THEOREM 4.58 (Stieltjes). *Let* $K(a_n z/1)$ *be an S-fraction, so that*

$$a_n > 0, \qquad n = 1, 2, 3, \dots . \tag{4.5.25}$$

(A) *Then the even and odd parts of* $K(a_n z/1)$ *converge to functions holomorphic in the cut plane*

$$R = [z : |\arg z| < \pi]. \tag{4.5.26}$$

Both even and odd parts converge uniformly on every compact subset of R.

(B) *The S-fraction* $K(a_n z/1)$ *converges to a function holomorphic in R iff at least one of the two series* (4.5.21) *diverges.*

(C) *If the S-fraction* $K(a_n z/1)$ *converges at a single point in R, then it converges at all points in R to a holomorphic function.*

(D) *A sufficient condition for an S-fraction to converge to a function holomorphic in R is that there exists a constant* $M > 0$ *such that*

$$|a_n| \leqslant M, \qquad n = 1, 2, 3, \dots . \tag{4.5.27}$$

Proof. If $g_n = \frac{1}{2}$ and $a_n > 0$ for all n, then the condition (4.5.18) holds for all $k > 0$. Let K be an arbitrary compact set contained in R. Then $K \subseteq D_k \subseteq R$ for some k sufficiently small, whose C_k is an open cardioid (4.5.20). Theorem 4.58 is thus an immediate consequence of Theorem 4.57. ∎

In Theorem 4.57 it is desirable to choose the g_n in such a manner that k can be as small as possible. If we assume that

$$|a_n| - \text{Re}(a_n) \leqslant h_n, \qquad n = 1, 2, 3, \dots ,$$

then the problem is to find that choice (if any) of $\{g_n\}$ and k for which

$$h_n = k g_{n-1}(1 - g_n), \qquad n = 1, 2, 3, \dots , \tag{4.5.28}$$

and k is as small as possible. No such minimal solution will exist in general. However, if

$$h_{2n} = h_0 \text{ and } h_{2n-1} = h_1, \qquad n = 1, 2, 3, \dots ,$$

then a minimal solution does exist. From (4.5.28)

$$h_0/k = g_1 - g_0 g_1, \qquad h_1/k = g_0 - g_0 g_1. \tag{4.5.29}$$

We assume that $h_0 > h_1$. (Similar arguments hold if $h_1 > h_0$.) Then

$$k = \frac{h_0 - h_1}{g_1 - g_0},$$

and so one wants to maximize $g_1 - g_0$. From (4.5.29)

$$\frac{h_0}{h_1} = \frac{g_1(1-g_0)}{g_0(1-g_1)} = \frac{(1/g_0)-1}{(1/g_1)-1}$$

and hence

$$h_0 t = (1/g_0) - 1 \quad \text{and} \quad h_1 t = (1/g_1) - 1$$

for some real $t \neq 0$. From this one obtains

$$g_1 - g_0 = \frac{1}{h_1 t + 1} - \frac{1}{h_0 t + 1} = \frac{(h_0 - h_1)t}{(h_0 t + 1)(h_1 t + 1)}.$$

The maximum of this expression is obtained for $t = 1/(h_0 h_1)^{1/2}$. This leads to

$$g_0 = \frac{\sqrt{h_1}}{\sqrt{h_0} + \sqrt{h_1}}, \qquad g_1 = \frac{\sqrt{h_0}}{\sqrt{h_0} + \sqrt{h_1}}$$

and hence

$$k = \left(\sqrt{h_0} + \sqrt{h_1} \right)^2.$$

We thus have established the following result due to [Thron, 1943a]:

COROLLARY 4.59. *Let* $K(a_n z/1)$ *be a regular C-fraction such that*

$$|a_{2n}| - \text{Re}(a_{2n}) \leq h_0, \quad |a_{2n-1}| - \text{Re}(a_{2n-1}) \leq h_1, \qquad n = 1, 2, 3, \ldots$$

where h_0 *and* h_1 *are positive real numbers.*

(A) *Then the even and odd parts of* $K(a_n z/1)$ *converge to functions holomorphic in the cardioid domain*

$$C(h_0, h_1) = \left[z : |z| < \frac{1 + \cos(\arg z)}{\left(\sqrt{h_0} + \sqrt{h_1} \right)^2} \right]. \qquad (4.5.30)$$

The convergence is uniform on every compact subset of $C(h_0, h_1)$.

(B) *The continued fraction* $K(a_n z/1)$ *converges iff at least one of the series* (4.5.21) *diverges.*

COROLLARY 4.60. *A g-fraction*

$$\frac{s_0}{1} + \frac{g_1 z}{1} + \frac{(1-g_1)g_2 z}{1} + \frac{(1-g_2)g_3 z}{1} + \cdots, \tag{4.5.31a}$$

$$s_0 > 0, \qquad 0 < g_n < 1, \quad n = 1, 2, 3, \ldots \tag{4.5.31b}$$

converges to a function holomorphic in the cut plane

$$Q_1 = [\, z : |\arg(z+1)| < \pi\,]. \tag{4.5.32}$$

The convergence is uniform on every compact subset of Q_1.

Remark. We adopt the convention that a g-fraction (4.5.31) and all of its approximants have value s_0 at $z = 0$.

Proof. By Theorem 4.58 the assertions of the corollary are valid for all z in the cut plane R [defined by (4.5.26)]. Therefore it suffices to show that these assertions are also valid in the disk $N_0(1) = [\, z : |z| < 1\,]$. An application of Corollary 4.36 (A), with $p_n = 1/g_n$, shows that the g-fraction (4.5.31) converges for all $z \in N_0(1)$. Theorem 4.35 implies that the approximants of (4.5.31) all lie in $N_0(1)$ if $|z| < 1$. Hence by Theorem 4.54, the convergence of (4.5.31) is uniform on every compact subset of $N_0(1)$. ∎

4.5.3   *Positive Definite J-fractions*

For positive definite *J*-fractions the following result can be derived from Theorem 4.44. A more detailed discussion of these continued fractions can be found in [Wall, 1948].

THEOREM 4.61. *A positive definite J-fraction*

$$\frac{1}{d_1 + z} - \frac{c_1^2}{d_2 + z} - \frac{c_2^2}{d_3 + z} - \frac{c_3^2}{d_4 + z} - \cdots, \qquad c_n^2 \neq 0, \tag{4.5.33a}$$

where

$$|c_n^2| - \mathrm{Re}(c_n^2) \leqslant 2\delta_n \delta_{n+1}(1 - g_{n-1})g_n, \qquad n = 1, 2, 3, \ldots, \tag{4.5.33b}$$

$$\delta_n = \mathrm{Im}(d_n) \geqslant 0, \quad 0 < g_{n-1} < 1, \qquad n = 1, 2, 3, \ldots, \tag{4.5.33c}$$

converges to a function holomorphic in

$$H = [\, z : \mathrm{Im}(z) > 0\,] \tag{4.5.34}$$

provided there exists an $M > 0$ such that

$$|c_n^2| < M, \qquad n = 1, 2, 3, \ldots , \qquad (4.5.35)$$

and

$$\delta_1 > 0.$$

The convergence is uniform on every compact subset of H.

Proof. Set

$$a_1 = 1; \qquad a_n = -c_{n-1}^2, \quad n = 2, 3, 4, \ldots ,$$
$$b_n = d_n + z, \qquad n = 1, 2, 3, \ldots ,$$
$$p_0 = \frac{1}{\delta_1 g_0} ; \qquad p_n = \operatorname{Im}(z) + \delta_n(1 - g_{n-1}), \quad n = 1, 2, 3, \ldots ,$$
$$\psi_n = \pi/2, \qquad n = 0, 1, 2, \ldots .$$

To apply Theorem 4.44 we must verify the inequality (4.4.53). It is easily seen that

$$|a_1| - \operatorname{Re}(a_1 e^{-i(\psi_1 + \psi_0)}) = 2 = 2p_0[\operatorname{Re}(b_1 e^{-i\psi_1}) - p_1].$$

For $n \geqslant 2$, we have

$$|a_n| - \operatorname{Re}(a_n e^{-i(\psi_n + \psi_{n-1})}) = |c_{n-1}^2| - \operatorname{Re}(c_{n-1}^2)$$

and

$$2p_{n-1}[\operatorname{Re}(b_n e^{-i\psi_n}) - p_n] = 2\delta_{n-1}\delta_n(1 - g_{n-2})g_{n-1} + 2\delta_n g_{n-1}\operatorname{Im}(z).$$

Thus if $\operatorname{Im}(z) > 0$, the inequality (4.4.53) holds for all $n \geqslant 1$. In view of (4.5.35), the sequence

$$\left\{ \frac{a_n}{p_n p_{n-1}} \right\}$$

is bounded for each $z \in H$. It follows from Theorem 4.44 that the positive definite J-fraction (4.5.33) converges to a finite value for each $z \in H$. Moreover, since

$$\operatorname{Re}(b_1 e^{-i\psi_1}) = \operatorname{Im}(z) + \delta_1 > p_1,$$

it also follows from Theorem 4.44 that, for every $z \in H$, the values of all approximants of (4.5.33) lie in the circular disk

$$\left[w : \left| w + \frac{i}{2\delta_1 g_0} \right| < \frac{1}{2\delta_1 g_0} \right].$$

Hence by Theorem 4.54 the convergence of (4.5.33) is uniform on every compact subset of H. Since each approximant of (4.5.33) is holomorphic in H, the limit function is holomorphic.   ∎

4.5.4   General T-fractions

Various criteria for the convergence of general T-fractions

$$\frac{z}{e_1 + d_1 z} + \frac{z}{e_2 + d_2 z} + \frac{z}{e_3 + d_3 z} + \cdots , \qquad e_n \neq 0, \qquad (4.5.36)$$

can be obtained by considering the equivalent continued fractions (see Theorem 2.6)

$$\frac{1/\zeta}{e_1 \zeta + \dfrac{d_1}{\zeta}} + \frac{1}{e_2 \zeta + \dfrac{d_2}{\zeta}} + \frac{1}{e_3 \zeta + \dfrac{d_3}{\zeta}} + \cdots , \qquad (4.5.37)$$

where

$$\zeta = z^{-1/2}, \qquad -\frac{\pi}{2} < \arg \zeta < \frac{\pi}{2}.$$

Throughout this book we adopt the convention that a general T-fraction (4.5.36) and all of its approximants have value zero at $z = 0$. From Theorem 4.35 with $a_n = 1$ it follows that (4.5.37) converges to a finite value if

$$\left| e_n \zeta + \frac{d_n}{\zeta} \right| \geqslant 2, \qquad n = 1, 2, 3, \dots . \qquad (4.5.38)$$

Moreover, from Van Vleck's criterion (Theorem 4.29), it follows that the even and odd parts of (4.5.37) converge to finite values if, for some $\epsilon > 0$,

$$\left| \arg\left(e_n \zeta + \frac{d_n}{\zeta} \right) \right| < \frac{\pi}{2} - \epsilon, \qquad n = 1, 2, 3, \dots . \qquad (4.5.39)$$

If (4.5.39) holds, then the continued fraction (4.5.37) converges to a finite value iff

$$\sum_{n=1}^{\infty} \left| e_n \zeta + \frac{d_n}{\zeta} \right| = \infty . \qquad (4.5.40)$$

Suppose that there exist positive constants k_1 and k_2 such that

$$|e_n| > k_1 \text{ and } |d_n| < k_2, \qquad n = 1, 2, 3, \ldots . \tag{4.5.41}$$

Then

$$\left| e_n \zeta + \frac{d_n}{\zeta} \right| > |e_n||\zeta| - \frac{|d_n|}{|\zeta|} > k_1|\zeta| - \frac{k_2}{|\zeta|}, \qquad n = 1, 2, 3, \ldots ,$$

so that (4.5.38) is satisfied if

$$k_1|\zeta| - \frac{k_2}{|\zeta|} > 2.$$

This is the case if

$$|\zeta| > \frac{1 + \sqrt{1 + k_1 k_2}}{k_1},$$

that is, if

$$|z| < \left(\frac{k_1}{1 + \sqrt{1 + k_1 k_2}} \right)^2 = \left(\frac{\sqrt{1 + k_1 k_2} - 1}{k_2} \right)^2 . \tag{4.5.42}$$

We further observe that (4.5.38) insures that the nth approximant $g_n(\zeta)$ of (4.5.37) satisfies

$$|\zeta g_n(\zeta)| < 1, \qquad n = 1, 2, 3, \ldots ,$$

(see Theorem 4.35). Hence if (4.5.41) and (4.5.42) hold, then the nth approximant $f_n(z)$ of (4.5.36), which is a rational function of z, satisfies

$$|f_n(z)| < |z|^{1/2} < \frac{\sqrt{1 + k_1 k_2} - 1}{k_2} . \tag{4.5.43}$$

Theorem 4.54 can therefore be invoked to assure the uniform convergence of $\{f_n(z)\}$ on compact subsets of (4.5.42). We have now proved:

THEOREM 4.62. *Let*

$$\frac{z}{e_1 + d_1 z} + \frac{z}{e_2 + d_2 z} + \frac{z}{e_3 + d_3 z} + \cdots \tag{4.5.44}$$

be a general T-fraction such that

$$|e_n| > k_1 \text{ and } |d_n| \leqslant k_2, \quad n = 1, 2, 3, \ldots, \qquad (k_1 > 0, \quad k_2 > 0). \qquad (4.5.45)$$

(A) *For all z in the closed circular disk*

$$G = \left[z : |z| \leqslant \left(\frac{\sqrt{1 + k_1 k_2} - 1}{k_2} \right)^2 \right], \qquad (4.5.46)$$

the continued fraction (4.5.44) converges to a function f(z) satisfying

$$|f(z)| \leqslant |z|^{1/2}. \qquad (4.5.47)$$

(B) *The convergence is uniform on every compact subset of* Int G, *and* f(z) *is holomorphic on* Int G.

An analogous result can be proved if we assume that for $k_1 > 0$, $k_2 > 0$,

$$|e_n| \leqslant k_1 \text{ and } |d_n| \geqslant k_2, \quad n = 1, 2, 3, \ldots. \qquad (4.5.48)$$

In this case (4.5.38) holds provided

$$|z| > \left(\frac{k_1}{\sqrt{1 + k_1 k_2} - 1} \right)^2 = \left(\frac{1 + \sqrt{1 + k_1 k_2}}{k_2} \right)^2 \qquad (4.5.49)$$

is satisfied. For the nth approximant $f_n(z)$ we have as before

$$|f_n(z)| < |z|^{1/2}, \qquad (4.5.50)$$

and hence if $|z| < M$, Theorem 4.54 can be used to conclude that $\{f_n(z)\}$ converges uniformly on compact subsets of

$$\left(\frac{1 + \sqrt{1 + k_1 k_2}}{k_2} \right)^2 < |z| < M$$

to a holomorphic function. Since M here is arbitrary, we have established the following result.

THEOREM 4.63. *Let*

$$\frac{z}{e_1 + d_1 z} + \frac{z}{e_2 + d_2 z} + \frac{z}{e_3 + d_3 z} + \cdots, \qquad e_n \neq 0, \qquad (4.5.51)$$

be a general T-fraction such that

$$|e_n| \le k_1 \text{ and } |d_n| \ge k_2, \quad n=1,2,3,\dots, \qquad (k_1>0, \quad k_2>0). \qquad (4.5.52)$$

(A) For z contained in the set

$$H = \left[z : \left(\frac{\sqrt{1+k_1 k_2}+1}{k_2} \right)^2 \le |z| < \infty \right], \qquad (4.5.53)$$

the continued fraction (4.5.51) converges to a function $f(z)$ satisfying

$$|f(z)| \le |z|^{1/2}. \qquad (4.5.54)$$

(B) The convergence is uniform on every compact subset of Int H, and $f(z)$ is holomorphic on Int H.

In the following theorem an angular opening is obtained for the domain of convergence of a general T-fraction.

THEOREM 4.64. Let

$$\frac{z}{e_1+d_1 z} + \frac{z}{e_2+d_2 z} + \frac{z}{e_3+d_3 z} + \cdots, \qquad e_n \ne 0, \quad d_n \ne 0, \quad (4.5.55)$$

be a general T-fraction such that

$$|\arg e_n| \le \alpha \text{ and } |\arg d_n| \le \alpha, \qquad n=1,2,3,\dots, \qquad (4.5.56a)$$

where

$$0 \le \alpha < \pi/2. \qquad (4.5.56b)$$

(A) The even and odd parts of (4.5.55) converge to functions $f(z)$ and $g(z)$, respectively, which are holomorphic and satisfy

$$|\arg f(z)| < \alpha + |\arg z|, \qquad |\arg g(z)| < \alpha + |\arg z| \qquad (4.5.57)$$

for all z in the domain

$$G_\alpha = [z : |\arg z| < \pi - 2\alpha]. \qquad (4.5.58)$$

(B) For each $z \in G_\alpha$, the continued fraction (4.5.55) converges to a finite value $F(z)$ iff

$$\sum_{n=1}^{\infty} (|e_n| + |d_n|) = \infty. \qquad (4.5.59)$$

(C) *If (4.5.59) holds, then the continued fraction (4.5.55) converges uniformly on compact subsets of G_α to a function $F(z)$ holomorphic in G_α satisfying*

$$|\arg F(z)| < \alpha + |\arg z|. \tag{4.5.60}$$

(D) *If the continued fraction (4.5.55) converges to a finite value for one point z_0 in G_α, then (4.5.59) holds and hence the conclusions of (C) are valid.*

Proof. Let $\zeta = z^{-1/2}$, $-\pi/2 < \arg \zeta < \pi/2$, and let $h_n(\zeta)$ denote the nth approximant of the continued fraction

$$\overset{\infty}{\underset{n=1}{\mathrm{K}}}\left[\frac{1}{e_n \zeta + \dfrac{d_n}{\zeta}}\right]. \tag{4.5.61}$$

If $f_n(z)$ denotes the nth approximant of the general T-fraction (4.5.55), then

$$f_n(z) = \frac{h_n(\zeta)}{\zeta}, \qquad n = 1, 2, 3, \dots . \tag{4.5.62}$$

The condition (4.5.39) will be satisfied with

$$\epsilon = \frac{\pi}{2} - \alpha - |\arg \zeta| \tag{4.5.63}$$

for each ζ in the domain

$$H_\alpha = \left[\zeta : |\arg \zeta| < \frac{\pi}{2} - \alpha\right]. \tag{4.5.64}$$

This assertion is easily verified by using the inequalities

$$|\arg(e_n \zeta)| \le |\arg e_n| + |\arg \zeta| < \frac{\pi}{2} - \epsilon, \qquad n = 1, 2, 3, \dots, \tag{4.5.65a}$$

and

$$\left|\arg \frac{d_n}{\zeta}\right| \le |\arg d_n| + |\arg \zeta| < \frac{\pi}{2} - \epsilon, \qquad n = 1, 2, 3, \dots . \tag{4.5.65b}$$

It follows from (4.5.39) and the remark preceding it that for each $z \in G_\alpha$, the even and odd parts of (4.5.55) converge to finite values $f(z)$ and $g(z)$, respectively. This proves the first part of (A). We shall deal with the second part of (A) later.

It also follows from (4.5.65) that for $\zeta \in H_\alpha$, ϵ defined by (4.5.63), and $n > 1$,

$$\mathrm{Re}(e_n\zeta) = |e_n\zeta|\cos\arg(e_n\zeta) > |e_n\zeta|\cos\left(\frac{\pi}{2} - \epsilon\right),$$

and

$$\mathrm{Re}\left(\frac{d_n}{\zeta}\right) = \left|\frac{d_n}{\zeta}\right|\cos\arg\frac{d_n}{\zeta} > \left|\frac{d_n}{\zeta}\right|\cos\left(\frac{\pi}{2} - \epsilon\right).$$

Thus, for $n > 1$,

$$\left(|e_n\zeta| + \left|\frac{d_n}{\zeta}\right|\right)\cos\left(\frac{\pi}{2} - \epsilon\right) < \mathrm{Re}\left(e_n\zeta + \frac{d_n}{\zeta}\right)$$

$$< \left|e_n\zeta + \frac{d_n}{\zeta}\right|$$

$$< |e_n\zeta| + \left|\frac{d_n}{\zeta}\right|. \qquad (4.5.66)$$

Hence (4.5.40) is equivalent to (4.5.59), provided $\zeta \in H_\alpha$. This, together with the remark preceding (4.5.40), proves (B).

By Van Vleck's theorem (Theorem 4.29), for each $\zeta \in H_\alpha$, every approximant $h_n(\zeta)$ is a finite complex number satisfying

$$|\arg h_n(\zeta)| < \frac{\pi}{2} - \epsilon, \qquad (4.5.67)$$

where ϵ is defined by (4.5.63). In view of (4.5.62), it follows from (4.5.67) that for all $z \in G_\alpha$, $\zeta \in H_\alpha$,

$$|\arg f_n(z)| = \left|\arg\frac{1}{\zeta} + \arg h_n(\zeta)\right|$$

$$< |\arg\zeta| + |\arg h_n(\zeta)|$$

$$< |\arg\zeta| + \frac{\pi}{2} - \epsilon$$

$$= \alpha + 2|\arg\zeta|$$

$$= \alpha + |\arg z|. \qquad (4.5.68)$$

The uniform convergence of the sequence $\{f_n(z)\}$ on compact subsets of G_α is then guaranteed by Theorem 4.54. The inequality (4.5.60) follows from (4.5.68). This proves (C). The proof of the uniform convergence in (A) is similar to that for (C) and hence is omitted. (D) is an immediate consequence of (B) and (C). ■

Of particular importance is the case $\alpha = 0$.

COROLLARY 4.65. *Let*

$$\frac{z}{e_1+d_1z} + \frac{z}{e_2+d_2z} + \frac{z}{e_3+d_3z} + \cdots \tag{4.5.69}$$

be a general T-fraction such that

$$e_n>0, \quad d_n>0, \qquad n=1,2,3,\ldots. \tag{4.5.70}$$

(A) *The even and odd parts of* (4.5.69) *converge to functions* $f(z)$ *and* $g(z)$, *respectively, which are holomorphic and satisfy*

$$|\arg f(z)| \leqslant |\arg z|, \qquad |\arg g(z)| \leqslant |\arg z|, \tag{4.5.71}$$

for all z in the domain

$$G_0=[\,z:|\arg z|<\pi\,]. \tag{4.5.72}$$

(B) *For each* $z\in G_0$, *the continued fraction* (4.5.69) *converges to a finite value* $F(z)$ *iff*

$$\sum_{n=1}^{\infty} (e_n+d_n)=\infty. \tag{4.5.73}$$

(C) *If* (4.5.73) *holds, then the continued fraction* (4.5.69) *converges uniformly on compact subsets of* G_0 *to a function* $f(z)$ *holomorphic in* G_0 *and satisfying*

$$|\arg F(z)| \leqslant |\arg z|. \tag{4.5.74}$$

(D) *If the continued fraction converges to a finite value for one point* $z_0\in G_0$, *then* (4.5.73) *holds and hence the conclusions of* (C) *are valid.*

We conclude this section with the following instructive example. From Theorem 3.2 one can easily show that the periodic T-fraction

$$\frac{z}{1-z} + \frac{z}{1-z} + \frac{z}{1-z} + \cdots \tag{4.5.75}$$

converges to $f(z)=z$ for $|z|<1$, diverges for all z in the set $[z:|z|=1, z\neq -1]$, and converges to $g(z)=-1$ for all other values of z. Thus it is possible that a general T-fraction, satisfying the condition of both Theorem 4.62 and Theorem 4.63, converges to a function $f(z)$ near $z=0$ and to another function $g(z)$ near $z=\infty$, such that $f(z)$ and $g(z)$ are not analytic continuations of each other.

CHAPTER 5

Methods for Representing Analytic Functions by Continued Fractions

This chapter deals with the representation of analytic functions by continued fractions. Two main approaches are considered. In the first approach a formal continued-fraction expansion is obtained by requiring that the Laurent expansion of the nth approximant agree term by term with a given Laurent series L up to the ν_n power of z, where ν_n tends to infinity with n. Continued fractions defined in this manner are said to correspond to the series L [or to the function $f(z)$ of which L is a Laurent expansion]. A general theory of correspondence is developed in Sections 5.1, 5.2, and 5.4 for sequences of functions $\{R_n(z)\}$ meromorphic at the origin. As a special case $R_n(z)$ can be the nth approximant of a continued fraction. A norm is introduced for the field $\mathcal{L}$ of all formal Laurent series, such that convergence with respect to the norm is equivalent to correspondence. Necessary and sufficient conditions for the existence of a Laurent series L to which a given sequence $\{R_n(z)\}$ corresponds are given by Theorem 5.1. A method for obtaining a sequence $\{R_n(z)\}$ (or continued fraction) corresponding to a given Laurent series L is provided by Theorems 5.2 and 5.3, in which three-term recurrence relations play an important role. As a consequence of the property of correspondence, it is shown (Section 5.4) that, with suitable restrictions, uniform convergence of a sequence $\{R_n(z)\}$ is equivalent to uniform boundedness, and that when a sequence $\{R_n(z)\}$ converges uniformly, its limit $f(z)$ is a function whose Laurent expansion is L. Although the basic ideas of correspondence go back to Gauss [1813], the general theory described here is based on [Jones and Thron, 1975, 1979]. Further work on correspondence of continued fractions is discussed in Chapters 6, 7 and 10.

ENCYCLOPEDIA OF MATHEMATICS and Its Applications, Gian-Carlo Rota (ed.).
Vol. 11: William B. Jones and W. J. Thron, Continued Fractions.   ISBN 0-201-13510-8

Three-term recurrence relations also play an important role in the second approach for representing analytic functions by continued fractions (Section 5.3). The method is based on a theorem of Pincherle [1894] concerning minimal solutions of three-term recurrence relations. In a paper on computational aspects of three-term recurrence relations, Gautschi [1967] recognized the importance of Pincherle's result and applied it extensively. A generalization of Pincherle's theorem for continued fractions over a normed field $\mathbb{F}$ is given in Theorem 5.7. As applications we obtain a new theorem (Corollary 5.9) on correspondence of continued fractions over the field $\mathcal{L}$ as well as Pincherle's original result. Another result relating continued fractions and three-term recurrence relations is a theorem due to Auric [1907] and independently formulated by Perron [1913]. A generalization of Auric's theorem for continued fractions over a normed field is given by Theorem 5.10, and this result is applied to obtain an independent proof of Theorem 5.5(A). Bessel functions of the first kind are used to illustrate the methods described in Sections 5.1, 5.2, 5.3 and 5.4. Many other examples are given in Chapter 6.

In Section 5.5 it is shown that the concept of correspondence can be used to define Padé approximants. A brief summary is given of some classical properties of the Padé table and recent generalizations such as the Newton-Padé and two-point Padé tables. Connections between certain classes of continued fractions and Padé tables are pointed out. Other relations of this type are developed in Chapter 7. It is clear from Theorems 2.2 and 2.7 that an arbitrary sequence of distinct Padé approximants is the sequence of approximants of a continued fraction. Thus the theory of continued fractions forms an integral part of the study of Padé tables.

5.1　Correspondence

Correspondence of sequences of meromorphic functions plays a key role in the theory of continued fractions and Padé approximants. Following Henrici [1974], we call

$$L = c_m z^m + c_{m+1} z^{m+1} + c_{m+2} z^{m+2} + \cdots, \qquad c_m \neq 0, \qquad (5.1.1)$$

a *formal Laurent series* (fLs), provided the c_k are complex numbers and m is an integer. $L = 0$ is also considered an fLs. The set $\mathcal{L}$ of all fLs forms a field with respect to addition and multiplication defined in the manner suggested by (5.1.1) (see, for example, [Henrici, 1974, Section 1.8]). If $m \geqslant 0$, then (5.1.1) is also called a *formal power series* (fps). If $f(z)$ is a function meromorphic at the origin (i.e. in an open disk containing the origin), its Laurent expansion will be denoted by $L(f)$. A sequence $\{R_n(z)\}$ of functions meromorphic at the origin will be said to *correspond to a fLs L (at z = 0)* if

$$\lim_{n \to \infty} \lambda(L - L(R_n)) = \infty, \qquad (5.1.2)$$

where $\lambda: \mathcal{L} \to \mathbb{R} \cup [\infty]$ is the function defined as follows: if $L = 0$, then $\lambda(L) = \infty$; if $L \neq 0$, then $\lambda(L) = m$ where m is defined by (5.1.1).

Every function $f(z)$ meromorphic at the origin has a unique fLs expansion $L(f)$. The one-to-one mapping L thus provides an embedding of the field $\mathfrak{M}$, of all functions meromorphic at the origin, in the field $\mathcal{L}$. If $\{R_n(z)\}$ corresponds to an fLs L, then the *order of correspondence* of $R_n(z)$ is defined to be

$$\nu_n = \lambda(L - L(R_n)).$$

Thus if $\{R_n(z)\}$ corresponds to L, it can be seen that $L(R_n)$ and L agree term by term up to and including the term involving $z^{\nu_n - 1}$. A continued fraction

$$\mathbf{K}\left(\frac{a_n(z)}{b_n(z)}\right)$$

will be said to *correspond to a fLs L* if each approximant $f_n(z)$ is a meromorphic function of z at the origin and if $\{f_n(z)\}$ corresponds to L. A sequence $\{R_n(z)\}$ [or a continued fraction $\mathbf{K}(a_n(z)/b_n(z))$] is said to *correspond at z = 0 to a function f(z)* meromorphic at the origin if it corresponds to the fLs expansion $L(f)$.

The following properties of λ are easily deduced: For every L_1 and L_2 in $\mathcal{L}$,

$$\lambda(L_1 L_2) = \lambda(L_1) + \lambda(L_2), \qquad (5.1.3)$$

$$\lambda(L_1/L_2) = \lambda(L_1) - \lambda(L_2), \qquad L_2 \neq 0, \qquad (5.1.4)$$

$$\lambda(L_1 \pm L_2) \geqslant \min[\lambda(L_1), \lambda(L_2)], \qquad (5.1.5)$$

$$\lambda(L_1 \pm L_2) = \min[\lambda(L_1), \lambda(L_2)] \quad \text{if} \quad \lambda(L_1) \neq \lambda(L_2). \quad (5.1.6)$$

We make two observations about correspondence (due to Baker [1975] for the case of Padé approximants): If $\{R_n(z)\}$ corresponds to L and if A, B, C, D are functions of z, meromorphic at the origin, such that

$$L(C) + L(D)L \neq 0, \qquad AD - BC \neq 0 \qquad (5.1.7)$$

and

$$\lambda(L(C) + L(D)L(R_n)) < k \qquad (5.1.8)$$

for all $n \geqslant 1$ and some fixed k, then

$$\left\{ \frac{A+BR_n}{C+DR_n} \right\} \text{ corresponds to } \frac{L(A)+L(B)L}{L(C)+L(D)L}. \tag{5.1.9}$$

Similarly

$$R_n\left(\frac{az}{cz+d} \right) \text{ corresponds to } L\left(\frac{az}{cz+d} \right), \tag{5.1.10}$$

where a, c, d are complex constants such that

$$ad \neq 0. \tag{5.1.11}$$

The definition of correspondence can be extended as follows. We call

$$L^* = c_{-m}z^{-m} + c_{-m-1}z^{-m-1} + c_{-m-2}z^{-m-2} + \cdots, \qquad c_{-m} \neq 0, \tag{5.1.12}$$

a *formal Laurent series at* $z = \infty$ provided the c_k are complex constants and m is an integer. In a similar manner the set $\mathcal{L}^*$ of all fLs at $z = \infty$ (including $L^* = 0$) forms a field. If $f(z)$ is a function meromorphic at $z = \infty$ (i.e. in an open subset of $\hat{C}$ containing ∞), then its Laurent expansion at $z = \infty$ (convergent in a deleted neighborhood of ∞) will be denoted by $L^*(f)$. A sequence $\{R_n(z)\}$ of functions meromorphic at $z = \infty$ will be said to *correspond to an fLs* L^* at $z = \infty$ if the sequence $\{R_n(1/w)\}$ of functions meromorphic at $w = 0$ corresponds to the fLs L obtained from L^* by replacing w with $1/w$. Similarly correspondence at $z = a$ ($a \in C$) can be defined by considering $z = w + a$.

To motivate further the definition of correspondence we observe that the function $\| L \|$ defined by

$$\| L \| = \begin{cases} 0 & \text{if} \quad L = 0 \\ 2^{-\lambda(L)} & \text{if} \quad L \neq 0 \end{cases}, \qquad L \in \mathcal{L} \tag{5.1.13}$$

is a norm for the field $\mathcal{L}$. In fact it is easily seen [by use of (5.1.5)] that for $L_1, L_2 \in \mathcal{L}$,

$$\| L_1 \| \geqslant 0, \tag{5.1.14a}$$

$$\| L_1 \| = 0 \qquad \text{iff} \quad L_1 = 0, \tag{5.1.14b}$$

$$\| L_1 L_2 \| = \| L_1 \| \| L_2 \|, \tag{5.1.14c}$$

$$\| L_1 + L_2 \| \leqslant \| L_1 \| + \| L_2 \|. \tag{5.1.14d}$$

In terms of the norm $\| L \|$ we define the function $\rho : \mathcal{L} \times \mathcal{L} \to \mathbb{R}^+ = [x : x \in \mathbb{R},$

$x \geqslant 0$] by

$$\rho(L_1, L_2) = \|L_1 - L_2\| \qquad \text{for all} \quad L_1, L_2 \in \mathcal{L}. \qquad (5.1.15)$$

One can verify that ρ satisfies the following properties: for all $L_1, L_2, L_3 \in \mathcal{L}$,

$$\rho(L_1, L_2) = 0 \quad \text{iff} \quad L_1 = L_2, \qquad (5.1.16a)$$

$$\rho(L_1, L_2) = \rho(L_2, L_1), \qquad (5.1.16b)$$

$$\rho(L_1, L_2) + \rho(L_2, L_3) \geqslant \rho(L_1, L_3). \qquad (5.1.16c)$$

Thus ρ is a *metric* on the field $\mathcal{L}$. In terms of the metric ρ (which has been used previously in a similar context by Franzen [1972]), the statement that "the sequence $\{R_n(z)\}$ corresponds to a fLs L" is equivalent to saying that "$\{L(R_n)\}$ converges to L with respect to the metric ρ."

Let $\mathfrak{M}$ be the field of functions meromorphic at the origin. Set $L(\mathfrak{M}) = [L : L = L(f), f \in \mathfrak{M}]$. We are interested in *Cauchy* sequences in $L(\mathfrak{M})$, that is, sequences $\{L(R_n)\}$ which satisfy the condition: given $\varepsilon > 0$, there exists an n_ε such that

$$\rho(L(R_{n+k}), L(R_n)) < \varepsilon \qquad \text{for} \quad n \geqslant n_\varepsilon \text{ and } k = 0, 1, 2, \dots. \qquad (5.1.17)$$

It can be shown that $\mathcal{L}$ is the completion of $L(\mathfrak{M})$ with respect to ρ and hence every Cauchy sequence in $L(\mathfrak{M})$ converges to some fLs in $\mathcal{L}$.

In this chapter we are concerned with convergence of sequences in $\mathcal{L}$ with respect to the metric ρ and with convergence of sequences in $\mathbb{C}$ with respect to the usual metric d defined by $d : \mathbb{C} \times \mathbb{C} \rightarrow \mathbb{R}^+$:

$$d(z_1, z_2) = |z_1 - z_2| \qquad \text{for} \quad z_1, z_2 \in \mathbb{C}.$$

In our first theorem we give a characterization of sequences $\{R_n(z)\}$ which correspond to formal Laurent series (i.e. of Cauchy sequences with respect to the metric ρ).

THEOREM 5.1.

(A) *Given a sequence $\{R_n(z)\}$ of functions meromorphic at the origin, there exists an fLs L such that $\{R_n(z)\}$ corresponds to L iff*

$$\lim_{n \to \infty} \lambda(L(R_{n+1}) - L(R_n)) = \infty. \qquad (5.1.18)$$

(B) *If (5.1.18) holds, then the L to which $\{R_n(z)\}$ corresponds is determined uniquely.*

(C) *If the sequence $\{\lambda(L(R_{n+1}) - L(R_n))\}$ tends monotonically to ∞, then the order of correspondence of $R_n(z)$ is given by*

$$\nu_n = \lambda(L(R_{n+1}) - L(R_n)). \qquad (5.1.19)$$

Proof. In view of the preceding discussion, to prove (A) it suffices to show that $\{L(R_n)\}$ is a Cauchy sequence with respect to the metric ρ [see (5.1.15)] iff (5.1.18) holds. We note that condition (5.1.17) is equivalent to

$$\lambda(L(R_{n+k})-L(R_n))>\mathrm{Log}_2(1/\varepsilon) \qquad \text{for} \quad n\geqslant n_\varepsilon \text{ and } k=0,1,2,\dots. \tag{5.1.20}$$

By (5.1.5) we see that for $L_1, L_2,\dots, L_k\in\mathcal{L}$,

$$\lambda\left(\sum_{j=1}^{k} L_j\right)> \min_{1\leqslant j\leqslant k} \lambda(L_j).$$

Thus

$$\lambda(L(R_{n+k})-L(R_n))=\lambda\left(\sum_{j=1}^{k} (L(R_{n+j})-L(R_{n+j-1}))\right)$$
$$> \min_{1\leqslant j\leqslant k} \lambda(L(R_{n+j})-L(R_{n+j-1})). \tag{5.1.21}$$

It follows from (5.1.20) and (5.1.21) that $\{L(R_n)\}$ is a Cauchy sequence with respect to ρ iff given $N>0$, there exists an n_N such that

$$\lambda(L(R_{n+1})-L(R_n))>N \qquad \text{for} \quad n\geqslant n_N.$$

Hence (5.1.18) holds iff $\{L(R_n)\}$ is a Cauchy sequence with respect to ρ. This proves (A). (B) follows immediately from properties of a complete metric space. (C) is a direct consequence of the definitions. ∎

Some applications of Theorem 5.1 will be considered after the following theorem:

THEOREM 5.2. *Let $\{a_n(z)\}$ and $\{b_n(z)\}$ be sequences of functions meromorphic at the origin, with*

$$a_n(z)\not\equiv 0, \qquad n=1,2,3,\dots, \tag{5.1.22}$$

and let L_0 be a fLs. Let $\{L_n\}$ be a sequence of fLs defined recursively as follows:

$$L_{n+1}=\frac{L(a_{n+1})}{L_n-L(b_n)}, \qquad n=0,1,2,\dots, \tag{5.1.23}$$

provided

$$L_n \neq L(b_n), \qquad n = 0, 1, 2, \ldots, \tag{5.1.24}$$

[*otherwise see* (B)].

(A) *The continued fraction*

$$b_0(z) + \frac{a_1(z)}{b_1(z)} + \frac{a_2(z)}{b_2(z)} + \frac{a_3(z)}{b_3(z)} + \cdots \tag{5.1.25}$$

corresponds to L_0 *provided that*

$$\lambda(L(b_n)) + \lambda(L(b_{n-1})) < \lambda(L(a_n)), \qquad n = 1, 2, 3, \ldots, \tag{5.1.26a}$$

and

$$\lambda(L_n) + \lambda(L(b_{n-1})) < \lambda(L(a_n)), \qquad n = 1, 2, 3, \ldots. \tag{5.1.26b}$$

If (5.1.24) *and* (5.1.26) *hold, then the order of correspondence of the nth approximant* $f_n(z)$ *is*

$$\nu_0 = \lambda(L(a_1)) - \lambda(L_1), \tag{5.1.27a}$$

$$\nu_n = \sum_{k=1}^{n+1} \lambda(L(a_k)) - 2 \sum_{k=1}^{n} \lambda(L(b_k)) - \lambda(L_{n+1}), \qquad n = 1, 2, 3, \ldots.$$
$$\tag{5.1.27b}$$

(B) *If in defining the sequence* $\{L_n\}$ *by* (5.1.23) *one obtains*

$$L_k \neq L(b_k) \qquad for \quad 0 \leqslant k \leqslant m - 1 \ and \ L_m = L(b_m), \tag{5.1.28}$$

then

$$L_0 = L\left(b_0(z) + \frac{a_1(z)}{b_1(z)} + \cdots + \frac{a_m(z)}{b_m(z)}\right). \tag{5.1.29}$$

Remarks.

(1) To insure correspondence of (5.1.25) it suffices for the conditions (5.1.26) to hold for all n sufficiently large.

(2) The conditions (5.1.26) are invariant under equivalence transformation of the continued fraction (5.1.25). By this we mean the following: Let $\{r_n(z)\}$ be a sequence of non-vanishing functions meromorphic at the

origin, and define

$$a_n^\dagger = r_n(z) r_{n-1}(z) a_n(z), \qquad n = 1,2,3,\ldots \quad [r_0(z) \equiv 1],$$

$$(5.1.30a)$$

$$b_n^\dagger(z) = r_n(z) b_n(z), \qquad\qquad n = 0,1,2,\ldots, \qquad (5.1.30b)$$

$$L_n^\dagger = L(r_n) L_n, \qquad\qquad n = 0,1,2,\ldots . \qquad (5.1.30c)$$

Then in view of (5.1.22), we have $a_n^\dagger(z) \not\equiv 0$. Moreover

$$L_n^\dagger \neq L(b_n^\dagger), \qquad n = 0,1,2,\ldots, \qquad (5.1.31)$$

iff (5.1.24) holds;

$$L_{n+1}^\dagger = \frac{L(a_{n+1}^\dagger)}{L_n^\dagger - L(b_n^\dagger)}, \qquad n = 0,1,2,\ldots, \qquad (5.1.32)$$

provided (5.1.31) holds; and the conditions

$$\lambda\big(L(b_n^\dagger)\big) + \lambda\big(L(b_{n-1}^\dagger)\big) < \lambda\big(L(a_n^\dagger)\big), \qquad n = 1,2,3,\ldots, \quad (5.1.33a)$$

and

$$\lambda(L_n^\dagger) + \lambda\big(L(b_{n-1}^\dagger)\big) < \big(L(a_n^\dagger)\big), \qquad n = 1,2,3,\ldots, \quad (5.1.33b)$$

hold iff (5.1.26a) and (5.1.26b) hold, respectively. In consequence the continued fraction

$$b_0^\dagger(z) + \mathbf{K}\left(\frac{a_n^\dagger(z)}{b_n^\dagger(z)} \right) \qquad (5.1.34)$$

is equivalent to (5.1.25); hence (5.1.34) corresponds to $L_0^\dagger$ provided (5.1.33) [or equivalently (5.1.26)] hold. The primary significance of these observations is that it is unnecessary to search for an "appropriate" equivalence transformation of a continued fraction for the purpose of making Theorem 5.2 applicable.

Proof of Theorem 5.2. (A): Suppose that (5.1.24) holds, and let $A_n(z)$ and $B_n(z)$ denote the nth numerator and denominator, respectively, of (5.1.25). From (5.1.23) we have

$$L_n = L(b_n) + \frac{L(a_{n+1})}{L_{n+1}}, \qquad n = 0,1,2,\ldots . \qquad (5.1.35)$$

Here division by L_{n+1} is possible, since $L_{n+1} \neq 0$ follows from (5.1.22) and (5.1.23). Thus

$$L_0 = L(b_0) + \frac{L(a_1)}{L(b_1)} + \cdots + \frac{L(a_{n-1})}{L(b_{n-1})} + \frac{L(a_n)}{L_n}, \qquad n = 1, 2, 3, \ldots,$$

and hence by (2.1.7)

$$L_0 = \frac{L(a_n)L(A_{n-2}) + L_n L(A_{n-1})}{L(a_n)L(B_{n-2}) + L_n L(B_{n-1})}, \qquad n = 2, 3, 4, \ldots. \qquad (5.1.36)$$

Using (5.1.36), the determinant formulas (2.1.9) and $L(f_{n-1}) = L(A_{n-1})/L(B_{n-1})$, one obtains

$$L_0 - L(f_{n-1}) = \frac{(-1)^{n-1} \prod\limits_{k=1}^{n} L(a_k)}{L(B_{n-1})[L(a_n)L(B_{n-2}) + L_n L(B_{n-1})]}, \qquad n = 2, 3, 4, \ldots.$$

Hence from (5.1.3) and (5.1.4)

$$\lambda(L_0 - L(f_{n-1})) = \sum_{k=1}^{n} \lambda(L(a_k)) - \lambda(L(B_{n-1}))$$
$$\qquad\qquad - \lambda(L(a_n)L(B_{n-2}) + L_n L(B_{n-1})), \qquad n = 2, 3, 4, \ldots.$$

A simple induction argument based on (5.1.6), (5.1.26) and the difference equations (2.1.6) can be used to establish the following relations:

$$\lambda(L(B_0)) = 0 \qquad\qquad\qquad\qquad\qquad\qquad (5.1.38a)$$

$$\lambda(L(B_n)) = \sum_{k=1}^{n} \lambda(L(b_k)), \qquad n = 1, 2, 3, \ldots. \qquad (5.1.38b)$$

Then by use of (5.1.6), (5.1.26) and (5.1.38) one can prove that

$$\lambda(L(a_n)L(B_{n-2}) + L_n L(B_{n-1})) = \lambda(L_n) + \sum_{k=1}^{n-1} \lambda(L(b_k)),$$

$$n = 2, 3, 4, \ldots. \qquad (5.1.39)$$

Now substituting (5.1.38) and (5.1.39) into (5.1.37) gives

$$\lambda(L_0 - L(f_{n-1})) = \sum_{k=1}^{n} \lambda(L(a_k)) - 2 \sum_{k=1}^{n-1} \lambda(L(b_k)) - \lambda(L_n),$$

$$n = 2, 3, 4, \ldots. \qquad (5.1.40)$$

Rearranging the terms in (5.1.40), one obtains

$$\lambda(L_0 - L(f_{n-1})) = \lambda(L(a_1)) - \lambda(L(b_1))$$

$$+ \sum_{k=2}^{n-1} \left[\lambda(L(a_k)) - \lambda(L(b_k)) - \lambda(L(b_{k-1})) \right]$$

$$+ \lambda(L(a_n)) - \lambda(L(b_{n-1})) - \lambda(L_n),$$

$$n = 2, 3, 4, \dots . \quad (5.1.41)$$

It follows from (5.1.26) that in (5.1.41) each term in the summation and the term following the summation is a positive integer. Hence

$$\lim_{n \to \infty} \lambda(L_0 - L(f_{n-1})) = \infty,$$

and so the continued fraction (5.1.25) corresponds to L_0. Equation (5.1.27a) is an immediate consequence of the definitions; (5.1.27b) follows from (5.1.40). This proves (A). To prove (B) it suffices to observe that from (5.1.23) and (5.1.28) one has

$$L_0 = L(b_0) + \frac{L(a_1)}{L(b_1) +} \cdots \frac{L(a_{m-1})}{+ L(b_{m-1}) +} \frac{L(a_m)}{L_m},$$

and from this, (5.1.29) follows, since [by (5.1.28)] $L_m = L(b_m)$. ∎

In the following two corollaries we apply Theorems 5.1 and 5.2 to two types of continued fractions. Leighton and Scott introduced C-fractions and proved Corollary 5.3 in [1939]. P-fractions were introduced and Corollary 5.4 proved by Magnus in [1962a, b, 1964, 1974]. Other applications of Theorems 5.1 and 5.2 will be given in Chapter 6.

COROLLARY 5.3.

(A) *Every C-fraction*

$$1 + \frac{a_1 z^{\alpha_1}}{1 +} \frac{a_2 z^{\alpha_2}}{1 +} \frac{a_3 z^{\alpha_3}}{1 +} \cdots, \qquad a_n \neq 0, \qquad (5.1.42)$$

where the α_n are positive integers, corresponds to a uniquely determined formal power series (fps) of the form

$$L_0 = 1 + c_1 z + c_2 z^2 + c_3 z^3 + \dots . \quad (5.1.43)$$

The order of correspondence of the n th approximant $f_n(z)$ is

$$\nu_n = \sum_{k=1}^{n+1} \alpha_k . \quad (5.1.44)$$

(B) *Let* (5.1.43) *be a given fps. Then either* (B1) *there exists a C-fraction* (5.1.42) *which corresponds to* L_0, *or else* (B2) *there exists a finite C-fraction*

$$f_m(z)=1+\frac{a_1 z^{\alpha_1}}{1}+\frac{a_2 z^{\alpha_2}}{1}+\cdots+\frac{a_m z^{\alpha_m}}{1}, \qquad a_n\neq 0, \quad (5.1.45)$$

such that

$$L_0=L(f_m). \qquad (5.1.46)$$

In the latter case L_0 *is the expansion at* $z=0$ *of the rational function* $f_m(z)$.
(C) *If* L_0 *is the Taylor series expansion at* $z=0$ *of a rational function* $f(z)$, *then there exists a finite C-fraction* $f_m(z)$ *of the form* (5.1.45) *such that* (5.1.46) *holds.*

Proof. (A): If $A_n(z)$ and $B_n(z)$ denote the nth numerator and denominator, respectively, of the C-fraction (5.1.42), then from the difference equations (2.1.6) one sees that $A_n(z)$ and $B_n(z)$ are polynomials in z and $B_n(0)=1$. From the determinant formulas (2.1.9) one obtains

$$\frac{A_{n+1}(z)}{B_{n+1}(z)}-\frac{A_n(z)}{B_n(z)}=\frac{(-1)^n\prod\limits_{k=1}^{n+1}a_k z^{\alpha_k}}{B_n(z)B_{n+1}(z)}.$$

It follows that

$$\nu_n=\lambda(L(f_{n+1})-L(f_n))=\sum_{k=1}^{n+1}\alpha_k$$

tends monotonically to ∞ as $n\to\infty$. Hence (A) is an immediate consequence of Theorem 5.1. To prove (B) we assume that an fps L_0 of the form (5.1.43) is given. Suppose that

$$L_0\neq L(b_0)=L(1)=1.$$

Let $c_{k_1}z^{k_1}$ denote the first non-vanishing term in L_0-1. Then define

$$a_1=c_{k_1}, \qquad \alpha_1=k_1\geqslant 1,$$

and

$$L_1=\frac{L(a_1 z^{\alpha_1})}{L_0-1}.$$

Then L_1 is a fps of the form

$$L_1=1+c_1^{(1)}z+c_2^{(1)}z^2+c_3^{(1)}z^3+\cdots.$$

If

$$L_{n-1} \neq L(b_{n-1}) = L(1) = 1, \qquad n = 1,2,3,\ldots, \qquad (5.1.47)$$

we can continue in the above manner to let, for each $n \geqslant 1$, $c_{k_n}^{(n)} z^{k_n}$ denote the first non-vanishing term in $L_{n-1} - 1$. Then define

$$a_n = c_{k_n}^{(n)}, \qquad \alpha_n = k_n$$

and

$$L_n = \frac{L(a_n z^{\alpha_n})}{L_{n-1} - 1}. \qquad (5.1.48)$$

Since

$$\lambda(a_n z^{\alpha_n}) = \alpha_n \geqslant 1, \qquad \lambda(b_n) = \lambda(1) = 0 \quad \text{and} \quad \lambda(L_n) = 0, \qquad (5.1.49)$$

the conditions of Theorem 5.2(A) are satisfied and hence the C-fraction (5.1.42) (formed by the construction described above) corresponds to L_0. On the other hand, suppose that (5.1.47) fails to hold, so that for some $n = m \geqslant 0$,

$$L_k \neq 1 \qquad \text{for} \quad 0 \leqslant k \leqslant m-1 \text{ and } L_m = 1.$$

Then (B2) follows from Theorem 5.2(B).

(C): In the construction used to prove (B), each fps L_n is the Taylor series expansion of a rational function. In fact, there exists a sequence of polynomials $\{P_n(z)\}$ such that

$$L_n = L\left(\frac{P_n}{P_{n+1}}\right) \quad \text{and} \quad P_n(0) = 1, \quad n = 0,1,2,\ldots.$$

Let q_n denote the degree of $P_n(z)$. From (5.1.48) it follows that

$$P_{n+2}(z) = \frac{P_n(z) - P_{n+1}(z)}{a_{n+1} z^{\alpha_{n+1}}}.$$

Hence

$$q_{n+2} \leqslant \max(q_n, q_{n+1}) - 1.$$

It follows that

$$q_2 \leqslant \max(q_0, q_1) - 1$$

and

$$q_3 \leqslant \max(q_1, q_2) - 1 \leqslant \max(q_0, q_1) - 1,$$

since

$$\max(q_1, q_2) \leqslant \max(q_0, q_1).$$

Thus

$$q_4 \leqslant \max(q_2, q_3) - 1 \leqslant \max(q_0, q_1) - 2.$$

Continuing in this manner we see that for some m

$$q_n \geqslant 1 \quad \text{for } n < m, \quad \text{and} \quad q_n = 0 \quad \text{for } n \geqslant m.$$

Applying (5.1.48) we obtain

$$L_0 = 1 + \frac{a_1 z^{\alpha_1}}{1} + \frac{a_2 z^{\alpha_2}}{1} + \cdots + \frac{a_m z^{\alpha_m}}{L_m}.$$

The proof is completed by noting that

$$L_m = L\left(\frac{P_m}{P_{m+1}}\right) = L\left(\frac{1}{1}\right) = 1. \qquad \blacksquare$$

A consequence of Corollary 5.3 is that if a non-terminating C-fraction corresponds to a fps L_0, then L_0 cannot be the Taylor series expansion at $z = 0$ of a rational function. By a proof very similar to that given for Corollary 5.3, we obtain the following:

COROLLARY 5.4.

(A) *Every P-fraction*

$$b_0(z) + \frac{1}{b_1(z)} + \frac{1}{b_2(z)} + \frac{1}{b_3(z)} + \cdots, \qquad (5.1.50a)$$

where each $b_n(z)$ *is a polynomial in* $1/z$,

$$b_n(z) = \sum_{k=-N_n}^{0} a_{-k}^{(n)} z^k,$$

$$N_0 \geqslant 0; \qquad N_n \geqslant 1 \text{ and } a_{-N_n}^{(n)} \neq 0 \text{ for } n = 1, 2, 3, \ldots, \qquad (5.1.50b)$$

corresponds to a uniquely determined fLs of the form

$$L_0 = \sum_{k=-N_0}^{\infty} a_{-k}^{(0)} z^k. \tag{5.1.51}$$

The order of correspondence of the n th approximant $f_n(z)$ is

$$\nu_n = 2 \sum_{k=1}^{n} N_k + N_{n+1}, \qquad n = 0, 1, 2, \ldots . \tag{5.1.52}$$

(B) *Conversely, let* (5.1.51) *be a given fLs. Then either* (B1) *there exists a P-fraction* (5.1.50) *which corresponds to L_0, or else* (B2) *there exists a finite P-fraction*

$$f_m(z) = b_0(z) + \frac{1}{b_1(z)} + \frac{1}{b_2(z)} + \cdots + \frac{1}{b_m(z)},$$

where the $b_n(z)$ are polynomials in $1/z$ of the form (5.1.50b), *such that*

$$L_0 = L(f_m) \tag{5.1.53}$$

5.2 Three-Term Recurrence Relations

There is an intimate connection between continued fractions and sequences that satisfy a system of three-term recurrence relations. This connection is brought out by three main results: Theorems 5.2, 5.5 and 5.7 (Section 5.3).

THEOREM 5.5. *Let $\{a_n(z)\}$ and $\{b_n(z)\}$ be sequences of functions meromorphic at the origin with*

$$a_n(z) \not\equiv 0, \qquad n = 1, 2, 3, \ldots . \tag{5.2.1}$$

Let $\{P_n\}$ be a sequence of non-zero fLs satisfying the three-term recurrence relations

$$P_n = L(b_n) P_{n+1} + L(a_{n+1}) P_{n+2}, \qquad n = 0, 1, 2, \ldots . \tag{5.2.2}$$

(A) *Then the continued fraction*

$$b_0(z) + \frac{a_1(z)}{b_1(z)} + \frac{a_2(z)}{b_2(z)} + \frac{a_3(z)}{b_3(z)} + \cdots \tag{5.2.3}$$

corresponds to the fLs

$$L = \frac{P_0}{P_1} \tag{5.2.4}$$

provided the following conditions are satisfied:

$$\lambda(L(b_n)) + \lambda(L(b_{n-1})) < \lambda(L(a_n)), \qquad n = 1, 2, 3, \ldots, \qquad (5.2.5a)$$

$$\lambda(P_n/P_{n+1}) + \lambda(L(b_{n-1})) < \lambda(L(a_n)), \qquad n = 1, 2, 3, \ldots. \qquad (5.2.5b)$$

(B) *The order of correspondence of the nth approximant* $f_n(z)$ *is*

$$\nu_0 = \lambda(L(a_1)) - \lambda(L_1) \qquad (5.2.6a)$$

$$\nu_n = \sum_{k=1}^{n+1} \lambda(L(a_k)) - 2 \sum_{k=1}^{n} \lambda(L(b_k)) - \lambda(L_{n+1}), \qquad n = 1, 2, 3, \ldots.$$

$$(5.2.6b)$$

(C) *Moreover, for each* $m = 0, 1, 2, \ldots$, *the continued fraction*

$$b_m(z) + \frac{a_{m+1}(z)}{b_{m+1}(z)} + \frac{a_{m+2}(z)}{b_{m+2}(z)} + \frac{a_{m+3}(z)}{b_{m+3}(z)} + \cdots \qquad (5.2.7)$$

corresponds to the fLs

$$L_m = \frac{P_m}{P_{m+1}}. \qquad (5.2.8)$$

Proof. Letting $L_n = P_n/P_{n+1}$ for $n \geqslant 0$, we obtain from (5.2.2) the relations

$$L_n - L(b_n) = \frac{L(a_{n+1})}{L_{n+1}}, \qquad n = 0, 1, 2, \ldots. \qquad (5.2.9)$$

The condition (5.2.1) implies that $L(a_{n+1}) \neq 0$ for all $n \geqslant 0$ and hence [from (5.2.9)]

$$L_n \neq L(b_n), \qquad n = 0, 1, 2, \ldots.$$

Thus $\{L_n\}$ satisfies all of the conditions of Theorem 5.2(A), and so Theorem 5.5 is an immediate consequence. ∎

Remark. The conditions (5.2.5) will hold provided either of the following sets of conditions holds:

$$\lambda(L(a_n)) \geqslant 1, \quad \lambda(L(b_{n-1})) \leqslant 0, \quad \lambda(L_n) \leqslant 0, \quad n = 1, 2, 3, \ldots$$

$$(5.2.10)$$

$$\lambda(L(a_n)) \geqslant 0, \quad \lambda(L(b_{n-1})) \leqslant -1, \quad \lambda(L_n) \leqslant 0, \quad n = 1, 2, 3, \ldots,$$

$$(5.2.11)$$

where $L_n = P_n/P_{n+1}$.

Many applications of Theorem 5.5 will be discussed in Chapter 6. For now we consider the following example. The Bessel function $J_\nu(z)$ of the first kind of order ν is defined by the infinite series

$$J_\nu(z) = \left(\frac{z}{2}\right)^\nu \sum_{k=0}^\infty \frac{(-1)^k \left(\frac{z}{2}\right)^{2k}}{k!\,\Gamma(\nu+k+1)}, \qquad \nu \in \mathbb{C}, \qquad (5.2.12)$$

convergent for all $z \in \mathbb{C}$. Here we assume that $1/\Gamma(t)$ is zero if t is a negative integer or zero (see [Copson, 1935, Chapter XII]). The $J_\nu(z)$ are easily shown to satisfy the three-term recurrence relations

$$J_\nu(z) = \frac{2(\nu+1)}{z} J_{\nu+1}(z) - J_{\nu+2}(z). \qquad (5.2.13)$$

For fixed $\nu \in \mathbb{C}$, we define the sequence $\{P_n(z)\}$ of convergent power series (or fLs) by

$$P_n(z) = \left(\frac{z}{2}\right)^{-\nu} J_{\nu+n}(z), \qquad n=0,1,2,\dots. \qquad (5.2.14)$$

Upon substitution of (5.2.14) into (5.2.13) and cancellation of $(z/2)^\nu$, we obtain

$$P_n(z) = \frac{2(\nu+n+1)}{z} P_{n+1}(z) - P_{n+2}(z), \qquad n=0,1,2,\dots. \qquad (5.2.15)$$

With $a_n(z) = -1$ and $b_n(z) = 2(\nu+n+1)/z$, we have

$$\lambda(L(a_n)) = 0, \qquad \lambda(L(b_n)) = -1 \quad \text{and} \quad L(P_n/P_{n+1}) = -1,$$

so that (5.2.11) is satisfied and hence Theorem 5.5 can be applied. It follows that the continued fraction

$$\frac{2(\nu+m+1)}{z} - \cfrac{1}{\dfrac{2(\nu+m+2)}{z}} - \cfrac{1}{\dfrac{2(\nu+m+3)}{z}} - \cfrac{1}{\dfrac{2(\nu+m+4)}{z}} - \cdots$$

$$(5.2.16)$$

corresponds to $P_m(z)/P_{m+1}(z) = J_{\nu+m}(z)/J_{\nu+m+1}(z)$. Taking the reciprocal of (5.2.16) and applying a simple equivalence transformation yields

COROLLARY 5.6. *Let ν be an arbitrary complex number. Then for each $m = 0, 1, 2, \dots$, the continued fraction*

$$\frac{z}{2(\nu+m+1)} - \frac{z^2}{2(\nu+m+2)} - \frac{z^2}{2(\nu+m+3)} - \frac{z^2}{2(\nu+m+4)} - \cdots$$

$$(5.2.17)$$

corresponds to $J_{\nu+m+1}(z)/J_{\nu+m}(z)$ at $z=0$. Here $J_\nu(z)$ is the Bessel function of the first kind of order ν. The order of correspondence of the nth approximant of (5.2.17) is

$$\nu_n = 2n - 1, \qquad n = 0, 1, 2, \ldots. \qquad (5.2.18)$$

5.3   Minimal Solutions of Three-Term Recurrence Relations

Many sequences of special functions can be computed by means of three-term recurrence relations that they are known to satisfy. However, for certain sequences (called minimal solutions) the algorithm based on the three-term recurrence relations is numerically unstable. Gautschi pointed out this difficulty in [1967] and showed that the problem could be dealt with by continued fractions based on a theorem of Pincherle [1894]. In this section we shall state and prove (Theorem 5.7) a generalization of Pincherle's theorem which applies not only to convergence in the usual metric of $\mathbb{C}$ but also to convergence in the normed field $\mathcal{L}$ of formal Laurent series [see (5.1.3) for the definition of the norm on $\mathcal{L}$]. We also give (Theorem 5.10) a generalization of a closely related theorem of Auric [1907]. Gautschi's method is illustrated by considering Bessel functions of the first kind. Other examples are described in Chapter 6. Additional results on minimal solutions of three-term recurrence relations are given in Appendix B.

We consider systems of three-term recurrence relations

$$y_{n+1} = b_n y_n + a_n y_{n-1}, \qquad a_n \neq 0, \qquad n = 1, 2, 3, \ldots, \qquad (5.3.1)$$

where the a_n, b_n and y_n are elements of an arbitrary normed field $\mathbb{F}$. For each element $y \in \mathbb{F}$ the norm of y will be denoted by $\|y\|_{\mathbb{F}}$. The set of all solutions $\{y_n\}$ of (5.3.1) forms a linear vector space V of dimension two over the field $\mathbb{F}$. If there exists a non-trivial solution $\{h_n\}$ (i.e. $h_n \neq 0$ for some n) and another solution $\{g_n\}$ of (5.3.1) such that

$$\lim_{n \to \infty} \frac{h_n}{g_n} = 0, \qquad (5.3.2)$$

then $\{h_n\}$ is called a *minimal solution* of (5.3.1). For any solution $\{y_n\}$ of (5.3.1) not proportional to $\{h_n\}$, it is easily shown that

$$\lim_{n \to \infty} \frac{h_n}{y_n} = 0. \qquad (5.3.3)$$

If the set of all solutions $\{h_n\}$ of (5.3.1) having the property (5.3.2) is not empty, then it is a one-dimensional subspace of V. A solution of (5.3.1) which is not minimal is called *dominant*. In general a system of recurrence relations (5.3.1) may or may not have a minimal solution.

The following theorem deals with continued fractions over an arbitrary normed field $\mathbb{F}$. Its approximants are in the extended field $\hat{\mathbb{F}} \cup [\infty]$ (see last part of Section 2.1.1).

THEOREM 5.7 (Pincherle generalized). *For each* $n = 1, 2, 3, \ldots$, *let* a_n *and* b_n *be elements of a normed field* $\mathbb{F}$, *with*

$$a_n \neq 0, \qquad n = 1, 2, 3, \ldots. \tag{5.3.4}$$

(A) *The system of three-term recurrence relations*

$$y_{n+1} = b_n y_n + a_n y_{n-1}, \qquad n = 1, 2, 3, \ldots, \tag{5.3.5}$$

has a minimal solution $\{h_n\}$, $h_n \in \mathbb{F}$, *iff the continued fraction over the field* $\mathbb{F}$

$$\frac{a_1}{b_1} + \frac{a_2}{b_2} + \frac{a_3}{b_3} + \cdots \tag{5.3.6}$$

converges (*to a finite value in* $\mathbb{F}$ *or to infinity*).

(B) *Suppose that* (5.3.5) *has a minimal solution* $\{h_n\}$, $h_n \in \mathbb{F}$. *Then for each* $m = 1, 2, 3, \ldots$,

$$\frac{h_m}{h_{m-1}} = -\frac{a_m}{b_m} + \frac{a_{m+1}}{b_{m+1}} + \frac{a_{m+2}}{b_{m+2}} + \cdots. \tag{5.3.7}$$

By (5.3.7) *we mean the following: If* $h_{m-1} = 0$, *then* $h_m \neq 0$ *and the continued fraction* (5.3.7) *converges to* $\infty = h_m / h_{m-1}$. *If* $h_{m-1} \neq 0$, *then the continued fraction* (5.3.7) *converges to the finite value* $h_m / h_{m-1} \in \mathbb{F}$.

Proof. Let A_n and B_n denote the nth numerator and denominator, respectively, of (5.3.6), with $A_0 = B_{-1} = 0$ and $A_{-1} = B_0 = 1$. Define sequences $\{\alpha_n\}$ and $\{\beta_n\}$ by

$$\alpha_n = A_{n-1}, \qquad \beta_n = B_{n-1}, \qquad n = 0, 1, 2, \ldots.$$

It follows from the difference equations (2.1.6) that $\{\alpha_n\}$ and $\{\beta_n\}$ are solutions of (5.3.5). Moreover, it is easily shown that $\{\alpha_n\}$ and $\{\beta_n\}$ are linearly independent and hence form a basis for the linear space U of solutions of (5.3.5). (A): Suppose that there exists a minimal solution $\{h_n\}$ of (5.3.5). Then there exist constants h_0 and h_1 (not both zero) such that

$$h_n = h_0 \alpha_n + h_1 \beta_n, \qquad n = 0, 1, 2, \ldots.$$

If $h_0 \neq 0$, then $\{\beta_n\}$ is not proportional to $\{h_n\}$; hence

$$\lim_{n \to \infty} \frac{h_n}{\beta_n} = 0.$$

This implies that

$$0= \lim_{n\to\infty} \frac{h_0\alpha_n+h_1\beta_n}{\beta_n} = h_0 \lim_{n\to\infty} \frac{\alpha_n}{\beta_n} +h_1,$$

so that

$$\lim_{n\to\infty} \frac{A_n}{B_n} = -\frac{h_1}{h_0}.$$

Thus the continued fraction (5.3.6) converges to the finite value $-h_1/h_0 \in$ F. On the other hand, if $h_0=0$, then $h_1\neq0$ and $\beta_n=h_n/h_1$, so that $\{\beta_n\}$ is a minimal solution. It follows that

$$\lim_{n\to\infty} \frac{A_n}{B_n} = \lim_{n\to\infty} \frac{\alpha_n}{\beta_n} = \infty,$$

that is, the continued fraction (5.3.6) converges to infinity. Conversely, suppose that

$$\lim_{n\to\infty} \frac{A_n}{B_n} =c\in F$$

exists and is finite. Define $\{y_n\}$ by

$$y_n=\alpha_n- c\beta_n, \qquad n=0,1,2,\dots .$$

Then

$$\lim_{n\to\infty} \frac{y_n}{\beta_n} = \lim_{n\to\infty} \frac{\alpha_n}{\beta_n} -c=c-c=0,$$

and hence $\{y_n\}$ is a minimal solution. If the continued fraction (5.3.6) converges to infinity, then

$$\lim_{n\to\infty} \frac{\beta_n}{\alpha_n} = \lim_{n\to\infty} \frac{B_n}{A_n} =0,$$

and so $\{\beta_n\}$ is a minimal solution. This completes the proof of (A).

To prove (B) note that for a non-trivial solution $\{y_n\}$ of (5.3.5) no two consecutive y_n can vanish. Hence

$$-\frac{y_n}{y_{n-1}} = \cfrac{a_n}{b_n - \cfrac{y_{n+1}}{y_n}} \tag{5.3.8}$$

is a valid consequence of (5.3.5) even if $y_n=0$ or $y_{n-1}=0$. In the latter case

$y_{n+1}/y_n = b_n$, and hence both sides of (5.3.8) are equal to ∞. It follows that, provided $y_0 \neq 0$,

$$-\frac{y_1}{y_0} = S_n\left(-\frac{y_{n+1}}{y_n}\right),$$

where $S_n(w)$ is defined by (2.1.1). We thus obtain

$$-\frac{y_1}{y_0} - \frac{A_n}{B_n} = S_n\left(-\frac{y_{n+1}}{y_n}\right) - \frac{A_n}{B_n}$$

and, using (2.1.7),

$$-\frac{y_1}{y_0} - \frac{A_n}{B_n} = \frac{A_n y_n - A_{n-1} y_{n+1}}{B_n y_n - B_{n-1} y_{n+1}} - \frac{A_n}{B_n}$$

$$= \frac{-(A_{n-1} B_n - A_n B_{n-1}) y_{n+1}}{(B_n y_n - B_{n-1} y_{n+1}) B_n}.$$

Since $y_{n+1} = y_0 A_n + y_1 B_n$, one obtains

$$B_n y_n - B_{n-1} y_{n+1} = B_n(y_0 A_{n-1} + y_1 B_{n-1}) - B_{n-1}(y_0 A_n + y_1 B_n)$$

$$= y_0(A_{n-1} B_n - A_n B_{n-1}). \tag{5.3.9}$$

Hence

$$-\frac{y_1}{y_0} - \frac{A_n}{B_n} = -\frac{y_{n+1}}{y_0 B_n}. \tag{5.3.10}$$

For a minimal solution $\{h_n\}$ of (5.3.5) for which $h_0 \neq 0$ we then have

$$\lim_{n\to\infty}\left(-\frac{h_1}{h_0} - \frac{A_n}{B_n}\right) = -\frac{1}{h_0}\lim\frac{h_{n+1}}{B_{n+1}} = 0.$$

If $h_0 = 0$ then $h_1 \neq 0$ and hence $\{\beta_n\}$ is a minimal solution, so that

$$\lim_{n\to\infty}\frac{\alpha_{n+1}}{\beta_{n+1}} = \lim_{n\to\infty}\frac{A_n}{B_n} = \infty = -\frac{h_1}{h_0}.$$

Hence (5.3.7) holds for $m = 1$. The proof of (5.3.7) for $m > 1$ is analogous to that given for $m = 1$. ∎

We consider now the problem of computing a minimal solution $\{h_n\}$ of a system of three-term recurrence relations (5.3.1), where $a_n, b_n, h_n \in \mathbb{C}$ for all n. Two algorithms will be discussed. The first algorithm, which will be shown to be numerically unstable, consists in starting with initial values h_0 and h_1 and then computing, successively, the numbers $h_2, h_3, h_4, \ldots$ (as far as we want) by the relations (5.3.1). The problem with this algorithm can be seen as follows. In practice, because there is rounding error, one will begin with approximate values y_0 and y_1 of h_0 and h_1, respectively. Then, even if no rounding errors were committed in the subsequent computations of $y_2, y_3, y_4, \ldots$ (where y_n is the approximation of h_n), one will obtain in general a solution $\{y_n\}$ of (5.3.1) which is linearly independent of $\{h_n\}$. Since $\{h_n\}$ is a minimal solution, one has $\lim h_n/y_n = 0$. Hence for the relative error $(y_n - h_n)/h_n$,

$$\lim_{n \to \infty} \left| \frac{y_n - h_n}{h_n} \right| = \lim_{n \to \infty} \left| \frac{1 - \dfrac{h_n}{y_n}}{\dfrac{h_n}{y_n}} \right| = \infty,$$

so that the algorithm is unstable.

An example (given by Gautschi [1967]) of the type of behavior one can expect is found in studying Bessel functions. Let $J_\nu(z)$ denote the Bessel function of the first kind of order ν [see (5.2.12)], and let $Y_\nu(z)$ denote the Bessel function of the second kind of order ν defined by

$$Y_\nu(z) = \frac{J_\nu(z) \cos \nu \pi - J_{-\nu}(z)}{\sin \nu \pi}, \qquad \nu \notin [0, \pm 1, \pm 2, \ldots], \quad (5.3.11a)$$

$$Y_n(z) = \lim_{\nu \to n} Y_\nu(z), \qquad n = 0, \pm 1, \pm 2, \ldots. \qquad (5.3.11b)$$

For fixed $\nu \in \mathbb{C}$ and $z \in \mathbb{C}$, we define $\{h_n\}$ and $\{g_n\}$ by

$$h_n = J_{\nu+n}(z), \quad g_n = Y_{\nu+n}(z), \qquad n = 0, 1, 2, \ldots. \qquad (5.3.12)$$

Then it is known that $\{h_n\}$ and $\{g_n\}$ are linearly independent solutions of the system of recurrence relations

$$y_{n+1} = \frac{2(\nu+n)}{z} y_n - y_{n-1}, \qquad n = 1, 2, 3, \ldots, \qquad (5.3.13)$$

(see, for example, [Copson, 1935] or [Watson, 1952]). With $\nu > 0$, $J_\nu(z)$ and $Y_\nu(z)$ have the asymptotic behavior

$$J_\nu(z) \sim \frac{1}{\sqrt{2\pi\nu}} \left(\frac{ez}{2\nu} \right)^\nu \qquad \text{as} \quad \nu \to \infty \quad (\nu > 0), \qquad (5.3.14a)$$

$$Y_\nu(z) \sim -\sqrt{\frac{2}{\pi\nu}} \left(\frac{ez}{2\nu} \right)^{-\nu} \qquad \text{as} \quad \nu \to \infty \quad (\nu > 0), \qquad (5.3.14b)$$

where the symbol $x_\nu \sim X_\nu$ as $\nu \to \infty$ means that $x_\nu / X_\nu \to 1$ as $\nu \to \infty$ (see, for example, [Abramowitz and Stegun, 1964, p. 365]). It follows that

$$\lim_{n \to \infty} \frac{h_n}{g_n} = 0 \quad \text{(if } \nu > 0\text{)}; \qquad (5.3.15)$$

hence $\{h_n\}$ is a minimal solution of (5.3.14) provided $\nu > 0$. For our numerical example we take $\nu = 0$ and $z = 1$, so that $h_n = J_n(1)$, $n \geqslant 0$. The starting values $J_0(1)$ and $J_1(1)$ are given by $y_0 = 0.7651976866$ and $y_1 = 0.4400505857$, respectively; these are correct to ten figures. Values of $y_2, y_3, y_4, \ldots$ computed by the recurrence relations (5.3.13) are given in Table 5.3.1; the computations were performed on a CDC 3600 computer, which yields approximately 12 decimal digits in floating-point arithmetic [Gautschi, 1967, p. 26]. For comparison the corresponding values of $J_n(1)$ (correct to ten figures) are also given in Table 5.3.1 (these values were taken from [Abramowitz and Stegun, 1964, p. 407]). It can be seen that the accuracy of y_n deteriorates rapidly; for $n \geqslant 7$ there are no correct digits in y_n. Thus the recurrence relation (5.3.14) is of little use in computing $J_n(1)$ for large n.

An alternative and preferred algorithm for computing values of Bessel functions $J_\nu(z)$ is based on the following corollary of Theorem 5.7.

COROLLARY 5.8. *Let* $0 \leqslant \nu < 1$ *be given and let m be a positive integer.*

(A) *At each* $z \in \mathbb{C}$ *such that* $J_{\nu+m-1}(z) \neq 0$, *the function* $J_{\nu+m}(z)/J_{\nu+m-1}(z)$ *is holomorphic and*

$$\frac{J_{\nu+m}(z)}{J_{\nu+m-1}(z)} = \frac{z}{2(\nu+m)} - \frac{z^2}{2(\nu+m+1)} - \frac{z^2}{2(\nu+m+2)} - \cdots . \qquad (5.3.16)$$

The continued fraction converges to the finite value on the left side of (5.3.16).
(B) *At each* $z \in \mathbb{C}$ *such that* $J_{\nu+m-1}(z) = 0$, *the function on the left of* (5.3.16) *has a simple pole and the continued fraction converges to infinity.*

Proof. Since, for $\nu > 0$, $\{h_n\}$ [where $h_n = J_{\nu+n}(z)$] is a minimal solution of (5.3.13), it follows from Theorem 5.7 (with $\mathbb{F} = \mathbb{C}$) that

$$\frac{J_{\nu+m}(z)}{J_{\nu+m-1}(z)} = \frac{1}{\dfrac{2(\nu+m)}{z}} - \frac{1}{\dfrac{2(\nu+m+1)}{z}} - \frac{1}{\dfrac{2(\nu+m+2)}{z}} - \cdots .$$

Applying an elementary equivalence transformation, we arrive at (5.3.16).   ∎

Let $f_{m,n}(z)$ denote the nth approximant of the continued fraction (5.3.16). Then since

$$J_{\nu+m}(z) = J_\nu(z) \prod_{k=1}^{m} \frac{J_{\nu+k}(z)}{J_{\nu+k-1}(z)}, \qquad (5.3.17)$$

we choose, for an nth approximation of $J_{\nu+m}(z)$, the function $G_{m,n}(z)$ defined by

$$G_{m,n}(z) = J_\nu(z) \prod_{k=1}^{m} f_{k,n}(z), \qquad n = 1, 2, 3, \ldots. \qquad (5.3.18)$$

Now for the numerical example considered above, we take $\nu = 0$, $z = 1$ and $n = 10$, to obtain $G_{m,10}(1)$ as an approximation of $J_m(1)$. Values of $G_{m,10}(1)$ are given in Table 5.3.1 for comparison; the computations were performed on a calculator which uses approximately ten decimal digits in floating-point arithmetic. The computations of the continued-fraction approximants $f_{k,10}(1)$ were made by the backward recurrence algorithm (Section 2.1.4). Only one initial value $J_0(1)$ is required in (5.3.18). As can be seen in

Table 5.3.1. Approximations of $J_m(1)$, Bessel Functions of the First Kind of Order m at $z = 1$. $J_m(1)$ (correct to ten figures) is taken from [Abramowitz and Stegun, 1964, p. 407]. y_m is computed by three-term recurrence relations. For $m > 7$, y_m has no correct digits. $G_{m,10}(1)$ is computed using continued-fraction approximants of order 10. (The numbers in parentheses denote powers of 10 by which the preceding numbers have to be multiplied.)

m	y_m	$G_{m,10}(1)$	$J_m(1)$
0	7.65197 6866(-1)	7.65197 6866(-1)	7.65197 6866(-1)
1	4.40050 5857(-1)	4.40050 5857(-1)	4.40050 5857(-1)
2	1.14903 4848(-1)	1.14903 4849(-1)	1.14903 4849(-1)
3	1.95633 5358(-2)	1.91633 5398(-2)	1.95633 5398(-2)
4	2.47633 6684(-3)	2.47633 8964(-3)	2.47663 8964(-3)
5	2.49739 8891(-4)	2.49757 7302(-4)	2.49757 7302(-4)
6	2.07622 0699(-5)	2.09383 3800(-5)	2.09383 3800(-5)
7	-5.93405 2751(-7)	1.50232 5817(-6)	1.50232 5817(-6)
8	-2.90698 8084(-5)	9.42234 4170(-8)	9.42234 4173(-8)
9	-4.64524 6881(-4)	5.24925 0178(-9)	5.24925 0180(-9)
10	-8.33237 4506(-3)	2.63061 5123(-10)	2.63061 5124(-10)
11	-1.66182 9654(-1)	1.19800 6746(-11)	1.19800 6746(-11)
12	-3.64769 2865(0)	4.99971 8179(-13)	4.99971 8179(-13)
13	-8.73784 4579(1)	1.92561 6765(-14)	1.92561 6764(-14)
$\vdots$	$\vdots$	$\vdots$	$\vdots$
20	-2.81859 0869(12)	3.87350 3006(-25)	3.87350 3009(-25)

Table 5.3.1, the $G_{m,10}(1)$ have remarkable accuracy. This accuracy is due in part to the rapid convergence of the continued fractions involved and in part to the numerical stability of the backward recurrence algorithm. These two factors are dealt with more fully in Chapters 8 and 10.

The following corollary of Theorem 5.7 is the interpretation of Pincherle's theorem in the normed field $\mathcal{L}$ of formal Laurent series. In the system of three-term recurrence relations (5.3.5), the a_n and b_n are elements of $\mathcal{L}$. A non-trivial solution $\{Q_n\}$, $Q_n \in \mathcal{L}$, of (5.3.5) will be a minimal solution if there exists another solution $\{P_n\}$ of (5.3.5) such that

$$\lim_{n \to \infty} \frac{Q_n}{P_n} = 0. \tag{5.3.19}$$

In the norm of $\mathcal{L}$ [see (5.1.3)], (5.3.19) is equivalent to

$$\lim_{n \to \infty} \lambda\left(\frac{Q_n}{P_n}\right) = \lim_{n \to \infty} \left[\lambda(Q_n) - \lambda(P_n)\right] = \infty. \tag{5.3.20}$$

As in the discussion in Section 2.1, we adjoin to $\mathcal{L}$ an additional element called infinity, and denoted by ∞. A sequence $\{R_n\}$ of elements of $\mathcal{L}$ *corresponds* to ∞ if

$$\lim_{n \to \infty} \lambda(R_n) = -\infty. \tag{5.3.21}$$

Since convergence in the norm of $\mathcal{L}$ is equivalent to correspondence, we arrive at

COROLLARY 5.9. *For each* $n = 1, 2, 3, \ldots$ *let* $a_n(z)$ *and* $b_n(z)$ *be elements of* $\mathfrak{M}$ (*i.e. functions meromorphic at the origin*), *with*

$$a_n(z) \not\equiv 0, \qquad n = 1, 2, 3, \ldots . \tag{5.3.22}$$

(A) *The continued fraction*

$$\frac{a_1(z)}{b_1(z)} + \frac{a_2(z)}{b_2(z)} + \frac{a_3(z)}{b_3(z)} + \cdots \tag{5.3.23}$$

corresponds to an element of the extended field $\hat{\mathcal{L}} = \mathcal{L} \cup [\infty]$ *iff the system of three-term recurrence relations in* $\mathcal{L}$

$$y_{n+1} = L(b_n)y_n + L(a_n)y_{n-1}, \qquad n = 1, 2, 3, \ldots, \tag{5.3.24}$$

has a minimal solution $\{Q_n\}$, $Q_n \in \mathcal{L}$.

(B) *Suppose that (5.3.24) has a minimal solution* $\{Q_n\}$, $Q_n \in \mathcal{L}$. *Then, for each* $m = 1,2,3,\ldots$, *the continued fraction*

$$\frac{a_m(z)}{b_m(z)} + \frac{a_{m+1}(z)}{b_{m+1}(z)} + \frac{a_{m+2}(z)}{b_{m+2}(z)} + \cdots \tag{5.3.25}$$

corresponds to $-Q_m/Q_{m-1} \in \hat{\mathcal{L}} = \mathcal{L} \cup [\infty]$.

As an example we shall now show how Pincherle's theorem in the field $\mathcal{L}$ (Corollary 5.9) can be used to prove the result on correspondence of C-fractions given by Corollary 5.3(B). Let P be a formal power series of the form

$$P = c_{q_1} z^{q_1} + c_{q_1+1} z^{q_1+1} + c_{q_1+2} z^{q_1+2} + \cdots, \tag{5.3.26}$$

where

$$c_{q_1} \neq 0 \quad \text{and} \quad q_1 \text{ is a positive integer.} \tag{5.3.27}$$

It suffices to show that either there exists a C-fraction $\mathrm{K}(a_n z^{\alpha_n}/1)$ which corresponds to P or there exists a finite C-fraction

$$\frac{a_1 z^{\alpha_1}}{1} + \frac{a_2 z^{\alpha_2}}{1} + \cdots + \frac{a_m z^{\alpha_m}}{1}, \tag{5.3.28}$$

which is identically equal to P. We set $Q_0 = 1$, $Q_1 = -P$ and will show inductively that a_n, α_n and Q_n can be determined so that the α_n are positive integers, the a_n are non-zero complex constants,

$$Q_{n+1} = Q_n + a_n z^{\alpha_n} Q_{n-1}, \qquad n = 1,2,3,\ldots, \tag{5.3.29}$$

and

$$\lambda(Q_{n+1}) = q_{n+1} > q_n = \lambda(Q_n). \tag{5.3.30}$$

Assume that, for all n such that $1 \leqslant n \leqslant m$,

$$Q_n = c^{(n)} z^{q_n} + (\text{higher powers of } z), \qquad c^{(n)} \neq 0, \tag{5.3.31a}$$

and

$$q_n > q_{n-1}. \tag{5.3.31b}$$

Set

$$\alpha_n = q_n - q_{n-1}, \quad a_n = -\frac{c^{(n)}}{c^{(n-1)}}, \qquad 1 \leqslant n \leqslant m. \tag{5.3.32}$$

Then

$$Q_{m+1} = Q_m + a_m z^{\alpha_m} Q_{m-1} \tag{5.3.33}$$

is a formal power series of the form

$$Q_{m+1} = c^{(m+1)} z^{q_{m+1}} + (\text{higher powers of } z) \tag{5.3.34}$$

and either

$$c^{(m+1)} \neq 0 \quad \text{and} \quad q_{m+1} > q_m \tag{5.3.35}$$

or

$$Q_{m+1} = 0 \text{ (the zero element of } \mathcal{L} \text{)}. \tag{5.3.36}$$

Suppose that (5.3.36) holds. Then from (5.3.29)

$$-\frac{Q_n}{Q_{n-1}} = \frac{a_n z^{\alpha_n}}{1 - \dfrac{Q_{n+1}}{Q_n}}, \quad 1 \leqslant n \leqslant m. \tag{5.3.37}$$

Upon application of (5.3.37) we obtain

$$P = -\frac{Q_1}{Q_0} = \frac{a_1 z^{\alpha_1}}{1} + \frac{a_2 z^{\alpha_2}}{1} + \cdots + \frac{a_m z^{\alpha_m}}{1 - \dfrac{Q_{m+1}}{Q_m}}. \tag{5.3.38}$$

Since $Q_{m+1} = 0$, (5.3.38) is a finite C-fraction of the form (5.3.28). Now suppose that $Q_n \neq 0$ for all n. Then in view of Corollary 5.8, it suffices to show that $\{Q_n\}$ is a minimal solution of (5.3.29). That this is the case can be seen as follows. Let $\{P_n\}$ be the solution of (5.3.29) with initial conditions $P_0 = 0$ and $P_1 = 1$. Then $\lambda(P_n) = 0$ for all $n > 1$ and

$$\lambda\left(\frac{Q_n}{P_n}\right) = \lambda(Q_n) - \lambda(P_n) = q_n \to \infty \quad \text{as} \quad n \to \infty.$$

Thus $\{Q_n\}$ is a minimal solution of (5.3.29) in $\mathcal{L}$ [see (5.3.20)]. It follows from Corollary 5.8 that the C-fraction $K(a_n z^{\alpha_n}/1)$ corresponds to $P = -Q_1/Q_0$, which was to be proved.

Auric's theorem, which was also independently formulated by Perron [1913], can be generalized to:

THEOREM 5.10 (Auric generalized). *For each* $n = 1, 2, 3, \ldots,$ *let* a_n *and* b_n *be elements of a normed field* $\mathbb{F}$, *with*

$$a_n \neq 0, \qquad n = 1, 2, 3, \ldots. \tag{5.3.39}$$

Let A_n, B_n *denote the* n *th numerator and denominator, respectively, of the continued fraction*

$$\frac{a_1}{b_1 +} \frac{a_2}{b_2 +} \frac{a_3}{b_3 +} \cdots \tag{5.3.40}$$

over the normed field $\mathbb{F}$. *Let* $\{y_n\}$, $y_n \in \mathbb{F}$, *be a solution of the system of three-term recurrence relations*

$$y_{n+1} = b_n y_n + a_n y_{n-1}, \qquad n = 1, 2, 3, \ldots. \tag{5.3.41}$$

If $y_n \neq 0$, $n = 0, 1, 2, \ldots,$ *then the continued fraction* (5.3.40) *over the normed field* $\mathbb{F}$ *converges to the finite limit*

$$-\frac{y_1}{y_0} = \lim_{n \to \infty} \frac{A_n}{B_n} \tag{5.3.42}$$

iff the series

$$\sum_{k=1}^{\infty} \frac{(-1)^k \displaystyle\prod_{m=1}^{k} a_m}{y_k y_{k+1}} \tag{5.3.43}$$

tends to ∞.

Proof. Since $y_0 \neq 0$, (5.3.10) can be used to conclude that (5.3.42) holds iff y_{n+1}/B_n tends to zero, that is, B_n/y_{n+1} tends to ∞. Now

$$\frac{B_n}{y_{n+1}} = \sum_{k=0}^{n} \left(\frac{B_k}{y_{k+1}} - \frac{B_{k-1}}{y_k} \right) = \sum_{k=0}^{n} \left(\frac{B_k y_k - B_{k-1} y_{k+1}}{y_k y_{k+1}} \right).$$

Using (5.3.9) and the determinant formulas (2.1.9) then leads to

$$\frac{B_n}{y_{n+1}} = \frac{y_0}{y_0 y_1} + y_0 \sum_{k=1}^{n} \frac{(-1)^k \displaystyle\prod_{m=1}^{k} a_m}{y_k y_{k+1}}. \qquad \blacksquare$$

If the recurrence relation (5.2.2) is generalized for an arbitrary normed field $\mathbb{F}$, it becomes

$$P_n = b_n P_{n+1} + a_{n+1} P_{n+2}, \qquad n = 0, 1, 2, \ldots, \tag{5.3.44}$$

where the a_n, b_n, P_n are elements of $\mathbb{F}$, with

$$a_n \neq 0, \qquad n=1,2,3,\ldots. \tag{5.3.45}$$

This is a different three-term recurrence relation than the one studied in this section [(5.3.1), (5.3.5) and (5.3.41)]. The bridge between the two is given by

$$y_0 = P_1; \qquad y_n = (-1)^n \prod_{k=1}^n a_k P_{n+1}, \quad n=1,2,3,\ldots. \tag{5.3.46}$$

By this we mean that if $\{y_n\}$ and $\{P_n\}$ are related by (5.3.46), then $\{y_n\}$ is a solution of (5.3.1) iff $\{P_n\}$ is a solution of (5.3.44). In view of this it is not surprising that Theorem 5.5 (as well as Theorem 5.2) can be derived from our present results. In terms of the P_n, $L(a_n)$ and $L(b_n)$ in (5.2.2), Auric's criterion (5.3.43) becomes that

$$\sum_{n=2}^{\infty} \frac{(-1)^n}{\left(\displaystyle\prod_{k=1}^n L(a_k)\right) P_n P_{n+1}} \tag{5.3.47}$$

tends to infinity in the normed field $\mathbb{F}$. Let d_n be defined by

$$d_n = \frac{(-1)^n}{\displaystyle\prod_{k=1}^n L(a_k) P_n P_{n+1}}, \qquad n=2,3,4,\ldots.$$

Then $\sum_{n=2}^m d_n \to \infty$ is equivalent to $\|1/\sum_{n=2}^m d_n\| \to 0$. This in turn is true iff

$$\lambda\!\left(\sum_{n=2}^m d_n\right) \to -\infty \qquad \text{as} \quad m\to\infty.$$

Now from (5.2.2) we have

$$\lambda\!\left(\frac{P_n}{P_{n+1}}\right) = \lambda\!\left(L(b_n) + L(a_{n+1})\frac{P_{n+2}}{P_{n+1}}\right). \tag{5.3.48}$$

From (5.2.5b) in Theorem 5.5 we deduce

$$\lambda\!\left(L(a_{n+1})\frac{P_{n+2}}{P_{n+1}}\right) > \lambda(L(b_n)). \tag{5.3.49}$$

Hence by (5.1.6), (5.3.48) and (5.3.49),

$$\lambda\left(\frac{P_n}{P_{n+1}}\right) = \lambda(L(b_n)),$$

or, equivalently,

$$\lambda(P_{n+1}) - \lambda(P_n) = -\lambda(L(b_n)).$$

This implies that

$$\lambda(P_{n+2}) - \lambda(P_n) = -\lambda(L(b_n)) - \lambda(L(b_{n+1})). \tag{5.3.50}$$

Applying (5.2.5a) to (5.3.50) yields

$$\lambda(P_n) - \lambda(P_{n+2}) - \lambda(L(a_{n+1})) < 0. \tag{5.3.51}$$

An easy calculation leads to

$$\lambda(d_{n+1}) - \lambda(d_n) = \lambda(P_n) - \lambda(P_{n+2}) - \lambda(L(a_{n+1})).$$

Making use of (5.3.51), we then conclude that

$$\lambda(d_{n+1}) < \lambda(d_n).$$

This allows us to use (5.1.6) to obtain

$$\lambda\left(\sum_{n=2}^{m} d_n\right) = \lambda(d_m),$$

and hence

$$\lambda\left(\sum_{n=2}^{m} d_n\right) = -\lambda(P_m) - \lambda(P_{m+1}) - \sum_{k=1}^{m} \lambda(L(a_k)). \tag{5.3.52}$$

Summing (5.3.51) over $n = 0, 1, \ldots, m-1$, and using the fact that the left side of (5.3.51) is an integer, we have the inequality

$$\lambda(P_m) + \lambda(P_{m+1}) + \sum_{k=1}^{m} \lambda(L(a_k)) \geqslant m + \lambda(P_0) + \lambda(P_1).$$

Thus the right side of (5.3.52) tends to $-\infty$ as $m \to \infty$, and therefore (5.3.47) tends to infinity. Hence Auric's theorem (Theorem 5.10) proves Theorem 5.5(A) as asserted.

5.4   Uniform Convergence

A sequence $\{R_n(z)\}$ of functions meromorphic in a domain D is said to *converge uniformly on a compact subset K of D* if:

1. there exists $N(K)$ such that $R_n(z)$ is holomorphic in some domain containing K for all $n \geqslant N(K)$, and
2. given $\varepsilon > 0$ there exists $N_\varepsilon > N(K)$ such that

$$\sup_{z \in K} |R_{n+k}(z) - R_n(z)| < \varepsilon \qquad \text{for} \quad n > N_\varepsilon, \quad k > 0. \qquad (5.4.1)$$

The sequence $\{R_n(z)\}$ is said to be *uniformly bounded on a compact subset K of D* if there exist $M(K)$ and $B(K)$ such that

$$\sup_{z \in K} |R_n(z)| < B(K) \qquad \text{for} \quad n \geqslant M(K). \qquad (5.4.2)$$

In Theorem 5.11 it is shown that, subject to certain restrictions, a sequence $\{R_n(z)\}$ which corresponds to an fLs L will be uniformly convergent iff it is uniformly bounded. First, however, we point out that correspondence of a sequence of functions $\{R_n(z)\}$ is not a necessary condition for uniform convergence. For example, the sequence $\{R_n(z)\}$ defined by

$$R_n(z) = 1 + \sum_{k=1}^{\infty} \left(\frac{1}{k!} + \frac{1}{n^2} \right) z^k, \qquad n = 1, 2, 3, \ldots \qquad (5.4.3)$$

converges uniformly to the function e^z on the closed neighborhood of the origin $[z : |z| \leqslant \rho < 1]$. However, $\{R_n(z)\}$ does not correspond to

$$L(e^z) = 1 + \frac{z}{1!} + \frac{z^2}{2!} + \frac{z^3}{3!} + \cdots, \qquad (5.4.4)$$

the Taylor series expansion of e^z at $z = 0$.

THEOREM 5.11. *Let $\{R_n(z)\}$ be a sequence of functions meromorphic at the origin and corresponding (at $z = 0$) to an fLs*

$$L = c_m z^m + c_{m+1} z^{m+1} + c_{m+2} z^{m+2} + \cdots, \qquad c_m \neq 0. \qquad (5.4.5)$$

Further suppose that there exists a deleted neighborhood of the origin $D^* = [z : 0 < |z| < \delta]$ such that each $R_n(z)$ is holomorphic in D^*. Let D be a domain containing D^*; if $m < 0$, we require that $0 \notin D$. Then:

(A) *$\{R_n(z)\}$ converges uniformly on every compact subset of D iff $\{R_n(z)\}$ is uniformly bounded on every compact subset of D.*

(B) *If* $\{R_n(z)\}$ *converges uniformly on every compact subset of D, then the function*

$$f(z) = \lim_{n \to \infty} R_n(z)$$

is holomorphic in D, and $L = L(f)$.

Proof. (A): That uniform convergence implies uniform boundedness follows from a standard argument which need not be repeated here. Now suppose that $\{R_n(z)\}$ is uniformly bounded on every compact subset of D. Let K be an arbitrary compact subset of D. Let K_0 be an open, connected, bounded subset of D such that

$$K \subset K_0 \subset c(K_0) \subset D,$$

and such that K_0 contains an annulus

$$A = [z : \eta < |z| < 5\eta < \delta].$$

Since K_0 is assumed to be bounded, $c(K_0)$ is a compact subset of D. Hence there exist an n_0 and an M, both depending upon $c(K_0)$, such that

$$\sup_{z \in c(K_0)} |R_n(z)| < M \qquad \text{for} \quad n > n_0. \tag{5.4.6}$$

Since $R_n(z)$ is meromorphic at the origin and holomorphic in D^*, it can be represented by its convergent Laurent series

$$L(R_n) = \sum_{k=m_n}^{\infty} \gamma_k^{(n)} z^k \qquad \text{for} \quad z \in A, \tag{5.4.7}$$

where

$$\gamma_k^{(n)} = \frac{1}{2\pi i} \int_C \frac{R_n(\zeta)}{\zeta^{k+1}} d\zeta, \quad k = m_n, m_n + 1, \ldots; \qquad \gamma_{m_n}^{(n)} \neq 0, \tag{5.4.8}$$

and where C is the circular path $|\zeta| = 4\eta$ traversed once in the counter-clockwise direction. It follows from (5.4.6) and (5.4.8) that

$$|\gamma_k^{(n)}| < \frac{M}{(4\eta)^k} \qquad \text{for} \quad k > m_n \text{ and } n > n_0. \tag{5.4.9}$$

We note in passing that the assumption that each $R_n(z)$ is holomorphic in D^* was made to insure that $L(R_n)$ would be the Laurent expansion of

$R_n(z)$ in the annulus A. From (5.4.7) we see that

$$\lambda(L(R_n)) = m_n, \qquad n = 0, 1, 2, \ldots. \qquad (5.4.10)$$

Further

$$\lambda(L(R_{j+n}) - L(R_n)) = \lambda(L(R_{j+n}) - L + L - L(R_n))$$

$$\geqslant \min\left[\lambda(L(R_{j+n}) - L), \lambda(L - L(R_n))\right],$$

$$\text{for} \quad n \geqslant 0 \text{ and } j > 0. \quad (5.4.11)$$

Therefore, since $\lambda(L - L(R_n)) \to \infty$ as $n \to \infty$, given any N, there exists an $n_1 > n_0$ such that

$$\lambda(L(R_{j+n}) - L(R_n)) > N \qquad \text{for} \quad n \geqslant n_1 \text{ and } j \geqslant 0. \quad (5.4.12)$$

It follows from (5.4.7), (5.4.9) and (5.4.12) that for $n \geqslant n_1$ and $j \geqslant 0$,

$$\sup_{\eta < |z| < 2\eta} |R_{j+n}(z) - R_n(z)| \leqslant \sup_{\eta < |z| < 2\eta} \sum_{k=N}^{\infty} |(\gamma_k^{(j+n)} - \gamma_k^{(n)}) z^k|$$

$$\leqslant \sum_{k=N}^{\infty} \frac{2M}{(4\eta)^k} (2\eta)^k = \frac{4M}{2^N}. \qquad (5.4.13)$$

Thus we see that $\{R_n(z)\}$ is a uniform Cauchy sequence on the set $[z: \eta < |z| < 2\eta]$. An application of the Stieltjes-Vitali theorem (Theorem 4.30) completes the proof that $\{R_n(z)\}$ converges uniformly on $c(K_0)$ and hence on K.

(B): Suppose now that $\{R_n(z)\}$ is uniformly convergent on all compact subsets of D. Define

$$L_n = c_m z^m + c_{m+1} z^{m+1} + \cdots + c_{m+n} z^{m+n}, \qquad n = 0, 1, 2, \ldots. \qquad (5.4.14)$$

Then

$$\lambda(L_n - L(R_n)) = \lambda((L_n - L) + (L - L(R_n)))$$

$$\geqslant \min\left[m + n + 1, \lambda(L - L(R_n))\right].$$

Hence

$$\lim_{n \to \infty} \lambda(L_n - L(R_n)) = \infty, \qquad (5.4.15)$$

and for every $k \geqslant m$ there exists l_k such that

$$c_k = \gamma_k^{(l)} \quad \text{for} \quad l > l_k. \tag{5.4.16}$$

Now let K^* be an arbitrary compact subset of $D^* = [z : 0 < |z| < \delta]$. Then there exist ε and η^* such that

$$0 < \varepsilon < |z| < \eta^* < \delta \quad \text{for} \quad z \in K^*.$$

The set $K_1 = [z : \varepsilon \leqslant |z| \leqslant \eta^*]$ is a compact subset of D^*. Let M_1 be the bound belonging to K_1; that is,

$$\sup_{z \in K_1} |R_n(z)| \leqslant M_1 \quad \text{for} \quad n \geqslant M(K_1). \tag{5.4.17}$$

The existence of the bound M_1 follows from part (A) already proved. Let ρ be chosen so that $\varepsilon < \rho < \eta^*$ and

$$\mu = \max_{z \in K^*} \left| \frac{z}{\rho} \right| < 1. \tag{5.4.18}$$

Let C_1 denote the path $|z| = \rho$ traversed one time in the counterclockwise direction. Then the coefficients $\gamma_k^{(n)}$ in (5.4.7) can be written as

$$\gamma_k^{(n)} = \frac{1}{2\pi i} \int_{C_1} \frac{R_n(\zeta)}{\zeta^{k+1}} d\zeta \quad \text{for} \quad k \geqslant m_n. \tag{5.4.19}$$

It follows from (5.4.16), (5.4.17) and (5.4.19) that for $l > l_k$,

$$|c_k| = |\gamma_k^{(l)}|,$$

$$= \frac{1}{2\pi} \left| \int_{C_1} \frac{R_l(\zeta)}{\zeta^{k+1}} d\zeta \right| < \frac{M_1}{\rho^k}. \tag{5.4.20}$$

Thus we have, for all $z \in K^*$,

$$|L_n(z)| \leqslant \sum_{k=m}^{m+n} |c_k z^k|$$

$$\leqslant \sum_{k=m}^{m+n} |\gamma_k^{(l)} z^k| \quad \text{for} \quad l > l_k$$

$$\leqslant M_1 \sum_{k=m}^{m+n} \left| \frac{z}{\rho} \right|^k. \tag{5.4.21}$$

If $m < 0$, then by (5.4.18) and (5.4.21),

$$|L_n(z)| < M_1 \left(\sum_{k=m}^{-1} \nu^k + \frac{1}{1-\mu} \right)$$

$$\text{for} \quad z \in K^*, \quad \nu = \min_{z \in K^*} \left| \frac{z}{\rho} \right| > 0, \quad (5.4.22a)$$

and if $m = 0$, then

$$|L_n(z)| < \frac{M_1}{1-\mu} \quad \text{for} \quad z \in K^*. \quad (5.4.22b)$$

We have shown that the sequence $\{L_n(z)\}$ is uniformly bounded on every compact subset of D^*. Since the $L_n(z)$ are rational functions, holomorphic in D^*, and since

$$\lambda(L - L_n) = m + n + 1 \to \infty \quad \text{as} \quad n \to \infty,$$

by part (A) of the theorem $\{L_n(z)\}$ converges uniformly on every compact subset of D^* to a function $f(z)$ holomorphic in D^*. Clearly $L = L(f)$. Now, with $\tau_n = \lambda(L_n - L(R_n))$

$$|L_n(z) - R_n(z)| < \sum_{k=\tau_n}^{\infty} |(c_k - \gamma_k^{(n)})z^k|$$

$$< \sum_{k=\tau_n}^{\infty} 2M_1 \left| \frac{z}{\rho} \right|^k, \quad \text{by (5.4.20)}$$

$$< \frac{2M_1 \mu^{\tau_n}}{1-\mu} \quad \text{for } z \in K^*, \quad \text{by (5.4.19).} \quad (5.4.23)$$

Since $0 < \mu < 1$ and $\tau_n \to \infty$ as $n \to \infty$ by (5.4.15), it follows from (5.4.23) that $\{L_n(z) - R_n(z)\}$ converges uniformly to 0 on K^*. Since

$$|f(z) - R_n(z)| < |f(z) - L_n(z)| + |L_n(z) - R_n(z)|,$$

we can conclude that $\{R_n(z)\}$ converges uniformly to $f(z)$ on all compact subsets of D^*. The extension to D can be obtained by analytic continuation. ∎

The following is an immediate corollary of Theorem 5.11.

COROLLARY 5.12. *Let $\{R_n(z)\}$ and $\{Q_n(z)\}$ be two sequences of functions meromorphic at the origin which correspond to the same fLs L. If both*

sequences converge uniformly on every compact subset of a domain D contain-
ing a deleted neighborhood of the origin, say D^*, and if $\lambda(L)<0$ implies
$0 \notin D$, and if all $R_n(z)$ and $Q_n(z)$ are holomorphic in D^*, then $\{R_n(z)\}$ and
$\{Q_n(z)\}$ converge to the same function $f(z)$ which is holomorphic in D and for
which $L=L(f)$.

For $\lambda(L)=m \geqslant 0$, the statement of Theorem 5.11 becomes sufficiently
simpler that it is worth stating separately as follows:

THEOREM 5.13. *Let $\{R_n(z)\}$ be a sequence of functions meromorphic at the*
origin which corresponds to a formal power series

$$P=c_0+c_1z+c_2z^2+\cdots.$$

Let D be a domain containing the origin. Then:

(A) *$\{R_n(z)\}$ converges uniformly on every compact subset of D iff $\{R_n(z)\}$*
is uniformly bounded on every compact subset of D.

(B) *If $\{R_n(z)\}$ converges uniformly on every compact subset of D, then*
$f(z)=\lim_{n\to\infty}R_n(z)$ is holomorphic on D and $P=L(f)$ is the Taylor series of
$f(z)$ at $z=0$.

The following example throws some light on what can happen if some of
the conditions in Theorems 5.11 and 5.13 are not met. Let $R_n(z)$ be
defined by

$$R_n(z)=\frac{1}{1-(nz)^n}, \qquad n=1,2,3,\ldots.$$

Then

$$L(R_n)=1+(nz)^n+(nz)^{2n}+\cdots,$$

so that $\{R_n(z)\}$ corresponds to $L=1$. Each $R_n(z)$ has n poles on the circle
$|z|=1/n$, but $\{R_n(z)\}$ is uniformly bounded on every compact subset of
$0<|z|$. Finally, $\{R_n(z)\}$ converges to 0 for $0<|z|$.

Part (A) of Theorem 5.13 was proved for regular C-fractions and
associated continued fractions by Pringsheim [1910]. For C-fractions, Theorem 5.13 and Corollary 5.12 were established by Leighton and Scott
[1939]; for T-fractions, the same three results can be found in [Thron,
1948]. The corresponding results for P-fractions were pointed out in [Jones
and Thron, 1979], where the more general result in Theorem 5.11 was
given. Further applications of Theorem 5.13 are given by the following two
theorems which are extensions of Theorems 4.55 and 4.56, respectively. We
recall the convention made in Section 4.5.2 that a regular C-fraction
$K(a_n z/1)$ and all of its approximants have value zero at $z=0$.

THEOREM 5.14. *Let* $1+K(a_n z/1)$ *be a regular C-fraction such that*

$$\lim_{n\to\infty} a_n = 0 \qquad (a_n \neq 0). \tag{5.4.24}$$

Then:

(A) $1+K(a_n z/1)$ *converges to a meromorphic function* $f(z)$.
(B) *The convergence is uniform on every compact subset K of* $\mathbb{C}$ *which contains no poles of* $f(z)$.
(C) $f(z)$ *is holomorphic at* $z=0$, *and* $f(0)=1$.

Proof. (A): Let K be an arbitrary compact subset of $\mathbb{C}$ which contains no poles of $f(z)$. Let M be chosen such that

$$K \subseteq G_M = [z : |z| < M].$$

By Theorem 4.55(A) there exists an n_M such that the continued fraction

$$\frac{a_{n_M} z}{1} + \frac{a_{n_M+1} z}{1} + \frac{a_{n_M+2} z}{1} + \cdots \tag{5.4.25}$$

converges uniformly on every compact subset of G_M to a function $F(z)$ holomorphic in G_M. By Corollary 5.3(A) the C-fraction (4.5.25) corresponds to a formal power series

$$L_0 = c_1 z + c_2 z^2 + c_3 z^3 + \cdots.$$

By Theorem 5.13(B), L_0 is the Taylor series expansion of $F(z)$ at $z=0$. Since the C-fraction (5.4.25) is non-terminating, it follows from Corollary 5.3 that $F(z)$ cannot be a rational function. It follows from this together with Theorem 4.55(C) that the continued fraction $1+K(a_n z/1)$ converges to a meromorphic function $f(z)$.

(B): Now let $f_k(z)$, $A_k(z)$ and $B_k(z)$ denote the kth approximant, numerator and denominator, respectively, of $1+K(a_n z/1)$ and let $F_k(z)$ denote the kth approximant of (5.4.25). Then by (2.1.7),

$$f(z) = \frac{A_{n_M}(z) + F(z) A_{n_M-1}(z)}{B_{n_M}(z) + F(z) B_{n_M-1}(z)} \tag{5.4.26}$$

and

$$f_{n_M+k}(z) = \frac{A_{n_M}(z) + F_k(z) A_{n_M-1}(z)}{B_{n_M}(z) + F_k(z) B_{n_M-1}(z)}, \qquad k = 0, 1, 2, \ldots. \tag{5.4.27}$$

It follows that

$$|f(z)-f_{n_M+k}(z)|=\frac{|A_{n_M}(z)B_{n_M-1}(z)-A_{n_M-1}(a)B_{n_M}(z)|\cdot|F(z)-F_k(z)|}{|B_{n_M}(z)+F(z)B_{n_M-1}(z)|\cdot|B_{n_M}(z)+F_k(z)B_{n_M-1}(z)|}$$

$$\text{for } k \geqslant n_M. \quad (5.4.28)$$

Since $\{F_k(z)\}$ converges uniformly to $F(z)$ on compact subsets of G_M and since the denominator of (5.4.26) vanishes only at the poles of $f(z)$, the denominator on the right side of (5.4.28) is uniformly bounded away from zero for all k sufficiently large and $z \in K$. Hence $\{f_k(z)\}$ converges uniformly on K to $f(z)$.

(C): Let $S = \sup[|a_n|: n=1,2,3,...]$. Then $|a_n z| < \frac{1}{4}$ if $|z| < 1/4S$. It follows from the convergence-neighborhood theorem (Theorem 4.45) that the regular C-fraction $1 + K(a_n z/1)$ converges uniformly on the set $[z:|z| \leqslant 1/4S]$. By Worpitzky's criterion [Corollary 4.36(B)] the approximants of the continued fraction are finite-valued and hence holomorphic functions of z. Thus the limit $f(z)$ is holomorphic in a neighborhood of $z=0$. ∎

By a similar argument one can prove the following:

THEOREM 5.15. *Let* $1 + K(a_n z/1)$ *be a regular C-fraction such that*

$$\lim_{n \to \infty} a_n = a \neq 0, \quad (5.4.29)$$

where a is a complex constant, and let

$$R_a = \left[z: \left| \arg\left(az + \tfrac{1}{4} \right) \right| < \pi \right] \quad (5.4.30)$$

(see Figure 5.4.1). Then:

(A) *The continued fraction converges to a function* $f(z)$ *meromorphic in* R_a.
(B) *The convergence is uniform on every compact subset* K *of* R_a *which contains no poles of* $f(z)$.
(C) $f(z)$ *is holomorphic at* $z=0$.

We consider now an application of Theorems 5.13 and 5.14 to obtain a continued-fraction representation of the ratios of Bessel functions.

THEOREM 5.16. *Let m be a non-negative integer, and let ν be a complex number not contained in the set* $[-(m+n): n=1,2,3,...]$. *Then:*

(A)

$$f(z) = \frac{J_{\nu+m+1}(z)}{J_{\nu+m}(z)} \quad (5.4.31)$$

is a meromorphic function of z, holomorphic at $z=0$. Here $J_\nu(z)$ denotes the Bessel function of the first kind of order ν, defined by (5.2.12).

  (B) *Let D be the domain containing all points of $\mathbb{C}$ which are not poles of $f(z)$. For every $z \in D$, the continued fraction*

$$\cfrac{z}{2(\nu+m+1)} - \cfrac{z^2}{2(\nu+m+2)} - \cfrac{z^2}{2(\nu+m+3)} - \cdots \qquad (5.4.32)$$

converges to $f(z)$; the convergence is uniform on every compact subset of D.

  Proof. (A) follows directly from the series definition of $J_\nu(z)$ in (5.2.12). (B): Let $w=z^2$, and consider the continued fraction

$$\cfrac{w}{2(\nu+m+1)} - \cfrac{w}{2(\nu+m+2)} - \cfrac{w}{2(\nu+m+3)} - \cdots, \qquad (5.4.33)$$

which is equivalent to the regular C-fraction

$$\cfrac{\dfrac{w}{2(\nu+m+1)}}{1} + \cfrac{\dfrac{-w}{4(\nu+m+1)(\nu+m+2)}}{1} + \cfrac{\dfrac{-w}{4(\nu+m+2)(\nu+m+3)}}{1} + \cdots. $$

$$(5.4.34)$$

By Theorem 5.14 the continued fraction (5.4.34) [and hence also (5.4.33)] converges to a meromorphic function $G(w)$, and the convergence is uniform on every compact subset of the w-plane which contains no poles of $G(w)$. Moreover $G(w)$ is holomorphic at $w=0$. Since $G(z^2)$ has a zero of order 2 at $z=0$, it follows that the continued fraction (5.4.32) converges to a meromorphic function $h(z)$, holomorphic at $z=0$, and the convergence is uniform on every compact subset of the z-plane which contains no poles of $h(z)$. By Corollary 5.6 the continued fraction (5.4.32) corresponds to the function $f(z)$ at $z=0$ and hence to the Taylor series expansion $L(f)$ of $f(z)$ at $z=0$. Thus by Theorem 5.13(B), $L(f)=L(h)$. Since $L(h)$ is the Taylor series expansion of $h(z)$ at $z=0$, $h(z)=f(z)$. ∎

  Theorem 5.16 illustrates a general method for obtaining continued-fraction representations of analytic functions. We note that the result given in Theorem 5.16 is more general than that derived from Pincherle's theorem in Corollary 5.8. In that case ν was restricted to the real interval $0 \leqslant \nu < 1$. Many more applications of the results developed in this chapter will be given in Chapter 6. However, first we shall define the Padé table and discuss some of its relations to the present chapter.

5.5 Padé Table

The concept of the Padé table originated with Frobenius [1881], who developed the early algorithmic aspects of the subject. The attribution to Padé seems to be due to work in his thesis [Padé, 1892], which came later, on certain abnormal situations that can arise. Because of its intimate connection with the analytic theory of continued fractions, the Padé table was dealt with briefly in the books by Wall [1948] and Perron [1957a]. The more recent stimulation of interest in Padé tables is the result of work by D. Shanks [1955] and P. Wynn [1956], which led to Wynn's epsilon algorithm for computing Padé approximants. In this section we define the Padé table and certain generalizations, summarize some of the classical properties and describe its connection with two classes of continued fractions. For more details on Padé tables the reader can refer to the review articles [Baker, 1965; Chui, 1976; Gragg, 1972], books [Baker and Gammel, 1970; Baker, 1975; Baker and Graves-Morris, to appear; Gilewicz, 1978], and conference proceedings [Jones and Thron, 1974c; Graves-Morris, 1973; Cabannes, 1976; Saff and Varga, 1977; and Wuytack, 1979]. An extensive bibliography on Padé approximants has recently been given by Brezinski [1977].

5.5.1 Padé Approximants

If $u(z)$ and $v(z)$ are polynomials in z, $v(z)$ not identically zero, then (u, v) is called a *rational expression*. Two rational expressions (u, v) and (u^*, v^*) are called *equivalent*, denoted by $(u, v) \sim (u^*, v^*)$, iff

$$u(z)v^*(z) \equiv u^*(z)v(z); \tag{5.5.1}$$

they are said to be *equal*, denoted by $(u, v) = (u^*, v^*)$, iff there exists a non-zero complex number a such that

$$a \cdot u(z) \equiv u^*(z) \quad \text{and} \quad a \cdot v(z) \equiv v^*(z). \tag{5.5.2}$$

Equivalence and equality of rational expressions are both equivalence relations. Equality of two rational expressions implies equivalence, but the converse is not true. The equivalence class of all rational expressions equivalent to a given rational expression (u, v) determine a unique *rational function R*, represented by

$$R(z) = \frac{P(z)}{Q(z)}, \tag{5.5.3}$$

where (P, Q) is the rational expression (uniquely determined up to equality) such that $(P, Q) \sim (u, v)$ and P and Q are relatively prime. A rational expression (u, v) is said to be of *type* $[m, n]$ if the degree of u is at most m and the degree of v is at most n. A rational function (5.5.3) is said to be of *type* $[m, n]$ if the rational expression (P, Q) is of type $[m, n]$.

We now consider the following *Hermite rational interpolation problem*: For a given

$$L = c_0 + c_1 z + c_2 z^2 + \cdots \qquad (c_0 \neq 0) \tag{5.5.4}$$

and a given pair of non-negative integers (m, n), can we find a rational function (5.5.3) of type $[m, n]$ such that the Taylor expansion of $R(z)$ at $z = 0$ has the form

$$L(R) = c_0 + c_1 z + \cdots + c_{m+n} z^{m+n} + \gamma_{m+n+1}^{(m, n)} z^{m+n+1} + \cdots; \tag{5.5.5}$$

that is, such that

$$\lambda(L - L(R)) \geqslant m + n + 1? \tag{5.5.6}$$

If L converges in an open neighborhood of $z = 0$ to a holomorphic function $f(z)$, then (5.5.6) is equivalent to the conditions

$$R^{(j)}(0) = f^{(j)}(0), \qquad j = 0, 1, \ldots, m + n. \tag{5.5.7}$$

This interpolation problem cannot always be solved, but it is closely related to the following problem which always has a solution.

For a given fps (5.5.4) and a given pair of non-negative integers (m, n), can we find a rational expression (u, v) of type $[m, n]$ such that

$$\lambda(vL - u) \geqslant m + n + 1? \tag{5.5.8}$$

A solution (u, v) to this problem always exists and determines a unique rational function

$$R_{m, n}(z) = \frac{P_{m, n}(z)}{Q_{m, n}(z)} \qquad (Q_{m, n}(0) \neq 0) \tag{5.5.9}$$

of type $[m, n]$, where $P_{m, n}$ and $Q_{m, n}$ are relatively prime polynomials with $(P_{m, n}, Q_{m, n}) \sim (u, v)$. This can be seen as follows: Let u and v be polynomials of the forms

$$u(z) = a_0 + a_1 z + \cdots + a_m z^m, \qquad v(z) = b_0 + b_1 z + \cdots + b_n z^n, \tag{5.5.10}$$

respectively. The condition (5.5.8) will hold iff the following equations are satisfied:

$$c_0 b_0 = a_0,$$
$$c_1 b_0 + c_0 b_1 = a_1,$$
$$\vdots \qquad\qquad\qquad\qquad (5.5.11a)$$
$$c_m b_0 + c_{m-1} b_1 + \cdots + c_0 b_m = a_m,$$
$$c_{m+1} b_0 + c_m b_1 + \cdots + c_{m-n+1} b_n = 0,$$
$$\vdots \qquad\qquad\qquad\qquad (5.5.11b)$$
$$c_{m+n} b_0 + c_{m+n-1} b_1 + \cdots + c_m b_n = 0.$$

Here we have set $c_k = 0$ if $k < 0$ and $b_k = 0$ if $k > n$. Clearly the homogeneous system (5.5.11b) has a non-trivial solution $b_0, b_1, \ldots, b_n$. Substituting these values into (5.5.11a) determines $a_0, a_1, \ldots, a_m$. Hence there exists a rational expression (u, v) of type $[m, n]$ satisfying (5.5.8). Now suppose that (u^*, v^*) is another rational expression of type $[m, n]$ satisfying

$$\lambda(v^* L - u^*) \geq m + n + 1.$$

It follows that

$$\lambda(v^* v L - v^* u) \geq m + n + 1.$$

These inequalities imply that

$$\lambda(u v^* - u^* v) \geq m + n + 1.$$

Since $u v^* - u^* v$ is a polynomial of degree at most $m + n$, (5.5.1) is satisfied and hence $(u^*, v^*) \sim (u, v)$. Thus the rational expressions (u, v) of type $[m, n]$ satisfying (5.5.8) form an equivalence class which determines the unique (up to equality) rational function (5.5.9) of type $[m, n]$. The rational function (5.5.9) is called the (m, n) *Padé approximant of L*. The two-dimensional array of functions

$$
\begin{array}{cccc}
R_{0,0} & R_{0,1} & R_{0,2} & \cdots \\
R_{1,0} & R_{1,1} & R_{1,2} & \cdots \\
R_{2,0} & R_{2,1} & R_{2,2} & \cdots \\
\vdots & \vdots & \vdots &
\end{array}
\qquad (5.5.12)
$$

is called the *Padé table of L*. That $Q_{m,n}(0) \neq 0$ follows from the fact that $P_{m,n}(z)$ and $Q_{m,n}(z)$ are relatively prime, and that $(P_{m,n}, Q_{m,n}) \sim (u, v)$ where (u, v) satisfies (5.5.8). It can also be seen that if the Hermite rational-interpolation problem has a solution, then it is given by the (m, n) Padé approximant $R_{m,n}$. In fact, suppose that (P, Q) is a rational expression of type $[m, n]$ with P and Q relatively prime, and suppose that the rational function

$$R(z) = \frac{P(z)}{Q(z)}$$

satisfies (5.5.6). It follows easily that

$$\lambda(QL - P) \geqslant m + n + 1.$$

Therefore, since P and Q are relatively prime, $R = R_{m,n}$ as asserted.

Some of the more significant known properties of Padé tables will now be stated. In general a given rational function can occur in more than one place in a Padé table; however, the set of all such places forms a square block. This property of the block structure is given by the following theorem. For a proof see [Gragg, 1972].

THEOREM 5.17 (Block theorem). *Suppose that a rational function*

$$R(z) = \frac{P(z)}{Q(z)},$$

where $P(z)$ and $Q(z)$ are relatively prime polynomials, occurs at some place in the Padé table of a fps L. Further suppose that the degrees of P and Q are m and n, respectively. Then:

(A) *The set of all places in the Padé table of L in which $R(z)$ occurs is a square block.*

(B) *If*

$$\lambda(QL - P) = m + n + r + 1,$$

then $r \geqslant 0$ and the square block consists of the $(r + 1)^2$ places

$$(m + r_1, n + r_2), \qquad r_1, r_2 = 0, 1, \ldots, r.$$

Here $r = \infty$ is possible, and this means that $QL - P = 0$, the zero fps.

The important special case in which a square block contains only one place is discussed in the following paragraph.

An (m, n) Padé approximant (5.5.9) is said to be *normal* if the degrees of $P_{m,n}$ and $Q_{m,n}$ (recall that $P_{m,n}$ and $Q_{m,n}$ are relatively prime) are exactly m and n, respectively, and

$$\lambda(Q_{m,n}L - P_{m,n}) = m + n + 1. \qquad (5.5.13)$$

We note that (5.5.13) means that $Q_{m,n}L - P_{m,n}$ has the form

$$Q_{m,n}L - P_{m,n} = d_{m+n+1}z^{m+n+1} + d_{m+n+2}z^{m+n+2} + \cdots, \quad (5.5.14a)$$

where

$$d_{m+n+1} \neq 0. \quad (5.5.14b)$$

The fps L and its Padé table are said to be *normal* if every Padé approximant is normal. It can be shown that a rational function occurs in exactly one place in a Padé table iff it is a normal Padé approximant. If the (m, n) approximant $R_{m,n} = P_{m,n}/Q_{m,n}$ is normal, then it solves the rational Hermite interpolation problem. In fact,

$$\lambda(L - L(R_{m,n})) = m+n+1$$

follows from (5.5.13) and the fact that $Q_{m,n}(0) \neq 0$. The determinants

$$c_{m,n} = \begin{vmatrix} c_m & c_{m-1} & \cdots & c_{m-n+1} \\ c_{m+1} & c_m & \cdots & c_{m-n+2} \\ \vdots & \vdots & & \vdots \\ c_{m+n-1} & c_{m+n-2} & \cdots & c_m \end{vmatrix},$$

$$m, n = 0, 1, 2, \ldots \quad (c_k = 0 \text{ if } k < 0), \quad (5.5.15a)$$

$$c_{m,0} = 1, \quad m = 0, 1, 2, \ldots, \quad (5.5.15b)$$

associated with a fps (5.5.4) are useful for characterizing normal Padé approximants. This is done by the following theorem, a proof of which can be found in [Gragg, 1972].

THEOREM 5.18.

(A) *An (m, n) Padé approximant $R_{m,n}(z)$ of a fps*

$$L = c_0 + c_1 z + c_2 z^2 + \cdots \quad (c_0 \neq 0)$$

is normal iff the determinants

$$c_{m,n}, \quad c_{m,n-1}, \\ c_{m+1,n}, \quad c_{m+1,n+1}$$

are all non-zero.

(B) *The fps L and its Padé table are normal iff*

$$c_{m,n} \neq 0, \quad m, n = 0, 1, 2, \ldots.$$

In particular, each $c_{m,1} = c_m$ must be non-zero.

(C) *The fps L is the Taylor series at $z=0$ of an irreducible rational function*

$$R(z)=\frac{P(z)}{Q(z)},$$

with P and Q of degrees M and N, respectively, iff

$$c_{m,N}\neq 0, \qquad m \geqslant M,$$
$$c_{M,n}\neq 0, \qquad n \geqslant N,$$
$$c_{m,n}=0, \qquad m > M \text{ and } n > N.$$

Following Gragg, we shall say that a fps L is (M,N)-*normal* if, in addition to the conditions of Theorem 5.18, part (C), we also have

$$c_{m,n}\neq 0, \qquad 0 \leqslant m \leqslant M \text{ or } 0 \leqslant n \leqslant N. \tag{5.5.16}$$

If a Padé table is normal, then the "staircase" sequence of approximants

$$
\begin{array}{cc}
R_{0,0} & \\
R_{1,0} & R_{1,1} \\
& R_{2,1} \quad R_{2,2} \\
& \qquad R_{3,1} \quad \ddots \\
& \qquad\qquad \ddots
\end{array}
$$

gives the successive approximants of a regular C-fraction. This connection between continued fractions and Padé approximants is described more fully by the following:

THEOREM 5.19.

(A) *Let*

$$L = 1 + c_1 z + c_2 z^2 + \cdots \tag{5.5.17}$$

be a fps for which the Padé approximants of the staircase sequence

$$R_{0,0}, R_{1,0}, R_{1,1}, R_{2,1}, R_{2,2}, R_{3,2}, \cdots \tag{5.5.18}$$

are all normal. Then there exists a regular C-fraction

$$1 + \frac{a_1 z}{1} + \frac{a_2 z}{1} + \frac{a_3 z}{1} + \cdots \qquad (a_n \neq 0) \tag{5.5.19}$$

with n *th approximant* f_n *satisfying*

$$f_{2m} = R_{m,m}, \quad f_{2m+1} = R_{m+1,m}, \quad m = 0, 1, 2, \ldots . \quad (5.5.20)$$

(B) *Conversely, let* (5.5.17) *be the fps to which a given regular C-fraction* (5.5.19) *corresponds* (*see Corollary* 5.3). *Then, for each Padé approximant* $R_{m,n} = P_{m,n}/Q_{m,n}$ *of L in the sequence* (5.5.18), *Equation* (5.5.20) *holds and the degrees of* $P_{m,n}$ *and* $Q_{m,n}$ *are m and n, respectively.*

Proof. (A): For each approximant $R_{m,n}$ in the sequence (5.5.18), let $R_{m,n} = P_{m,n}/Q_{m,n}$, where $P_{m,n}$ and $Q_{m,n}$ are relatively prime polynomials of degrees m and n, respectively. Define $\{A_n\}$ and $\{B_n\}$ by

$$A_{2m} = P_{m,m}, \qquad B_{2m} = Q_{m,m}, \qquad m = 0, 1, 2, \ldots, \quad (5.5.21a)$$
$$A_{2m+1} = P_{m+1,m}, \quad B_{2m+1} = Q_{m+1,m}, \qquad m = 0, 1, 2, \ldots . \quad (5.5.21b)$$

Clearly the elements of the sequence $\{f_n\}_{n=0}^{\infty}$, where $f_n = A_n/B_n$, are all distinct and satisfy (2.1.10). Hence by Theorem 2.2, there exists a continued fraction

$$b_0(z) + \overset{\infty}{\underset{n=1}{\mathrm{K}}} \left(\frac{a_n(z)}{b_n(z)} \right) \qquad (5.5.22)$$

with elements defined by (2.1.11), or equivalently,

$$b_0(z) = 1, \qquad a_1(z) = c_1 z, \qquad b_1(z) = 1, \qquad (5.5.23a)$$

$$a_{2m}(z) = \frac{P_{m,m-1}Q_{m,m} - P_{m,m}Q_{m,m-1}}{P_{m,m-1}Q_{m-1,m-1} - P_{m-1,m-1}Q_{m,m-1}}, \qquad (5.5.23b)$$

$$b_{2m}(z) = \frac{P_{m,m}Q_{m-1,m-1} - P_{m-1,m-1}Q_{m,m}}{P_{m,m-1}Q_{m-1,m-1} - P_{m-1,m-1}Q_{m,m-1}},$$

and

$$a_{2m+1}(z) = \frac{P_{m,m}Q_{m+1,m} - P_{m+1,m}Q_{m,m}}{P_{m,m}Q_{m,m-1} - P_{m,m-1}Q_{m,m}}, \qquad (5.5.23c)$$

$$b_{2m+1}(z) = \frac{P_{m+1,m}Q_{m,m-1} - P_{m,m-1}Q_{m+1,m}}{P_{m,m}Q_{m,m-1} - P_{m,m-1}Q_{m,m}},$$

for $m = 1, 2, 3, \ldots$. We shall show that there exist non-zero constants d_n and e_n such that

$$a_n(z) = d_n z \quad \text{and} \quad b_n(z) = e_n. \qquad (5.5.24)$$

Then by an equivalence transformation (Theorem 2.6), it is easily shown that (5.5.22) is equivalent to a regular C-fraction. We prove (5.5.24) for $n=2m$; a similar proof holds for $n=2m+1$. Since the Padé approximants are all normal,

$$\lambda(L-L(R_{m,m-1}))=2m \quad \text{and} \quad \lambda(L-L(R_{m,m}))=2m+1.$$

Hence

$$\lambda(P_{m,m-1}Q_{m,m}-P_{m,m}Q_{m,m-1})=2m. \tag{5.5.25}$$

Therefore, since the polynomial in parentheses on the left side of (5.5.25) is of degree exactly $2m$, there exists a non-zero constant k_{2m} such that

$$P_{m,m-1}Q_{m,m}-P_{m,m}Q_{m,m-1}=k_{2m}z^{2m}. \tag{5.5.26}$$

Similarly

$$\lambda(L-L(R_{m,m-1}))=2m \quad \text{and} \quad \lambda(L-L(R_{m-1,m-1}))=2m-1$$

imply that

$$\lambda(P_{m,m-1}Q_{m-1,m-1}-P_{m-1,m-1}Q_{m,m-1})=2m-1. \tag{5.5.27}$$

Since the polynomial on the left side in (5.5.27) is of degree exactly $2m-1$, there exists a non-zero constant l_{2m} such that

$$P_{m,m-1}Q_{m-1,m-1}-P_{m-1,m-1}Q_{m,m-1}=l_{2m}z^{2m-1}. \tag{5.5.28}$$

It follows from (5.5.26) and (5.5.28) that

$$a_{2m}(z)=\frac{k_{2m}}{l_{2m}}z=d_{2m}z \quad (d_{2m}\neq 0).$$

A similar argument shows that $b_{2m}(z)=e_{2m}\neq 0$. This completes the proof of (A).

To prove (B) we let A_n, B_n, f_n denote the nth numerator, denominator and approximant of the regular C-fraction (5.5.19). From the difference equations (2.1.6) it is easily shown that the A_n and B_n are polynomials of the forms given by

$$A_0=B_0=1, \tag{5.5.29a}$$

$$A_{2m}=1+\alpha_{m,1}z+\cdots+\alpha_{m,m}z^m, \qquad m=1,2,3,\ldots, \tag{5.5.29b}$$

$$B_{2m}=1+\beta_{m,1}z+\cdots+\beta_{m,m}z^m, \qquad m=1,2,3,\ldots, \tag{5.5.29c}$$

$$A_{2m-1}=1+\gamma_{m,1}z+\cdots+\gamma_{m,m}z^m, \qquad m=1,2,3,\ldots, \tag{5.5.29d}$$

$$B_{2m-1}=1+\delta_{m,1}z+\cdots+\delta_{m,m-1}z^{m-1}, \qquad m=1,2,3,\ldots, \tag{5.5.29e}$$

where

$$\beta_{m,m} = a_2 a_4 \cdots a_{2m} \neq 0, \qquad m = 1, 2, 3, \ldots, \qquad (5.5.29f)$$

$$\gamma_{m,m} = a_1 a_3 \cdots a_{2m-1} \neq 0, \qquad m = 1, 2, 3, \ldots. \qquad (5.5.29g)$$

Thus, for $m \geqslant 0$, $f_{2m+1} = A_{2m+1}/B_{2m+1}$ is a rational function of type $[m+1, m]$ and $f_{2m} = A_{2m}/B_{2m}$ is a rational function of type $[m, m]$. It follows from Corollary 5.3 that the regular C-fraction (5.5.19) corresponds to a fps L of the form (5.5.17) and that

$$\lambda(L - L(f_n)) = n + 1, \qquad n = 0, 1, 2, \ldots.$$

Thus

$$\lambda(L - L(f_{2m})) = 2m + 1 \quad \text{and} \quad \lambda(L - L(f_{2m+1})) = 2m + 2,$$

so that (5.5.20) holds. That the degrees of the $P_{m,n}$ and $Q_{m,n}$ are as asserted follows from (5.5.29). This proves (B). ■

Another class of continued fractions related to Padé tables consists of the P-fractions (see Corollary 5.4). The following result is due to Magnus [1974, Theorem 6; 1962a, Theorem 6].

THEOREM 5.20. *Let*

$$L = c_0 + c_1 z + c_2 z^2 + \cdots \qquad (c_0 \neq 0) \qquad (5.5.30)$$

be an fps, let s be a fixed integer $(s = 0, \pm 1, \pm 2, \ldots)$, *and let*

$$b_{s,0}(z) + \cfrac{1}{b_{s,1}(z)} + \cfrac{1}{b_{s,2}(z)} + \cdots, \qquad (5.5.31)$$

with n th approximant $f_{s,n}(z)$, *denote the P-fraction corresponding to the fps*

$$z^s L = c_0 z^s + c_1 z^{s+1} + c_2 z^{s+2} + \cdots \qquad (5.5.32)$$

in the sense of Corollary 5.4. Then $\{f_{s,n}(z)/z^s\}_{n=0}^{\infty}$ *is the sequence of consecutive distinct Padé approximants of L along the diagonal* $\{(m, m-s)\}_{m=0}^{\infty}$ *if* $s \leqslant 0$, *and along the diagonal* $\{(m+s, m)\}_{m=0}^{\infty}$ *if* $s \geqslant 0$.

From the above theorem it is clear that every Padé approximant of L is one of the rational functions $f_{s,n}(z)/z^s$.

Other continued fractions related to the Padé table will be discussed in Chapter 7. The close relationship between Padé tables and continued fractions makes it possible to apply the convergence theory of continued fractions to obtain convergence results for sequences of Padé approximants.

The following theorem due to Montessus de Ballore [1902] is a classical result for Padé tables associated with meromorphic functions. Modern proofs can be found, for example, in [Gragg, 1972, Theorem 8.3a] and in [Perron, 1957a, Satz 5.12].

THEOREM 5.21 (Montessus de Ballore). *Let $f(z)$ be a function holomorphic in the disk*

$$[z:|z|\leqslant R]$$

except for n simple poles at points $\pi_1, \pi_2, \ldots, \pi_n$, where

$$0<|\pi_1|\leqslant|\pi_2|\leqslant \cdots \leqslant|\pi_n|<R.$$

Let

$$L=c_0+c_1z+c_2z^2+\cdots \qquad (c_0\neq 0)$$

denote the Taylor series expansion of $f(z)$ at $z=0$, and let $R_{m,n}(z)$ denote the (m, n) Padé approximant of L. Then for all z in the set

$$D_R=[z:|z|\leqslant R, z\neq\pi_k, k=1,2,\ldots,n],$$

$$\lim_{m\to\infty} R_{m,n}(z)=f(z),$$

and the convergence is uniform on every compact subset of D_R.

We conclude this section with a convergence result for Padé approximants which is obtained by an application of Theorem 5.13.

THEOREM 5.22. *Let $\{m_\nu\}$ and $\{n_\nu\}$ be sequences of non-negative integers such that*

$$\lim_{\nu\to\infty} \max[m_\nu, n_\nu]=\infty. \qquad (5.5.33)$$

Let $R_\nu(z)$ denote the (m_ν, n_ν) Padé approximant of the fps

$$L=c_0+c_1z+c_2z^2+\cdots,$$

and let D be a domain containing the origin. Then:

(A) $\{R_\nu(z)\}$ converges uniformly on every compact subset of D iff $\{R_\nu(z)\}$ is uniformly bounded on every compact subset of D.

(B) If $\{R_\nu(z)\}$ converges uniformly on every compact subset of D, then $f(z)=\lim_{\nu\to\infty}R_\nu(z)$ is holomorphic in D, and P is the Taylor series expansion of $f(z)$ about $z=0$.

Proof. Let

$$R_{m,n}(z) = \frac{A_{m,n}(z)}{B_{m,n}(z)}$$

denote the (m, n) Padé approximant of L, where $(A_{m,n}, B_{m,n})$ is a rational expression of type $[m, n]$, but $A_{m,n}$ and $B_{m,n}$ are not necessarily relatively prime. Since

$$\lambda(B_{m,n}L - A_{m,n}) \geqslant m+n+1, \qquad (5.5.34)$$

it follows that

$$\lambda\left(L - L\left(\frac{A_{m,n}}{B_{m,n}}\right)\right) \geqslant m+n+1 - \min[m, n] = 1 + \max[m, n].$$

$$(5.5.35)$$

In the transition from $B_{m,n}L - A_{m,n}$ to $L - L(A_{m,n}/B_{m,n})$ the value of λ may decrease by an integer value r which is such that z^r is the highest power of z contained as a factor of $B_{m,n}$. Hence $r \leqslant n$. If z^r is a factor of $B_{m,n}$, then by (5.5.34) it must also be a factor of $A_{m,n}$, so that $r \leqslant m$. It follows from (5.5.33) and (5.5.35) that the sequence of rational functions $\{R_\nu(z)\}$ corresponds to the fps L. Hence the present theorem is an immediate consequence of Theorem 5.13.   ∎

5.5.2   *Multiple-Point Padé Tables*

The concept of the Padé table has been generalized to give rational approximants for formal Newton series (called Newton-Padé approximants) and for approximation alternately at 0 and ∞ (called two-point Padé approximants). In the case of the Newton-Padé table the approximation is at a sequence of (not necessarily distinct) points in the finite complex plane. Because of their connection with continued fractions, we shall discuss very briefly some of the recent results in these areas. Greater details can be found in references given below.

A series of the form

$$L = c_0 + \sum_{k=1}^{\infty} c_k \prod_{i=1}^{k} (z - \beta_i) \qquad (5.5.36)$$

is called a *formal Newton series* (fNs) *with* sequence of (not necessarily distinct) interpolation points $\{\beta_i\}$ in the finite complex plane. The (m, n) *Newton-Padé approximant* $R_{m,n}(L, z)$ of L is a rational function of type

$[m, n]$ defined in a manner completely analogous to that of the (m, n) Padé approximant of a power series (see, for example, [Gallucci and Jones, 1976] and [Warner, 1974]). Existence and uniqueness proofs are given in the preceding references. If all $\beta_i = 0$, then (5.5.36) reduces to an fps and $R_{m,n}(L, z)$ becomes its (m, n) Padé approximant. Many of the classical properties of the Padé table have been extended to the more general Newton-Padé table. We shall cite references where some of the principal results can be found. A large number of algebraic identities and algorithms that can be used to construct (numerically) Newton-Padé approximants are contained in the following: [Claessens, 1976, 1977/78], [Larkin, 1967], [Stoer, 1961], [Thacher and Tukey, 1960], [Wuytack, 1973, 1975]. The most complete sets of these are contained in the doctoral theses by Warner [1974] and Claessens [1976]. A Newton-Padé approximant

$$R_{m,n}(L, z) = \frac{P_{m,n}(L, z)}{Q_{m,n}(L, z)}$$

for which $P_{m,n}$ and $Q_{m,n}$ are relatively prime polynomials is said to be *normal* if the degrees of $P_{m,n}$ and $Q_{m,n}$ are m and n, respectively, and if $R_{m,n}(L, z)$ occurs only at one place in the table. Necessary and sufficient conditions for normality, similar to those given by Theorem 5.18, can be found in [Gallucci and Jones, 1976]. It has also been shown that certain "staircase" sequences of Newton-Padé approximants are the approximants of a Thiele-type continued fraction of the form

$$a_0 + \frac{a_1(z - \beta_1)}{1} + \frac{a_2(z - \beta_2)}{1} + \frac{a_3(z - \beta_3)}{1} + \cdots ;$$

see, for example, [Thiele, 1909], [Nörlund, 1924], and [Claessens, 1977/78]. The classical convergence theorem of Montessus de Ballore (Theorem 5.21) was generalized by Saff in [1972] and further extended by Warner [1974, 1976]. Karlsson [1976] has extended the theorems on convergence in measure by Nuttall and in capacity by Pommerenke. Theorem 5.22 was extended to Newton-Padé tables by Gallucci and Jones [1976]. In the same paper they showed that with suitable restrictions (insured by normality) the approximant $R_{m,n}(L, z)$ is a continuous function of the c_k and β_i in (5.5.36). Bounds for Newton-Padé approximants of series of Stieltjes have been given by Baker [1969] and Barnsley [1974].

Corresponding to a pair of fLs

$$L = c_0 + c_1 z + c_2 z^2 + \cdots \quad \text{(increasing powers)},$$

$$L^* = c_\mu^* z^\mu + c_{\mu-1}^* z^{\mu-1} + c_{\mu-2}^* z^{\mu-2} + \cdots \quad \text{(decreasing powers)},$$

rational-function approximants (*called two-point Padé approximants*) can be formed in a manner analogous to that for Padé and Newton-Padé approximants. In the present case the sequence of interpolation points is given by $\{0, \infty, 0, \infty, 0, \infty, \cdots \}$, so that the approximation takes place alternately at 0 and ∞. The (m, n) two-point Padé approximant of (L, L^*) will be denoted by $R_{m,n}(L, L^*, z)$. Applications of two-point Padé approximants in theoretical physics have been described in the papers [Baker et al., 1964], [Isihara and Montroll, 1971], [Sheng, 1974] and [Sheng and Dow, 1971]. Recently it has been found that certain diagonal sequences of two-point Padé approximants form the approximants of general T-fractions (4.5.36) (see [McCabe, 1975], [McCabe and Murphy, 1976], [Jones and Thron, 1977], [Jones, 1977] and [Thron, 1977]. This connection will be described in Section 7.3.

Other generalizations of the Padé table include Laurent, Fourier and Chebyschev-Padé tables (see [Gragg and Johnson, 1974] and [Gragg, 1977]) and Padé approximants involving several complex variables (see, for example, [Chisholm, 1977], [Graves-Morris, 1977], [Karlsson and Wallin, 1977] and references contained therein).

CHAPTER 6

Representations of Analytic Functions by Continued Fractions

A large number of analytic functions are known to have continued-fraction representations. Frequently a given function will be represented by several different continued fractions, each with its own convergence behavior. The purpose of this chapter is to give a wide selection of examples of continued-fraction representations of functions based on the methods of Chapter 5. Our selection is far from exhaustive. In Section 6.1 the examples consist of continued fractions of Gauss associated with various hypergeometric and confluent hypergeometric series. Section 6.2 deals with examples associated with minimal solutions of three-term recurrence relations. Further methods and examples will be discussed in Chapters 7 and 10. Additional examples can be found, among others, in the following books and articles as well as in the references contained therein: [Abramowitz and Stegun, 1964], [Andrews, 1968], [Askey and Ismail, to appear], [de Bruin, 1977], [Cody et al., 1970], [Erdelyi et al., 1953, Vols. 1, 2, 3], [Frank, 1958, 1960a, 1960b], [Gautschi, 1967, 1970, 1977], [Henrici, 1977, Vol. 2, Chapter 12], [Ince, 1919], [Khovanskii, 1963], [Luke, 1969], [Maurer, 1966], [McCabe, 1974], [Miller, 1975], [Murphy, 1971], [Murphy and O'Donohoe, 1976], [Perron, 1957a], [Phipps, 1971], [Schlömilch, 1871], [Sobhy, 1973], [Stegun and Zucker, 1970], [Tannery, 1882], [Thacher, 1967 and Tech. Rpt. 38-77], [Wall, 1948], [Widder, 1968], [Wynn, 1959].

6.1 Continued Fractions of Gauss

In this section we consider a number of examples of continued-fraction representations of functions associated with hypergeometric and confluent hypergeometric series. The method employed is described in Sections 5.2

ENCYCLOPEDIA OF MATHEMATICS and Its Applications, Gian-Carlo Rota (ed.). Vol. 11: William B. Jones and W. J. Thron, Continued Fractions. ISBN 0-201-13510-8

and 5.4. In Section 6.1.1 the series converge only in the unit disk $|z|<1$ (and possibly on its boundary), while the corresponding continued fraction of Gauss converges throughout the z-plane cut along the real axis from 1 to $+\infty$ and provides the analytic continuation in the cut plane. In Sections 6.1.2 and 6.1.3 the functions considered are meromorphic and are represented in $\mathbb{C}$ (even at the poles) by the continued fractions. In Section 6.1.4 the confluent hypergeometric series are divergent except at $z=0$; nevertheless, the corresponding continued fractions can sometimes be proved to converge to analytic functions for which the divergent series are asymptotic expansions. We are thus provided with a summability method for divergent series. The final step of the summability method is treated in Chapter 10.

6.1.1 *Hypergeometric Functions $F(a, b; c; z)$*

The hypergeometric function $F(a, b; c; z)$ is defined by the power series

$$F(a, b; c; z) = 1 + \frac{ab}{c}\frac{z}{1!} + \frac{a(a+1)b(b+1)}{c(c+1)}\frac{z^2}{2!} + \cdots, \quad (6.1.1)$$

where a, b, c are complex constants and $c \notin [0, -1, -2, \ldots]$. If a or b is contained in the set $[0, -1, -2, \ldots]$, then $F(a, b; c; z)$ is a polynomial. Otherwise the power series (6.1.1) has radius of convergence equal to 1. The function $F(a, b; c; z)$ is sometimes denoted by $_2F_1(a, b; c; z)$.

THEOREM 6.1. *Let $\{a_n\}$ be a sequence of complex numbers defined by*

$$a_{2n+1} = -\frac{(a+n)(c-b+n)}{(c+2n)(c+2n+1)}, \quad n=0,1,2,\ldots, \quad (6.1.2a)$$

$$a_{2n} = -\frac{(b+n)(c-a+n)}{(c+2n-1)(c+2n)}, \quad n=1,2,3,\ldots, \quad (6.1.2b)$$

where a, b, c are constants such that

$$a_n \neq 0, \quad n=1,2,3,\ldots. \quad (6.1.3)$$

Then:

(A) *The regular C-fraction*

$$1 + \mathop{K}_{n=1}^{\infty}\left(\frac{a_n z}{1}\right) \quad (6.1.4)$$

converges to a function $f(z)$ meromorphic in the domain

$$D = [z : 0 < \arg(z-1) < 2\pi]. \quad (6.1.5)$$

(B) *The convergence is uniform on every compact subset of D which contains no poles of $f(z)$.*
(C) *$f(z)$ is holomorphic at $z=0$, and $f(0)=1$.*
(D) *For all z such that $|z|<1$,*

$$f(z) = \frac{F(a,b;c;z)}{F(a,b+1;c+1;z)},$$ (6.1.6)

and hence $f(z)$ provides the analytic continuation of the function on the right side of (6.1.6) into the cut plane D.

The regular C-fraction (6.1.4) is called the *continued fraction of Gauss* [1813]. We note that (6.1.5) is the plane cut along the real axis from 1 to $+\infty$.

Proof of Theorem 6.1.   The formal identities

$$F(a,b;c;z) = F(a,b+1;c+1;z) - \frac{a(c-b)}{c(c+1)} zF(a+1,b+1;c+2;z)$$ (6.1.7a)

and

$$F(a,b+1;c+1;z) = F(a+1,b+1;c+2;z)$$
$$- \frac{(b+1)(c-a+1)}{(c+1)(c+2)} zF(a+1,b+2;c+3;z)$$ (6.1.7b)

are easily verified from (6.1.1). If we let

$$P_{2n} = F(a+n,b+n;c+2n;z), \qquad n=0,1,2,\ldots,$$ (6.1.8a)
$$P_{2n+1} = F(a+n,b+n+1;c+2n+1;z), \qquad n=0,1,2,\ldots,$$ (6.1.8b)

then it follows from (6.1.7) that

$$P_n = P_{n+1} + a_{n+1} z P_{n+2}, \qquad n=0,1,2,\ldots.$$ (6.1.9)

Clearly

$$\lambda(a_n z) = 1, \qquad \lambda(b_n) = \lambda(1) = 0$$

and

$$\lambda(P_n/P_{n+1}) = 0, \qquad n=1,2,3,\ldots,$$

where λ is the function defined in Section 5.1. Thus by Theorem 5.5, the regular C-fraction (6.1.4) corresponds at $(z=0)$ to the function

$$\frac{P_0}{P_1} = \frac{F(a,b;c;z)}{F(a,b+1;c+1;z)}.$$

From (6.1.2) we see that

$$\lim_{n\to\infty} a_n = -\tfrac{1}{4}.$$

Therefore (A), (B) and (C) follow from Theorem 5.15. (D) is an immediate consequence of Theorem 5.13. ∎

Setting $b=0$ and replacing c by $c-1$ in Theorem 6.1, we obtain

COROLLARY 6.2. *Let a and c be complex constants such that $\{a_n\}$, defined by*

$$a_{2n+1} = -\frac{(a+n)(c+n-1)}{(c+2n-1)(c+2n)}, \qquad n=0,1,2,\ldots, \quad (6.1.10a)$$

$$a_{2n} = -\frac{(n)(c-a+n-1)}{(c+2n-2)(c+2n-1)}, \qquad n=1,2,3,\ldots, \quad (6.1.10b)$$

is a sequence of non-zero complex numbers. Then:

(A) *For all z such that $|z|<1$,*

$$F(a,1;c;z) = \cfrac{1}{1 + \overset{\infty}{\underset{n=1}{\mathrm{K}}}\left(\dfrac{a_n z}{1}\right)}. \qquad (6.1.11)$$

(B) *The continued fraction on the right side of (6.1.11) converges to a function $f(z)$ meromorphic in the domain D of (6.1.5), and $f(z)$ is the analytic continuation in D of $F(a,1;c;z)$. $f(z)$ is holomorphic at $z=0$, and $f(0)=1$. The continued fraction converges uniformly on compact subsets of D.*

The initial part of the continued fraction in (6.1.11) is given by

$$\frac{1}{1} - \frac{\dfrac{a}{c}z}{1} - \frac{\dfrac{1(c-a)}{c(c+1)}z}{1} - \frac{\dfrac{(a+1)c}{(c+1)(c+2)}z}{1} - \frac{\dfrac{2(c-a+1)}{(c+2)(c+3)}z}{1}$$

$$- \frac{\dfrac{(a+2)(c+1)}{(c+3)(c+4)}z}{1} - \frac{\dfrac{3(c-a+2)}{(c+4)(c+5)}z}{1} - \cdots, \qquad (6.1.12)$$

which is equivalent to

$$\frac{1}{1-}\ \frac{az}{c-}\ \frac{1(c-a)z}{c+1}\ -\ \frac{(a+1)cz}{c+2}\ -\ \frac{2(c-a+1)z}{c+3}$$
$$-\ \frac{(a+2)(c+1)z}{c+4}\ -\ \frac{3(c-a+2)z}{c+5}\ -\ \cdots. \tag{6.1.13}$$

We shall now consider a few examples of elementary functions represented by such continued fractions.

Example 1. Arctangent.

$$\arctan z = zF\left(\tfrac{1}{2},1;\tfrac{3}{2};-z^2\right)$$
$$= \frac{z}{1+}\ \frac{1^2z^2}{3}\ +\ \frac{2^2z^2}{5}\ +\ \frac{3^2z^2}{7}\ +\ \frac{4^2z^2}{9}\ +\cdots, \tag{6.1.14}$$

as given by (2.1.16). The continued fraction converges and represents a single-valued branch of the analytic function $\arctan z$ in the cut z-plane with cuts along the imaginary axis extending from i to $+i\infty$ and from $-i$ to $-i\infty$. The points $z = \pm i$ are branch points of $\arctan z$. The Taylor series expansion of $\arctan z$ at $z=0$,

$$\arctan z = z - \frac{z^3}{3} + \frac{z^5}{5} - \frac{z^7}{7} + \frac{z^9}{9} - \cdots, \tag{6.1.15}$$

converges for $|z| < 1$, $z \neq \pm i$. To compute $\pi/4 = \arctan 1$ with an error not exceeding 10^{-7} by the series (6.1.15), one would need to sum the terms out to (approximately) the 10^6 power of z. It can be shown from Theorem 4.28 that $\arctan 1$ can be computed with an error less than 10^{-7} by using the 9th approximant of the continued fraction (6.1.14). A continued-fraction representation for $\operatorname{arctanh} z$ can be obtained from (6.1.14) by using the relation $\operatorname{arctanh} z = -i \arctan iz$.

Example 2. General binomial.

$$(1+z)^\alpha = F(-\alpha,1;1;-z)$$
$$= \frac{1}{1+}\ \frac{(-\alpha)z}{1}\ +\ \frac{1(1+\alpha)z}{2}$$
$$+\ \frac{1(1-\alpha)z}{3}\ +\ \frac{2(2+\alpha)z}{4}\ +\ \frac{2(2-\alpha)z}{5}\ +\cdots, \tag{6.1.16}$$

where

$$\alpha \in [\, w : w \in \mathbb{C},\ w \neq 0, \pm 1, \pm 2, \cdots \,].$$

The continued fraction in (6.1.16) converges and represents a single-valued branch of the analytic function $(1+z)^\alpha$ throughout the cut z-plane with cut along the real axis from -1 to $-\infty$.

Example 3. Natural logarithm.

$$\log(1+z)=zF(1,1;2;-z)$$

$$=\frac{z}{1+}\frac{1^2z}{2+}\frac{1^2z}{3+}\frac{2^2z}{4+}\frac{2^2z}{5+}\frac{3^2z}{7+}\frac{3^2z}{9+}\cdots. \qquad (6.1.17)$$

The continued fraction in (6.1.17) converges and represents a single-valued branch of the analytic function $\log(1+z)$ for all z in the cut z-plane with cut along the real axis from -1 to $-\infty$. We also have

$$\log\left(\frac{1+z}{1-z}\right)=2zF\left(\frac{1}{2},1;\frac{3}{2};z^2\right)$$

$$=\frac{2z}{1-}\frac{1^2z^2}{3-}\frac{2^2z^2}{5-}\frac{3^2z^2}{7-}\frac{4^2z^2}{9-}\cdots, \qquad (6.1.18)$$

where the continued fraction in (6.1.18) converges and represents a single-valued branch of the analytic function $\log[(1+z)/(1-z)]$ for all z in the cut z-plane with cuts along the real axis from 1 to $+\infty$ and from -1 to $-\infty$.

Example 4. A generalization.

$$\int_0^z \frac{dt}{1+t^n}\,dt=zF\left(\frac{1}{n},1;1+\frac{1}{n};-z^n\right)$$

$$=\frac{z}{1+}\frac{1^2\cdot z^n}{n+1+}\frac{(1\cdot n)^2z^n}{2n+1+}\frac{(n+1)^2z^n}{3n+1}$$

$$+\frac{(2n)^2z^n}{4n+1+}\frac{(2n+1)^2z^n}{5n+1+}\cdots,$$

$$n\in[1,2,3,\dots], \qquad (6.1.19)$$

where the continued fraction converges and represents the analytic function on the left of (6.1.19) for all z in the $w=z^n$ plane cut along the real axis from -1 to $-\infty$. If $n=1$, (6.1.19) reduces to (6.1.17), and if $n=2$ it reduces to (6.1.14).

Example 5. Arcsine. By Theorem 6.1 we obtain

$$\frac{\arcsin z}{\sqrt{1-z^2}}=\frac{zF\left(\frac{1}{2},\frac{1}{2};\frac{3}{2};z^2\right)}{F\left(\frac{1}{2},-\frac{1}{2};\frac{1}{2};z^2\right)}$$

$$=\frac{z}{1-}\frac{1\times2z^2}{3-}\frac{1\times2z^2}{5-}\frac{3\times4z^2}{7}$$

$$-\frac{3\times4z^2}{9-}\frac{5\times6z^2}{11-}\frac{5\times6z^2}{13-}\cdots, \qquad (6.1.20)$$

where the continued fraction converges and represents a single-valued branch of the analytic function $(\arcsin z)/\sqrt{1-z^2}$ for all z in the cut z-plane with cuts along the real axis from 1 to $+\infty$ and from -1 to $-\infty$. A continued-fraction representation for

$$\frac{\text{arcsinh}\, z}{\sqrt{1+z^2}}$$

can be obtained from (6.1.20) by using the relation $\text{arcsinh}\, z = -i \arcsin iz$.

Example 6. Legendre functions of the second kind of degree α and order m are defined by

$$Q_\alpha^m(z) = K_n z^{-\alpha-m-1}(z^2-1)^{m/2} F\left(\tfrac{1}{2}\alpha+\tfrac{1}{2}m+1, \tfrac{1}{2}\alpha+\tfrac{1}{2}m+\tfrac{1}{2}; \alpha+\tfrac{3}{2}; z^{-2}\right)$$

$$\text{for}\quad |z|>1, \quad (6.1.21a)$$

where

$$K_n = \frac{e^{im\pi}\sqrt{\pi}}{2^{\alpha+1}} \frac{\Gamma(\alpha+m+1)}{\Gamma\left(\alpha+m+\tfrac{3}{2}\right)}, \qquad (6.1.21b)$$

and

$$|\arg(z-1)|<\pi \quad \text{and} \quad (z^2-1)^{m/2}=(z-1)^{m/2}(z+1)^{m/2}. \quad (6.1.21c)$$

Here m is a non-negative integer and α is an arbitrary complex number. When extended by analytic continuation, $Q_\alpha^m(z)$ is single-valued and analytic in the z-plane cut along the real axis from 1 to $-\infty$ (see [Erdelyi et al., 1953, Vol. 1, Chapter 3]).

Let m and α be fixed, let $w=1/z$, and let $\{y_n\}$ be defined by

$$y_n(w) = Q_{\alpha+n}^m(z), \qquad n=0,1,2,\dots. \qquad (6.1.22)$$

It is easily shown that the $y_n(w)$ satisfy the three-term recurrence relations

$$y_n(w) = \frac{b_n}{w} y_{n+1}(w) + a_{n+1} y_{n+2}(w), \qquad n=0,1,2,\dots, \quad (6.1.23a)$$

where

$$a_{n+1} = -\frac{n+\alpha+m+2}{n+\alpha+m+1}, \qquad n=0,1,2,\dots, \qquad (6.1.23b)$$

$$b_n = \frac{2n+2\alpha+3}{n+\alpha+m+1}, \qquad n=0,1,2,\dots. \qquad (6.1.23c)$$

Since

$$\lambda(a_{n+1})=0, \quad \lambda\left(\frac{b_n}{w}\right)=-1, \quad \lambda\left(\frac{y_n}{y_{n+1}}\right)=-1, \quad n=0,1,2,\ldots,$$

it follows from Theorem 5.5 that for each $n=0,1,2,\ldots$, the continued fraction

$$\frac{b_n}{w} + \overset{\infty}{\underset{k=1}{\mathrm{K}}}\left(\frac{a_{n+k}}{b_{n+k}/w}\right) \tag{6.1.24}$$

corresponds (at $w=0$) to $y_n(w)/y_{n+1}(w)$. The continued fraction (6.1.24) is equivalent to

$$\frac{b_n}{w} + \cfrac{\dfrac{a_{n+1}}{b_{n+1}}w}{1} \ + \ \cfrac{\dfrac{a_{n+2}}{b_{n+1}b_{n+2}}w^2}{1} \ + \ \cfrac{\dfrac{a_{n+3}}{b_{n+2}b_{n+3}}w^2}{1} \ + \cdots, \tag{6.1.25}$$

which can be written as

$$\frac{b_n}{w}\left[1+\overset{\infty}{\underset{k=1}{\mathrm{K}}}\left(\frac{a_{n+k}^* w^2}{1}\right)\right], \tag{6.1.26a}$$

where

$$a_{n+k}^* = \frac{a_{n+k}}{b_{n+k-1}b_{n+k}}, \quad n,k=0,1,2,\ldots. \tag{6.1.26b}$$

Since $\lim_{k\to\infty} a_{n+k}^* = -\tfrac{1}{4}$, an application of Theorems 5.13 and 5.15 shows that

$$\frac{Q_{\alpha+n}^m(z)}{Q_{\alpha+n+1}^m(z)} = b_n z\left[1+\overset{\infty}{\underset{k=1}{\mathrm{K}}}\left(\frac{a_{n+k}^*/z^2}{1}\right)\right], \quad m,n=0,1,2,\ldots, \tag{6.1.27}$$

for all $\alpha\in\mathbb{C}$ such that $a_{n+k}^*\in\mathbb{C}$, $a_{n+k}^*\neq 0$ for all $k>0$. The right side of (6.1.27) converges and represents the analytic function $Q_{\alpha+n}^m(z)/Q_{\alpha+n+1}^m(z)$ for all z in the cut z-plane with cut along the real axis from -1 to $+1$. Other continued-fraction representations of Legendre functions are considered in Section 6.2.

6.1.2 Confluent Hypergeometric Functions $\Phi(b;c;z)$.

The entire function defined by

$$\Phi(b;c;z)=1+\frac{b}{c}\frac{z}{1!}+\frac{b(b+1)}{c(c+1)}\frac{z^2}{2!}+\frac{b(b+1)(b+2)}{c(c+1)(c+2)}\frac{z^3}{3!}+\cdots, \tag{6.1.28}$$

where b and c are complex constants with $c \notin [0, -1, -2, -3, \ldots]$, is called a *confluent hypergeometric function*. The function $\Phi(b; c; z)$ is sometimes denoted by $_1F_1(b; c; z)$ or $M(b; c; z)$ and is called Kummer's function.

THEOREM 6.3. *Let $\{a_n\}$ be a sequence of complex numbers defined by*

$$a_{2n} = \frac{b+n}{(c+2n-1)(c+2n)}, \qquad n = 1, 2, 3, \ldots, \qquad (6.1.29a)$$

$$a_{2n+1} = -\frac{c-b+n}{(c+2n)(c+2n+1)}, \qquad n = 0, 1, 2, \ldots, \qquad (6.1.29b)$$

where b and c are complex constants chosen such that

$$a_n \neq 0, \qquad n = 1, 2, 3, \ldots. \qquad (6.1.30)$$

Then:

(A) *The regular C-fraction*

$$1 + \overset{\infty}{\underset{n=1}{\mathrm{K}}} \left(\frac{a_n z}{1} \right) \qquad (6.1.31)$$

converges to the meromorphic function

$$f(z) = \frac{\Phi(b; c; z)}{\Phi(b+1; c+1; z)} \qquad (6.1.32)$$

for all $z \in \mathbb{C}$.

(B) *The convergence is uniform on every compact subset of $\mathbb{C}$ which contains no poles of $f(z)$.*

(C) *$f(z)$ is holomorphic at $z = 0$, and $f(0) = 1$.*

Proof. Let

$$P_{2n} = \Phi(b+n; c+2n; z), \qquad n = 0, 1, 2, \ldots,$$

$$P_{2n+1} = \Phi(b+n+1; c+2n+1; z), \qquad n = 0, 1, 2, \ldots.$$

Then it can be shown that the power series P_n satisfy the three-term recurrence relations

$$P_n = P_{n+1} + a_{n+1} z P_{n+2}, \qquad n = 0, 1, 2, \ldots.$$

Since

$$\lambda(a_{n+1} z) = 1, \quad \lambda(1) = 0, \quad \lambda(P_n / P_{n+1}) = 0, \qquad n = 0, 1, 2, \ldots,$$

by Theorem 5.5 the continued fraction (6.1.31) corresponds (at $z=0$) to P_0/P_1. Further, since

$$\lim_{n\to\infty} a_n = 0,$$

the results stated in the theorem follow directly from Theorems 5.13 and 5.14. ∎

Setting $b=0$ in (6.1.29), then replacing $c+1$ by c and taking the reciprocal of the regular C-fraction (6.1.30), we obtain

$$\Phi(1;c;z) = \frac{1}{1} - \frac{\dfrac{1}{c}z}{1} + \frac{\dfrac{1}{c(c+1)}z}{1} - \frac{\dfrac{c}{(c+1)(c+2)}z}{1} + \frac{\dfrac{2}{(c+2)(c+3)}z}{1}$$

$$- \frac{\dfrac{c+1}{(c+3)(c+4)}z}{1} + \frac{\dfrac{3}{(c+4)(c+5)}z}{1} - \cdots, \tag{6.1.33}$$

which is equivalent to

$$\Phi(1;c;z) = \frac{1}{1} - \frac{z}{c} + \frac{1\cdot z}{c+1} - \frac{cz}{c+2} + \frac{2\cdot z}{c+3} - \frac{(c+1)z}{c+4} + \frac{3\cdot z}{c+5} - \cdots. \tag{6.1.34}$$

The continued fractions in (6.1.33) and (6.1.34) converge and represent the entire function $\Phi(1;c;z)$ for all $z\in\mathbb{C}$, provided that $c\notin[0,-1,-2,-3,\ldots]$. Now setting $c=1$, we obtain

Example 7. Exponential function.

$$e^z = \Phi(1;1;z)$$

$$= \frac{1}{1} - \frac{z}{1} + \frac{1\cdot z}{2} - \frac{1\cdot z}{3} + \frac{2\cdot z}{4} - \frac{2\cdot z}{5} + \frac{3\cdot z}{6} - \frac{3\cdot z}{7} + \cdots, \tag{6.1.35}$$

is valid for all $z\in\mathbb{C}$. It is easily seen that (6.1.35) is equivalent to

$$e^z = \frac{1}{1} - \frac{z}{1} + \frac{z}{2} - \frac{z}{3} + \frac{z}{2} - \frac{z}{5} + \frac{z}{2} - \frac{z}{7} + \cdots \tag{6.1.36}$$

[see (2.1.21)]. Taking the reciprocal of (6.1.36) and replacing z by $-z$ gives the representation

$$e^z = 1 + \frac{z}{1} - \frac{z}{2} + \frac{z}{3} - \frac{z}{2} + \frac{z}{5} - \frac{z}{2} + \frac{z}{7} - \cdots. \tag{6.1.37}$$

Example 8. Error function.

$$\operatorname{erf} z = \frac{2}{\sqrt{\pi}} \int_0^z e^{-t^2}\, dt = \frac{2z}{\sqrt{\pi}} e^{-z^2} \Phi\left(1; \tfrac{3}{2}; z^2\right)$$

$$= \frac{2}{\sqrt{\pi}} e^{-z^2} \left[\frac{z}{1} - \frac{z^2}{\frac{3}{2}} + \frac{1 \cdot z^2}{\frac{5}{2}} - \frac{\frac{3}{2} \cdot z^2}{\frac{7}{2}} + \frac{2 \cdot z^2}{\frac{9}{2}} - \frac{\frac{5}{2} \cdot z^2}{\frac{11}{2}} + \frac{3 \cdot z^2}{\frac{13}{2}} - \cdots\right]$$

$$(6.1.38)$$

is valid for all $z \in \mathbb{C}$. Closely related to the error function is *Dawson's integral*

$$\int_0^z e^{t^2}\, dt = \frac{i\sqrt{\pi}}{2}\operatorname{erf}(-iz) = ze^{z^2}\Phi\left(1; \tfrac{3}{2}; -z^2\right)$$

$$= ze^{z^2}\left[\frac{1}{1} + \frac{z^2}{\frac{3}{2}} - \frac{1 \cdot z^2}{\frac{5}{2}} + \frac{\frac{3}{2} \cdot z^2}{\frac{7}{2}} - \frac{2 \cdot z^2}{\frac{9}{2}} + \frac{\frac{5}{2} \cdot z^2}{\frac{11}{2}} - \frac{3 \cdot z^2}{\frac{13}{2}} + \cdots\right],$$

$$(6.1.39)$$

valid for all $z \in \mathbb{C}$.

Example 9. Fresnel integrals are defined by

$$C(z) = \int_0^z \cos\left(\frac{\pi}{2}t^2\right) dt, \qquad S(z) = \int_0^z \sin\left(\frac{\pi}{2}t^2\right) dt. \qquad (6.1.40)$$

These functions are related to the error function and confluent hypergeometric functions by

$$C(z) + iS(z) = \frac{1+i}{2}\operatorname{erf}\left(\frac{\sqrt{\pi}}{2}(1-i)z\right) = ze^{(i\pi/2)z^2}\Phi\left(1; \tfrac{3}{2}; -\frac{i\pi}{2}z^2\right).$$

$$(6.1.41)$$

Thus by (6.1.34) we obtain

$$C(z) + iS(z) = e^{i\pi z^2/2}\left(\frac{z}{1} + \frac{i\pi z^2/2}{\frac{3}{2}} - \frac{i\pi z^2/2}{\frac{5}{2}} + \frac{i3\pi z^2/4}{\frac{7}{2}}\right.$$

$$\left. - \frac{i\pi z^2}{\frac{9}{2}} + \frac{i5\pi z^2/4}{\frac{11}{2}} - \frac{i3\pi z^2/2}{\frac{13}{2}} + \cdots\right), \qquad (6.1.42)$$

valid for all $z \in \mathbb{C}$.

Example 10. Incomplete gamma functions are defined by

$$\gamma(a, z) = \int_0^z e^{-t} t^{a-1} dt, \qquad \text{Re } a > 0. \tag{6.1.43}$$

These functions are related to confluent hypergeometric functions by

$$\gamma(a, z) = a^{-1} z^a e^{-z} \Phi(1; a+1; z). \tag{6.1.44}$$

Thus by (6.1.33)

$$\gamma(a, z) = a^{-1} z^a e^{-z} \left(\frac{1}{1} - \frac{z}{a+1} + \frac{1 \cdot z}{a+2} - \frac{(a+1)z}{a+3} \right.$$
$$\left. + \frac{2 \cdot z}{a+4} - \frac{(a+2)z}{a+5} + \frac{3 \cdot z}{a+6} - \cdots \right) \tag{6.1.45}$$

is valid for all $z \in \mathbb{C}$. We note that (6.1.38), (6.1.39) and (6.1.42) can be obtained from (6.1.45).

6.1.3 *Confluent Hypergeometric Functions* $\Psi(c; z)$

The entire function defined by

$$\Psi(c; z) = 1 + \frac{1}{c} \frac{z}{1!} + \frac{1}{c(c+1)} \frac{z^2}{2!} + \frac{1}{c(c+1)(c+2)} \frac{z^3}{3!} + \cdots, \tag{6.1.46}$$

where c is a complex constant with $c \notin [0, -1, -2, -3, \ldots]$, is also a confluent hypergeometric function. $\Psi(c; z)$ is sometimes denoted by $_0F_1(c; z)$.

THEOREM 6.4. *Let* $\{a_n\}$ *be a sequence of complex numbers defined by*

$$a_n = \frac{1}{(c+n-1)(c+n)}, \qquad n = 1, 2, 3, \ldots, \tag{6.1.47}$$

where c *is a complex constant such that* $c \notin [0, -1, -2, -3, \ldots]$. *Then:*

(A) *The regular C-fraction*

$$1 + \mathop{\mathrm{K}}_{n=1}^{\infty} \left(\frac{a_n z}{1} \right) \tag{6.1.48}$$

converges to the meromorphic function

$$f(z) = \frac{\Psi(c; z)}{\Psi(c+1; z)} \qquad (6.1.49)$$

for all $z \in \mathbb{C}$.

(B) *The convergence is uniform on every compact subset of C which contains no poles of $f(z)$.*

(C) *$f(z)$ is holomorphic at $z=0$, and $f(0)=1$.*

Proof. Let

$$P_n = \Psi(c+n; z), \qquad n = 0,1,2,\dots .$$

Then it is easily shown that the power series P_n satisfy the three-term recurrence relations

$$P_n = P_{n+1} + a_{n+1} z P_{n+2}, \qquad n = 0,1,2,\dots .$$

Since

$$\lambda(a_{n+1} z) = 1, \quad \lambda(1) = 0, \quad \lambda(P_n/P_{n+1}) = 0, \qquad n = 0,1,2,\dots,$$

by Theorem 5.5 the regular C-fraction (6.1.40) corresponds (at $z=0$) to P_0/P_1. Further, since

$$\lim_{n \to \infty} a_n = 0,$$

the assertions of the theorem follow immediately from Theorems 5.13 and 5.14.

The regular C-fraction (6.1.48) can be expressed by

$$\frac{\Psi(c; z)}{\Psi(c+1; z)} = 1 + \cfrac{\dfrac{z}{c(c+1)}}{1} + \cfrac{\dfrac{z}{(c+1)(c+2)}}{1} + \cfrac{\dfrac{z}{(c+2)(c+3)}}{1} + \cdots,$$

$$(6.1.50)$$

which is equivalent to

$$\frac{\Psi(c; z)}{\Psi(c+1; z)} = 1 + \frac{z/c}{c+1} + \frac{z}{c+2} + \frac{z}{c+3} + \frac{z}{c+4} + \cdots . \qquad (6.1.51)$$

Both (6.1.50) and (6.1.51) are valid for all $z \in \mathbb{C}$.

Example 11. Bessel functions of first kind. From (5.2.12) it is easily seen that the Bessel function $J_\nu(z)$ of the first kind of order ν can be expressed by

$$J_\nu(z) = \frac{(z/2)^\nu}{\Gamma(\nu+1)} \Psi\left(\nu+1; -\frac{z^2}{4}\right), \qquad z \in \mathbb{C}, \qquad (6.1.52)$$

provided $\nu \notin [-1, -2, -3, \ldots]$. Thus

$$\frac{J_\nu(z)}{J_{\nu+1}(z)} = \frac{2(\nu+1)}{z} \frac{\Psi(\nu+1; -z^2/4)}{\Psi(\nu+2; -z^2/4)}, \qquad (6.1.53)$$

and hence, by Theorem 6.4 and an equivalence transformation, we obtain

$$\frac{J_{\nu+1}(z)}{J_\nu(z)} = \frac{z}{2(\nu+1)} - \frac{z^2}{2(\nu+2)} - \frac{z^2}{2(\nu+3)} - \frac{z^2}{2(\nu+4)} - \cdots, \qquad (6.1.54)$$

in agreement with Theorem 5.16. The representation (6.1.54) is valid for all $z \in \mathbb{C}$ provided $\nu \notin [-1, -2, -3, \ldots]$.

Example 12. Tangent function. Since

$$\sin z = z \Psi\left(\frac{3}{2}; -\frac{z^2}{4}\right) \quad \text{and} \quad \cos z = \Psi\left(\frac{1}{2}; -\frac{z^2}{4}\right),$$

we obtain from Theorem 6.4 and an equivalence transformation

$$\tan z = z \frac{\Psi(\frac{3}{2}; -z^2/4)}{\Psi(\frac{1}{2}; -z^2/4)}$$

$$= \frac{z}{1} - \frac{z^2}{3} - \frac{z^2}{5} - \frac{z^2}{7} - \frac{z^2}{9} - \cdots. \qquad (6.1.55)$$

The continued fraction in (6.1.55) converges and represents $\tan z$ for all $z \in \mathbb{C}$. Replacing z by iz in (6.1.55) gives

$$\tanh z = \frac{e^z - e^{-z}}{e^z + e^{-z}} = \frac{z}{1} + \frac{z^2}{3} + \frac{z^2}{5} + \frac{z^2}{7} + \frac{z^2}{9} + \cdots, \qquad (6.1.56)$$

valid for all $z \in \mathbb{C}$.

6.1.4 *Confluent Hypergeometric Functions* $\Omega(a, b; z)$.

The formal power series

$$\Omega(a, b; z) = 1 + ab\frac{z}{1!} + a(a+1)b(b+1)\frac{z^2}{2!} + \cdots, \qquad (6.1.57)$$

where $a, b \notin [0, -1, -2, -3, \ldots]$, converges only at $z = 0$. For certain of these divergent series, continued fractions provide a method of summability. Theorem 6.5 asserts that there exist regular C-fractions which correspond to certain ratios of the divergent Ω series. That the continued fractions converge to analytic functions is stated in Theorem 6.6. The final argument, that the divergent series (obtained as ratios of two Ω series) are, under additional conditions, asymptotic expansions of the analytic functions, is given in Chapter 10.

THEOREM 6.5. *Let* $\{a_n\}$ *be a sequence of complex numbers defined by*

$$a_{2m-1} = a + m - 1, \quad a_{2m} = b + m, \qquad m = 1, 2, 3, \ldots, \qquad (6.1.58)$$

where a *and* b *are complex constants such that* $a \notin [0, -1, -2, -3, \ldots]$ *and* $b \notin [-1, -2, -3, \ldots]$. *Let* $\{P_n\}$ *be a sequence of fps defined by*

$$P_{2m} = \Omega(a+m, b+m; -z), \qquad m = 0, 1, 2, \ldots, \qquad (6.1.59a)$$

$$P_{2m+1} = \Omega(a+m, b+m+1; -z), \qquad m = 0, 1, 2, \ldots. \qquad (6.1.59b)$$

Then for each $n = 0, 1, 2, \ldots$, *the regular* C-*fraction*

$$1 + \overset{\infty}{\underset{k=1}{\mathbf{K}}}\left(\frac{a_{n+k}z}{1}\right) \qquad (6.1.60)$$

corresponds (*at* $z = 0$) *to the fps* P_n/P_{n+1}.

The proof of Theorem 6.5 is analogous to the proofs of Theorems 6.1, 6.3 and 6.4; hence it is omitted. We note, in particular, that the regular C-fraction

$$1 + \overset{\infty}{\underset{n=1}{\mathbf{K}}}\left(\frac{a_n z}{1}\right) \qquad (6.1.61)$$

corresponds (at $z = 0$) to the fps

$$\frac{P_0}{P_1} = \frac{\Omega(a, b; -z)}{\Omega(a, b+1; -z)}. \qquad (6.1.62)$$

The regular C-fraction (6.1.61) can be expressed by

$$1 + \frac{az}{1} + \frac{(b+1)z}{1} + \frac{(a+1)z}{1} + \frac{(b+2)z}{1} + \frac{(a+2)z}{1} + \cdots. \quad (6.1.63)$$

THEOREM 6.6. *Let the sequence* $\{a_n\}$ *be defined by (6.1.58), where a and b are complex constants satisfying* $a \notin [0, -1, -2, \ldots]$ *and* $b \notin [-1, -2, -3, \ldots]$. *Then:*

(A) *For each positive constant* $k > 0$ *there exists a positive integer* $n(k)$ *such that the regular C-fraction*

$$1 + \mathop{K}_{m=1}^{\infty} \left(\frac{a_{n(k)+m} z}{1} \right) \quad (6.1.64)$$

converges uniformly on every compact subset of the cardioid domain

$$C_k = \left[z : |z| < \frac{1 + \cos \arg z}{k} \right] \quad (6.1.65)$$

to a function holomorphic in C_k.

(B) *The regular C-fraction*

$$1 + \mathop{K}_{n=1}^{\infty} \left(\frac{a_n z}{1} \right) \quad (6.1.66)$$

converges to a function $f(z)$ *meromorphic in the cut plane*

$$R = [z : |\arg z| < \pi], \quad (6.1.67)$$

and the convergence is uniform on every compact subset of R which contains no poles of $f(z)$.

Proof. To prove (A) we use the cardioid theorem (Theorem 4.57) with $g_n = \frac{1}{2}$, for all n. Let a, b and $k > 0$ be given. Then clearly there exists a positive integer $n(k)$ such that for all $m > n(k)$, the a_m will all be in the parabolic region

$$|a_m| - \mathrm{Re}(a_m) < \frac{k}{4}.$$

Thus it remains to investigate the convergence behavior of the series (4.5.21) in the case where the a_n satisfy (6.1.58). Suppose that $\sum_{n=1}^{\infty} d_n < \infty$, where

$$d_n = \left| \frac{a_1 a_3 \cdots a_{2n-1}}{a_2 a_4 \cdots a_{2n}} \right|, \quad n = 1, 2, 3, \ldots;$$

that is, the first series in (4.5.21) converges. Then the second series in (4.5.21) can be written as

$$\sum_{n=2}^{\infty} \frac{1}{d_n |a_{2n+1}|}.$$
(6.1.68)

Since $d_n \to 0$ and $|a_{2n+1}| = O(n)$ as $n \to \infty$, the series (6.1.68) must diverge. Hence at least one of the two series in (4.5.21) diverges, which proves (A).

The proof of (B) is similar to the proof of Theorem 5.14. Since R is the union of all cardioid domains C_k, $k > 0$, and since one shows easily that the meromorphic functions to which (6.1.66) converges on the various C_k are analytic continuations of each other, the existence of one function $f(z)$ meromorphic on R follows. ∎

6.2   Representations from Minimal Solutions

In this section we give a number of examples of continued fraction representations of functions derived from Gautschi's method based on Pincherle's theorem (Section 5.3, Theorem 5.7). Among the functions considered are the Legendre functions of the first kind, Coulomb wave functions, incomplete beta and gamma functions, and repeated integrals of the error function. From these last functions we derive a well-known continued fraction representation of the plasma dispersion function. For proofs that sequences of functions are minimal solutions of three-term recurrence relations, we refer to [Gautschi, 1967]. Other continued-fraction representations considered in the above reference include an example arising in the numerical computation of Fourier coefficients and Sturm-Liouville boundary-value problem with one boundary condition at infinity.

Example 1. Legendre functions of the first kind of degree α and order m are defined by

$$P_\alpha^m(z) = \frac{1}{\Gamma(1-m)} \left(\frac{z+1}{z-1} \right)^{m/2} F\left(-\alpha, \alpha+1; 1-m; \tfrac{1}{2} - \tfrac{1}{2}z\right), \qquad |1-z| < 2$$
(6.2.1)

where m is a non-negative integer, α is an arbitrary complex number and $|\arg(z-1)| < \pi$. Here F denotes the hypergeometric function (6.1.1). The factor $\Gamma(1-m)$ is included so that the right side of (6.2.1) has meaning when m is a positive integer. In this case, (6.2.1) should be interpreted as the limiting value as m (complex) tends to a positive integer. The analytic continuation of $P_\alpha^m(z)$ is single-valued analytic in the z-plane cut along the real axis from 1 to $-\infty$ (see [Erdelyi et al., 1953, Vol. 1, Chapter 3]).

Let z and α be fixed, α not an integer, and define $\{y_m\}$ by

$$y_m = P_\alpha^m(z), \qquad m = 0, 1, 2, \ldots . \tag{6.2.2}$$

Gautschi [1967, Section 6] has shown that, for $\mathrm{Re}(z) > 0$, $\{y_m\}$ is a minimal solution of the system of three-term recurrence relations

$$y_{m+1} = -\frac{2mz}{(z^2 - 1)^{1/2}} y_m - (m + \alpha)(m - \alpha - 1) y_{m-1} = 0, \qquad m = 1, 2, 3, \ldots . \tag{6.2.3}$$

Thus by Theorem 5.7 (Pincherle)

$$\frac{P_\alpha^m(z)}{P_\alpha^{m-1}(z)} = - \mathop{\mathrm{K}}_{k=0}^{\infty} \left(\frac{a_{m+k}}{b_{m+k}} \right), \qquad m = 1, 2, 3, \ldots, \tag{6.2.4a}$$

where

$$a_n = -(n + \alpha)(n - \alpha - 1), \quad b_n = -\frac{2nz}{(z^2 - 1)^{1/2}}, \qquad n = 1, 2, 3, \ldots . \tag{6.2.4b}$$

Another continued-fraction representation of the ratios $P_\alpha^m(z)/P_\alpha^{m-1}(z)$, for complex $\alpha = -\frac{1}{2} + i\tau$, can be obtained from [Gautschi, 1967, (6.8)]. Continued-fraction representations of Legendre functions of the second kind $Q_\alpha^m(z)$ can also be obtained from results of the preceding reference. However, we have already derived these continued fractions in Section 6.1.1, Example 6.

Example 2. Coulomb wave functions play an important role in the study of nuclear interactions. The *regular Coulomb wave function* $F_L(\eta, \rho)$ is defined

$$F_L(\eta, \rho) = \rho^{L+1} e^{-i\rho} C_L(\eta) \Phi(L + 1 - i\eta; 2L + 2; 2i\rho), \tag{6.2.5a}$$

where η is a real parameter, L a non-negative integer, $\rho > 0$ and

$$C_L(\eta) = \frac{2^L e^{-\pi\eta/2} |\Gamma(L + 1 + i\eta)|}{(2L + 1)!}. \tag{6.2.5b}$$

Here Φ denotes the confluent hypergeometric function (6.1.28). Gautschi [1967, Section 7] has shown that for fixed η and ρ, $\{F_L(\eta, \rho)\}$ is a minimal solution of the system of three-term recurrence relations

$$L\left[(L+1)^2+\eta^2\right]^{1/2}y_{L+1}=(2L+1)\left(\eta+\frac{L(L+1)}{\rho}\right)y_L$$

$$-(L+1)(L^2+\eta^2)^{1/2}y_{L-1}, \qquad L=1,2,3,\dots. \qquad (6.2.6)$$

Thus it follows from Theorem 5.7 (Pincherle) that

$$\frac{F_L(\eta,\rho)}{F_{L-1}(\eta,\rho)}=-\overset{\infty}{\underset{k=0}{\mathrm{K}}}\left(\frac{a_{L+k}}{b_{L+k}}\right), \qquad L=1,2,3,\dots, \qquad (6.2.7a)$$

where

$$b_L=\frac{(2L+1)\left[\eta+L(L+1)/\rho\right]}{L\left[(L+1)^2+\eta^2\right]^{1/2}}, \qquad L=1,2,3,\dots, \qquad (6.2.7b)$$

and

$$a_L=\frac{-(L+1)(L^2+\eta^2)^{1/2}}{L\left[(L+1)^2+\eta^2\right]^{1/2}}, \qquad L=1,2,3,\dots. \qquad (6.2.7c)$$

The preceding reference and [Gautschi, 1969b] contain helpful discussions on the computation of $F_L(\eta,\rho)$.

Example 3. Incomplete beta functions are defined by

$$B_x(p,q)=\int_0^x t^{p-1}(1-t)^{q-1}dt, \qquad p>0, \quad q>0, \quad 0\leqslant x\leqslant 1. \qquad (6.2.8)$$

It can be shown that

$$B_x(p,q)=p^{-1}x^p F(p,1-q;p+1;x), \qquad (6.2.9)$$

where F denotes the hypergeometric function (6.1.1), (see [Erdelyi et al., 1953, Vol. 1, p. 87]). In probability-distribution theory and mathematical statistics the ratio

$$I_x(p,q)=\frac{B_x(p,q)}{B_1(p,q)} \qquad (6.2.10)$$

is important (see, for example, [Abramowitz and Stegun, 1964, p. 944]). We note here that

$$B_1(p,q) = \frac{\Gamma(p)\Gamma(q)}{\Gamma(p+q)}. \tag{6.2.11}$$

Let p, q and x be fixed, and let $\{g_n\}$ be defined by

$$g_n = I_x(p+n,q), \qquad 0<p\leqslant 1, \quad q>0, \quad n=0,1,2,\ldots. \tag{6.2.12}$$

Gautschi [1967, Section 8] has shown that $\{g_n\}$ is a minimal solution of the system of three-term recurrence relations

$$g_{n+1} = \left(1 + \frac{n+p+q-1}{n+p}x\right)g_n - \frac{n+p+q-1}{n+p}xg_{n-1}, \qquad n=1,2,3,\ldots.$$

$$\tag{6.2.13}$$

It follows from this and Theorem 5.7 (Pincherle) that, for $m=1,2,3,\ldots$, $0<p\leqslant 1$, $q>0$, $0\leqslant x<1$,

$$\frac{I_x(p+m,q)}{I_x(p+m-1,q)} = -\overset{\infty}{\underset{n=1}{\mathbf{K}}}\left(\frac{F_n x}{1-F_n x}\right), \tag{6.2.14a}$$

where

$$F_n = -\frac{n+m+p+q-2}{n+m+p-1}, \qquad n=1,2,3,\ldots. \tag{6.2.14b}$$

We note here that the continued fraction on the right side of (6.2.14a) is equivalent to a general T-fraction [see (4.5.36) and Section 7.3].

Example 4. Incomplete gamma function. We consider here the incomplete gamma function

$$P(a,x) = \frac{1}{\Gamma(a)}\int_0^x e^{-t}t^{a-1}dt, \tag{6.2.15a}$$

with the restrictions

$$a>0, \qquad x>0. \tag{6.2.15b}$$

We note that $\Gamma(a)P(a,x)=\gamma(a,x)$, where $\gamma(a,z)$ is defined by (6.1.43). For fixed $a>0$ and $x>0$ we define $\{h_n\}$ by

$$h_n = P(a+n,x), \qquad n=0,1,2,\ldots. \tag{6.2.16}$$

Gautschi [1967, Section 8] has shown that $\{h_n\}$ is a minimal solution of the system of three-term recurrence formulas

$$y_{n+1}=\left(1+\frac{x}{a+n}\right)y_n-\frac{x}{a+n}y_{n-1}, \qquad n=1,2,3,\dots. \qquad (6.2.17)$$

Hence by Theorem 5.7 (Pincherle), for each $m=1,2,3,\dots$, $a>0$, $x>0$,

$$\frac{P(a+m,x)}{P(a+m-1,x)}=-\mathop{\mathbf{K}}_{n=1}^{\infty}\left(\frac{F_n x}{1-F_n x}\right), \qquad (6.2.18a)$$

where

$$F_n=-\frac{1}{n+a+m-1}, \qquad n=1,2,3,\dots. \qquad (6.2.18b)$$

Thus again we obtain a continued fraction of the same form as that in (6.2.14a) which is equivalent to a general T-fraction.

Example 5. Repeated integrals of the error function. The *complementary error function* erfc z is defined by

$$\operatorname{erfc} z=\frac{2}{\sqrt{\pi}}\int_z^{\infty}e^{-t^2}\,dt=1-\operatorname{erf} z, \qquad (6.2.19)$$

[see (6.1.38)]. Here the path of integration is subject to the restriction $\arg t\to\alpha$ with $|\arg\alpha|<\pi/4$ as $t\to\infty$ along the path (see [Abramowitz and Stegun, 1964, p. 297]). erfc z is an entire function of z. Repeated integrals of erfc z are defined by

$$I^{-1}\operatorname{erfc} z=\frac{2}{\sqrt{\pi}}e^{-z^2}, \qquad I^0\operatorname{erfc} z=\operatorname{erf} z, \qquad (6.2.20a)$$

$$I^n\operatorname{erfc} z=\int_z^{\infty}I^{n-1}\operatorname{erfc} t\,dt, \qquad n=1,2,3,\dots. \qquad (6.2.20b)$$

Further let $\{h_n\}$ and $\{g_n\}$ be defined by

$$h_n=e^{z^2}I^n\operatorname{erfc} z, \qquad\qquad n=-1,0,1,2,\dots, \qquad (6.2.21a)$$

$$g_n=(-1)^n e^{z^2}I^n\operatorname{erfc}(-z), \qquad n=-1,0,1,2,\dots. \qquad (6.2.21b)$$

Gautschi [1967, Section 9] has shown that $\{h_n\}$ is a minimal solution of the system of three-term recurrence relations

$$y_{n+1}=-\frac{z}{n+1}y_n+\frac{1}{2(n+1)}y_{n-1}, \qquad n=0,1,2,\dots, \qquad (6.2.22)$$

if $\mathrm{Re}(z)>0$, and $\{g_n\}$ is a minimal solution if $\mathrm{Re}(z)<0$. It follows from Theorem 5.7 (Pincherle) and an equivalence transformation that for each $m=0,1,2,\ldots,$

$$\frac{I^m\,\mathrm{erfc}\,z}{I^{m-1}\,\mathrm{erfc}\,z}=\frac{\frac{1}{2}}{z}+\frac{\frac{m+1}{2}}{z}+\frac{\frac{m+2}{2}}{z}+\frac{\frac{m+3}{2}}{z}+\cdots,\qquad \mathrm{Re}(z)>0.$$

$$(6.2.23)$$

In the special case with $m=0$ we obtain

$$2e^{z^2}\int_z^\infty e^{-t^2}\,dt=\frac{1}{z}+\frac{\frac{1}{2}}{z}+\frac{1}{z}+\frac{\frac{3}{2}}{z}+\frac{2}{z}+\frac{\frac{5}{2}}{z}+\frac{3}{z}+\cdots\qquad\text{for } \mathrm{Re}(z)>0,$$

$$(6.2.24)$$

and

$$\mathrm{erfc}\,z=\frac{e^{-z^2}}{\sqrt{\pi}}\left(\frac{1}{z}+\frac{\frac{1}{2}}{z}+\frac{1}{z}+\frac{\frac{3}{2}}{z}+\frac{2}{z}+\frac{\frac{5}{2}}{z}+\frac{3}{z}+\cdots\right)\qquad\text{for } \mathrm{Re}(z)>0.$$

$$(6.2.25)$$

By taking the even part of the continued fraction in (6.2.25) (see Theorem 2.10) one obtains the following representation of the *plasma dispersion function*:

$$P(\zeta)=\frac{1}{\sqrt{\pi}}\int_{-\infty}^\infty\frac{e^{-t^2}\,dt}{t-\zeta}=i\sqrt{\pi}\,e^{-\zeta^2}\mathrm{erfc}(-i\zeta)$$

$$=\mathop{\mathbf{K}}_{n=1}^\infty\left(\frac{a_n(\zeta)}{b_n(\zeta)}\right),\qquad \mathrm{Im}(\zeta)>0,\qquad\qquad(6.2.26a)$$

where

$$a_1(\zeta)=\zeta,\qquad a_n(\zeta)=\frac{-(n-1)(2n-3)}{2},\qquad n=2,3,4,\ldots,\qquad(6.2.26b)$$

$$b_n(\zeta)=-\zeta^2+2n-\tfrac{3}{2},\qquad n=1,2,3,\ldots.\qquad\qquad(6.2.26c)$$

It can be seen that the continued fraction in (6.2.26a) is the product of ζ and a real J-fraction in the variable $-\zeta^2$ [see (4.5.4)]. This continued fraction was used by Fried and Conte [1961] for computing a table of values of $P(\zeta)$ with complex ζ.

CHAPTER 7

Types of Corresponding Continued Fractions and Related Algorithms

The concept of correspondence of continued fractions with formal Laurent series (fLs) was introduced in Chapter 5. Some general theory of correspondence was developed there, and two types of corresponding continued fractions were discussed, the C-fractions and the P-fractions. Some applications of correspondence to obtain continued-fraction representations of analytic functions were given in Chapters 5 and 6. The present chapter deals with special properties of the three most important types of corresponding continued fractions: regular C-fractions (Section 7.1), associated continued fractions (Section 7.2) and general T-fractions (Section 7.3). J-fractions are discussed in connection with associated continued fractions, since they are essentially of the same type. Also g-fractions are considered with regular C-fractions.

A great deal of this chapter is devoted to algorithms. For each type of continued fraction an algorithm is given to compute the coefficients of the continued fraction in terms of the corresponding fLs. The quotient-difference algorithm (Section 7.1.2) and the FG algorithms (Section 7.3.2) can also be used to compute zeros and poles of analytic functions. Among the power series to which the quotient-difference algorithm can be applied are all normal series. Subsumed among these are the Pólya frequency series discussed in Section 7.1.2. Some examples of analytic functions represented by general T-fractions are given in Section 7.3.3; these consist of ratios of confluent hypergeometric functions and include as special cases the error function and Fresnel integrals. The close connection between J-fractions and the general theory of orthogonal polynomials is described briefly in Section 7.2.2. Stable polynomials are discussed in Section 7.4 and

ENCYCLOPEDIA OF MATHEMATICS and Its Applications, Gian-Carlo Rota (ed.). Vol. 11: William B. Jones and W. J. Thron, Continued Fractions.   ISBN 0-201-13510-8

a method based on continued fractions is given to determine whether a polynomial is stable.

There are other types of corresponding continued fractions which space does not permit us to describe in detail. One of these consists of the continued fractions (introduced by Frank [1952]) of the form

$$k_0\gamma_0 + \frac{k_0(1-\gamma_0\bar{\gamma}_0)z}{\bar{\gamma}_0 z} - \frac{1}{k_1\gamma_1 +} \frac{k_1(1-\gamma_1\bar{\gamma}_1)z}{\bar{\gamma}_1 z} - \frac{1}{k_2\gamma_2 +} \cdots,$$

where the k_n and γ_n are non-zero complex constants and $|\gamma_n| \neq 1$ for all n. The special case with all $k_n = 1$ was first discussed by Schur [1917] with $|\gamma_n| < 1$ in connection with functions bounded on the unit disk (see also [Thron, 1977]). Frank [1952] showed that every formal power series (fps) corresponds to a continued fraction of the above form and, conversely, every such continued fraction corresponds to a fps L; in fact, the Taylor series at $z = 0$ of the $2n$th and $(2n+1)$th approximants agree with L up to and including the terms involving z^n and z^{n-1}, respectively. An algorithm for the computation of the γ_n, when the power series and the k_n are known, is also given in the preceding reference. In [Frank and Perron, 1954] it is shown by examples that a continued fraction of the above type can converge to one function in part of the plane and to a different function in another part.

Algorithms for computing C-fractions from their corresponding power series have been given by Frank [1946a] and by Murphy and O'Donohoe [1977].

7.1 Regular C-Fractions

Perhaps the most important known type of continued fraction is the regular C-fraction

$$1 + \frac{a_1 z}{1 +} \frac{a_2 z}{1 +} \frac{a_3 z}{1 +} \cdots, \qquad a_n \neq 0. \tag{7.1.1}$$

Some general convergence theorems for regular C-fractions have been given in Section 4.5.2 and in Theorems 5.14, 5.15 and 6.6. Some correspondence properties for C-fractions were given in Corollary 5.3, and in Theorem 5.19 it was shown that the approximants of a regular C-fraction form a "staircase" sequence in the Padé table of the corresponding power series. Numerous examples of analytic functions represented by regular C-fractions were given in Section 6.1.

This section deals with further properties of regular C-fractions. The principal results center on the quotient-difference (qd) algorithm of Rutishauser [1954a]. It is shown that the qd algorithm gives a convenient numerical method for computing coefficients a_n of the regular C-fraction (7.1.1.) corresponding to a given formal power series (Theorems 7.6 and

7.7). In Theorems 7.8 and 7.9 it is shown that the qd algorithm can be used to compute poles of analytic functions. The qd algorithm can always be applied to a normal fps. Some other sufficient conditions to insure the applicability of the qd algorithm are given in Theorems 7.10 and 7.11. The latter result deals with Pólya frequency series and a class of functions introduced by Schoenberg [1948]. Finally in Theorem 7.12 we give a recent result of Arms and Edrei [1970] on the convergence of sequences of Padé approximants of Pólya frequency series. The application to regular C-fractions and associated continued fractions is briefly described.

7.1.1   Correspondence of Regular C-Fractions

THEOREM 7.1.

(A) *Every regular C-fraction* (7.1.1) *corresponds to a uniquely determined fps*

$$L = 1 + c_1 z + c_2 z^2 + c_3 z^3 + \cdots. \tag{7.1.2}$$

The order of correspondence of the nth approximant $f_n(z)$ is $v_n = n + 1$, so that the Taylor expansion of $f_n(z)$ agrees with L up to and including the term $c_n z^n$; that is,

$$f_n(z) = 1 + c_1 z + \cdots + c_n z^n + \gamma_{n+1}^{(n)} z^{n+1} + \cdots. \tag{7.1.3}$$

(B) *If two regular C-fractions* (7.1.1) *and*

$$1 + \frac{a_1^\dagger z}{1} + \frac{a_2^\dagger z}{1} + \frac{a_3^\dagger z}{1} + \cdots, \qquad a_n^\dagger \neq 0, \tag{7.1.4}$$

correspond to the same power series (7.1.2), *then*

$$a_n = a_n^\dagger, \qquad n = 1, 2, 3, \ldots. \tag{7.1.5}$$

Proof. Part (A) is an immediate consequence of Corollary 5.3 (A). To prove (B) we let $A_n(z)$ and $B_n(z)$ denote the nth numerator and denominator, respectively, of (7.1.1), and let $A_n^\dagger(z)$ and $B_n^\dagger(z)$ denote the nth numerator and denominator, respectively, of (7.1.4). It is easily verified that $a_1 = c_1 = a_1^\dagger$. Proceeding by induction, we assume that

$$a_m = a_m^\dagger, \qquad m = 1, 2, \ldots, n. \tag{7.1.6}$$

Then

$$\frac{A_{n+1}^\dagger(z)}{B_{n+1}^\dagger(z)} - \frac{A_n^\dagger(z)}{B_n^\dagger(z)} = \frac{(-1)^n a_1 a_2 \cdots a_n a_{n+1}^\dagger}{B_{n+1}^\dagger(z) B_n^\dagger(z)} z^{n+1}$$

$$= (-1)^n a_1 a_2 \cdots a_n a_{n+1}^\dagger z^{n+1}$$

$$+ \left(\gamma_{n+2}^{(n+1)^\dagger} - \gamma_{n+2}^{(n)^\dagger} \right) z^{n+2} + \cdots, \tag{7.1.7}$$

and

$$\frac{A_{n+1}(z)}{B_{n+1}(z)} - \frac{A_n(z)}{B_n(z)} = \frac{(-1)^n a_1 a_2 \cdots a_n a_{n+1}}{B_{n+1}(z) B_n(z)} z^{n+1}$$

$$= (-1)^n a_1 a_2 \cdots a_n a_{n+1} z^{n+1}$$

$$+ \left(\gamma_{n+2}^{(n+1)} - \gamma_{n+2}^{(n)} \right) z^{n+2} + \cdots . \qquad (7.1.8)$$

In view of (7.1.6) we have $A_n(z) = A_n^\dagger(z)$ and $B_n(z) = B_n^\dagger(z)$. Hence subtracting (7.1.7) from (7.1.8) gives

$$\frac{A_{n+1}(z)}{B_{n+1}(z)} - \frac{A_{n+1}^\dagger(z)}{B_{n+1}^\dagger(z)} = \frac{(-1)^n a_1 a_2 \cdots a_n \left[a_{n+1} B_{n+1}^\dagger(z) - a_{n+1}^\dagger B_{n+1}(z) \right] z^{n+1}}{B_{n+1}(z) B_{n+1}^\dagger(z) B_n(z)}$$

$$= (-1)^n a_1 a_2 \cdots a_n \left(a_{n+1} - a_{n+1}^\dagger \right) z^{n+1} + \left(\gamma_{n+2}^{(n+1)} - \gamma_{n+2}^{(n+1)\dagger} \right) z^{n+2} + \cdots .$$

$$\qquad (7.1.9)$$

Since the Taylor series at $z=0$ of A_{n+1}/B_{n+1} and $A_{n+1}^\dagger/B_{n+1}^\dagger$ both agree with (7.1.2) up to and including the term $c_{n+1} z^{n+1}$, it follows from (7.1.9) that $a_{n+1} = a_{n+1}^\dagger$. ∎

The following theorem gives necessary and sufficient conditions such that for a given fps (7.1.2) there will exist a corresponding regular *C*-fraction. It also provides explicit formulas for the coefficients of the regular *C*-fraction in terms of the coefficients of the fps. The conditions are formulated in terms of the *Hankel determinants* $H_k^{(n)}$ (of dimension k) associated with the fps

$$L = c_0 + c_1 z + c_2 z^2 + c_3 z^3 + \cdots ,$$

defined by

$$H_0^{(n)} = 1;$$

$$H_k^{(n)} = \begin{vmatrix} c_n & c_{n+1} & \cdots & c_{n+k-1} \\ c_{n+1} & c_{n+2} & \cdots & c_{n+k} \\ \vdots & \vdots & & \vdots \\ c_{n+k-1} & c_{n+k} & \cdots & c_{n+2k-2} \end{vmatrix}, \qquad k = 1,2,3,\ldots . \qquad (7.1.10)$$

THEOREM 7.2.

(A) *If for a given fps*

$$L = 1 + c_1 z + c_2 z^2 + c_3 z^3 + \cdots \qquad (7.1.11)$$

there exists a regular C-fraction

$$1 + \overset{\infty}{\underset{n=1}{\mathbf{K}}} \left(\frac{a_n z}{1} \right), \qquad a_n \neq 0, \tag{7.1.12}$$

which corresponds to L (at z = 0), then

$$H_k^{(1)} \neq 0 \text{ and } H_k^{(2)} \neq 0, \qquad k = 1, 2, 3, \ldots, \tag{7.1.13}$$

and

$$a_1 = H_1^{(1)};$$

$$a_{2m} = -\frac{H_{m-1}^{(1)} H_m^{(2)}}{H_m^{(1)} H_{m-1}^{(2)}}, \quad a_{2m+1} = -\frac{H_{m+1}^{(1)} H_{m-1}^{(2)}}{H_m^{(1)} H_m^{(2)}}, \quad m = 1, 2, 3, \ldots . \tag{7.1.14}$$

(B) *Conversely, if* (7.1.13) *holds, then the regular C-fraction* (7.1.12) *with coefficients a_n defined by* (7.1.14) *corresponds to* (7.1.11).

Proof. For convenience we adopt the notation

$$\varphi_m = H_m^{(1)}, \quad \psi_m = H_{m-1}^{(2)}, \qquad m = 1, 2, 3, \ldots . \tag{7.1.15}$$

(A): First we investigate the conditions that must be satisfied if there exists a corresponding regular C-fraction (7.1.12). We shall use the terminology for the nth numerator $A_n(z)$ and denominator $B_n(z)$ of (7.1.12) defined in (5.5.29), and we shall write

$$\frac{A_n(z)}{B_n(z)} = 1 + c_1 z + \cdots + c_n z^n + \gamma_{n+1}^{(n)} z^{n+1} + \cdots \tag{7.1.16}$$

for the Taylor series (at $z = 0$) of $A_n(z)/B_n(z)$. From (5.5.29b, c) and (7.1.16) we have

$$(1 + \beta_{m,1} z + \cdots + \beta_{m,m} z^m)(1 + c_1 z + \cdots + c_{2m} z^{2m} + \gamma_{2m+1}^{(2m)} z^{2m+1} + \cdots)$$
$$= 1 + \alpha_{m,1} z + \cdots + \alpha_{m,m} z^m,$$

from which we obtain, upon equating coefficients of like powers of z,

$$c_1 \beta_{m,m} + c_2 \beta_{m,m-1} + \cdots + c_m \beta_{m,1} = -c_{m+1},$$
$$c_2 \beta_{m,m} + c_3 \beta_{m,m-1} + \cdots + c_{m+1} \beta_{m,1} = -c_{m+2},$$
$$\vdots \tag{7.1.17}$$
$$c_m \beta_{m,m} + c_{m+1} \beta_{m,m-1} + \cdots + c_{2m-1} \beta_{m,1} = -c_{2m}.$$

It follows that

$$\varphi_m \beta_{m,m} = (-1)^m \psi_{m+1}, \qquad m = 1, 2, 3, \ldots. \qquad (7.1.18)$$

Similarly, from (5.5.29d, e) and (7.1.16) we have

$$(1 + \delta_{m,1} z + \cdots + \delta_{m,m-1} z^{m-1})$$
$$\times (1 + c_1 z + \cdots + c_{2m-1} z^{2m-1} + \gamma_{2m}^{(2m-1)} z^{2m} + \cdots)$$
$$= 1 + \gamma_{m,1} z + \cdots + \gamma_{m,m} z^m,$$

and upon equating coefficients of like powers of z we obtain

$$c_1 \delta_{m,m-1} + c_2 \delta_{m,m-2} + \cdots + c_{m-1} \delta_{m,1} - \gamma_{m,m} = -c_m,$$
$$c_2 \delta_{m,m-1} + c_3 \delta_{m,m-2} + \cdots + c_m \delta_{m,1} = -c_{m+1},$$
$$\vdots \qquad (7.1.19)$$
$$c_m \delta_{m,m-1} + c_{m+1} \delta_{m,m-2} + \cdots + c_{2m-2} \delta_{m,1} = -c_{2m-1}.$$

It follows that

$$\psi_m \gamma_{m,m} = (-1)^{m-1} \varphi_m, \qquad m = 1, 2, 3, \ldots. \qquad (7.1.20)$$

Now combining (5.5.29f, g), (7.1.18) and (7.1.20) gives

$$a_2 a_4 \cdots a_{2m} \varphi_m = (-1)^m \psi_{m+1}, \qquad m = 1, 2, 3, \ldots, \qquad (7.1.21a)$$

and

$$a_1 a_3 \cdots a_{2m-1} \psi_m = (-1)^{m-1} \varphi_m, \qquad m = 1, 2, 3, \ldots. \qquad (7.1.21b)$$

Taking $m = 1$ and noting that $\psi_1 = 1$, we obtain

$$a_1 = \varphi_1 \quad \text{and} \quad a_2 \varphi_1 = -\psi_2.$$

Since $a_1 \neq 0$, we must have $\varphi_1 \neq 0$, and since $a_2 \neq 0$, we must have $\psi_2 \neq 0$. Continuing in this manner we conclude that (7.1.13) and (7.1.14) must hold. This proves (A).

To prove (B) we assume that (7.1.13) holds and let (7.1.12) be the regular C-fraction with coefficients defined by (7.1.14). It follows from Theorem 7.1 that the regular C-fraction (7.1.12) corresponds to some fps

$$\hat{L} = 1 + \hat{c}_1 z + \hat{c}_2 z^2 + \hat{c}_3 z^3 + \cdots$$

at $z = 0$. Now the procedure employed in proving part (A) to define the a_n in terms of the c_n can be applied also to the coefficients $\hat{c}_n$, and this will give the same a_n. It is easily shown that the sequence $\{a_n\}$ satisfying (7.1.14) uniquely determines the sequence $\{c_n\}$. Thus $\hat{c}_n = c_n$, $n = 1, 2, 3, \ldots$. ∎

From (5.5.15) and (7.1.10) we see that the determinants $c_{m,n}$ and $H_k^{(n)}$ are related by

$$c_{m,n} = (-1)^{n(n-1)/2} H_n^{(m-n+1)}, \qquad m, n = 0, 1, 2, \ldots, \qquad (7.1.22)$$

if we set $c_0 = 1$ and $c_k = 0$ for $k < 0$. From (7.1.22) one can see a clear connection between Theorems 7.2 and (5.19). In particular, for a given fps (7.1.11), there will exist a regular C-fraction (7.1.12) whose approximants form a staircase sequence in the Padé table of L as in Theorem 5.19, provided

$$c_{n,n} = (-1)^{n(n+1)/2} H_n^{(1)} \neq 0, \qquad n = 1, 2, 3, \ldots, \qquad (7.1.23a)$$

and

$$c_{n+1,n} = (-1)^{n(n+1)/2} H_n^{(2)} \neq 0, \qquad n = 1, 2, 3, \ldots. \qquad (7.1.23b)$$

The following corollary is an immediate consequence of Theorem 7.2:

COROLLARY 7.3.

(A) *If for a given fps*

$$L_n = 1 + c_n z + c_{n+1} z^2 + c_{n+2} z^3 + \cdots \qquad (7.1.24)$$

there exists a regular C-fraction

$$1 + \frac{a_1^{(n)} z}{1} + \frac{a_2^{(n)} z}{1} + \frac{a_3^{(n)} z}{1} + \cdots, \qquad a_k^{(n)} \neq 0, \qquad (7.1.25)$$

which corresponds to L_n (at $z = 0$), then

$$H_k^{(n)} \neq 0 \quad \text{and} \quad H_k^{(n+1)} \neq 0, \qquad k = 1, 2, 3, \ldots, \qquad (7.1.26)$$

and

$$a_1^{(n)} = H_1^{(n)} = c_n;$$

$$a_{2m}^{(n)} = -\frac{H_{m-1}^{(n)} H_m^{(n+1)}}{H_m^{(n)} H_{m-1}^{(n+1)}}, \qquad m = 1, 2, 3, \ldots, \qquad (7.1.27a)$$

$$a_{2m+1}^{(n)} = -\frac{H_{m+1}^{(n)} H_{m-1}^{(n+1)}}{H_m^{(n)} H_m^{(n+1)}}, \qquad m = 1, 2, 3, \ldots. \qquad (7.1.27b)$$

(B) *Conversely, if (7.1.26) holds, then the regular C-fraction defined by (7.1.25) and (7.1.27) corresponds to L_n at $z=0$.*

The following result follows easily from the proof of Theorem 7.2(A).

THEOREM 7.4. *If*

$$L = 1 + c_1 z + c_2 z^2 + c_3 z^3 + \cdots$$

is the Taylor series at $z=0$ of a finite regular C-fraction

$$1 + \frac{a_1 z}{1} + \frac{a_2 z}{1} + \cdots + \frac{a_n z}{1}, \qquad a_k \neq 0, \quad k = 1, 2, \ldots, n,$$

then the coefficients $a_1, a_2, \ldots, a_n$ satisfy (7.1.14) and the Hankel determinants involved are all non-zero.

7.1.2.   Quotient-Difference Algorithm

We shall describe now the *quotient-difference* algorithm of H. Rutishauser [1954a, b, c]. For a given fps

$$L = c_0 + c_1 z + c_2 z^2 + c_3 z^3 + \cdots, \tag{7.1.28}$$

we define sequences $\{e_m^{(n)}\}$ and $\{q_m^{(n)}\}$ as follows:

$$e_0^{(n)} = 0, \qquad\qquad n = 1, 2, 3, \ldots, \tag{7.1.29a}$$

$$q_1^{(n)} = \frac{c_{n+1}}{c_n}, \qquad n = 0, 1, 2, \ldots, \tag{7.1.29b}$$

$$e_m^{(n)} = q_m^{(n+1)} - q_m^{(n)} + e_{m-1}^{(n+1)}, \qquad m = 1, 2, 3, \ldots, \quad n = 0, 1, 2, \ldots, \tag{7.1.29c}$$

$$q_{m+1}^{(n)} = \frac{e_m^{(n+1)}}{e_m^{(n)}} q_m^{(n+1)}, \qquad m = 1, 2, 3, \ldots, \quad n = 0, 1, 2, \ldots. \tag{7.1.29d}$$

It is tacitly assumed that no $e_m^{(n)}$ vanishes, so that the $\{e_m^{(n)}\}$ and $\{q_m^{(n)}\}$ are defined by (7.1.29). The two-dimensional array shown in Table 7.1.1 is called the *quotient-difference table* (or *qd table*). Any four elements of the table forming a rhombus (as illustrated) are related by the equations (7.1.29); hence these equations are sometimes called the *rhombus rules*. Any element of each such rhombus can be computed in terms of the other three elements by means of the rhombus rules. Thus, for example, if one knows the first two columns of a qd table, the remaining columns can be computed column by column using the rhombus rules. Also if one knows

Table 7.1.1. Quotient-Difference Table

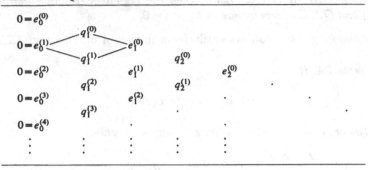

the elements of a diagonal, such as

$$0 = e_0^{(0)}, \ q_1^{(0)}, \ e_1^{(0)}, \ q_2^{(0)}, \ e_2^{(0)}, \ldots,$$

the remaining elements can be computed diagonal by diagonal by the rhombus rules. A quotient-difference algorithm (qd algorithm) is a computational procedure of this type based on the rhombus rules (7.1.29). Both procedures described above are known to be numerically unstable. However, the computation by columns is found to be more unstable than that by diagonals (see, for example, [Henrici, 1974, Section 7.6]).

The significance of the qd algorithm for continued fractions is made clear, in part, by the following

THEOREM 7.5.

(A) *If the Hankel determinants associated with a fps*

$$L = 1 + c_1 z + c_2 z^2 + c_3 z^3 + \cdots \qquad (7.1.30)$$

satisfy the conditions

$$H_m^{(n)} \neq 0, \qquad m = 1, 2, \ldots, k, \quad n = 0, 1, 2, \ldots, \qquad (7.1.31)$$

then in the qd table associated with L the columns $q_m^{(n)}$, $m = 1, 2, \ldots, k$, *exist, and for* $n = 0, 1, 2, \ldots$,

$$q_m^{(n)} = \frac{H_{m-1}^{(n)} H_m^{(n+1)}}{H_m^{(n)} H_{m-1}^{(n+1)}}, \qquad m = 1, 2, \ldots, k, \qquad (7.1.32a)$$

$$e_m^{(n)} = \frac{H_{m+1}^{(n)} H_{m-1}^{(n+1)}}{H_m^{(n)} H_m^{(n+1)}}, \qquad m = 1, 2, \ldots, k. \qquad (7.1.32b)$$

(B) *If the fps is normal, then (7.1.31) is satisfied for all $k \geqslant 1$.*

Proof. (A): Let $e_m^{(n)}$ and $q_m^{(n)}$ denote the elements of the qd table of L. Define sequences $\{\hat{e}_m^{(n)}\}$ and $\{\hat{q}_m^{(n)}\}$ by

$$\hat{e}_0^{(n)} = 0, \qquad n = 0, 1, 2, \ldots,$$

$$\hat{q}_m^{(n)} = \frac{H_{m-1}^{(n)} H_m^{(n+1)}}{H_m^{(n)} H_{m-1}^{(n+1)}}, \qquad \hat{e}_m^{(n)} = \frac{H_{m+1}^{(n)} H_{m-1}^{(n+1)}}{H_m^{(n)} H_m^{(n+1)}},$$

$$m = 1, 2, \ldots, k, \quad n = 0, 1, 2, \ldots.$$

It follows that for $n = 0, 1, 2, \ldots,$

$$\hat{e}_0^{(n)} = 0 = e_0^{(n)} \quad \text{and} \quad \hat{q}_1^{(n)} = \frac{c_{n+1}}{c_n} = q_1^{(n)},$$

where $c_0 = 1$. Thus it remains to show that the $\hat{e}_m^{(n)}$ and $\hat{q}_m^{(n)}$ satisfy the rhombus rules (7.1.29c, d). It is easy to verify the identity

$$\frac{\hat{e}_m^{(n+1)}}{\hat{e}_m^{(n)}} \hat{q}_m^{(n+1)} = \hat{q}_{m+1}^{(n)}.$$

By using Jacobi's identity [Henrici, 1974, Theorem 7.5a]

$$\left[H_m^{(n)} \right]^2 - H_m^{(n-1)} H_m^{(n+1)} + H_{m+1}^{(n-1)} H_{m-1}^{(n+1)} = 0, \tag{7.1.33}$$

one can also show that

$$\hat{q}_m^{(n+1)} - \hat{q}_m^{(n)} + \hat{e}_{m-1}^{(n+1)} = \hat{e}_m^{(n)},$$

which proves (A). (B) is an immediate consequence of Theorem 5.18 and (7.1.22). ∎

Combining Corollary 7.3 with Theorem 7.5 we obtain

THEOREM 7.6. *Let*

$$L = 1 + c_1 z + c_2 z^2 + c_3 z^3 + \cdots \tag{7.1.34}$$

be a normal fps, and let $e_m^{(n)}, q_m^{(n)}$ denote the elements of the qd table of L. Then for each $n = 1, 2, 3, \ldots$, the regular C-fraction

$$1 + \frac{c_n z}{1} - \frac{q_1^{(n)} z}{1} - \frac{e_1^{(n)} z}{1} - \frac{q_2^{(n)} z}{1} - \frac{e_2^{(n)} z}{1} - \cdots \tag{7.1.35}$$

corresponds (at $z=0$) to the fps

$$L_n = 1 + c_n z + c_{n+1} z^2 + c_{n+2} z^3 + \cdots . \qquad (7.1.36)$$

In Chapter 6 a number of functions were represented by continued fractions of the form

$$g(z) = \frac{a_1}{1} + \frac{a_2 z}{1} + \frac{a_3 z}{1} + \cdots , \qquad a_n \neq 0 \qquad (7.1.37)$$

[see, for example, (6.1.11) and 6.1.32)]. The continued fraction (7.1.37) is obtained from the regular C-fraction

$$f(z) = 1 + \underset{n=1}{\overset{\infty}{\mathbf{K}}} \left(\frac{a_n z}{1} \right), \qquad a_n \neq 0, \qquad (7.1.38)$$

by first subtracting 1 and then dividing by z. If $g_n(z)$ and $f_n(z)$ denote the n approximants of (7.1.37) and (7.1.38), respectively, then

$$f_n(z) = 1 + z g_n(z), \qquad n = 0, 1, 2, \ldots . \qquad (7.1.39)$$

Since (7.1.38) corresponds to a fps

$$\hat{L} = 1 + \hat{c}_1 z + \hat{c}_2 z^2 + \hat{c}_3 z^3 + \cdots , \qquad (7.1.40)$$

in the sense that the Taylor series at $z=0$ of $f_n(z)$ has the form

$$f_n(z) = 1 + \hat{c}_1 z + \cdots + \hat{c}_n z^n + \gamma_{n+1}^{(n)} z^{n+1} + \cdots , \qquad (7.1.41)$$

the continued fraction (7.1.37) corresponds at $z=0$ to the fps

$$L = c_0 + c_1 z + c_2 z^2 + c_3 z^3 + \cdots ,$$

where

$$\hat{c}_n = c_{n-1}, \qquad n = 1, 2, 3, \ldots ,$$

in the sense that $g_n(z)$ has a Taylor series at $z=0$ of the form

$$g_n(z) = \hat{c}_1 + \hat{c}_2 z + \cdots + \hat{c}_n z^{n-1} + \gamma_{n+1}^{(n)} z^n + \cdots$$
$$= c_0 + c_1 z + \cdots + c_{n-1} z^{n-1} + \gamma_{n+1}^{(n)} z^n + \cdots . \qquad (7.1.42)$$

Applying this discussion together with Theorem 7.6, we arrive at

THEOREM 7.7. *Let the fps*

$$L = c_0 + c_1 z + c_2 z^2 + c_3 z^3 + \cdots , \qquad c_0 \neq 0, \qquad (7.1.43)$$

be normal, and let $e_m^{(n)}$ and $q_m^{(n)}$ denote the elements of the qd table associated with L. Then the continued fraction

$$\frac{c_0}{1} - \frac{q_1^{(0)}z}{1} - \frac{e_1^{(0)}z}{1} - \frac{q_2^{(0)}z}{1} - \frac{e_2^{(0)}z}{1} - \cdots \qquad (7.1.44)$$

corresponds to L at $z=0$ in the sense that the n th approximant $g_n(z)$ has a Taylor series at $z=0$ of the form (7.1.42).

The qd algorithm provides a convenient numerical procedure for computing the coefficients of continued fractions corresponding to a given normal formal power series. We shall consider now a few examples.

Example 1. Arctangent. The fps

$$L = 1 + z - \frac{z^2}{3} + \frac{z^3}{5} - \frac{z^4}{7} + \frac{z^5}{9} - \cdots \qquad (7.1.45)$$

converges to the function

$$f(z) = 1 + z^{1/2} \arctan z^{1/2} \qquad (7.1.46)$$

at least for $|z| \leqslant 1$, $z \neq -1$. The initial part of the qd table for L is given in Table 7.1.2. From this (and Theorem 7.6) we can see that the regular C-fraction

$$1 + \frac{z}{1} + \frac{\frac{1}{3}z}{1} + \frac{\frac{2^2}{3 \times 5}z}{1} + \frac{\frac{3^2}{5 \times 7}z}{1} + \frac{\frac{4^2}{7 \times 9}z}{1} + \cdots \qquad (7.1.47)$$

corresponds (at $z=0$) to the fps (7.1.37). The continued fraction (7.1.47) is equivalent to

$$1 + \frac{z}{1} + \frac{1^2 \cdot z}{3} + \frac{2^2 \cdot z}{5} + \frac{3^2 \cdot z}{7} + \frac{4^2 \cdot z}{9} + \cdots, \qquad (7.1.48)$$

which agrees with the results of Example 1 in Section 6.1.

Example 2. Hypergeometric series. For the series

$$F(a, 1; c; z) = 1 + \frac{a}{c}z + \frac{a(a+1)}{c(c+1)}z^2 + \frac{a(a+1)(a+2)}{c(c+1)(c+2)}z^3 + \cdots, \qquad (7.1.49)$$

with $c \notin \{0, -1, -2, \ldots\}$, one obtains for the elements $e_m^{(n)}$, $q_m^{(n)}$ of the

Table 7.1.2.　Qd *table for* $1+z^{1/2}\arctan z^{1/2}$.

$e_0^{(n)}$	$q_1^{(n)}$	$e_1^{(n)}$	$q_2^{(n)}$	$e_2^{(n)}$	$q_3^{(n)}$	$e_3^{(n)}$	$\cdots$
$e_0^{(0)}=0$							
	1						
$e_0^{(1)}=0$		$-\dfrac{2^2}{1\times3}$					
	$-\dfrac13$		$-\dfrac{1^2}{3\times5}$				
$e_0^{(2)}=0$		$-\dfrac{2^2}{3\times5}$		$-\dfrac{4^2}{5\times7}$			
	$-\dfrac35$		$-\dfrac{3^2}{5\times7}$		$-\dfrac{3^2}{7\times9}$		
$e_0^{(3)}=0$		$-\dfrac{2^2}{5\times7}$		$-\dfrac{4^2}{7\times9}$		$-\dfrac{6^2}{9\times11}$	
	$-\dfrac57$		$-\dfrac{5^2}{7\times9}$		$-\dfrac{5^2}{9\times11}$	$\vdots$	$\ddots$
$e_0^{(4)}=0$		$-\dfrac{2^7}{7\times9}$		$-\dfrac{4^2}{9\times11}$	$\vdots$	$\vdots$	
	$-\dfrac79$		$-\dfrac{7^2}{9\times11}$	$\vdots$			
$e_0^{(5)}=0$		$-\dfrac{2^2}{9\times11}$	$\vdots$				
	$-\dfrac{9}{11}$	$\vdots$					
$e_0^{(6)}=0$	$\vdots$						
$\vdots$							

associated qd table

$$q_1^{(n)}=\frac{a+n}{c+n},\qquad n=0,1,2,\ldots,\tag{7.1.50a}$$

$$q_m^{(n)}=\frac{(a+n+m-1)(c+n+m-2)}{(c+n+2m-3)(c+n+2m-2)},\qquad m=2,3,4,\ldots\ ;n\geqslant0,\tag{7.1.50b}$$

$$e_m^{(n)}=\frac{m(c-a+m-1)}{(c+n+2m-2)(c+n+2m-1)},\qquad m=1,2,3,\ldots\ ;n\geqslant0.\tag{7.1.50c}$$

Thus by Theorem 7.7 the continued fraction

$$\cfrac{1}{1}\ -\ \cfrac{\dfrac{a}{c}z}{1}\ -\ \cfrac{\dfrac{1(c-a)}{c(c+1)}z}{1}\ -\ \cfrac{\dfrac{(a+1)(c)}{(c+1)(c+2)}z}{1}\ -\cdots\tag{7.1.51}$$

corresponds (at $z=0$) to (7.1.49), in agreement with (6.1.12).

Example 3. Confluent hypergeometric series. For the series

$$\Phi(1;c;z)=1+\frac{1}{c}z+\frac{1}{c(c+1)}z^2+\frac{1}{c(c+1)(c+2)}z^3+\cdots, \quad (7.1.52)$$

with $c\notin[0,-1,-2,...]$, one obtains for the elements $e_m^{(n)}$, $q_m^{(n)}$ of the associated qd table

$$q_1^{(n)}=\frac{1}{c+n}, \quad n=0,1,2,..., \qquad\qquad (7.1.53a)$$

$$q_m^{(n)}=\frac{c+n+m-2}{(c+n+2m-3)(c+n+2m-2)}, \quad m=2,3,4,...;n\geqslant 0, \qquad\qquad (7.1.53b)$$

$$e_m^{(n)}=\frac{-m}{(c+n+2m-2)(c+n+2m-1)}, \quad m=1,2,3,...;n\geqslant 0. \qquad\qquad (7.1.53c)$$

Thus by Theorem 7.7 the continued fraction

$$\frac{1}{1}-\frac{\frac{1}{c}z}{1}+\frac{\frac{1}{c(c+1)}z}{1}-\frac{\frac{c}{(c+1)(c+2)}z}{1}+\frac{\frac{2}{(c+2)(c+3)}z}{1}-\cdots$$

$$(7.1.54)$$

corresponds (at $z=0$) to (7.1.52), in agreement with (6.1.32).

The qd algorithm can also be used to compute zeros and poles of certain functions, as is shown by the following

THEOREM 7.8. *Let*

$$L=c_0+c_1z+c_2z^2+c_3z^3+\cdots \qquad\qquad (7.1.55)$$

be the Taylor series at $z=0$ of a function $f(z)$ holomorphic at the origin and meromorphic in the disk $D_r=[z:|z|<r]$. Let the poles z_j of $f(z)$ in D_r be arranged such that

$$0<|z_1|\leqslant|z_2|\leqslant|z_3|\leqslant\cdots<r, \qquad\qquad (7.1.56)$$

where each pole occurs as many times in the sequence as its order. Let the Hankel determinants associated with L satisfy

$$H_m^{(n)}\neq 0, \quad m=1,2,...,k, \quad n=0,1,2,.... \qquad\qquad (7.1.57)$$

Then:

(A) *For each m such that* $0 < m \leqslant k$ *and*

$$|z_{m-1}| < |z_m| < |z_{m+1}|, \qquad (7.1.58)$$

we have

$$\lim_{n \to \infty} q_m^{(n)} = \frac{1}{z_m}. \qquad (7.1.59)$$

Here $z_0 = 0$, *and if f has only k poles,* $z_{k+1} = \infty$.
(B) *For each m such that* $0 < m \leqslant k$ *and*

$$|z_{m-1}| < |z_m| < |z_{m+1}| < |z_{m+2}|, \qquad (7.1.60)$$

we have

$$\lim_{n \to \infty} e_m^{(n)} = 0. \qquad (7.1.61)$$

Proof. Both (A) and (B) are immediate consequences of (7.1.32) and the following property of Hankel determinants (see, for example, [Henrici, 1974, Theorem 7.5b]). When the hypotheses of the theorem are satisfied, there exists a constant $b_m \neq 0$ independent of n such that

$$H_m^{(n)} = b_m(\zeta_1\zeta_2 \cdots \zeta_m)^n \left[1 + O\left(\left(\frac{\rho}{|\zeta_m|} \right)^n \right) \right] \qquad (7.1.62)$$

for each ρ such that $|\zeta_m| > \rho > |\zeta_{m+1}|$. Here we have set $\zeta_i = 1/z_j$. ∎

The preceding theorem gives conditions by which the qd algorithm can be used to compute poles of meromorphic functions. However, the poles must be simple and the distance from the pole to the origin must differ from that of other poles. A treatment of the more general case in which several poles have the same modulus can be found in [Henrici, 1974, Section 7.9]. Also in that reference it is shown that the qd algorithm can be used to compute zeros of entire functions and polynomials in particular. The following result describes the manner in which the qd algorithm can be used to compute poles of functions represented by regular C-fractions.

THEOREM 7.9. *Let*

$$1 + \overset{\infty}{\underset{n=1}{\mathbf{K}}} \left(\frac{a_n z}{1} \right), \qquad a_n \neq 0, \qquad (7.1.63)$$

be a regular C-fraction which converges uniformly on compact subsets of a domain D containing the origin to a function $f(z)$ holomorphic at $z=0$ and meromophic in the disk $D_r=[z:|z|<r]$. Further suppose that (7.1.63) corresponds at $z=0$ to a fps

$$L=1+c_1z+c_2z^2+c_3z^3+\cdots,$$

whose associated Hankel determinants $H_m^{(n)}$ satisfy

$$H_m^{(n)}\neq0,\qquad m=1,2,\ldots,k,\quad n=0,1,2,\ldots. \qquad (7.1.64)$$

Let the poles z_j of $f(z)$ in D_r be arranged such that

$$0<|z_1|\leqslant|z_2|\leqslant|z_3|\leqslant\cdots<r, \qquad (7.1.65)$$

where each pole occurs as many times in the sequence as its order. Then for each m such that $0<m\leqslant k$ and

$$|z_{m-1}|<|z_m|<|z_{m+1}|, \qquad (7.1.66)$$

we have

$$\lim_{n\to\infty} q_m^{(n)}=\frac{1}{z_m}, \qquad (7.1.67)$$

where the $q_m^{(n)}$ are obtained from the qd algorithm (7.1.29) starting with the elements of the diagonal

$$q_m^{(1)}=-a_{2m},\quad e_m^{(1)}=-a_{2m+1},\qquad m=1,2,3,\ldots. \qquad (7.1.68)$$

Here $z_0=0$, and if f has only k poles in D_r, then $z_{k+1}=\infty$.

Proof. The theorem is a direct consequence of Theorems 5.13, 7.2, 7.5 and 7.8. ∎

A further discussion of these and other applications of the qd algorithm can be found in the following references: [Anderson, 1964], [Bandemer, 1964], [Bauer, 1954, 1959, 1967], [Gargantini and Henrici, 1967], [Henrici, 1958, 1963a,b, 1967, 1977], [Henrici and Watkins, 1965], [Householder, 1970, 1971], [Rutishauser, 1954a,b,c, 1955, 1956, 1957, 1962, 1963].

Theorem 7.9 can be modified for computing poles of rational functions represented by finite regular C-fractions, using Theorem 7.4. We conclude this section by stating some sufficient conditions to insure that the Hankel determinants will satisfy (7.1.57) in Theorem 7.8. The first can be found in [Henrici, 1974, Theorem 7.5g; 1977, Chapter 12].

THEOREM 7.10. *Let $\psi(t)$ be a real-valued, non-decreasing, bounded function on the interval $0 \leqslant t < \infty$, and let the Stieltjes integrals*

$$c_n = \int_0^\infty t^n \, d\psi(t), \qquad n = 0, 1, 2, \ldots \tag{7.1.69}$$

exist. Let $H_m^{(n)}$ denote the Hankel determinants associated with the fps

$$L = c_0 + c_1 z + c_2 z^2 + c_3 z^3 + \cdots. \tag{7.1.70}$$

Then:

(A) *If $\psi(t)$ has at least k points of increase, then*

$$H_m^{(n)} \neq 0, \qquad m = 1, 2, \ldots, k, \quad n = 0, 1, 2, \ldots. \tag{7.1.71}$$

(B) *If $\psi(t)$ has infinitely many points of increase, then (7.1.71) holds for all $k = 1, 2, 3, \ldots$.*

From Theorem 5.18 and (7.1.22) if can be seen that a fps

$$L = c_0 + c_1 z + c_2 z^2 + \cdots \tag{7.1.72}$$

is normal iff its associated Hankel determinants satisfy the conditions

$$H_m^{(n)} \neq 0 \qquad \text{for} \quad m = 0, 1, 2, \ldots \text{ and } n \geqslant -m + 1. \tag{7.1.73}$$

Clearly this condition insures that (7.1.57) holds for all $k > 0$. Also from (7.1.22) and (5.5.16) one sees that if the fps L is (M, N)-normal, then

$$H_m^{(n)} \neq 0 \qquad \text{if either} \quad 0 \leqslant m \leqslant M \quad \text{or} \quad -m + 1 \leqslant n \leqslant -m + N + 1. \tag{7.1.74}$$

Thus if L is (M, N)-normal, then (7.1.57) is satisfied with $k = M$.

A sequence $\{c_n\}_{n=0}^\infty$ of real numbers is called *totally positive* if the infinite matrix

$$C = \begin{bmatrix} c_0 & 0 & 0 & \cdots \\ c_1 & c_0 & 0 & \cdots \\ c_2 & c_1 & c_0 & \cdots \\ \vdots & \vdots & \vdots & \end{bmatrix} \tag{7.1.75}$$

has only non-negative minors (of all finite orders, with any choice of rows and columns). If $\{c_n\}$ is a totally positive sequence, then the fps

$$L = c_0 + c_1 z + c_2 z^2 + \cdots \tag{7.1.76}$$

is said to be a *Polya frequency* (PF) *series*. Closely related to PF series is the class $\mathfrak{S}$ of all analytic functions $f(z)$ representable in the form

$$f(z) = c_0 e^{\gamma z} \frac{\displaystyle\prod_{j=1}^{\infty} (1 + \alpha_j z)}{\displaystyle\prod_{j=1}^{\infty} (1 - \beta_j z)} \tag{7.1.77a}$$

where

$$c_0 > 0, \qquad \gamma \geqslant 0, \tag{7.1.77b}$$

$$\alpha_j \geqslant 0, \quad \beta_j \geqslant 0 \quad \text{for} \quad j = 1, 2, 3, \ldots, \tag{7.1.77c}$$

and

$$\sum_{j=1}^{\infty} (\alpha_j + \beta_j) < \infty. \tag{7.1.77d}$$

Schoenberg [1948] introduced the class of functions $\mathfrak{S}$ and proved that *if $f(z)$ is a member of $\mathfrak{S}$, then the Taylor series L of $f(z)$ at $z = 0$ is a PF series*. The converse of this theorem was conjectured by Schoenberg and later proved by Edrei [1953a, b]. A proof of the following result can be found in [Gragg, 1972, Theorem 7.1].

THEOREM 7.11 (Schoenberg, Karlin). *Let $f(z)$ be a member of the class $\mathfrak{S}$, and let M and N denote the number of positive α_j and β_j, respectively, in* (7.1.77). *Then*:

(A) *The Taylor series*

$$L = c_0 + c_1 z + c_2 z^2 + \cdots$$

of $f(z)$ at $z = 0$ is a PF series.

(B) *If $\gamma = 0$ and $M + N < \infty$, then L is (M, N)-normal and*

$$c_{m,n} > 0 \quad \text{for} \quad 0 \leqslant m \leqslant M \text{ and/or } 0 \leqslant n \leqslant M, \tag{7.1.78a}$$

$$c_{m,n} = 0 \quad \text{for} \quad m > M \text{ and } n > N. \tag{7.1.78b}$$

(C) *If $\gamma > 0$ or $M + N = \infty$, then L is normal and*

$$c_{m,n} > 0 \quad \text{for} \quad m, n = 0, 1, 2, \ldots. \tag{7.1.79}$$

Here $c_{m,n}$ is the determinant defined by (5.5.15).

The following convergence theorem for functions of class $\mathfrak{S}$ was proved by Arms and Edrei [1970].

THEOREM 7.12 (Arms and Edrei). *Let $f(z)$ be a function of class $\mathfrak{S}$, let L denote its Taylor series at $z=0$, and let*

$$R_{m,n}(z) = \frac{P_{m,n}(z)}{Q_{m,n}(z)} \qquad (P_{m,n}, Q_{m,n} \text{ relatively prime})$$

denote the (m, n) Padé approximant of L. Let $\{m_\nu\}_{\nu=1}^\infty$ and $\{n_\nu\}_{\nu=1}^\infty$ be sequences of positive integers such that

$$\lim_{\nu\to\infty} m_\nu = \lim_{\nu\to\infty} n_\nu = \infty \tag{7.1.80a}$$

and

$$\lim_{\nu\to\infty} \frac{m_\nu}{n_\nu} = \omega, \qquad \text{where} \quad 0 \leqslant \omega \leqslant +\infty. \tag{7.1.80b}$$

Then

$$\lim_{\nu\to\infty} P_{m_\nu, n_\nu}(z) = c_0 e^{\omega\gamma z/(1+\omega)} \prod_{j=1}^\infty (1+\alpha_j z), \tag{7.1.81a}$$

and

$$\lim_{\nu\to\infty} Q_{m_\nu, n_\nu}(z) = e^{-\gamma z/(1+\omega)} \prod_{j=1}^\infty (1-\beta_j z), \tag{7.1.81b}$$

and the convergence is uniform on every bounded region of C. In the special case $\gamma=0$, the assumption (7.1.80b) may be omitted.

An extension to Theorem 7.12 was given by Edrei [1974]. In Section 4.5 it was pointed out that the even part of a regular C-fraction is a continued fraction of the form

$$1 + \frac{k_1 z}{1+l_1 z} - \frac{k_2 z^2}{1+l_2 z} - \frac{k_3 z^2}{1+l_3 z} - \frac{k_4 z^2}{1+l_4 z} - \cdots, \qquad k_n \neq 0, \quad (7.1.82)$$

called an associated continued fraction. By Theorem 5.19, for a given normal Padé table, there exists a regular C-fraction whose approximants form a staircase sequence in the table. Thus there exists an associated continued fraction (7.1.82) whose nth approximant $f_n(z)$ is the (n, n) Padé approximant $R_{n,n}(z)$. Let $A_n(z)$ and $B_n(z)$ denote the nth numerator and denominator, respectively, of the associated continued fraction (7.1.82) whose approximants form a diagonal in the Padé table of a PL series

$$L = 1 + c_1 z + c_2 z^2 + c_3 z^3 + \cdots,$$

which is the Taylor series at $z=0$ of a function $f(z)$ of class $\mathcal{S}$. Then by Theorem 7.12

$$\lim_{n\to\infty} A_n(z)=e^{\gamma z/2}\prod_{j=1}^{\infty}(1+\alpha_j z) \tag{7.1.83a}$$

and

$$\lim_{n\to\infty} B_n(z)=e^{-\gamma z/2}\prod_{j=1}^{\infty}(1-\beta_j z), \tag{7.1.83b}$$

and the convergence is uniform on every bounded region of C. Moreover, Arms and Edrei [1970] have shown that if

$$A_n(z)=1+a_{n,1}z+a_{n,2}z^2+\cdots+a_{n,n}z^n \tag{7.1.84a}$$

and

$$B_n(z)=1+b_{n,1}z+b_{n,2}z^2+\cdots+b_{n,n}z^n, \tag{7.1.84b}$$

then

$$a_{n,j}>0 \text{ and } (-1)^n b_{n,j}>0, \qquad j=1,2,\ldots,n, \tag{7.1.85}$$

$$k_1>0; \qquad k_j<0, \ j=2,3,4,\ldots, \tag{7.1.86}$$

$$\lim_{j\to\infty} k_j=0, \tag{7.1.87}$$

$$-c_1<\sum_{j=1}^{n} l_j<0, \qquad n=1,2,3,\ldots, \tag{7.1.88}$$

and

$$\sum_{j=1}^{\infty} l_j=-\frac{\gamma}{2}-\sum_{j=1}^{\infty}\beta_j. \tag{7.1.89}$$

Many important functions of mathematical physics are members of class $\mathcal{S}$. An example is the function

$$z^{\nu/2}\left[J_\nu(2\sqrt{z})\right]^{-1}, \qquad \nu>-1, \tag{7.1.90}$$

where J_ν is the Bessel function of the first kind of order ν [see (5.2.12)]. Some other examples of functions of class $\mathcal{S}$ whose Taylor series at $z=0$

are normal include

$$\frac{\sin\sqrt{-z}}{\sqrt{-z}} = \prod_{n=1}^{\infty}\left(1+\frac{z}{n^2\pi^2}\right),$$ (7.1.91)

$$\frac{\sqrt{z}}{\sin\sqrt{z}} = \frac{1}{\displaystyle\prod_{n=1}^{\infty}\left(1-\frac{z}{n^2\pi^2}\right)},$$ (7.1.92)

and

$$\cos\sqrt{-z} = \prod_{n=1}^{\infty}\left(1+\frac{z}{\pi^2\left(n-\frac{1}{2}\right)^2}\right).$$ (7.1.93)

7.1.3 g-Fractions

We conclude this section by stating an algorithm for *g-fractions*

$$\frac{s_0}{1} + \frac{g_1 z}{1} + \frac{(1-g_1)g_2 z}{1} + \frac{(1-g_2)g_3 z}{1} + \cdots,$$ (7.1.94)

with

$$s_0 > 0, \qquad 0 < g_n < 1 \quad \text{for } n=1,2,3,\ldots.$$

A g-fraction corresponds to a fps L of the form

$$L = s_0 - s_1 z + s_2 z^2 - s_3 z^3 + \cdots.$$ (7.1.95)

In fact, if $f_n(z)$ denotes the nth approximant of (7.1.94), then the Taylor series at $z=0$ of $f_n(z)$ has the form

$$f_n(z) = s_0 - s_1 z + s_2 z^2 - \cdots + (-1)^{n-1}s_{n-1}z^{n-1} + \gamma_n^{(n)}z^n + \cdots.$$
   (7.1.96)

The coefficients g_n are the diagonal elements $g_n^{(0)}$ of the *g-table*

$$\begin{array}{ccccccc}
 & & g_1^{(0)} & & & & \\
 & g_0^{(1)} & & g_2^{(0)} & & & \\
 & & g_1^{(1)} & & g_3^{(0)} & & \\
 & g_0^{(2)} & & g_2^{(1)} & & g_4^{(0)} & \\
 & & g_1^{(2)} & & g_3^{(1)} & & \ddots \\
 & g_0^{(3)} & & g_2^{(2)} & & g_4^{(1)} & \\
 & \vdots & g_1^{(3)} & \vdots & g_3^{(2)} & \vdots & \ddots
\end{array}$$ (7.1.97)

whose entries are defined by the initial conditions

$$g_0^{(m)}=0, \qquad g_1^{(m)}=\frac{s_{m+1}}{s_m}, \tag{7.1.98a}$$

and the rhombus rules

$$\left(1-g_{2n+1}^{(m)}\right)\left(1-g_{2n+2}^{(m)}\right)=\left(1-g_{2n}^{(m+1)}\right)\left(1-g_{2n+1}^{(m+1)}\right), \tag{7.1.98b}$$

$$g_{2n}^{(m)}g_{2n+1}^{(m)}=g_{2n-1}^{(m+1)}g_{2n}^{(m+1)} \tag{7.1.98c}$$

[Bauer, 1959, 1960, 1965]. By Corollary 4.60 the g-fraction always converges to a function $f(z)$ holomorphic in the cut plane

$$Q_1=\left[z:|\arg(1+z)|<\pi\right]. \tag{7.1.99}$$

Wall [1940; 1948, Theorem 74.1 and (67.5)] has shown that the *class* $\mathfrak{W}$ of all functions $f(z)$ represented by g-fractions can be characterized in two ways:

(A) $f \in \mathfrak{W}$ iff $f(z)$ *is a non-rational function, holomorphic in* Q_1, *real-valued for real z, and which satisfies*

$$\mathrm{Re}\left(\sqrt{1+z}\,f(z)\right)>0 \qquad for \quad z \in Q_1. \tag{7.1.100}$$

(B) $f \in \mathfrak{W}$ iff *there exists a bounded non-decreasing function* $\psi(t)$ *with infinitely many points of increase on* $(0,1)$ *such that*

$$f(z)=\int_0^1 \frac{d\psi(t)}{1+tz} \qquad for \quad z \in Q_1. \tag{7.1.101}$$

An example of a function represented by a g-fraction is given in Chapter 8, Example 5, equation (8.3.56).

7.2   Associated Continued Fractions and J-Fractions

Continued fractions of the form

$$1+\frac{k_1 z}{1+l_1 z}-\frac{k_2 z^2}{1+l_2 z}-\frac{k_3 z^2}{1+l_3 z}-\frac{k_4 z^2}{1+l_4 z}-\cdots, \qquad k_n \neq 0, \tag{7.2.1}$$

where the k_n and l_n are complex constants, are called *associated continued fractions*. In Section 4.5.1 it was seen that the even part of a regular C-fraction is an associated continued fraction. However, there exist associated continued fractions which cannot be obtained as the even part of a regular C-fraction. Given a fps L, the existence of an associated continued fraction corresponding to L can be insured under weaker conditions than for regular C-fractions (see Theorem 7.14). Theorem 7.14 also gives the basic machinery used in Algorithm 7.2.1 for computing the coefficients k_n and l_n in (7.2.1). The connection between associated continued fractions and J-fractions is described by Theorem 7.15. Finally, Theorem 7.16 shows, in part, the important relationship between J-fractions and orthogonal polynomials.

7.2.1 Correspondence of Associated Continued Fractions

It follows from the difference equations (2.1.6) that the nth numerator $A_n(z)$ and denominator $B_n(z)$ of (7.2.1) are polynomials in z of the form

$$A_n(z)=1+a_{n,1}z+a_{n,2}z^2+\cdots+a_{n,n}z^n, \qquad n=0,1,2,\ldots, \qquad (7.2.2a)$$

$$B_n(z)=1+b_{n,1}z+b_{n,2}z^2+\cdots+b_{n,n}z^n, \qquad n=0,1,2,\ldots. \qquad (7.2.2b)$$

Thus the nth approximant $f_n(z)=A_n(z)/B_n(z)$ is holomorphic at the origin, and its Taylor series at $z=0$

$$\frac{A_n(z)}{B_n(z)}=1+\gamma_1^{(n)}z+\gamma_2^{(n)}z^2+\cdots \qquad (7.2.3)$$

has a positive radius of convergence.

THEOREM 7.13.

(A) *Every associated continued fraction (7.2.1) corresponds to a uniquely determined fps*

$$L=1+c_1z+c_2z^2+c_3z^3+\cdots. \qquad (7.2.4)$$

The order of correspondence of the nth approximant $A_n(z)/B_n(z)$ is

$$\nu_n=2n+1, \qquad (7.2.5)$$

and hence the Taylor series at $z=0$ of $A_n(z)/B_n(z)$ has the form

$$\frac{A_n(z)}{B_n(z)}=1+c_1z+c_2z^2+\cdots+c_{2n}z^{2n}+\gamma_{2n+1}^{(n)}z^{2n+1}+\cdots \qquad (7.2.6)$$

(B) *If two associated continued fractions (7.2.1) and*

$$1+\cfrac{k_1^\dagger z}{1+l_1^\dagger z}-\cfrac{k_2^\dagger z^2}{1+l_2^\dagger z}-\cfrac{k_3^\dagger z^2}{1+l_3^\dagger z}-\cdots \qquad (7.2.7)$$

correspond to the same fps (7.2.4), *then*

$$k_n = k_n^\dagger \quad \text{and} \quad l_n = l_n^\dagger, \qquad n = 1, 2, 3, \ldots. \tag{7.2.8}$$

Proof. (A): By using the determinant formulas (2.1.9) we obtain

$$\frac{A_{n+1}(z)}{B_{n+1}(z)} - \frac{A_n(z)}{B_n(z)} = \frac{k_1 k_2 \cdots k_{n+1}}{B_{n+1}(z)B_n(z)} z^{2n+1}$$

$$= k_1 k_2 \cdots k_{n+1} z^{2n+1} + \left(\gamma_{2n+2}^{(n+1)} - \gamma_{2n+2}^{(n)} \right) z^{2n+2} + \cdots. \tag{7.2.9}$$

Thus

$$\lambda \left(\frac{A_{n+1}(z)}{B_{n+1}(z)} - \frac{A_n(z)}{B_n(z)} \right) = 2n+1, \qquad n = 1, 2, 3, \ldots,$$

and hence part (A) follows immediately from Theorem 5.1.

To prove (B) we let $A_n^\dagger(z)$ and $B_n^\dagger(z)$ denote the nth numerator and denominator, respectively, of (7.2.7). It is easily verified that $k_1 = k_1^\dagger$ and $l_1 = l_1^\dagger$. We now assume that, for some integer n,

$$k_m = k_m^\dagger \quad \text{and} \quad l_m = l_m^\dagger \qquad \text{for} \quad m = 1, 2, \ldots, n \tag{7.2.10}$$

Then

$$\frac{A_{n+1}^\dagger(z)}{B_{n+1}^\dagger(z)} - \frac{A_n^\dagger(z)}{B_n^\dagger(z)} = \frac{k_1 k_2 \cdots k_n k_{n+1}^\dagger}{B_{n+1}^\dagger(z)B_n^\dagger(z)} z^{2n+1}$$

$$= k_1 k_2 \cdots k_n k_{n+1}^\dagger z^{2n+1} + \left(\gamma_{2n+2}^{(n+1)\dagger} - \gamma_{2n+2}^{(n)\dagger} \right) z^{2n+2} + \cdots. \tag{7.2.11}$$

It follows from (7.2.10) that $A_n^\dagger(z) = A_n(z)$ and $B_n^\dagger(z) = B_n(z)$. Thus by subtracting (7.2.11) from (7.2.9) we obtain

$$\frac{A_{n+1}(z)}{B_{n+1}(z)} - \frac{A_{n+1}^\dagger(z)}{B_{n+1}^\dagger(z)} = \frac{k_1 k_2 \cdots k_n \left[k_{n+1} B_{n+1}^\dagger(z) - k_{n+1}^\dagger B_{n+1}(z) \right]}{B_n(z)B_{n+1}(z)B_{n+1}^\dagger(z)} z^{2n+1}$$

$$= k_1 k_2 \cdots k_n \left(k_{n+1} - k_{n+1}^\dagger \right) z^{2n+1}$$

$$+ \left(\gamma_{n+2}^{(n+1)} - \gamma_{n+2}^{(n+1)\dagger} \right) z^{2n+2} + \cdots. \tag{7.2.12}$$

Since both continued fractions (7.2.1) and (7.2.7) correspond to L, it follows from part (A) that the Taylor series at $z = 0$ of $A_{n+1}(z)/B_{n+1}(z)$

and $A_{n+1}^\dagger(z)/B_{n+1}^\dagger(z)$ agree with L up to and including the term $c_{2n+2}z^{2n+2}$. Hence, by (7.2.12),

$$k_{n+1}=k_{n+1}^\dagger \quad \text{and} \quad \gamma_{2n+2}^{(n+1)}=\gamma_{2n+2}^{(n+1)\dagger}=c_{2n+2}. \tag{7.2.13}$$

By the difference equations (2.1.6) we have

$$B_{n+1}(z)=(1+l_{n+1}z)B_n(z)-k_{n+1}z^2B_{n-1}(z),$$

$$B_{n+1}^\dagger(z)=(1+l_{n+1}^\dagger z)B_n^\dagger(z)-k_{n+1}^\dagger z^2B_{n-1}^\dagger(z).$$

Therefore

$$B_{n+1}^\dagger(z)-B_{n+1}(z)=(l_{n+1}^\dagger-l_{n+1})zB_n(z). \tag{7.2.14}$$

Applying (7.2.14) and (7.2.13) in (7.2.12), we have

$$\frac{A_{n+1}(z)}{B_{n+1}(z)}-\frac{A_{n+1}^\dagger(z)}{B_{n+1}^\dagger(z)}=\frac{k_1k_2\cdots k_{n+1}(l_{n+1}^\dagger-l_{n+1})}{B_{n+1}(z)B_{n+1}^\dagger(z)}z^{2n+2}$$

$$=k_1k_2\cdots k_{n+1}(l_{n+1}^\dagger-l_{n+1})z^{2n+2}+(\gamma_{2n+3}^{(n+1)}-\gamma_{2n+3}^{(n+1)\dagger})z^{2n+3}+\cdots.$$

Hence $l_{n+1}^\dagger=l_{n+1}$, and so part (B) follows by induction on n. ■

The following theorem provides necessary and sufficient conditions such that for a given fps L there exists a corresponding associated continued fraction. Also explicit formulas are given for the coefficients of the continued fraction, and it is shown that the approximants of the continued fraction form the main diagonal in the Padé table of L.

THEOREM 7.14.

(A) *If for a given fps*

$$L=1+c_1z+c_2z^2+c_3z^3+\cdots, \tag{7.2.15}$$

there exists an associated continued fraction

$$1+\frac{k_1z}{1+l_1z}-\frac{k_2z^2}{1+l_2z}-\frac{k_3z^2}{1+l_3z}-\cdots, \quad k_n\neq0 \tag{7.2.16}$$

which corresponds to L, then

$$H_m^{(1)}\neq0, \quad m=1,2,3,\ldots, \tag{7.2.17}$$

where $H_m^{(1)}$ is the Hankel determinant associated with L. Moreover,

$$k_m = \frac{H_m^{(1)} H_{m-2}^{(1)}}{\left(H_{m-1}^{(1)}\right)^2}, \quad m = 1, 2, 3, \ldots \quad \left(H_{-1}^{(1)} = H_0^{(1)} = 1\right), \quad (7.2.18a)$$

and

$$l_m = \frac{X_{m-1}}{H_{m-1}^{(1)}} - \frac{X_m}{H_m^{(1)}}, \quad m = 1, 2, 3, \ldots, \quad (7.2.18b)$$

where

$$X_0 = 0, \quad X_1 = c_1, \quad (7.2.19a)$$

$$X_n = \begin{vmatrix} c_1 & c_2 & \cdots & c_{n-1} & c_{n+1} \\ c_2 & c_3 & \cdots & c_n & c_{n+2} \\ \vdots & \vdots & & \vdots & \vdots \\ c_n & c_{n+1} & \cdots & c_{2n-2} & c_{2n} \end{vmatrix}, \quad n = 2, 3, 4, \ldots. \quad (7.2.19b)$$

(B) *Suppose that (7.2.17) holds. Then the associated continued fraction (7.2.16) with coefficients (7.2.18) corresponds to L at $z = 0$. Let $f_n(z)$ and $B_n(z)$ denote the n th approximant and denominator of (7.2.16), respectively, with*

$$B_n(z) = 1 + b_{n,1} z + b_{n,2} z^2 + \cdots + b_{n,n} z^n. \quad (7.2.20)$$

Let σ_n and τ_n be defined by $(b_{n,0} = 1)$

$$\sigma_n = \sum_{j=0}^{n} b_{n,j} c_{2n+1-j}, \quad \sigma_n \tau_n = \sum_{j=0}^{n} b_{n,j} c_{2n+2-j}, \quad (7.2.21)$$

and let $R_{m,n}(z)$ denote the (m, n) Padé approximant of L. (It will be seen that $\sigma_n \neq 0$.) Then

$$f_n(z) = R_{n,n}(z), \quad n = 0, 1, 2, \ldots, \quad (7.2.22a)$$

$$b_{n,j} = b_{n-1,j} + l_n b_{n-1,j-1} - k_n b_{n-2,j-2}, \quad (7.2.22b)$$

$$b_{n,1} = b_{n-1,1} + l_n = (l_1 + l_2 + \cdots + l_n) = -\frac{X_n}{H_n^{(1)}}, \quad (7.2.22c)$$

$$\sigma_n = k_1 k_2 \cdots k_{n+1} = \frac{H_{n+1}^{(1)}}{H_n^{(1)}}, \quad (7.2.22d)$$

$$\sigma_n \tau_n = \frac{X_{n+1}}{H_n^{(1)}}, \quad (7.2.22e)$$

$$\tau_n = -(l_1 + l_2 + \cdots + l_{n+1}) = \frac{X_{n+1}}{H_{n+1}^{(1)}}, \quad (7.2.22f)$$

and

$$B_n(z) = \frac{1}{H_n^{(1)}} \begin{vmatrix} c_1 & c_2 & \cdots & c_{n+1} \\ c_2 & c_3 & \cdots & c_{n+2} \\ \vdots & \vdots & & \vdots \\ c_n & c_{n+1} & \cdots & c_{2n} \\ z^n & z^{n-1} & \cdots & 1 \end{vmatrix}. \tag{7.2.22g}$$

Proof. (A): Let the nth numerator $A_n(z)$ be written in the form (7.2.2a). Then by (7.2.6)

$$(1 + b_{n,1}z + \cdots + b_{n,n}z^n)(1 + c_1 z + \cdots + c_{2n}z^{2n} + \gamma_{2n+1}^{(n)}z^{2n+1} + \cdots)$$
$$= 1 + a_{n,1}z + \cdots + a_{n,n}z^n, \tag{7.2.23}$$

and upon equating coefficients of like powers of z, we obtain the system of equations

$$\begin{aligned} c_1 b_{n,n} + c_2 b_{n,n-1} + \cdots + c_n b_{n,1} &= -c_{n+1}, \\ c_2 b_{n,n} + c_3 b_{n,n-1} + \cdots + c_{n+1} b_{n,1} &= -c_{n+2}, \\ &\vdots \\ c_n b_{n,n} + c_{n+1} b_{n,n-1} + \cdots + c_{2n-1} b_{n,1} &= -c_{2n}, \end{aligned} \tag{7.2.24}$$

and

$$c_{n+1} b_{n,n} + c_{n+2} b_{n,n-1} + \cdots + c_{2n} b_{n,1} = -\gamma_{2n+1}^{(n)}. \tag{7.2.25}$$

From (7.2.6) and (7.2.9) one finds that

$$c_{2n+1} - \gamma_{2n+1}^{(n)} = k_1 k_2 \cdots k_{n+1}, \tag{7.2.26}$$

and hence by (7.2.25)

$$c_{n+1} b_{n,n} + c_{n+2} b_{n,n-1} + \cdots + c_{2n} b_{n,1} = k_1 k_2 \cdots k_{n+1} - c_{2n+1}. \tag{7.2.27}$$

By Cramer's rule, the system (7.2.24) yields

$$b_{n,n} H_n^{(1)} = - \begin{vmatrix} c_{n+1} & c_2 & \cdots & c_n \\ c_{n+2} & c_3 & \cdots & c_{n+1} \\ \vdots & \vdots & & \vdots \\ c_{2n} & c_{n+1} & \cdots & c_{2n-1} \end{vmatrix},$$

$$b_{n,n-1}H_n^{(1)} = -\begin{vmatrix} c_1 & c_{n+1} & c_3 & \cdots & c_n \\ c_2 & c_{n+2} & c_4 & \cdots & c_{n+1} \\ \vdots & \vdots & \vdots & & \vdots \\ c_n & c_{2n} & c_{n+1} & \cdots & c_{2n-1} \end{vmatrix}, \qquad (7.2.28)$$

$$\vdots$$

$$b_{n,1}H_n^{(1)} = -\begin{vmatrix} c_1 & c_2 & \cdots & c_{n-1} & c_{n+1} \\ c_2 & c_3 & \cdots & c_n & c_{n+2} \\ \vdots & \vdots & & \vdots & \vdots \\ c_n & c_{n+1} & \cdots & c_{2n-2} & c_{2n} \end{vmatrix}.$$

Multiplying both sides of (7.2.27) by $H_n^{(1)}$ and substituting (7.2.28) gives

$$H_{n+1}^{(1)} = k_1 k_2 \cdots k_{n+1} H_n^{(1)}, \qquad n = 0,1,2,\dots. \qquad (7.2.29)$$

That $H_1^{(1)} \neq 0$ follows from $H_1^{(1)} = c_1 = k_1 \neq 0$. The formulas (7.2.17) and (7.2.18a) can be proved by induction from (7.2.29). From (7.2.20) and the difference equations (2.1.6), we have

$$b_{1,1} = l_1; \qquad b_{n,1} = b_{n-1,1} + l_n, \qquad n = 2,3,4,\dots, \qquad (7.2.30)$$

and (7.2.28) implies that

$$b_{n,1} = -\frac{X_n}{H_n^{(1)}}. \qquad (7.2.31)$$

The formula (7.2.18b) follows immediately from (7.2.30) and (7.2.31), which completes the proof of (A).

To prove (B)' we assume that (7.2.17) holds and let the k_n and l_n be defined by (7.2.18). Since $k_n \neq 0$ for all n, the associated continued fraction (7.2.16) corresponds to some fps

$$\hat{L} = 1 + \hat{c}_1 z + \hat{c}_2 z^2 + \hat{c}_3 z^3 + \cdots,$$

by Theorem 7.13. The procedure used in part (A) to determine the k_n and l_n can now be applied to the coefficients $\hat{c}_n$, and this will give the same k_n and l_n. However, it is easily shown that the sequences $\{k_n\}$ and $\{l_n\}$ uniquely determine the sequence $\{c_n\}$ by (7.2.18). Hence $c_n = \hat{c}_n$ for all n, and so (7.2.16) corresponds to the fps (7.2.15). To prove (7.2.22a) it suffices to use (7.2.6) and the fact that $f_n(z)$ is a rational function of type $[n, n]$. The recurrence formulas (7.2.22b) follow from the difference equations (2.1.6). (7.2.22c) follows easily from (7.2.30) and (7.2.31). One can prove (7.2.22d) by (7.2.21), (7.2.27) and (7.2.29). The proof of (7.2.22e) can be seen by expanding the determinant χ_{n+1} by elements of the last column and then using (7.2.28) and (7.2.21). (7.2.22f) follows from (7.2.22c, d and e). Finally (7.2.22g) can be shown from (7.2.24).   ∎

The following algorithm, which is based on Theorem 7.14, is due to Gragg [1974]. The basic ideas apparently go back to Tchebycheff [1858] and can be found in [Wall, 1948, Chapter 11].

ALGORITHM 7.2.1.   Let the coefficients of the fps

$$L = 1 + c_1 z + c_2 z^2 + c_3 z^3 + \cdots$$

satisfy the conditions

$$H_n^{(1)} \neq 0, \qquad n = 1, 2, 3, \ldots, \tag{7.2.32}$$

where the $H_n^{(1)}$ are Hankel determinants associated with L. Then the coefficients k_n and l_n in the associated continued fraction (7.2.16) corresponding to L can be computed as follows: Set

$$\sigma_{-1} = 1, \qquad \tau_{-1} = 0, \qquad b_{0,0} = 1,$$

and compute, for $n = 0, 1, 2, \ldots,$

$$\sigma_n = \sum_{j=0}^{n} b_{n,j} c_{2n+1-j},$$

$$\tau_n = \frac{\sum_{j=0}^{n} b_{n,j} c_{2n+2-j}}{\sigma_n},$$

$$k_{n+1} = \frac{\sigma_n}{\sigma_{n-1}}, \qquad l_{n+1} = \tau_{n-1} - \tau_n,$$

$$b_{n-1,-1} = b_{n,n+1} = 0, \qquad b_{n+1,0} = 1,$$

and for $j = 1, 2, \ldots, n+1,$

$$b_{n+1,j} = b_{n,j} + l_{n+1} b_{n,j-1} - k_{n+1} b_{n-1,j-2}.$$

Algorithm 7.2.1 is more general than the qd algorithm, which serves the same purpose but also requires that all $H_m^{(n)} \neq 0$. Gragg [1974] warns that in practical experience the functions $k_n = k_n(c_1, c_2, \ldots, c_{2n-1})$ and $l_n = l_n(c_1, c_2, \ldots, c_{2n})$ are ill-conditioned functions of the c_k, although this has not been precisely quantified. Hence he recommends that the algorithm should be used only if the c_n are rational and rational arithmetic is

Table 7.2.1.   Results of Algorithm 7.2.1 Applied to $f(z)=1+\sqrt{z}\ \arctan\sqrt{z}$

n	k_n	l_n	σ_n	τ_n	$b_{n,0}$	$b_{n,1}$	$b_{n,2}$
-1			1	0			
0			1	$-\frac{1}{3}$	1		
1	1	$\frac{1}{3}$	$\frac{4}{45}$	$-\frac{6}{7}$	1	$\frac{1}{3}$	
2	$\frac{4}{45}$	$\frac{11}{21}$	$\frac{64}{11025}$	$-\frac{15}{11}$	1	$\frac{6}{7}$	$\frac{3}{35}$
3	$\frac{16}{245}$	$\frac{39}{77}$					

employed. (See [Sack and Donovan, 1972] for a related treatment of "modified moments," [Gautschi, 1969a, 1970] for a related analysis, and [Wheeler, 1974] for an application).

We shall consider the following elementary example.

Example 1.   The function

$$f(z)=1+\sqrt{z}\ \arctan\sqrt{z} \qquad (7.2.33)$$

has Taylor series at $z=0$ given by

$$L=1+z-\frac{z^2}{3}+\frac{z^3}{5}-\frac{z^4}{7}+\frac{z^5}{9}-\cdots, \qquad (7.2.34)$$

convergent for $|z|\leqslant 1$, $z\neq -1$. Applying Algorithm 7.2.1, we obtain the coefficients k_n, l_n of the corresponding associated continued fraction given in Table 7.2.1.

7.2.2   *J-Fractions and Orthogonal Polynomials*

If in the associated continued fraction (7.2.1) we let $z=1/\zeta$, omit the initial term 1 and make an equivalence transformation, we obtain the J-fraction

$$\frac{k_1}{l_1+\zeta}-\frac{k_2}{l_2+\zeta}-\frac{k_3}{l_3+\zeta}-\cdots, \qquad k_n\neq 0 \qquad (7.2.35)$$

[see (4.5.4)]. A convergence result for positive definite J-fractions was given in Theorem 4.61. The following theorem summarizes the connections between associated continued fractions and J-fractions. The proof is a simple application of Theorem 7.14.

THEOREM 7.15.   *Let $A_n(z)$ and $B_n(z)$ ($P_n(\zeta)$ and $Q_n(\zeta)$) denote the nth numerator and denominator, respectively, of the associated continued fraction*

(7.2.1) (*J-fraction* (7.2.35)), *where* $z = 1/\zeta$. *Further let*

$$L = 1 + c_1 z + c_2 z^2 + c_3 z^3 + \cdots. \qquad (7.2.36)$$

be the fps to which (7.2.1) *corresponds* (*at* $z = 0$). *Then*:

(A) *For* $n = 1, 2, 3, \ldots,$

$$B_n(z) = z^n Q_n(1/z), \qquad Q_n(\zeta) = \zeta^n B_n(1/\zeta), \qquad (7.2.37)$$

and

$$\frac{A_n(z)}{B_n(z)} = 1 + \frac{P_n(\zeta)}{Q_n(\zeta)}. \qquad (7.2.38)$$

(B) *The formal Laurent expansion of* $P_n(\zeta)/Q_n(\zeta)$ *at* $\zeta = \infty$ *has the form*

$$\frac{P_n(\zeta)}{Q_n(\zeta)} = \frac{c_1}{\zeta} + \frac{c_2}{\zeta^2} + \cdots + \frac{c_{2n}}{\zeta^{2n}} + \frac{\gamma_{2n+1}^{(n)}}{\zeta^{2n+1}} + \cdots, \qquad (7.2.39)$$

and hence the J-fraction (7.2.35) *corresponds at* $z = \infty$ *to the fLs*

$$L^* = \frac{c_1}{\zeta} + \frac{c_2}{\zeta^2} + \frac{c_3}{\zeta^3} + \cdots. \qquad (7.2.40)$$

(C) *For* $n = 1, 2, 3, \ldots,$ $Q_n(\zeta)$ *is a polynomial in* ζ *of degree* n *and can be expressed by the determinant formula*

$$Q_n(\zeta) = \frac{1}{H_n^{(1)}} \begin{vmatrix} c_1 & c_2 & \cdots & c_n & c_{n+1} \\ c_2 & c_3 & \cdots & c_{n+1} & c_{n+2} \\ \vdots & \vdots & & \vdots & \vdots \\ c_n & c_{n+1} & \cdots & c_{2n-1} & c_{2n} \\ 1 & \zeta & \cdots & \zeta^{n-1} & \zeta^n \end{vmatrix}$$

$$= \zeta^n + b_{n,1} \zeta^{n-1} + b_{n,2} \zeta^{n-2} + \cdots + b_{n,n}, \qquad (7.2.41)$$

where the $b_{n,j}$ *are defined as in* (7.2.20).

Historically the general theory of orthogonal polynomials originated in the theory of continued fractions (see [Tchebycheff, 1858] and [Stieltjes, 1894]). According to Szegö [1959, Section 3.5], the connection between orthogonal polynomials and continued fractions is of great importance and is one of the possible starting points for the treatment of orthogonal polynomials. We shall indicate this connection briefly.

Following Gragg [1974] (see also [Wall, 1948, Section 50]), we consider real sequences $\{c_n\}_{n=1}^{\infty}$ and the associated linear functional c^* defined on the vector space V of real polynomials and determined by

$$c^*(\zeta^n) = c_{n+1}, \qquad n = 0, 1, 2, \ldots.$$

We shall derive conditions for $\{c_n\}$ which insure that $(P, Q) = c^*(PQ)$ defines an inner product on V, and then show (Theorem 7.16) that the polynomials (7.2.41) form an orthogonal system with respect to this inner product. For two polynomials

$$P(\zeta) = \sum_{j=0}^{m} a_j \zeta^j \quad \text{and} \quad Q(\zeta) = \sum_{j=0}^{m} b_j \zeta^j,$$

the Cauchy product PQ is defined by

$$PQ(\zeta) = \sum_{j=0}^{2m} \left(\sum_{k=0}^{j} a_k b_{j-k} \right) \zeta^j,$$

where $a_k = 0$ and $b_k = 0$ if $k > m$. Hence

$$c^*(PQ) = \sum_{j=0}^{2m} \left(\sum_{k=0}^{j} a_k b_{j-k} \right) c_{j+1} = \mathbf{a}^T H \mathbf{b}, \tag{7.2.42}$$

where $\mathbf{a}^T = (a_0, a_1, \ldots, a_m, 0, \ldots)$ and $\mathbf{b}^T = (b_0, b_1, \ldots, b_m, 0, \ldots)$ are row vectors and H is the infinite matrix

$$H = \begin{pmatrix} c_1 & c_2 & c_3 & \cdots \\ c_2 & c_3 & c_4 & \cdots \\ c_3 & c_4 & c_5 & \cdots \\ \vdots & \vdots & \vdots & \end{pmatrix}. \tag{7.2.43}$$

Here T denotes the transpose of column vectors. The functional $c^*(PQ) = (P, Q)$ is an inner product iff H is positive definite or, equivalently,

$$H_m^{(1)} > 0 \qquad \text{for} \quad m = 1, 2, 3, \ldots, \tag{7.2.44}$$

where $H_m^{(1)}$ denotes the Hankel determinant associated with $\{c_n\}$ [see (7.1.10)]. It is well known (see, for example, [Perron, 1957a, Satz 4.15, p. 233]) that (7.2.44) will hold iff there exists a bounded non-decreasing function $\psi(t)$ with infinitely many points of increase such that

$$c_n = \int_{-\infty}^{\infty} t^{n-1} d\psi(t), \qquad n = 1, 2, 3, \ldots \tag{7.2.45}$$

(see the discussion of the Hamburger moment problem in Chapter 9, Theorem 9.8.B, with slightly different notation). In terms of the inner product $c^*(PQ)$, we say that two polynomials P and Q are *orthogonal with* respect to $\{c_n\}$ if

$$c^*(PQ) = 0. \tag{7.2.46}$$

That the denominators $Q_n(\zeta)$ of a J-fraction are orthogonal with respect to $\{c_n\}$, provided (7.2.44) holds, is given in the following:

THEOREM 7.16. *Let*

$$\frac{k_1}{l_1 + \zeta} - \frac{k_2}{l_2 + \zeta} - \frac{k_3}{l_3 + \zeta} - \cdots, \qquad k_n \neq 0, \tag{7.2.47}$$

be a J-fraction corresponding (at $z = \infty$) to a fLs

$$L^* = \frac{c_1}{\zeta} + \frac{c_2}{\zeta^2} + \frac{c_3}{\zeta^3} + \cdots, \tag{7.2.48}$$

such that the c_n are all real and (7.2.44) holds. Let $c^*(PQ)$ denote the inner product with respect to $\{c_n\}$, and let $Q_n(\zeta)$ denote the nth denominator of (7.2.47). Then for each $n = 0, 1, 2, \ldots,$

$$c^*(Q_m Q_n) = \begin{cases} 0 & \text{if } 0 \leqslant m \leqslant n-1, \\ \dfrac{H_{m+1}^{(1)}}{H_m^{(1)}} & \text{if } n = m, \end{cases} \tag{7.2.49a}$$

$$c^*(\zeta^m Q_n) = 0, \qquad 0 \leqslant m \leqslant n-1, \quad n \geqslant 1, \tag{7.2.49b}$$

$$c^*(\zeta^n Q_n) = c^*(Q_n^2) = \frac{H_{n+1}^{(1)}}{H_n^{(1)}}, \tag{7.2.49c}$$

and

$$c^*(\zeta^{n+1} Q_n) = \frac{X_{n+1}}{H_n^{(1)}}. \tag{7.2.49d}$$

Here $H_m^{(1)}$ and X_n denote the determinants (7.1.10) and (7.2.19b), respectively.

Proof. Our proof is by induction. One can easily verify (7.2.49) for $n = 0, 1$. Assuming that (7.2.49) is true for $0, 1, \ldots, n$, we shall show that it is also true for $n + 1$. Equations (7.2.49c) and (7.2.49d) are immediate consequences of (7.2.21), (7.2.22d) and (7.2.22e) for all n. By the difference equations (2.1.6) we have

$$Q_{n+1}(\zeta) = (l_{n+1} + \zeta)Q_n(\zeta) - k_{n+1}Q_{n-1}(\zeta),$$

and hence

$$c^*(Q_m Q_{n+1}) = l_{n+1} c^*(Q_m Q_n) + c^*(\zeta Q_m Q_n) - k_{n+1} c^*(Q_m Q_{n-1}).$$

$$(7.2.50)$$

Therefore applying the induction hypotheses yields

$$c^*(Q_m Q_{n+1}) = 0 \quad \text{for} \quad 0 \leqslant m \leqslant n-2.$$

Another application of the induction hypotheses (7.2.49a) and (7.2.49c) together with (7.2.18a) and (7.2.50) gives

$$c^*(Q_{n-1} Q_{n+1}) = 0.$$

From (7.2.50) with $m = n$ we have, after applying (7.2.49a, b) and (7.2.41),

$$c^*(Q_n Q_{n+1}) = l_{n+1} \frac{H_{n+1}^{(1)}}{H_n^{(1)}} + c^*(\zeta^{n+1} Q_n) + b_{n,1} c^*(\zeta^n Q_n).$$

Hence by (7.2.18b), (7.2.22c) and (7.2.49c, d),

$$c^*(Q_n Q_{n+1}) = \left(\frac{X_n}{H_n^{(1)}} - \frac{X_{n+1}}{H_{n+1}^{(1)}} \right) \frac{H_{n+1}^{(1)}}{H_n^{(1)}} + \frac{X_{n+1}}{H_{n+1}^{(1)}} - \frac{X_n}{H_n^{(1)}} \cdot \frac{H_{n+1}^{(1)}}{H_n^{(1)}} = 0.$$

Finally, that

$$c^*(\zeta^m Q_{n+1}) = 0 \quad \text{for} \quad 0 \leqslant m \leqslant n$$

follows from what has been proven and the fact that each ζ^m can be expressed in the form

$$\zeta^m = \alpha_0 Q_0(\zeta) + \alpha_1 Q_1(\zeta) + \cdots + \alpha_{m-1} Q_{m-1}(\zeta) + Q_m(\zeta). \qquad \blacksquare$$

For each pair (a, b) such that $-\infty \leqslant a < b \leqslant +\infty$, we let $\Phi(a, b)$ denote the family of all real-valued, bounded, non-decreasing functions $\psi(t)$ with infinitely many points of increase in $a \leqslant t \leqslant b$. In Theorem 7.16 it is assumed that the determinant conditions (7.2.44) hold and hence that there exists a function $\psi \in \Phi(a, b)$ for some $-\infty \leqslant a < b \leqslant +\infty$, such that (7.2.45) is valid. These conditions will be assumed throughout the remainder of this section. It is then easily seen that

$$c^*(q_m q_n) = \int_a^b q_m(t) q_n(t) \, d\psi(t)$$

for arbitrary polynomials $q_m(\zeta)$ and $q_n(\zeta)$. Thus Theorem 7.16 asserts that the denominators $Q_n(\zeta)$ of a J-fraction (7.2.47), with all $k_n > 0$ and l_n real, form an orthogonal polynomial system with respect to $\psi(t)$; that is, for each $n = 0, 1, \ldots,$ $Q_n(t)$ is a polynomial in t of degree n and

$$\int_a^b Q_m(t) Q_n(t) \, d\psi(t) = \begin{cases} 0 & \text{if} \quad m \neq n, \\ d_n \neq 0 & \text{if} \quad m = n. \end{cases} \tag{7.2.51}$$

From the difference equations (2.1.6), the $Q_n(\zeta)$ satisfy the system of three-term recurrence relations

$$Q_0(\zeta) = 1, \quad Q_1(\zeta) = l_1 + \zeta, \tag{7.2.52a}$$

$$Q_n(\zeta) = (l_n + \zeta) Q_{n-1}(\zeta) - k_n Q_{n-2}(\zeta), \quad n = 2, 3, 4, \ldots, \tag{7.2.52b}$$

and by (7.2.18) and (7.2.44) we have

$$k_n > 0 \quad \text{and} \quad l_n \text{ is real for } n = 1, 2, 3, \ldots. \tag{7.2.52c}$$

It is well known that every system $\{Q_n(\zeta)\}$ of polynomials orthogonal with respect to a distribution $\psi \in \Phi(a, b)$ satisfies a system of three-term recurrence relations of the form (7.2.52), provided that each Q_n is normalized to have its leading coefficients equal to one (see, for example, [Szegö, 1959, Theorem 3.2.1]). The proof is a simple consequence of the orthogonality conditions (7.2.51). The l_n and k_n in (7.2.52) determine a IJ-fraction (7.2.47) with nth denominator Q_n. *Thus every system of orthogonal polynomials can be obtained from the nth denominators of J-fractions.*

It can be shown that *if $\{Q_n\}$ is a sequence of polynomials satisfying a system of three-term recurrence relations of the form* (7.2.52)*, then there exists a $\psi \in \Phi(a, b)$ such that $\{Q_n\}$ is a system of polynomials orthogonal with respect to $\psi(t)$.* To prove this we note first that the l_n and k_n in (7.2.52) determine a J-fraction (7.2.47) which corresponds at $z = \infty$ (by Theorem 7.15) to a fLs

$$L^* = \frac{c_1}{\zeta} + \frac{c_2}{\zeta^2} + \frac{c_3}{\zeta^3} + \cdots. \tag{7.2.53}$$

Since $k_n > 0$ for $n \geqslant 1$, it follows from (7.2.18a) that (7.2.44) holds. Hence by Theorem 9.8(B) (with a slight change in notation), there exists a $\psi \in \Phi(a, b)$ such that

$$c_n = \int_a^b t^{n-1} \, d\psi(t), \quad n = 1, 2, 3, \ldots. \tag{7.2.54}$$

Therefore by Theorem 7.16 the Q_n form an orthogonal polynomial system with respect to ψ. This result has been attributed to Favard [1935], although it can be deduced from [Perron, 1929, Section 67].

The preceding remarks justify the claim that the theory of orthogonal polynomials can be developed from continued fractions. Actually, to some extent the implication goes both ways. Further results on the connection can be found in [Chihara, 1978] and [Szegö, 1968]. We note also the book by Brezinski [1980] and the article [Brezinski, 1977a] which uses orthogonal polynomials to derive the qd and epsilon algorithms.

We conclude this section with a brief discussion of connections between Gaussian quadrature and continued fractions. Let $\psi \in \Phi(a, b)$ be such that the moments (7.2.54) all exist. Then by Theorem 9.8(B) (with a slight difference in notation), $H_n^{(1)} > 0$ for $n \geqslant 1$, where $H_n^{(1)}$ is the Hankel determinant associated with the sequence of moments $\{c_n\}$. Therefore Theorems 7.14 and 7.15 imply that there exists a *J*-fraction (7.2.47) corresponding at $z = \infty$ to the fLs (7.2.53) and the coefficients k_n and l_n satisfy (7.2.52c). Let $P_n(\zeta)$ and $Q_n(\zeta)$ denote the nth numerator and denominator, respectively, of the *J*-fraction, so that the $Q_n(\zeta)$ form a system of polynomials orthogonal with respect to ψ. By Theorem 9.3(B) the zeros ζ_{kn} of $Q_n(\zeta)$ are real and distinct and satisfy

$$-\infty \leqslant a < \zeta_{1n} < \zeta_{2n} \cdots < \zeta_{nn} < b \leqslant +\infty.$$

If $f(\zeta)$ is continuous on $[a, b]$, then the Gaussian quadrature formula with error term $E_n(f)$ is given by

$$\int_a^b f(t)\,d\psi(t) = \sum_{k=1}^n \lambda_{kn} f(\zeta_{kn}) + E_n(f), \qquad (7.2.55a)$$

where the constants λ_{kn}, called Christoffel numbers, are given by

$$\lambda_{kn} = \frac{1}{Q_n'(\zeta_{kn})} \int_a^b \frac{Q_n(t)}{t - \zeta_{kn}}\,d\psi(t). \qquad (7.2.55b)$$

It is well known that the error term $E_n(f)$ vanishes if f is any polynomial of degree not exceeding $2n - 1$ (see, for example, [Hildebrand, 1956] or [Davis, 1963, Theorem 14.2.1]). By a theorem due to Stieltjes one has that if a and b are both finite and if $f(\zeta)$ is continuous on $[a, b]$, then

$$\lim_{n \to \infty} \sum_{k=1}^n \lambda_{kn} f(\zeta_{kn}) = \int_a^b f(t)\,d\psi(t) \qquad (7.2.56)$$

(see, for example, [Davis, 1963, Corollary 14.4.7]). Using the difference equations (2.1.6) for the *J*-fraction, one can easily show that

$$P_n(\zeta) = \int_a^b \frac{Q_n(\zeta) - Q_n(t)}{\zeta - t}\,d\psi(t). \qquad (7.2.57)$$

Thus we obtain the expression for the Christoffel numbers

$$\lambda_{kn} = \frac{P_n(\zeta_{kn})}{Q'_n(\zeta_{kn})}, \qquad k = 1, 2, \ldots, n, \qquad (7.2.58)$$

and

$$\frac{P_n(\zeta)}{Q_n(\zeta)} = \sum_{k=1}^{n} \frac{1}{\zeta - \zeta_{kn}} \frac{P_n(\zeta_{kn})}{Q'_n(\zeta_{kn})} = \sum_{k=1}^{n} \lambda_{kn} f(\zeta_{kn}). \qquad (7.2.59)$$

If now we set $f(t) = 1/(\zeta - t)$ and apply (7.2.56) and (7.2.59), we obtain

$$\int_a^b \frac{d\psi(t)}{\zeta - t} = \lim_{n \to \infty} \sum_{k=1}^{n} \frac{\lambda_{kn}}{\zeta - \zeta_{kn}} = \lim_{n \to \infty} \frac{P_n(\zeta)}{Q_n(\zeta)} \qquad (7.2.60)$$

for all complex values of ζ not in $[a, b]$. The result (7.2.60) is due to Markoff (see Theorem 9.9). We have shown that Markoff's theorem can be deduced from Stieltjes's result for Gaussian quadrature. The ideas for this proof were given by Szegö [1968].

7.3   General T-Fractions

Continued fractions of the form

$$1 + d_0 z + \frac{z}{1 + d_1 z} + \frac{z}{1 + d_2 z} + \frac{z}{1 + d_3 z} + \cdots, \qquad (7.3.1)$$

in which the d_n are arbitrary complex constants, were introduced by Thron in [1948]. He proved that there exists a one-to-one correspondence between continued fractions of the form (7.3.1) and fps

$$L = 1 + c_1 z + c_2 z^2 + c_3 z^3 + \cdots, \qquad (7.3.2)$$

such that the Taylor series of the nth approximant of (7.3.1) agrees with (7.3.2) term by term up to and including the term $c_n z^n$. He also gave a number of convergence results for continued fractions of this type. The general theory of T-fractions has also been studied in [Jones and Thron, 1966], [Jefferson, 1969a, b], [Hovstad, 1975], [Waadeland, 1964, 1966, 1967] and [Hag, 1970, 1972] (see Section 12.1 and 12.2).

In his book, Perron [1957a, Section 31] considered the more general continued fractions of the form

$$e_0 + d_0 z + \frac{z}{e_1 + d_1 z} + \frac{z}{e_2 + d_2 z} + \frac{z}{e_3 + d_3 z} + \cdots, \qquad e_n \neq 0, \qquad (7.3.3)$$

in which the e_n and d_n are arbitrary complex constants with $e_n \neq 0$ for all n. Perron called (7.3.3) the *Thronschen Kettenbrüche*. Hence, for abbreviation, (7.3.1) is said to be a *T-fraction* and (7.3.3) a *general T-fraction*. Consistent with a convention made in Section 4.5.1, if an arbitrary function of z is added to (7.3.1) or (7.3.3), the continued fraction remains of the same type. Also a continued fraction will remain of the same type if it undergoes an equivalence transformation. Thus a general *T*-fraction

$$\mathop{K}_{n=1}^{\infty}\left(\frac{z}{e_n + d_n z}\right), \qquad e_n \neq 0, \tag{7.3.4}$$

can also be expressed in the useful equivalent form

$$\mathop{K}_{n=1}^{\infty}\left(\frac{F_n z}{1 + G_n z}\right), \qquad F_n \neq 0, \tag{7.3.5}$$

where the coefficients are related by

$$F_1 = \frac{1}{e_1}; \qquad F_n = \frac{1}{e_{n-1}e_n}, \qquad n = 2,3,4,\ldots, \tag{7.3.6a}$$

$$G_n = \frac{d_n}{e_n}, \qquad n = 1,2,3,\ldots. \tag{7.3.6b}$$

Perron [1957a, Section 31] observed that a general *T*-fraction (7.3.3) not only corresponds to a fps (7.3.2) at $z = 0$ but, if

$$d_n \neq 0, \qquad n = 1,2,3,\ldots, \tag{7.3.7}$$

it also corresponds to a fLs

$$L^* = c_1^* z + c_0^* + c_{-1}^* z^{-1} + c_{-2}^* z^{-2} + \cdots$$

at $z = \infty$. However, Murphy and McCabe appear to be the first to have recognized that the approximants of (7.3.3), with (7.3.7) satisfied, belong to the two-point Padé table of the pair of series (L, L^*) (see [McCabe, 1975] and [McCabe and Murphy, 1976], where they preferred to work with the closely connected *M*-fractions defined in Section 4.5.1). This result was discovered independently by Jones and Thron [1977] (see also [Jones, 1977] and [Thron, 1977]).

Some convergence results for general *T*-fractions have already been given in Section 4.5.4. This section deals with further properties and applications of these continued fractions. Included are basic correspondence properties (Theorems 7.17 and 7.18), and an algorithm similar to the qd algorithm (called the *FG* algorithm) which can be used to compute coefficients of general T-fractions and also zeros and poles of analytic functions (Theorems 7.19, 7.21; Corollary 7.22).

Ince in [1919] has stated that, in certain specific cases, the general T-fraction converges near $z=0$ to a ratio of confluent hypergeometric functions and near $z=\infty$ to a ratio of different confluent hypergeometric functions. Other expansions were also considered by Wynn [1959] in work on converging factors, and by McCabe [1974]. In Section 7.3.3 we give the basic convergence theory for the representation of these functions by general T-fractions. Other examples of functions represented by general T-fractions can be found in [Drew and Murphy, 1977], [Frank, 1958, 1960a, b], [Gautschi, 1967, 1969b, 1977] (see Section 6.2, Examples 3 and 4), [Gautschi and Slavik, 1978], [Grundy, 1977, 1978a, b], [Maurer, 1966] and [Murphy, 1971].

Additional results on general T-fractions will be found in later parts of this book. Truncation-error analysis is discussed in Chapter 8, and a theory of moments based on general T-fractions is described briefly in Chapter 9.

7.3.1   Correspondence of General T-Fractions

We shall consider general T-fractions of the form

$$\sum_{k=1}^{\mu} c_k^* z^k + \sum_{k=-\nu}^{0} c_k z^k + \mathop{\mathrm{K}}_{n=1}^{\infty}\left(\frac{F_n z}{1+G_n z}\right), \qquad F_n \neq 0, \quad \mu \geqslant 0, \ \nu \geqslant 0,$$

$$(7.3.8)$$

or, equivalently,

$$\sum_{k=1}^{\mu} c_k^* z^k + \sum_{k=-\nu}^{0} c_k z^k + \mathop{\mathrm{K}}_{n=1}^{\infty}\left(\frac{z}{e_n+d_n z}\right), \qquad e_n \neq 0, \quad \mu \geqslant 0, \ \nu \geqslant 0,$$

$$(7.3.9)$$

where the coefficients F_n, G_n and e_n, d_n are related by (7.3.6). Let $A_n(z)$ and $B_n(z)$ denote the nth numerator and denominator, respectively, of (7.3.8) and (7.3.9). Then from the difference equations (2.1.6), we have

$$A_{-1}=1, \qquad B_{-1}=0, \tag{7.3.10a}$$

$$A_0(z)=\sum_{k=1}^{\mu} c_k^* z^k + \sum_{k=-\nu}^{0} c_k z^k, \qquad B_0(z)=1, \tag{7.3.10b}$$

and for $n \geqslant 1$, $A_n(z)$ and $B_n(z)$ have the form

$$A_n(z)=a_{n,-\nu}z^{-\nu}+a_{n,-\nu+1}z^{-\nu+1}+\cdots+a_{n,n+\mu}z^{n+\mu}, \tag{7.3.10c}$$

$$B_n(z)=b_{n,0}+b_{n,1}z+\cdots+b_{n,n}z^n, \tag{7.3.10d}$$

where

$$a_{n,-\nu} = c_{-\nu} e_1 e_2 \cdots e_n \qquad (7.3.10e)$$

$$a_{n,n+\mu} = \begin{cases} (1 + c_0 d_1) d_2 \cdots d_n & \text{if } \mu = 0 \\ c_\mu^* d_1 d_2 \cdots d_n & \text{if } \mu > 0 \text{ and } c_\mu^* \neq 0, \end{cases} \qquad (7.3.10f)$$

$$b_{n,0} = e_1 e_2 \cdots e_n, \qquad (7.3.10g)$$

$$b_{n,n} = d_1 d_2 \cdots d_n. \qquad (7.3.10h)$$

We are now ready for our first theorem on correspondence.

THEOREM 7.17.

(A) *Every general T-fraction*

$$\sum_{k=1}^{\mu} c_k^* z^k + \sum_{k=-\nu}^{0} c_k z^k + \mathop{\mathbf{K}}_{n=1}^{\infty} \left(\frac{z}{e_n + d_n z} \right), \qquad e_n \neq 0, \qquad (7.3.11)$$

corresponds at $z = 0$ to a uniquely determined fLs

$$L = \sum_{k=-\nu}^{\infty} c_k z^k. \qquad (7.3.12)$$

The order of correspondence of the n th approximant $f_n(z)$ is $\nu_n = n + 1$, and the Taylor series at $z = 0$ of $f_n(z)$ has the form

$$f_n(z) = \sum_{k=-\nu}^{-1} c_k z^k + c_0 + c_1 z + \cdots + c_n z^n + \gamma_{n+1}^{(n)} z^{n+1} + \cdots, \qquad (7.3.13)$$

where

$$c_1 - \gamma_1^{(0)} = \frac{1}{e_1}, \qquad (7.3.14a)$$

$$c_{n+1} - \gamma_{n+1}^{(n)} = \frac{(-1)^n}{(e_1 e_2 \cdots e_n)^2 e_{n+1}}, \qquad n = 1, 2, 3, \ldots. \qquad (7.3.14b)$$

(B) *If*

$$d_n \neq 0, \qquad n = 1, 2, 3, \ldots, \qquad (7.3.15)$$

then the general T-fraction (7.3.11) corresponds at $z = \infty$ to a uniquely determined fLs

$$L^* = \sum_{k=-\infty}^{\mu} c_k^* z^k. \qquad (7.3.16)$$

The order of correspondence of $f_n(z)$ is $v_n = -n$, and the Laurent series at $z = \infty$ of $f_n(z)$ has the form

$$f_n(z) = \sum_{k=1}^{\mu} c_k^* z^k + c_0^* + c_{-1}^* z^{-1} + \cdots$$

$$+ c_{-(n-1)}^* z^{-(n-1)} + \gamma_{-n}^{(n)^{\bullet}} z^{-n} + \cdots, \qquad (7.3.17)$$

where

$$c_0^* - \gamma_0^{(0)^{\bullet}} = \frac{1}{d_1}, \qquad (7.3.18a)$$

$$c_{-n}^* - \gamma_{-n}^{(n)^{\bullet}} = \frac{(-1)^n}{(d_1 d_2 \cdots d_n)^2 d_{n+1}}, \qquad n = 1, 2, 3, \ldots. \quad (7.3.18b)$$

Proof. (A): Let $A_n(z)$ and $B_n(z)$ denote the nth numerator and denominator, respectively, of (7.3.11). Then by (7.3.10) and the determinant formulas (2.1.9) we have

$$\frac{A_{n+1}(z)}{B_{n+1}(z)} - \frac{A_n(z)}{B_n(z)} = \frac{(-1)^n z^{n+1}}{B_n(z) B_{n+1}(z)}. \qquad (7.3.19)$$

Expanding (7.3.19) in increasing powers of z, we obtain a fps of the form

$$\frac{A_{n+1}(z)}{B_{n+1}(z)} - \frac{A_n(z)}{B_n(z)} = \frac{(-1)^n z^{n+1}}{(e_1 e_2 \cdots e_n)^2 e_{n+1}}$$

$$+ \alpha_{n+2} z^{n+2} + \alpha_{n+3} z^{n+3} + \cdots. \qquad (7.3.20)$$

Thus an application of Theorem 5.1 proves part (A). To prove part (B) we consider the expansion of (7.3.19) in decreasing powers of z and obtain a fLs of the form

$$\frac{A_{n+1}(z)}{B_{n+1}(z)} - \frac{A_n(z)}{B_n(z)} = \frac{(-1)^n z^{-n}}{(d_1 d_2 \cdots d_n)^2 d_{n+1}}$$

$$+ \beta_{-(n+1)} z^{-(n+1)} + \beta_{-(n+2)} z^{-(n+2)} + \cdots. \qquad (7.3.21)$$

Hence the proof of part (B) is completed by an application of Theorem 5.1.

∎

We consider now a pair (L, L^*) of fLs

$$L = \sum_{k=-\nu}^{\infty} c_k z^k, \quad L^* = \sum_{k=-\infty}^{\mu} c_k^* z^k, \quad \mu \geqslant 0, \quad \nu \geqslant 0, \quad (7.3.22)$$

and let

$$\delta_k = c_k^* - c_k, \quad k = 0, \pm 1, \pm 2, \ldots, \quad (7.3.23a)$$

where

$$c_k = 0 \text{ if } k < -\nu \quad \text{and} \quad c_k^* = 0 \text{ if } k > \mu. \quad (7.3.23b)$$

Associated with the pair (L, L^*) are the Hankel determinants $\mathcal{H}_k^{(n)}$ defined by

$$\mathcal{H}_0^{(n)} = 1;$$

$$\mathcal{H}_k^{(n)} = \begin{vmatrix} \delta_n & \delta_{n+1} & \cdots & \delta_{n+k-1} \\ \delta_{n+1} & \delta_{n+2} & \cdots & \delta_{n+k} \\ \vdots & \vdots & & \vdots \\ \delta_{n+k-1} & \delta_{n+k} & \cdots & \delta_{n+2k-2} \end{vmatrix}, \quad n = 0, \pm 1, \pm 2, \cdots.$$

$$(7.3.24)$$

If a general T-fraction (7.3.11) corresponds both to L given by (7.3.12) and to L^* given by (7.3.16), then we say that it *corresponds to the pair of fLs* (L, L^*). The following theorem gives necessary and sufficient conditions for there to exist a general T-fraction corresponding to a pair of fLs (L, L^*). It also gives explicit expressions for the coefficients of the continued fraction in terms of the Hankel determinants $\mathcal{H}_k^{(n)}$ and shows that the continued-fraction approximants belong to the two-point Padé table of (L, L^*). Parts (A) and (C) of the theorem were obtained by Murphy and McCabe (See [McCabe, 1975] and [McCabe and Murphy, 1976]). Theorem 7.18 was obtained independently by Jones and Thron [1977] (see also [Jones, 1977] and [Thron, 1977]).

THEOREM 7.18.

(A) *If for a given pair (L, L^*) of fLs*

$$L = \sum_{k=-\nu}^{\infty} c_k z^k \text{ and } L^* = \sum_{k=-\infty}^{\mu} c_k^* z^k, \quad \mu \geqslant 0, \quad \nu \geqslant 0, \quad (7.3.25)$$

there exists a general T-fraction

$$\sum_{k=1}^{\mu} c_k^* z^k + \sum_{k=-\nu}^{0} c_k z^k + \mathop{\mathbf{K}}_{n=1}^{\infty} \left(\frac{F_n z}{1 + G_n z} \right),$$ (7.3.26a)

with

$$F_n \neq 0 \quad \text{and} \quad G_n \neq 0, \qquad n = 1, 2, 3, \ldots,$$ (7.3.26b)

which corresponds to (L, L^*), *then*

$$\mathcal{H}_k^{(-k+1)} \neq 0, \quad \mathcal{H}_k^{(-k+2)} \neq 0, \qquad k = 1, 2, 3, \ldots,$$ (7.3.27)

and

$$F_n = -\frac{\mathcal{H}_{n-2}^{(-n+3)} \mathcal{H}_n^{(-n+2)}}{\mathcal{H}_{n-1}^{(-n+2)} \mathcal{H}_{n-1}^{(-n+3)}}, \quad G_n = -\frac{\mathcal{H}_{n-1}^{(-n+2)} \mathcal{H}_n^{(-n+2)}}{\mathcal{H}_n^{(-n+1)} \mathcal{H}_{n-1}^{(-n+3)}}, \qquad n = 1, 2, 3, \ldots.$$

(7.3.28)

Here we set $\mathcal{H}_{-1}^{(n)} = 1$.

(B) *Conversely, if, for a given pair* (L, L^*) *of fLs (7.3.25) the conditions (7.3.27) hold, then the general T-fraction (7.3.26a) with coefficients defined by (7.3.28) corresponds to* (L, L^*) *and satisfies (7.3.26b).*

(C) *If* $\nu = 0$ *and (7.3.27) and (7.3.28) hold, then the nth numerator $A_n(z)$ and denominator $B_n(z)$ of (7.3.26a) are polynomials in z of degrees not greater than $n + \mu$ and n, respectively, and for $n \geqslant 1$,*

$$\frac{A_n(z)}{B_n(z)} = R_{n+\mu, n}(L, L^*, z),$$ (7.3.29)

the $(n + \mu, n)$ *two-point Padé approximant of* (L, L^*).

Proof. (A): In our proof we shall use the equivalent form (7.3.9) for the general T-fraction (7.3.26). Then by (7.3.6) and (7.3.26b), we have

$$e_n \neq 0 \quad \text{and} \quad d_n \neq 0, \qquad n = 1, 2, 3, \ldots.$$ (7.3.30)

It follows from (7.3.13) and (7.3.10) that $LB_n - A_n$ is a fLs of the form

$$LB_n(z) - A_n(z) = (e_1 e_2 \cdots e_n)(c_{n+1} - \gamma_{n+1}^{(n)}) z^{n+1}$$

$$+ \lambda_{n+2} z^{n+2} + \lambda_{n+3} z^{n+3} + \cdots.$$

Equating coefficients of like powers of z on both sides and using (7.3.10), we obtain the equations

$$c_0 b_{n,0} + c_{-1} b_{n,1} + \cdots + c_{-n} b_{n,n} = a_{n,0}, \qquad (7.3.31a)$$

$$\left.\begin{array}{l} c_1 b_{n,0} + c_0 b_{n,1} + \cdots + c_{-(n-1)} b_{n,n} = a_{n,1}, \\ c_2 b_{n,0} + c_1 b_{n,1} + \cdots + c_{-(n-2)} b_{n,n} = a_{n,2}, \\ \qquad\qquad\vdots \\ c_n b_{n,0} + c_{n-1} b_{n,1} + \cdots + c_0 b_{n,n} = a_{n,n}, \end{array}\right\} \qquad (7.3.31b)$$

and

$$c_{n+1} b_{n,0} + c_n b_{n,1} + \cdots + c_1 b_{n,n} = a_{n,n+1} + (e_1 e_2 \cdots e_n)(c_{n+1} - \gamma_{n+1}^{(n)}). \qquad (7.3.31c)$$

Similarly, it follows from (7.3.17) and (7.3.10) that $L^* B_n - A_n$ is a fLs of the form

$$L^* B_n(z) - A_n(z) = (d_1 d_2 \cdots d_n)\big(c_{-n}^* - \gamma_{-n}^{(n)'}\big) z^0$$

$$+ \kappa_{-(n+1)} z^{-1} + \kappa_{-(n+2)} z^{-2} + \cdots.$$

Equating coefficients of like powers of z on both sides and using (7.3.10), we arrive at the equations

$$c_1^* b_{n,n} + c_2^* b_{n,n-1} + \cdots + c_{n+1}^* b_{n,0} = a_{n,n+1}, \qquad (7.3.32a)$$

$$\left.\begin{array}{l} c_0^* b_{n,n} + c_1^* b_{n,n-1} + \cdots + c_n^* b_{n,0} = a_{n,n}, \\ c_{-1}^* b_{n,n} + c_0^* b_{n,n-1} + \cdots + c_{n-1}^* b_{n,0} = a_{n,n-1}, \\ \qquad\qquad\vdots \\ c_{-(n-1)}^* b_{n,n} + c_{-(n-2)}^* b_{n,n-1} + \cdots + c_1^* b_{n,0} = a_{n,1}, \end{array}\right\} \qquad (7.3.32b)$$

and

$$c_{-n}^* b_{n,n} + c_{-(n-1)}^* b_{n,n-1} + \cdots + c_0^* b_{n,0} = a_{n,0} + (d_1 d_2 \cdots d_n)\big(c_{-n}^* - \gamma_{-n}^{(n)'}\big). \qquad (7.3.32c)$$

Eliminating the coefficients $a_{n,1}, a_{n,2}, \ldots, a_{n,n}$ from (7.3.31b) and (7.3.32b) gives the system

$$\left.\begin{array}{l} \delta_{-(n-1)} b_{n,n} + \delta_{-(n-2)} b_{n,n-1} + \cdots + \delta_0 b_{n,1} = -\delta_1 b_{n,0}, \\ \delta_{-(n-2)} b_{n,n} + \delta_{-(n-3)} b_{n,n-1} + \cdots + \delta_1 b_{n,1} = -\delta_2 b_{n,0}, \\ \qquad\qquad\vdots \\ \delta_0 b_{n,n} + \delta_1 b_{n,n-1} + \cdots + \delta_{n-1} b_{n,1} = -\delta_n b_{n,0}. \end{array}\right\} \qquad (7.3.33)$$

The determinant of this system is $\mathfrak{H}_n^{(-n+1)}$. Thus an application of Cramer's rule yields the equations

$$\mathfrak{H}_n^{(-n+1)}b_{n,1} = -b_{n,0}\begin{vmatrix} \delta_{-(n-1)} & \cdots & \delta_{-1} & \delta_1 \\ \delta_{-(n-2)} & \cdots & \delta_0 & \delta_2 \\ \vdots & & \vdots & \vdots \\ \delta_0 & \cdots & \delta_{n-2} & \delta_n \end{vmatrix},$$

$$\mathfrak{H}_n^{(-n+1)}b_{n,2} = b_{n,0}\begin{vmatrix} \delta_{-(n-1)} & \cdots & \delta_{-2} & \delta_0 & \delta_1 \\ \delta_{-(n-2)} & \cdots & \delta_{-1} & \delta_1 & \delta_2 \\ \vdots & & \vdots & \vdots & \vdots \\ \delta_0 & \cdots & \delta_{n-3} & \delta_{n-1} & \delta_n \end{vmatrix}, \qquad (7.3.34)$$

$$\vdots$$

$$\mathfrak{H}_n^{(-n+1)}b_{n,n} = (-1)^n b_{n,0}\begin{vmatrix} \delta_{-(n-2)} & \cdots & \delta_0 & \delta_1 \\ \delta_{-(n-3)} & \cdots & \delta_1 & \delta_2 \\ \vdots & & \vdots & \vdots \\ \delta_1 & \cdots & \delta_{n-1} & \delta_n \end{vmatrix}.$$

Subtracting (7.3.31a) from (7.3.32c) gives

$$\delta_{-n}b_{n,n} + \delta_{-(n-1)}b_{n,n-1} + \cdots + \delta_0 b_{n,0} = (d_1 d_2 \cdots d_n)(c_{-n}^* - \gamma_{-n}^{(n)^*}). \tag{7.3.35}$$

Then expanding $\mathfrak{H}_{n+1}^{(-n)}$ by cofactors along the first row and applying (7.3.34), we obtain

$$\mathfrak{H}_{n+1}^{(-n)} = \frac{(-1)^n \mathfrak{H}_n^{(-n+1)}}{b_{n,0}}\left(\delta_{-n}b_{n,n} + \delta_{-(n-1)}b_{n,n-1} + \cdots + \delta_0 b_{n,0}\right). \tag{7.3.36}$$

Combining (7.3.35), (7.3.36), (7.3.18b) and (7.3.10g) yields

$$\mathfrak{H}_1^{(0)} = \frac{\mathfrak{H}_0^{(-1)}}{d_1} = \frac{1}{d_1} \tag{7.3.37a}$$

and

$$\mathfrak{H}_{n+1}^{(-n)} = \frac{\mathfrak{H}_n^{(-n+1)}}{(e_1 e_2 \cdots e_n)(d_1 d_2 \cdots d_n d_{n+1})}, \qquad n = 1, 2, 3, \ldots . \tag{7.3.37b}$$

It is easily shown by induction using (7.3.37) that

$$\mathcal{K}_k^{(-k+1)} \neq 0, \qquad k = 1, 2, 3, \ldots. \tag{7.3.38}$$

Now subtracting (7.3.31c) from (7.3.32a) yields

$$\delta_1 b_{n,n} + \delta_2 b_{n,n-1} + \cdots + \delta_{n+1} b_{n,0} = -e_1 e_2 \cdots e_n \left(c_{n+1} - \gamma_{n+1}^{(n)} \right). \tag{7.3.39}$$

If the determinant $\mathcal{K}_{n+1}^{(-n+1)}$ is expanded by cofactors along the last column, and (7.3.34) is applied, we obtain

$$\mathcal{K}_{n+1}^{(-n+1)} = \frac{\mathcal{K}_n^{(-n+1)}}{b_{n,0}} \left(\delta_1 b_{n,n} + \delta_2 b_{n,n-1} + \cdots + \delta_{n+1} b_{n,0} \right). \tag{7.3.40}$$

Combining (7.3.39), (7.3.40), (7.3.14b) and (7.3.10g) gives

$$\mathcal{K}_1^{(1)} = -\frac{1}{e_1} \tag{7.3.41a}$$

and

$$\mathcal{K}_{n+1}^{(-n+1)} = \frac{(-1)^{n-1} \mathcal{K}_n^{(-n+1)}}{(e_1 e_2 \cdots e_n)^2 e_{n+1}}, \qquad n = 1, 2, 3, \ldots. \tag{7.3.41b}$$

It follows from (7.3.41) and (7.3.38) that

$$\mathcal{K}_n^{(-n+2)} \neq 0, \qquad n = 1, 2, 3, \ldots.$$

Using (7.3.6), (7.3.37) and (7.3.41), one can easily derive the formulas (7.3.28). This proves (A).

(B): Suppose that (7.3.27) is satisfied for a given pair of fLs (L, L^*). Then the general *T*-fraction defined by (7.3.26a) and (7.3.28) satisfies (7.3.36b). Thus by Theorem 7.17, the continued fraction (7.3.26a) corresponds to a uniquely determined pair $(\hat{L}, \hat{L}^*)$ of fLs

$$\hat{L} = \sum_{k=-\nu}^{\infty} \hat{c}_k z^k \quad \text{and} \quad \hat{L}^* = \sum_{k=-\infty}^{\mu} \hat{c}_k^* z^k, \qquad \mu \geq 0, \quad \nu \geq 0.$$

Now the procedure used in the proof of part (A) to define the coefficients F_n and G_n in terms of the coefficients c_k and c_k^* can be applied to the $\hat{c}_k$ and $\hat{c}_k^*$, and this will yield the same F_n and G_n. It is readily shown that given sequences $\{F_n\}$, $\{G_n\}$ uniquely determine sequences $\{c_k\}$ and $\{c_k^*\}$ by means of the relations (7.3.28). Thus we conclude that $c_k = \hat{c}_k$ and $c_k^* = \hat{c}_k^*$ for all k, which proves part (B).

(C): Suppose that $\nu=0$. Since $A_n(z)/B_n(z)$ is a rational function of type $[n+\mu, n]$, whose Taylor series at $z=0$ agrees with L in $n+1$ terms [see (7.3.13)] and whose Laurent series at $z=\infty$ agrees with L^* in $n+\mu$ terms [see (7.3.17)], it follows that $A_n(z)/B_n(z)$ is the $(n+\mu, n)$ two-point Padé approximant of (L, L^*).   ■

7.3.2   FG Algorithms

Let (L, L^*) be a pair of fLs of the form

$$L=\sum_{k=1}^{\infty} c_k z^k, \qquad L^*=\sum_{k=-\infty}^{0} c_k^* z^k. \qquad (7.3.42)$$

This corresponds to setting $\nu=0$, $c_0=0$ and $\mu=0$ in (7.3.25). In this section we describe an algorithm, similar to the qd algorithm, which can be used to compute the general T-fraction corresponding to the given pair (L, L^*) (see Algorithm 7.3.1 and Theorem 7.19 with $r=0$). The algorithm is based on rhombus rules introduced by Murphy and McCabe ([McCabe, 1975] and [McCabe and Murphy, 1976]). The rhombus rules also provide a procedure for computing zeros and poles of analytic functions whose Laurent expansions at $z=0$ and ∞ are L and L^*, respectively (Theorems 7.21 and Corollary 7.22).

For this purpose we define in terms of the given pair of fLs (7.3.42) a double sequence of pairs of fLs $\{(L_r, L_r^*)\}_{r=-\infty}^{\infty}$ as follows: For $r=0$,

$$(L_0, L_0^*)=(L, L^*); \qquad (7.3.43a)$$

for $r \geqslant 1$,

$$L_r = \frac{L-(c_1 z+\cdots+c_r z^r)}{z^r}, \qquad (7.3.43b)$$

$$L_r^* = \frac{L^*-(c_1 z+\cdots+c_r z^r)}{z^r};$$

and for $r \leqslant 1$,

$$L_r = \frac{L-(c_0^*+c_{-1}^* z^{-1}+\cdots+c_{r+1}^* z^{r+1})}{z^r}, \qquad (7.3.43c)$$

$$L_r^* = \frac{L^*-(c_0^*+c_{-1}^* z^{-1}+\cdots+c_{r+1}^* z^{r+1})}{z^r}.$$

It is easily seen that in all cases the fLs L_r and L_r^* can be written in the form

$$L_r=\sum_{k=1}^{\infty} c_k^{(r)} z^k, \qquad L_r^*=\sum_{k=-\infty}^{0} c_k^{(r)'} z^k. \qquad (7.3.44)$$

Moreover, it can be shown that for all r,

$$\delta_{r+k}=c_k^{(r)^*}-c_k^{(r)}, \qquad k=0,\pm1,\pm2,\dots, \qquad (7.3.45)$$

where δ_k is defined by (7.3.23), and where $c_k^{(r)}=0$ if $k\leqslant0$ and $c_k^{(r)^*}=0$ if $k\geqslant1$. Using these results together with Theorem 7.18, we arrive at the following

THEOREM 7.19.

(A) *Let* (L,L^*) *be a given pair of fLs of the form* (7.3.42), *let* r *be an arbitrary integer, and let* (L_r,L_r^*) *be the pair of fLs defined by* (7.3.43). *If there exists a general T-fraction*

$$\frac{h^{(r)}z}{1+G_1^{(r)}z} + \frac{F_2^{(r)}z}{1+G_2^{(r)}z} + \frac{F_3^{(r)}z}{1+G_3^{(r)}z} + \cdots \qquad (7.3.46a)$$

with

$$h^{(r)}\neq0; \qquad F_n^{(r)}\neq0, \quad n=2,3,4,\dots, \qquad (7.3.46b)$$

$$G_n^{(r)}\neq0, \qquad n=1,2,3,\dots, \qquad (7.3.46c)$$

which corresponds to (L_r,L_r^*), *then*

$$\mathcal{H}_k^{(r-k+1)}\neq0, \quad \mathcal{H}_k^{(r-k+2)}\neq0, \qquad k=1,2,3,\dots, \qquad (7.3.47a)$$

$$h^{(r)}=-\delta_{r+1}, \qquad F_n^{(r)}=-\frac{\mathcal{H}_{n-2}^{(r-n+3)}\mathcal{H}_n^{(r-n+2)}}{\mathcal{H}_{n-1}^{(r-n+2)}\mathcal{H}_{n-1}^{(r-n+3)}}, \qquad n=2,3,4,\dots, \qquad (7.3.47b)$$

and

$$G_n^{(r)}=-\frac{\mathcal{H}_{n-1}^{(r-n+2)}\mathcal{H}_n^{(r-n+2)}}{\mathcal{H}_n^{(r-n+1)}\mathcal{H}_{n-1}^{(r-n+3)}}, \qquad n=1,2,3,\dots. \qquad (7.3.47c)$$

Here the Hankel determinants $\mathcal{H}_k^{(n)}$ *are defined by* (7.3.24).

(B) *Conversely, if for a given pair* (L_r,L_r^*) *of fLs* (7.3.43) *the conditions* (7.3.47a) *hold, then the general T-fraction* (7.3.46a) *with coefficients defined by* (7.3.47b,c) *corresponds to* (L_r,L_r^*) *and satisfies* (7.3.46b,c).

The coefficients $F_n^{(r)}$ and $G_n^{(r)}$ defined by (7.3.47) have been shown by Murphy and McCabe [McCabe, 1975; McCabe and Murphy, 1976] to satisfy rhombus rules analogous to those of Rutishauser in the qd algorithm (7.1.29). In fact, they proved that if we set

$$F_1^{(r)}=0, \qquad r=0,\pm1,\pm2,\dots, \qquad (7.3.48a)$$

Table 7.3.1. *FG* Table

$\vdots$	$\vdots$	$\vdots$	$\vdots$	$\vdots$	
$0 = F_1^{(-2)}$	$G_1^{(-2)}$	$F_2^{(-2)}$	$G_2^{(-2)}$	$F_3^{(-2)}$	$\ldots$
$0 = F_1^{(-1)}$	$G_1^{(-1)}$	$F_2^{(-1)}$	$G_2^{(-1)}$	$F_3^{(-1)}$	$\ldots$
$0 = F_1^{(0)}$	$G_1^{(0)}$	$F_2^{(0)}$	$G_2^{(0)}$	$F_3^{(0)}$	$\ldots$
$0 = F_1^{(1)}$	$G_1^{(1)}$	$F_2^{(1)}$	$G_2^{(1)}$	$F_3^{(1)}$	$\ldots$
$0 = F_1^{(2)}$	$G_1^{(2)}$	$F_2^{(2)}$	$G_2^{(2)}$	$F_3^{(2)}$	$\ldots$
$\vdots$	$\vdots$	$\vdots$	$\vdots$	$\vdots$	

then for $r = 0, \pm 1, \pm 2, \ldots$

$$G_1^{(r)} = -\frac{\delta_{r+1}}{\delta_r}, \tag{7.3.48b}$$

$$F_{n+1}^{(r)} + G_n^{(r)} = F_n^{(r+1)} + G_n^{(r+1)}, \qquad n = 1, 2, 3, \ldots, \tag{7.3.48c}$$

$$F_{n+1}^{(r)} \cdot G_{n+1}^{(r+1)} = F_{n+1}^{(r+1)} \cdot G_n^{(r)}, \qquad n = 1, 2, 3, \ldots. \tag{7.3.48d}$$

Equations (7.3.48b, d) can be verified by direct substitution from (7.3.47). The proof of (7.3.48d) can be made by a straightforward argument using Jacobi's identity for the Hankel determinants (7.1.33) with H replaced by $\mathcal{H}$. We shall call the two-dimensional array of coefficients indicated by Table 7.3.1 the *FG* table. If two adjacent columns are known, the remaining elements can be computed in terms of these using (7.3.48). Also if a row is known, then by using (7.3.48) one can compute the other rows. A computational procedure for this based on (7.3.48) will be called an *FG* algorithm.

Some examples of *FG* algorithms will now be described. In the first we assume that a pair of fLs (7.3.42) is given. Thus we know the first two columns of the *FG* table. The following algorithm can be used to compute the remaining columns successively to the right. Thus we obtain, in addition to the other elements of the *FG* table, the coefficients

$$G_1^{(0)}, F_2^{(0)}, G_2^{(0)}, F_3^{(0)}, G_3^{(0)}, \ldots$$

of the general *T*-fraction (7.3.46a) corresponding to (L, L^*).

ALGORITHM 7.3.1 (*FG* algorithm for computing successive columns). Suppose we are given

$$F_1^{(r)} = 0, \quad G_1^{(r)} = -\frac{\delta_{r+1}}{\delta_r}, \qquad r = 0, \pm 1, \pm 2, \ldots.$$

Then compute the remaining columns of the *FG* table as follows:
For $n = 1, 2, 3, \ldots,$

$$F_{n+1}^{(r)} = F_n^{(r+1)} + G_n^{(r+1)} - G_n^{(r)}, \qquad r = 0, \pm 1, \pm 2, \ldots,$$

$$G_{n+1}^{(r)} = \frac{F_{n+1}^{(r)}}{F_{n+1}^{(r-1)}} G_n^{(r-1)}, \qquad r = 0, \pm 1, \pm 2, \ldots.$$

In the following two algorithms we assume that the row with $r = 0$ in the *FG* table is given, and we wish to compute other rows successively.

ALGORITHM 7.3.2 (*FG* algorithm for computing successive rows descending). Suppose we are given

$$0 = F_1^{(0)}, \ G_1^{(0)}, F_2^{(0)}, G_2^{(0)}, F_3^{(0)}, \ldots.$$

Then the succeeding rows can be computed as follows: For $r = 0, 1, 2, \ldots,$ set

$$F_1^{(r+1)} = 0, \qquad G_1^{(r+1)} = G_1^{(r)} + F_2^{(r)},$$

$$F_n^{(r+1)} = \frac{F_n^{(r)}(G_n^{(r)} + F_{n+1}^{(r)})}{G_{n-1}^{(r)} + F_n^{(r)}}, \qquad n = 2, 3, 4, \ldots$$

and

$$G_n^{(r+1)} = G_n^{(r)} + F_{n+1}^{(r)} - F_n^{(r+1)}, \qquad n = 2, 3, 4, \ldots.$$

ALGORITHM 7.3.3 (*FG* algorithm for computing successive rows ascending). Suppose we are given

$$0 = F_1^{(0)}, \ G_1^{(0)}, F_2^{(0)}, G_2^{(0)}, F_3^{(0)}, \ldots.$$

Then the preceding rows can be computed as follows: For $r = 0, -1, -2, \ldots,$ set

$$F_1^{(r-1)} = 0, \qquad G_1^{(r-1)} = \frac{G_1^{(r)} G_2^{(r)}}{F_2^{(r)} + G_2^{(r)}},$$

$$F_n^{(r-1)} = F_{n-1}^{(r)} + G_{n-1}^{(r)} - G_{n-1}^{(r-1)}, \qquad n = 2, 3, 4, \ldots,$$

and

$$G_n^{(r-1)} = \frac{G_{n+1}^{(r)}(F_n^{(r)} + G_n^{(r)})}{F_{n+1}^{(r)} + G_{n+1}^{(r)}}, \qquad n = 2, 3, 4, \ldots,$$

In all of these algorithms it is, of course, assumed that the Hankel determinants involved do not vanish. For otherwise the computational procedure can break down. Before considering applications of *FG* algorithms, we give the following

THEOREM 7.20.

(A) *A general T-fraction*

$$\overset{\infty}{\underset{n=1}{\mathrm{K}}}\left(\frac{F_n z}{1+G_n z}\right), \qquad \text{with} \quad F_n \neq 0, \; G_n \neq 0, \quad n=1,2,3,\ldots, \qquad (7.3.49)$$

corresponds to a fLs $L^* = -1$ *at* $z = \infty$ *iff*

$$F_n = -G_n, \qquad n=1,2,3,\ldots. \qquad (7.3.50)$$

(B) *A general T-fraction of the form*

$$\overset{\infty}{\underset{n=1}{\mathrm{K}}}\left(\frac{F_n z}{1-F_n z}\right), \qquad F_n \neq 0, \quad n=1,2,3,\ldots, \qquad (7.3.51)$$

corresponds at $z=0$ *to the fLs*

$$L = \frac{d_0}{\displaystyle\sum_{n=0}^{\infty} d_n z^n} - 1, \qquad (7.3.52\text{a})$$

where

$$d_n = F_1 F_2 \cdots F_n d_0, \qquad n=1,2,3,\ldots. \qquad (7.3.52\text{b})$$

In fact, for each $m=1,2,3,\ldots$

$$\frac{d_0}{\displaystyle\sum_{n=1}^{m} d_n z^n} - 1 = \overset{m}{\underset{n=1}{\mathrm{K}}}\left(\frac{F_n z}{1-F_n z}\right) \qquad (7.3.53)$$

provided that (7.3.52b) *holds.*

Remark. Theorem 7.20(B) is equivalent to the statement that *if* $\{d_k\}$ *is an arbitrary sequence of non-zero complex numbers, then the general T-fraction*

$$\cfrac{-\dfrac{d_1}{d_0}z}{1+\dfrac{d_1}{d_0}z +} \cfrac{-\dfrac{d_2}{d_1}z}{1+\dfrac{d_2}{d_1}z +} \cfrac{-\dfrac{d_3}{d_2}z}{1+\dfrac{d_3}{d_2}z +} \cdots \qquad (7.3.51')$$

corresponds to the fLs (7.3.52a) *at* $z=0$.

Proof. (A): Suppose that (7.3.50) holds. Then a simple application of Theorem 5.2 shows that (7.3.49) corresponds to $L^* = -1$ at $z = \infty$. In fact, we let $w = 1/z$, $a_n(w) = F_n/w$, $b_n(w) = 1 - F_n/w$, $L_0 = L^* = -1$, $L_n = -F_n/w$. Then (5.1.23), (5.1.24) and (5.1.26) are satisfied, and hence Theorem 5.2 can be applied. Now suppose that (7.3.49) corresponds to $L^* = -1$ at $z = \infty$ and to

$$L = c_1 z + c_2 z^2 + c_3 z^3 + \cdots$$

at $z = 0$. Then

$$\delta_k = \begin{cases} -c_k & \text{if } k \geqslant 1, \\ -1 & \text{if } k = 0, \\ 0 & \text{if } k \leqslant -1. \end{cases} \tag{7.3.54}$$

It follows from (7.3.24) that

$$\mathcal{H}_n^{(-n+1)} = (-1)^{n(n+1)/2}, \qquad n = 1, 2, 3, \dots. \tag{7.3.55}$$

Thus by (7.3.28)

$$F_n = (-1)^n \frac{\mathcal{H}_n^{(-n+2)}}{\mathcal{H}_{n-1}^{(-n+3)}} = -G_n, \qquad n = 1, 2, 3, \dots. \tag{7.3.56}$$

This proves (A). To prove (B) it suffices to verify the identities

$$\frac{c_0}{\sum\limits_{n=0}^{m} c_n z^n} = 1 + \cfrac{-\dfrac{c_1}{c_0} z}{1 + \dfrac{c_1}{c_0} z +} \cdots \cfrac{-\dfrac{c_m}{c_{m-1}} z}{+ 1 + \dfrac{c_m}{c_{m-1}} z}, \qquad m = 1, 2, 3, \dots, \tag{7.3.57}$$

which follow from (2.3.29).   ∎

General *T*-fractions of the type considered in Theorem 7.20 are not of great interest as means of representing analytic functions, since the *m*th approximant is obtained by taking the *m*th partial sum of the series [see (7.3.53)]. However, the general *T*-fractions (7.3.51) are useful in the problem of computing zeros and poles of functions that they represent. This fact is made clear in part by Theorem 7.21, Corollary 7.22 and related results contained in [Jones and Magnus, 1980]. Some examples of continued fractions of the type considered in Theorem 7.20 were treated in Section 6.2 [see (6.2.14) and (6.2.18)]. Another example is given in the following.

Example 1. The exponential e^{-z}. Here we consider the convergent series

$$L = \sum_{k=0}^{\infty} \frac{(-z)^k}{k!} = \frac{1}{\displaystyle\sum_{k=0}^{\infty} \frac{z^k}{k!}} = e^{-z}. \tag{7.3.58}$$

By Theorem 7.20 the general T-fraction

$$1 + \frac{-z}{1+z} + \frac{-\frac{1}{2}z}{1+\frac{1}{2}z} + \frac{-\frac{1}{3}z}{1+\frac{1}{3}z} + \cdots \tag{7.3.59}$$

corresponds to L at $z=0$ and to $L^*=0$ at $z=\infty$. Note that this is essentially the same continued fraction as given by (6.2.18) with $m=0$ and $a=1$. We shall also derive (7.3.59) by means of the FG algorithm (Algorithm 7.3.1) applied to the pair of fLs (L, L^*). For this case we have

$$\delta_k = \begin{cases} \dfrac{(-1)^{k-1}}{k!}, & k \geqslant 1, \\[2mm] 0, & k \leqslant 0. \end{cases}$$

Applying Algorithm 7.3.1, we obtain the lower half of the FG table shown in Table 7.3.2. We note that the elements $G_n^{(0)}$ with $n \geqslant 2$ are not obtained by Algorithm 7.3.1. However since $G_n^{(0)} = -F_n^{(0)}$ (by Theorem 7.20), this is not a problem. Further examples of general T-fractions dealing with correspondence and convergence are given in Section 7.3.3. Before going to that, we describe some applications of FG algorithms in computing zeros and poles of analytic functions. The principal result for this application is the following theorem of Jones and Magnus [1980], which is similar to Theorem 7.8.

Table 7.3.2. FG Table for $e^{-z} - 1$

r	$F_1^{(r)}$	$G_1^{(r)}$	$F_2^{(r)}$	$G_2^{(r)}$	$F_3^{(r)}$	$G_3^{(r)}$	$\cdots$
0	0	1	$-\dfrac{1}{1\times2}$	$\dfrac{1}{2}$	$-\dfrac{2}{2\times3}$	$\dfrac{1}{3}$	$\cdots$
1	0	$\dfrac{1}{2}$	$-\dfrac{1}{2\times3}$	$\dfrac{1}{3}$	$-\dfrac{2}{3\times4}$	$\dfrac{1}{4}$	$\cdots$
2	0	$\dfrac{1}{3}$	$-\dfrac{1}{3\times4}$	$\dfrac{1}{4}$	$-\dfrac{2}{4\times5}$	$\dfrac{1}{5}$	$\cdots$
3	0	$\dfrac{1}{4}$	$-\dfrac{1}{4\times5}$	$\dfrac{1}{5}$	$-\dfrac{2}{5\times6}$	$\dfrac{1}{6}$	$\cdots$
4	0	$\dfrac{1}{5}$	$-\dfrac{1}{5\times6}$	$\dfrac{1}{6}$	$-\dfrac{2}{6\times7}$	$\dfrac{1}{7}$	$\cdots$
$\vdots$	$\vdots$	$\vdots$	$\vdots$	$\vdots$	$\vdots$	$\vdots$	

THEOREM 7.21. *Let*

$$L = c_0 + c_1 z + c_2 z^2 + \cdots \qquad (7.3.60)$$

be the Taylor series at $z=0$ of a function $f(z)$ analytic at $z=0$ and meromorphic in the disk $D_R = [z : |z| < R]$. Let the poles z_j of $f(z)$ in D_R be arranged such that

$$0 < |z_1| \leqslant |z_2| \leqslant |z_3| \leqslant \cdots < R, \qquad (7.3.61)$$

where each pole occurs as many times in the sequence as its order. Similarly let

$$L^* = c_0^* + c_{-1}^* z^{-1} + c_{-2}^* z^{-2} + \cdots \qquad (7.3.62)$$

be the Laurent series at $z = \infty$ of a function $f^*(z)$ analytic at $z = \infty$ and meromorphic in the domain $\Delta_{R^*} = [z : |z| > R^*]$. Let the poles of z_j^* of $f^*(z)$ in Δ_{R^*} be arranged such that

$$\infty > |z_1^*| \geqslant |z_2^*| \geqslant |z_3^*| \geqslant \cdots, \qquad (7.3.63)$$

where each pole occurs as many times in the sequence as its order. Then:

(A) *If, for some k,*

$$|z_{k-1}| < |z_k| < |z_{k+1}|, \qquad (7.3.64)$$

then $\mathcal{H}_{k-1}^{(m)} \neq 0$ and $\mathcal{H}_k^{(m)} \neq 0$ for all sufficiently large positive integers m, and

$$\lim_{m \to +\infty} G_k^{(m)} = -\frac{1}{z_k}. \qquad (7.3.65)$$

(B) *If, for some k,*

$$|z_{k-1}^*| > |z_k^*| > |z_{k+1}^*|, \qquad (7.3.66)$$

then $\mathcal{H}_{k-1}^{(m)} \neq 0$ and $\mathcal{H}_k^{(m)} \neq 0$ for all sufficiently large negative integers m, and

$$\lim_{m \to -\infty} G_k^{(m)} = -\frac{1}{z_k^*}. \qquad (7.3.67)$$

Here the $\mathcal{H}_k^{(m)}$ are Hankel determinants associated with (L, L^*) and the $G_k^{(m)}$ are defined by (7.3.47b). Also $z_0 = 0$ and $z_0^* = \infty$; if $f(z)$ has only n poles, then $z_{n+1} = \infty$, and if $f^*(z)$ has only n poles, then $z_{n+1}^* = 0$.

Since the $G_k^{(m)}$ can be computed using Algorithm 7.3.1 (assuming that the $F_k^{(m)}$ do not vanish), we have here a convenient method for locating poles of analytic functions. The following corollary of Theorem 7.21

applies to the problem of computing zeros of polynomials (for proof see [Jones and Magnus, 1980]).

COROLLARY 7.22. *Let*

$$P(z) = b_0 + b_1 z + \cdots + b_n z^n \tag{7.3.68a}$$

such that

$$b_k \neq 0, \qquad k = 0, 1, \ldots, n. \tag{7.3.68b}$$

let the zeros z_j of $P(z)$ be arranged so that

$$0 < |z_1| < |z_2| < \cdots < |z_n|. \tag{7.3.69}$$

Let

$$F_1^{(0)} = F_{n+1}^{(0)} = G_{n+1}^{(0)} = 0, \tag{7.3.70a}$$

$$F_k^{(0)} = \frac{b_{n-k}}{b_{n-k+1}}, \qquad k = 2, 3, \ldots, n, \tag{7.3.70b}$$

$$G_k^{(0)} = -\frac{b_{n-k}}{b_{n-k+1}}, \qquad k = 1, 2, \ldots, n. \tag{7.3.70c}$$

For $k = 1, 2, \ldots, n$ and $m = 1, 2, 3, \ldots$, let $F_k^{(m)}$ and $G_k^{(m)}$ be defined by Algorithm 7.3.2, assuming that

$$G_{k-1}^{(m)} + F_k^{(m)} \neq 0, \qquad k = 2, 3, \ldots, n, \quad m = 0, 1, 2, \ldots. \tag{7.3.71}$$

If for some k

$$|z_{k-1}| < |z_k| < |z_{k+1}|, \tag{7.3.72}$$

then

$$\lim_{m \to \infty} G_k^{(m)} = z_{n-k+1}. \tag{7.3.73}$$

Here $z_0 = 0$ and $z_{n+1} = \infty$.

Example 2. Zeros of polynomials.   For the polynomial

$$P(z) = (z-1)(z-2)(z-3) = z^3 - 6z^2 + 11z - 6,$$

set

$$F_1^{(0)} = 0, \qquad G_1^{(0)} = 6, \qquad F_2^{(0)} = -G_2^{(0)} = -\tfrac{11}{6}, \qquad F_3^{(0)} = -G_3^{(0)} = -\tfrac{6}{11}.$$

Table 7.3.3.   Values of $G_k^{(m)}$ from Algorithm 7.3.2 for $P(z)=(z-1)(z-2)(z-3)$

m	$G_1^{(m)}$	$G_2^{(m)}$	$G_3^{(m)}$
0	6.00000	1.83333	0.54545
1	4.16667	1.85455	0.77647
2	3.60000	1.87908	0.88696
3	3.34444	1.90374	0.94237
4	3.20930	1.92616	0.97062
5	3.13147	1.94509	0.98506
⋮	⋮	⋮	⋮
10	3.01565	1.99058	0.99952
20	3.00027	1.99982	1.00000
30	3.00000	2.00000	1.00000

Then an application of Algorithm 7.3.2 gives the results in Table 7.3.3. The computation was carried out using 10-decimal-digit floating-point arithmetic, and the results were then rounded to 5 decimal places. It can be seen that all three zeros are obtained correct to 5 decimal places at $m=30$.

The conditions (7.3.64) and (7.3.66) insure that the z_k and z_k^*, respectively, are simple poles of modulus different from all other poles. These conditions can be weakened, as has been done for the qd algorithm (see, for example, [Henrici, 1974, Section 7.9]). This will not be done here.

7.3.3   Representation of Analytic Functions

This section deals with the representation by general T-fractions of functions defined by ratios of the confluent hypergeometric functions $\Phi(b;c;z)$. First, however, we give a useful convergence theorem and recall the convention made in Section 4.5.4 that a general T-fraction (7.3.74) and all of its approximants have value equal to zero at $z=0$.

THEOREM 7.23.   *Let*

$$\overset{\infty}{\underset{n=1}{\mathrm{K}}}\left(\frac{F_n z}{1+G_n z}\right), \qquad F_n \neq 0 \tag{7.3.74}$$

be a general T-fraction with the property that for each $M>0$ there exists an n_M such that

$$\left|\frac{F_n z}{(1+G_{n-1}z)(1+G_n z)}\right| < \frac{1}{4} \qquad \text{for all} \quad n \geqslant n_M \text{ and } |z| < M. \tag{7.3.75}$$

Then:

(A) *The continued fraction* (7.3.74) *converges to a function* $f(z)$ *meromorphic in* $\mathbb{C}$.

(B) *The convergence is uniform on every compact subset of* $\mathbb{C}$ *which contains no poles of* $f(z)$.

(C) $f(z)$ *is holomorphic at* $z=0$, *and* $f(0)=0$.

(D) *If*

$$L = c_1 z + c_2 z^2 + c_3 z^3 + \cdots$$

is the power series to which the continued fraction (7.3.74) *corresponds at* $z=0$, *then* L *is the Taylor series of* $f(z)$ *at* $z=0$.

(E) *The conditions* (7.3.75) *of the theorem are satisfied if*

$$\lim_{n \to \infty} F_n = \lim_{n \to \infty} G_n = 0. \tag{7.3.76}$$

Proof. Parts (A), (B) and (C) can be proved in a manner very similar to the proof of Theorem 5.14. It will suffice to note that in the present case the denominator analogous to that on the right of (5.4.26) cannot vanish identically, since $F(0)=0$ and $B_n(0)=1$ for all n. Part (D) is an immediate consequence of Theorem 5.13, and part (E) is easily verified.   ∎

We shall now derive a general T-fraction expansion for a ratio of confluent hypergeometric functions. The method is of some interest in itself, since it can serve as a guide for the study of other differential equations. A similar derivation can be found in [Perron, 1957a, p. 278]. Many of the specific continued fractions that emerge have been discussed by [Wynn, 1959, pp. 297–299] and others. We also determine the fLs to which the continued fractions correspond at $z=\infty$, and in some cases (Corollary 7.25) we are able to determine functions for which the fLs are asymptotic expansions as $z \to \infty$.

We start by considering a differential equation of the form

$$y = (\alpha + \beta x) y' + \gamma x y'' \tag{7.3.77a}$$

where

$$\alpha, \beta, \gamma \neq 0 \quad \text{and} \quad \frac{1}{\beta}, \frac{1}{\gamma} \notin [1, 2, 3, \dots]. \tag{7.3.77b}$$

Successive differentiation and rearrangement of (7.3.77a) gives the sequence of three-term recurrence relations

$$y^{(n)} = (\alpha_n + \beta_n x) y^{(n+1)} + \gamma_n x y^{(n+2)}, \qquad n = 0, 1, 2, \dots, \tag{7.3.78a}$$

where

$$\alpha_n = \frac{\alpha + n\gamma}{1 - n\beta}, \quad \beta_n = \frac{\beta}{1 - n\beta}, \quad \gamma_n = \frac{\gamma}{1 - n\gamma}, \quad n = 1, 2, 3, \ldots.$$

$$(7.3.78\text{b})$$

By a simple change of variable $x = (-\gamma/\beta)z$, $y(x) = w(z)$, we obtain from (7.3.77) the equation

$$w(z) = (\alpha' + \beta z)w'(z) - \beta z w''(z), \quad \alpha' = -\frac{\alpha\gamma}{\beta}, \quad (7.3.79)$$

in which the coefficient of zw' is the negative of that of zw''. Successive differentiation of (7.3.79) leads to the system of three-term recurrence relations

$$w^{(n)}(z) = (\alpha'_n + \beta_n z)w^{(n+1)}(z) - \beta_n z w^{(n+2)}(z), \quad n = 0, 1, 2, \ldots,$$

$$(7.3.80\text{a})$$

where

$$\alpha'_n = \frac{\alpha' - n\beta}{1 - n\beta}, \quad \beta_n = \frac{\beta}{1 - n\beta}, \quad n = 1, 2, 3, \ldots. \quad (7.3.80\text{b})$$

This gives the following set of formal identities

$$\frac{w}{w'} = \alpha' + \beta z + \frac{-\beta\alpha}{\alpha'_1 + \beta_1 z} + \frac{-\beta_1 z}{\alpha'_2 + \beta_2 z} + \cdots + \frac{-\beta_{n-1} z}{w^{(n)}/w^{(n-1)}}, \quad (7.3.81)$$

so that the ratio w/w' is in some way associated with the general T-fraction

$$\alpha' + \beta z + \frac{-\beta z}{\alpha'_1 + \beta_1 z} + \frac{-\beta_1 z}{\alpha'_2 + \beta_2 z} + \frac{-\beta_2 z}{\alpha'_3 + \beta_3 z} + \cdots. \quad (7.3.82)$$

Kummer's confluent hypergeometric differential equation is usually written in the form

$$z\frac{d^2 w}{dz^2} + (c - z)\frac{dw}{dz} - bw = 0. \quad (7.3.83)$$

If $b \neq 0$ this can be written as

$$w = \left(\frac{c}{b} - \frac{1}{b}z\right)w' + \frac{1}{b}zw'', \quad (7.3.84\text{a})$$

which is the same as (7.3.79) if we take

$$\alpha' = \frac{c}{b} \quad \text{and} \quad \beta = -\frac{1}{b}. \tag{7.3.84b}$$

Thus the continued fraction (7.3.82) can be written as

$$\frac{c}{b} - \frac{1}{b}z + \cfrac{\frac{1}{b}z}{\frac{c+1}{b+1} - \frac{1}{b+1}z} + \cfrac{\frac{1}{b+1}z}{\frac{c+2}{b+2} - \frac{1}{b+2}z} + \cfrac{\frac{1}{b+2}z}{\frac{c+3}{b+3} - \frac{1}{b+3}z} + \cdots,$$

which is equivalent to

$$\frac{1}{b}\left(c - z + \frac{(b+1)z}{c+1-z} + \frac{(b+2)z}{c+2-z} + \frac{(b+3)z}{c+3-z} + \cdots \right). \tag{7.3.85}$$

It is well known and not difficult to show that the confluent hypergeometric function $w = \Phi(b; c; z)$ is a solution of the differential equation (7.3.84a). The following theorem shows the connection between the continued fractions (7.3.85) and the solutions $\Phi(b; c; z)$. It also tells what fLs (7.3.85) corresponds to at $z = \infty$.

THEOREM 7.24. *Let b and c be complex constants such that*

$$b \notin [-1, -2, -3, \dots], \qquad c \notin [0, -1, -2, \dots]. \tag{7.3.86}$$

Then:

(A) *The continued fraction*

$$\frac{c}{c-z} + \frac{(b+1)z}{c+1-z} + \frac{(b+2)z}{c+2-z} + \frac{(b+3)z}{c+3-z} + \cdots \tag{7.3.87}$$

converges to the meromorphic function

$$f(z) = \frac{\Phi(b+1; c+1; z)}{\Phi(b; c; z)} \tag{7.3.88}$$

for all $z \in \mathbb{C}$.

(B) *The convergence is uniform on every compact subset of $\mathbb{C}$ which contains no poles of $f(z)$.*

(C) *$f(z)$ is holomorphic at $z = 0$ and $f(0) = 1$.*

(D) *The continued fraction (7.3.87) corresponds at $z = \infty$ to the fLs*

$$L^* = -\frac{c}{z} \frac{\Omega(b+1, b-c+1; -1/z)}{\Omega(b, b-c+1; -1/z)}, \tag{7.3.89}$$

where Ω is the divergent confluent hypergeometric series defined by (6.1.56).

Proof. It is easily seen that

$$\frac{d\Phi(b;c;z)}{dz} = \frac{b}{c}\Phi(b+1;c+1;z)$$

and

$$\frac{d^2\Phi(b;c;z)}{dz^2} = \frac{b}{c}\left(\frac{b+1}{c+1}\right)\Phi(b+2;c+2;z).$$

Since $\Phi(b;c;z)$ is a solution of (7.3.84a), it follows that

$$\frac{c}{b}\frac{\Phi(b;c;z)}{\Phi(b+1;c+1;z)} = \frac{c}{b} - \frac{1}{b}z + \cfrac{\frac{1}{b}z}{\frac{c+1}{b+1}\frac{\Phi(b+1;c+1;z)}{\Phi(b+2;c+2;z)}}.$$

Thus if we let L_n denote the fps

$$L_n = \frac{c+n}{b+n}\frac{\Phi(b+n;c+n;z)}{\Phi(b+n+1;c+n+1;z)}, \qquad n=0,1,2,\ldots, \qquad (7.3.90)$$

then

$$L_n = \frac{c+n}{b+n} - \frac{1}{b+n}z + \cfrac{\frac{1}{b+n}z}{L_{n+1}},$$

or

$$L_{n+1} = \cfrac{\frac{1}{b+n}z}{L_n - \left(\frac{c+n}{b+n} - \frac{1}{b+n}z\right)}, \qquad n=0,1,2,\ldots. \qquad (7.3.91)$$

Now clearly

$$\lambda\left(\frac{1}{b+n}z\right) = 1, \qquad \lambda\left(\frac{c+n}{b+n} - \frac{1}{b+n}z\right) = 0, \qquad (7.3.92)$$

$$\lambda(L_n) = 0, \qquad n=0,1,2,\ldots.$$

Thus by Theorem 5.2 the continued fraction

$$\frac{c}{b} - \frac{1}{b}z + \cfrac{\frac{1}{b}z}{\frac{c+1}{b+1} - \frac{1}{b+1}z} + \cfrac{\frac{1}{b+1}z}{\frac{c+2}{b+2} - \frac{1}{b+2}z} + \cfrac{\frac{1}{b+2}z}{\frac{c+3}{b+3} - \frac{1}{b+3}z} + \cdots$$

$$(7.3.93)$$

corresponds to the fps

$$L_0 = \frac{c}{b} \frac{\Phi(b;c;z)}{\Phi(b+1;c+1;z)}. \qquad (7.3.94)$$

The continued fraction (7.3.93) is equivalent to the general T-fraction

$$\frac{c}{b} - \frac{1}{b}z + \mathop{\mathbf{K}}_{n=1}^{\infty}\left(\frac{F_n z}{1+G_n z}\right), \qquad (7.3.95a)$$

where

$$F_1 = \frac{b+1}{b(c+1)}, \qquad F_n = \frac{b+n}{(c+n-1)(c+n)}, \qquad n=2,3,4,\ldots, \qquad (7.3.95b)$$

$$G_n = -\frac{1}{c+n}, \qquad n=1,2,3,\ldots. \qquad (7.3.95c)$$

Since (7.3.76) is satisfied, Theorem 7.23 can be applied to (7.3.95). Parts (A), (B) and (C) of our theorem are therefore easily verified.

To prove (D) we shall study the differential equation (7.3.84a) at $z = \infty$ or, equivalently, let $z = 1/\zeta$, and study the resulting equation at $\zeta = 0$. It has an irregular singular point at $\zeta = 0$ and a formal-power-series (fps) solution starting with ζ^b. (b is the only solution of the indicial equation.) Expressed in terms of z, the original differential equation (7.3.84a) has a fps solution

$$G(b;c;z) = z^{-b} + \frac{b(c-b-1)}{1!}z^{-b-1} + \frac{b(b+1)(c-b-1)(c-b-2)}{2!}z^{-b-2}$$

$$+ \frac{b(b+1)(b+2)(c-b-1)(c-b-2)(c-b-3)}{3!}z^{-b-3} + \cdots. \qquad (7.3.96)$$

Since this series is divergent for all $z \in \mathbb{C}$, it is a solution of (7.3.84a) only in a formal sense. It is easily seen that

$$G(b;c;z) = z^{-b}\Omega\left(b, b-c+1; -\frac{1}{z}\right), \qquad (7.3.97)$$

where Ω is the confluent hypergeometric series defined by (6.1.56). Since $G(b;c;z)$ is a formal solution of the differential equation (7.3.84a), it can be shown that

$$-\frac{G(b;c;z)}{bG(b+1;c+1;z)} = \frac{c}{b} - \frac{1}{b}z + \cfrac{\frac{1}{b}z}{-\frac{1}{b+1}\frac{G(b+1;c+1;z)}{G(b+2;c+2;z)}}.$$

(7.3.98)

Hence, letting L_n^* denote the fLs

$$L_n^* = -\frac{1}{b+n}\frac{G(b+n;c+n;z)}{G(b+n+1;c+n+1;z)}$$

$$= -\frac{z}{b+n}\frac{\Omega(b+n,b-c+1;-1/z)}{\Omega(b+n+1,b-c+1;-1/z)}, \qquad n=0,1,2,\dots,$$

(7.3.99)

one can show that

$$L_{n+1}^* = \cfrac{\frac{1}{b+n}z}{L_n^* - \left(\frac{c+n}{b+n} - \frac{1}{b+n}z\right)}, \qquad n=0,1,2,\dots. \qquad (7.3.100)$$

In view of (7.3.92), it follows from Theorem 5.2 that the continued fraction (7.3.93) corresponds at $z=\infty$ to the fLs

$$L_0^* = -\frac{z}{b}\frac{\Omega(b,b-c+1;-1/z)}{\Omega(b+1,b-c+1;-1/z)}, \qquad (7.3.101)$$

from which (D) is an immediate consequence. ∎

Before considering examples, we state the following important corollary of the previous theorem.

COROLLARY 7.25. *Let c be a complex constant such that*

$$c \notin [0,-1,-2,\dots]. \qquad (7.3.102)$$

Then:

(A) *The continued fraction*

$$\frac{c}{c-z} + \frac{1\cdot z}{c+1-z} + \frac{2\cdot z}{c+2-z} + \frac{3\cdot z}{c+3-z} + \cdots \qquad (7.3.103)$$

converges to the entire function

$$f(z) = \Phi(1; c+1; z) = cz^{-c}e^z\gamma(c, z) \tag{7.3.104}$$

for all $z \in \mathbb{C}$, and the convergence is uniform on every compact subset of $\mathbb{C}$.
   (B) The continued fraction corresponds to the power series $\Phi(1; c+1; z)$ at $z=0$ and to the divergent fLs

$$L^* = -\frac{c}{z}\Omega\left(1, 1-c; -\frac{1}{z}\right) \tag{7.3.105}$$

at $z = \infty$.

Example 1. Error function.   The error function erf z, defined by (6.1.37), is represented by the equation

$$\frac{\sqrt{\pi}}{2z}e^{z^2}\operatorname{erf} z = \Phi\left(1; \tfrac{3}{2}; z^2\right)$$

$$= \frac{\tfrac{1}{2}}{\tfrac{1}{2}-z^2} + \frac{1 \cdot z^2}{\tfrac{3}{2}-z^2} + \frac{2 \cdot z^2}{\tfrac{5}{2}-z^2} + \frac{3 \cdot z^2}{\tfrac{7}{2}-z^2} + \cdots \qquad \text{for all} \quad z \in \mathbb{C}.$$

$$\tag{7.3.106}$$

The continued fraction on the right side of (7.3.106) corresponds to the fLs

$$L^* = -\frac{1}{2z^2}\Omega\left(1, \frac{1}{2}; -\frac{1}{z^2}\right) \tag{7.3.107}$$

at $z = \infty$.

Example 2. Dawson's integral [see (6.1.38)].

$$\int_0^z e^{t^2}\,dt = \frac{i\sqrt{\pi}}{2}\operatorname{erf}(-iz) = ze^{z^2}\Phi\left(1; \tfrac{3}{2}; -z^2\right)$$

$$= ze^{z^2}\left[\frac{\tfrac{1}{2}}{\tfrac{1}{2}+z^2} - \frac{z^2}{\tfrac{3}{2}+z^2} - \frac{2z^2}{\tfrac{5}{2}+z^2} - \frac{3z^2}{\tfrac{7}{2}+z^2} - \cdots\right] \tag{7.3.108}$$

is valid for all $z \in \mathbb{C}$.

Example 3. Fresnel integrals [see (6.1.39)].

$$C(z) + iS(z) = \frac{1+i}{2}\operatorname{erf}\left(\frac{\sqrt{\pi}}{2}(1-i)z\right) = ze^{i(\pi/2)z^2}\Phi\left(1; \frac{3}{2}; -\frac{i\pi}{2}z^2\right)$$

$$= ze^{i(\pi/2)z^2}\left[\frac{\tfrac{1}{2}}{\tfrac{1}{2}+\tfrac{i\pi}{2}z^2} - \frac{\tfrac{i\pi}{2}z^2}{\tfrac{3}{2}+\tfrac{i\pi}{2}z^2} - \frac{2\left(\tfrac{i\pi}{2}\right)z^2}{\tfrac{5}{2}+\tfrac{i\pi}{2}z^2} - \frac{3\left(\tfrac{i\pi}{2}\right)z^2}{\tfrac{7}{2}+\tfrac{i\pi}{2}z^2} - \cdots\right]$$

$$\tag{7.3.109}$$

is valid for all $z \in \mathbb{C}$.

7.4 Stable Polynomials

In this section we are concerned with determining whether a polynomial has all of its zeros in the left half of the complex plane [Re(z)<0]. Such a polynomial is called a *stable* (or *Hurwitz*) polynomial in honor of A. Hurwitz, who solved the problem for real polynomials in [1895] in terms of determinants and quadratic forms (see Theorem 7.34). The problem was suggested to Hurwitz by his colleague Stodola at the ETH in Zürich in connection with a problem on the regulation of turbines. Stable polynomials are of considerable importance in the study of eigenvalues associated with mechanical and electrical systems. The principal result of the section is Theorem 7.33, which gives necessary and sufficient conditions for a complex polynomial to be stable. The special case in which the polynomial has real coefficients is stated separately in Corollary 7.33. These results were obtained for real polynomials by Wall [1945] and extended to complex polynomials by Frank [1946b]. Our proof is based on an adaption of results of Levinson and Redheffer [1970, Section 5.5] on positive para-odd functions. Properties of these functions are developed in Theorems 7.26, 7.27, 7.28 and 7.29. Several numerical examples are given to illustrate the procedure for determining the stability of a polynomial. Hurwitz's criterion for the stability of real polynomials is stated without proof in Theorem 7.34. Finally in Theorem 7.35 we give sufficient conditions to insure that a complex polynomial of degree n has m zeros with negative real part and $n-m$ with positive real part. Additional material on this subject can be found in [Henrici, 1974, Section 6.7], [Henrici, 1977, Section 12.7] and [Wall, 1948, Chapter 10].

A non-constant polynomial f is said to be *stable* (or *Hurwitz*) if $f(z_k)=0$ implies that Re(z_k)<0. In the following we shall use the operation of paraconjugation. The *paraconjugate* $f^*(z)$ of a rational function $f(z)$ is defined by

$$f^*(z)=\overline{f(-\bar z)}. \tag{7.4.1}$$

It is easily seen that

$$f^{**}(z)=f(z). \tag{7.4.2}$$

If $f(z)$ is a polynomial

$$f(z)=a_0+a_1z+\cdots+a_nz^n, \tag{7.4.3}$$

then

$$f^*(z)=\bar a_0-\bar a_1z+\cdots+(-1)^n\bar a_nz^n, \tag{7.4.4}$$

whereas if $f(z)$ is written in the form

$$f(z)=c(z-z_1)(z-z_2)\cdots(z-z_n), \tag{7.4.5}$$

then

$$f^*(z) = (-1)^n \bar{c}(z + \bar{z}_1)(z + \bar{z}_2) \cdots (z + \bar{z}_n). \tag{7.4.6}$$

From (7.4.5) and (7.4.6) we see that if z_n is a zero of $f(z)$, then $-\bar{z}_k$ is a zero of $f^*(z)$. Clearly, z_k and $-\bar{z}_k$ are symmetric with respect to the imaginary axis. From (7.4.1) we have

$$f^*(iy) = \overline{f(iy)} \qquad \text{for } y \text{ real,}$$

so that

$$|f^*(iy)| = |f(iy)| \qquad \text{for } y \text{ real.} \tag{7.4.7}$$

Hence $w = r(z) = f^*(z)/f(z)$ maps the imaginary axis $\mathrm{Re}(z) = 0$ into the unit circle $|w| = 1$. In the following theorem it will be shown that $w = r(z)$ maps the upper half plane $\mathrm{Im}(z) > 0$ into the disk $|w| < 1$, provided f is a stable polynomial. It is also convenient to introduce another concept: a rational function h will be called *positive* if $\mathrm{Re}(h(z)) > 0$ when $\mathrm{Re}(z) > 0$.

THEOREM 7.26. *Let f be a non-constant polynomial such that f and f^* have no common zeros. Let r and h be rational functions defined by*

$$r(z) = \frac{f^*(z)}{f(z)}, \qquad h(z) = \frac{f(z) - f^*(z)}{f(z) + f^*(z)}. \tag{7.4.8}$$

Then the following statements are equivalent:

(i) *f is a stable polynomial.*
(ii) *$|r(z)| < 1$ for $\mathrm{Re}(z) > 0$.*
(iii) *h is positive.*

Proof. First we show that (i) implies (ii). If f is stable and is written in the form (7.4.5), then each zero z_k satisfies $\mathrm{Re}(z_k) < 0$. Moreover, if $\mathrm{Re}(z) > 0$, then it is easily shown that

$$|z + \bar{z}_k| < |z - z_k|$$

(note that z will be closer to $-\bar{z}_k$ than to z_k). Thus if f^* is written in the form (7.4.6), it follows that

$$|f^*(z)| < |f(z)| \qquad \text{for} \quad \mathrm{Re}(z) > 0, \tag{7.4.9}$$

so that (ii) holds. Next we show that (ii) implies (i). Clearly (ii) implies (7.4.9), from which we conclude that f can have no zeros in the right half plane. If z_k is an imaginary zero of f, then z_k is also a zero of f^*. However, this situation is not permitted by the hypothesis of the theorem. Thus all zeros of f lie in the left half plane, so that f is stable.

It remains now to show that (ii) and (iii) are equivalent. For this purpose it is sufficient to note that h can be written in the form

$$h(z) = \frac{1 - r(z)}{1 + r(z)},$$

and hence that h satisfies $\operatorname{Re}(h(z)) > 0$ iff r satisfies $|r(z)| < 1$. ∎

Example 1.   The polynomial

$$f(z) = (1 + z)(1 - z^2)$$

is not stable even though the function $h(z) = z$ [defined by (7.4.8)] is positive. Thus we see that the condition (in Theorem 7.26) that f and f^* have no common zeros is essential. In this case

$$f^*(z) = (1 - z)(1 - z^2).$$

A rational function $h(z)$ of the form (7.4.8) satisfies the condition

$$h^*(z) = -h(z). \tag{7.4.10}$$

We shall say that a function h satisfying (7.4.10) is *para-odd*. Theorem 7.26 reduces the study of stable polynomials to the study of positive para-odd rational functions. In the following we investigate some useful properties of these functions. In particular, positive para-odd functions h satisfying $\lim_{z \to \infty} h(z) = \infty$ are characterized in terms of continued fractions in Theorem 7.30. This is the key result used in the proof of Theorem 7.32, which gives a necessary and sufficient condition for a polynomial to be stable.

THEOREM 7.27. *Let a and b be constants satisfying $\operatorname{Re}(a) = 0$ and $b > 0$.*

(A) *If h_1 and h_2 are positive rational functions, then*

$$h_1 + h_2, \quad a + h_1, \quad bh_1 \text{ and } 1/h_1 \tag{7.4.11}$$

are all positive.

(B) *If h_1 and h_2 are para-odd, then the functions in (7.4.11) are all para-odd.*

Proof. The assertions of the theorem are simple consequences of the definition of positive and para-odd. ∎

THEOREM 7.28.

(A) *Let h be a positive para-odd rational function. Then $h(z)$ can be written in the partial-fraction form*

$$h(z)=a+b_0 z+\frac{b_1}{z-i\omega_1}+\frac{b_2}{z-i\omega_2}+\cdots+\frac{b_n}{z-i\omega_n}, \qquad (7.4.12)$$

where $\operatorname{Re}(a)=0$, $b_k \geqslant 0$ for $k=0,1,\ldots,n$, and the ω_i are distinct real numbers. Moreover, $h(z)$ is not identically constant, and hence $b_k>0$ for at least one k such that $0\leqslant k\leqslant n$.

(B) *Conversely, every non-constant function of the form (7.4.12) is positive para-odd.*

Proof. (A): We recall that $h(z_k)=\overline{h^*(-\bar{z}_k)}$, so that if z_k is a zero of h, then $-\bar{z}_k$ is a zero of h^*. The points z_k and $-\bar{z}_k$ are symmetric with respect to the imaginary axis. Therefore, since $-h^*$ is positive, h^* has no zeros in $\operatorname{Re}(z)>0$ and so h has no zeros in $\operatorname{Re}(z)<0$. Also, since h is positive, h can have no zeros in $\operatorname{Re}(z)>0$. Thus all zeros of h lie on the imaginary axis $\operatorname{Re}(z)=0$. Moreover, the zeros of h are all simple. To see this, suppose that h has a zero of order m at $z=\alpha$. Then there exists a function $H(z)$ holomorphic at $z=\alpha$ such that

$$h(z)=(z-\alpha)^m H(z) \quad \text{and} \quad H(\alpha)\neq 0.$$

Now let θ_0, ρ and δ be such that

$$H(\alpha)=|H(\alpha)|e^{i\theta_0} \quad \text{and} \quad H(z)=\rho e^{i(\delta+\theta_0)}.$$

Here ρ and δ are functions of z, with $\rho=|H(z)|\to|H(\alpha)|$ and $\delta\to 0$ as $z\to\alpha$. If we write $z-\alpha=re^{i\theta}$, then

$$h(z)=(z-\alpha)^m H(z)=r^m\rho e^{i(\delta+m\theta+\theta_0)}$$

and hence

$$\operatorname{Re}(h(z))=r^m\rho\cos(\delta+m\theta+\theta_0).$$

If $m\geqslant 2$, then we see that $\operatorname{Re}(h(z))$ changes sign in the interval $-\pi/2<\theta<\pi/2$. Since α is imaginary, this means that $\operatorname{Re}(h(z))$ changes sign in the right half plane $\operatorname{Re}(z)>0$. This is impossible, since h is positive; thus $0\leqslant m\leqslant 1$ as asserted.

By Theorem 7.27 the function $1/h$ is positive para-odd. Hence by the argument given above, the zeros of $1/h$ are all simple and imaginary. This implies that the poles of h are all simple and imaginary, so that $h(z)$ can be written in the form (7.4.12), except that the polynomial part $a+b_0 z$ may be replaced by a polynomial of higher degree. If the degree were 2 or higher, then one can easily see that for large $|z|$, $\text{Re}(h(z))$ would change sign in the right half plane. Similarly by considering large $|z|$ one concludes that $b_0 > 0$, and by considering z close to $i\omega_k$ one sees that $b_k > 0$ for $1 \leqslant k \leqslant n$. For z on the imaginary axis the only contribution to $\text{Re}(h(z))$ is $\text{Re}(a)$. Hence by letting $z \to iy$ we obtain $\text{Re}(a) = 0$, which completes the proof of (A).

(B): To prove the converse it suffices to apply Theorem 7.27, noting that $h(z)$ in (7.4.12) is a sum of positive para-odd functions plus the constant a. ∎

The term $a + b_0 z$ in (7.4.12) will be called the *integral part* of the positive para-odd function h, and it will be denoted by $[\![h]\!]$.

THEOREM 7.29. *If h is a positive para-odd function which does not coincide with $[\![h]\!]$, then*

$$\tilde{h} = h - [\![h]\!]$$

is positive para-odd.

Proof. By Theorem 7.28(A), $\tilde{h}$ is a sum of positive para-odd functions, and hence, by Theorem 7.27, $\tilde{h}$ is itself positive para-odd. ∎

THEOREM 7.30.

(A) *Let h_1 be a positive para-odd function satisfying*

$$\lim_{z \to \infty} h_1(z) = \infty. \tag{7.4.13}$$

Then h_1 can be expressed as a terminating continued fraction of the form

$$h_1(z) = c_1 + d_1 z + \cfrac{1}{c_2 + d_2 z} + \cfrac{1}{c_3 + d_3 z} + \cdots + \cfrac{1}{c_m + d_m z}, \tag{7.4.14a}$$

where

$$\text{Re}(c_j) = 0 \text{ and } d_j > 0, \quad j = 1, 2, \ldots, m. \tag{7.4.14b}$$

(B) *Conversely, if h_1 is a rational function of the form (7.4.14), then h_1 is positive para-odd and satisfies (7.4.13).*

Proof. (A): Since h_1 is positive para-odd, it can be expressed in the partial-fraction form (7.4.12), and in view of (7.4.13), we have

$$[\![h_1]\!](z) = c_1 + d_1 z, \quad \text{Re}(c_1) = 0, \quad d_1 > 0. \tag{7.4.15}$$

If $h_1 = [\![h_1]\!]$, then clearly h_1 has the form (7.4.14) as asserted. Otherwise we obtain a rational function h_2 defined by

$$h_2 = \frac{1}{h_1 - [h_1]},$$

so that

$$h_1 = [h_1] + \frac{1}{h_2} \qquad\qquad (7.4.16)$$

and

$$\lim_{z \to \infty} h_2(z) = \infty. \qquad\qquad (7.4.17)$$

It follows from Theorems 7.27 and 7.29 that h_2 is positive para-odd. Hence by Theorem 7.28(A), $[h_2]$ can be written in the form

$$[h_2](z) = c_2 + d_2 z, \qquad \mathrm{Re}(c_2) = 0, \quad d_2 > 0. \qquad (7.4.18)$$

That $d_2 > 0$ is a consequence of (7.4.17). If $h_2 = [h_2]$, then, by (7.4.15), (7.4.16) and (7.4.18), the function h_2 has the form of (7.4.14) as asserted. On the other hand, if $h_2 \neq [h_2]$, then the process described above can be continued. In general, suppose that for $j = 2, 3, \ldots, k$, h_{j-1} is a positive para-odd function satisfying

$$h_{j-1} = [h_{j-1}] + \frac{1}{h_j}, \qquad\qquad (7.4.19)$$

where

$$h_{j-1} \neq [h_{j-1}] = c_{j-1} + d_{j-1} z, \qquad \mathrm{Re}(c_{j-1}) = 0, \quad d_{j-1} > 0, \qquad (7.4.20)$$

and

$$\lim_{z \to \infty} h_{j-1}(z) = \infty. \qquad\qquad (7.4.21)$$

Then by Theorems 7.27 and 7.29, h_k is positive para-odd. Also, by Theorem 7.28(A),

$$\lim_{z \to \infty} h_k(z) = \infty \qquad\qquad (7.4.22)$$

and

$$[h_k](z) = c_k + d_k z, \qquad \mathrm{Re}(c_k) = 0, \quad d_k > 0. \qquad (7.4.23)$$

Eventually, for some $k = m$, we shall have $h_m = [h_m]$, so that the process will end and h_1 will be expressed in the form (7.4.14). To see this, we write

$$h_{j-1} = \frac{P_{j-1}}{Q_{j-1}}, \qquad (7.4.24)$$

where P_{j-1} and Q_{j-1} are polynomials with $\deg Q_{j-1} < \deg P_{j-1}$, since (7.4.21) holds. In view of (7.4.19) we see that

$$\frac{P_{j-1}}{Q_{j-1}} = c_{j-1} + d_{j-1} z + \frac{Q_j}{P_j}. \qquad (7.4.25)$$

Since the right side of (7.4.25) is obtained by long division of P_{j-1} by Q_{j-1}, it follows that

$$P_j = Q_{j-1} \quad \text{and} \quad \deg Q_j < \deg Q_{j-1} = \deg P_j.$$

Thus for some $m \geqslant 0$ we must have $\deg Q_m = 0$ and so $h_m = h_m = c_m + d_m z$. This completes the proof of (A).

(B): Suppose now that h_1 is a rational function in the form (7.4.14). Define $h_2, h_3, \ldots, h_m$ by

$$h_{j-1}(z) = c_{j-1} + d_{j-1} z + \frac{1}{h_j(z)}, \qquad j = 2, 3, \ldots, m-1, \qquad (7.4.26a)$$

and

$$h_m(z) = c_m + d_m z. \qquad (7.4.26b)$$

Since $\operatorname{Re}(c_j) = 0$ and $d_j > 0$ for $j = 1, 2, \ldots, m$, it follows from Theorem 7.27 that each $h_j, j = 1, 2, \ldots, m$, is positive para-odd. Moreover, it can be seen that

$$\lim_{z \to \infty} h_j(z) = \infty, \qquad j = m, m-1, \ldots, 2, 1. \qquad \blacksquare$$

The following corollary is an immediate consequence of Theorem 7.30.

COROLLARY 7.31. *A real rational function h_1 is positive para-odd and satisfies*

$$\lim_{z \to \infty} h_1(z) = \infty \qquad (7.4.27)$$

iff h_1 can be expressed as a terminating continued fraction of the form

$$h_1(z) = d_1 z + \frac{1}{d_2 z} + \frac{1}{d_3 z} + \cdots + \frac{1}{d_m z} \qquad (7.4.28a)$$

where

$$d_j > 0, \qquad j = 1, 2, \ldots, m. \qquad (7.4.28b)$$

Proof. Since h_1 is real (i.e., its coefficients are all real), $c_j = 0$ for $j = 1, 2, \ldots, m$ in (7.4.14). ■

The continued fraction (7.4.14) is equivalent to a terminating J-fraction (4.5.4), whereas (7.4.28) is closely related to a terminating S-fraction [see (4.5.1) and (8.3.41)]. Theorems 7.30 and Corollary 7.31 provide function-theoretic characterizations for the terminating continued fractions in-volved. Theorem 7.30 will now be used to prove the main result of this section.

THEOREM 7.32. *Let*

$$f(z) = z^n + a_{n-1} z^{n-1} + a_{n-2} z^{n-2} + \cdots + a_1 z + a_0 \qquad (7.4.29)$$

be a polynomial with complex coefficients

$$a_k = \alpha_k + i\beta_k, \qquad k = 0, 1, \ldots, n-1. \qquad (7.4.30)$$

Let

$$g(z) = \alpha_{n-1} z^{n-1} + i\beta_{n-2} z^{n-2} + \alpha_{n-3} z^{n-3} + i\beta_{n-4} z^{n-4} + \cdots. \qquad (7.4.31)$$

Then $f(z)$ is a stable polynomial iff the test fraction $t = g/f$ can be expressed as a terminating continued fraction of the form

$$t(z) = \frac{g(z)}{f(z)} = \frac{1}{1 + c_1 + d_1 z} + \frac{1}{c_2 + d_2 z} + \frac{1}{c_3 + d_3 z} + \cdots + \frac{1}{c_n + d_n z},$$

$$(7.4.32a)$$

where

$$\mathrm{Re}(c_j) = 0 \text{ and } d_j > 0, \qquad j = 1, 2, \ldots, n. \qquad (7.4.32b)$$

Proof. Suppose that $t(z)$ can be written in the form (7.4.32). Let $z = x + iy$ be a fixed complex number with $x = \mathrm{Re}(z) \geqslant 0$. We shall consider the linear fractional transformations

$$s_1(w) = \frac{1}{1 + c_1 + d_1 z + w}; \qquad s_j(w) = \frac{1}{c_j + d_j z + w}, \qquad j = 2, 3, \ldots, n,$$

and the half plane

$$H = [w : \mathrm{Re}(w) \geqslant 0].$$

It is easily shown that $s_1(H)$ is the circular disk

$$s_1(H)=\left[w:\left|w-\frac{1}{2(1+d_1x)}\right|\leqslant\frac{1}{2(1+d_1x)}\right]$$

and hence that

$$s_1(H)\subseteq\left[w:|w-\tfrac{1}{2}|\leqslant\tfrac{1}{2}\right].$$

Moreover, one can readily show that

$$s_j(H)\subseteq H \qquad \text{for } j=2,3,\ldots,n.$$

It follows that

$$s_1\circ s_2\circ\cdots\circ s_n(H)\subseteq\left[w:|w-\tfrac{1}{2}|\leqslant\tfrac{1}{2}\right],$$

and since

$$t(z)=s_1\circ s_2\circ\cdots\circ s_n(0),$$

we have

$$\left|\frac{g(z)}{f(z)}-\frac{1}{2}\right|=\left|t(z)-\frac{1}{2}\right|\leqslant\frac{1}{2}. \tag{7.4.33}$$

Since $t(z)=g(z)/f(z)$ is the nth approximant of a continued fraction (7.4.32), we conclude from the determinant formula (2.1.9) that $g(z)$ and $f(z)$ cannot vanish simultaneously. Therefore (7.4.33) implies that $f(z)\neq0$. Since z is an arbitrary point satisfying $\text{Re}(z)\geqslant0$, the polynomial $f(z)$ must be stable.

   Suppose now that $f(z)$ is stable. We shall show that the test fraction $t(z)$ can be expressed in the form (7.4.33). Let

$$h(z)=\frac{f(z)-f^*(z)}{f(z)+f^*(z)}.$$

Since f is stable, f and f^* can have no common zeros. Hence by Theorem 7.26 h is a positive para-odd function. By (7.4.29) and (7.4.30) we have

$$h(z)=\frac{i\beta_0+\alpha_1z+i\beta_2z^2+\alpha_3z^3+\cdots+\left(\dfrac{1-(-1)^n}{2}\right)z^n}{a_0+i\beta_1z+\alpha_2z^2+i\beta_3z^3+\cdots+\left(\dfrac{1+(-1)^n}{2}\right)z^n}. \tag{7.4.34}$$

From (7.4.4), (7.4.29) and (7.4.30) we see that

$$g(z) = \frac{f(z) + f^*(z)}{2} \qquad \text{if } n \text{ is odd}$$

and

$$g(z) = \frac{f(z) - f^*(z)}{2} \qquad \text{if } n \text{ is even.}$$

Hence we must distinguish the two separate cases.

Case (a). Suppose that n is odd. Then it can be seen that

$$\frac{1}{t(z)} - h(z) = \frac{2f(z)}{f(z) + f^*(z)} - \frac{f(z) - f^*(z)}{f(z) + f^*(z)} = 1,$$

so that

$$t(z) = \frac{1}{1 + h(z)}. \qquad (7.4.35)$$

Since n is odd, it follows from (7.4.34) that

$$\lim_{z \to \infty} h(z) = \infty.$$

Therefore Theorem 7.30(A) implies that $h(z)$ can be expressed as a terminating continued fraction of the form (7.4.14) for some integer $m > 0$. Since the numerator of $h(z)$ is a polynomial of degree n, an application of the difference equations (2.1.6) shows that $m = n$. Substitution of the continued fraction for $h(z)$ into (7.4.35) establishes the required representation of $t(z)$ in the form (7.4.32).

Case (b). The situation with n even can be handled in a similar manner. However, in this case

$$\frac{1}{t(z)} - \frac{1}{h(z)} = \frac{2f(z)}{f(z) - f^*(z)} - \frac{f(z) + f^*(z)}{f(z) - f^*(z)} = 1,$$

so that

$$t(z) = \frac{1}{1 + h_1(z)}, \qquad \text{where} \quad h_1(z) = \frac{1}{h(z)}.$$

Theorem 7.27 implies that $h_1(z)$ is positive para-odd, and from (7.4.34) we see that

$$\lim_{z \to \infty} h_1(z) = \infty.$$

Therefore Theorem 7.30(A) insures that $h_1(z)$ can be expressed as a continued fraction of the form (7.4.14), and we deduce that $t(z)$ has the required form (7.4.32).   ∎

COROLLARY 7.33.   *Let*

$$f(z)=z^n+a_{n-1}z^{n-1}+a_{n-2}z^{n-2}+\cdots+a_1z+a_0 \qquad (7.4.36)$$

be a polynomial with real coefficients a_k. *Let*

$$g(z)=a_{n-1}z^{n-1}+a_{n-3}z^{n-3}+a_{n-5}z^{n-5}+\cdots. \qquad (7.4.37)$$

Then $f(z)$ *is a stable polynomial iff the test fraction* $t=g/f$ *can be expressed as a terminating continued fraction of the form*

$$t(z)=\frac{g(z)}{f(z)}=\frac{1}{1+d_1z}+\frac{1}{d_2z}+\frac{1}{d_3z}+\cdots+\frac{1}{d_nz}, \qquad (7.4.38a)$$

where

$$d_j>0, \qquad j=1,2,\ldots,n. \qquad (7.4.38b)$$

The division process described in the proof of Theorem 7.30(A) can be used (when possible) to obtain the continued-fraction representation of $t(z)$ of the form (7.4.32) or (7.4.38). The process involves only the division of polynomials and is easily programmable. The procedure is illustrated by the following examples.

Example 2.   Determine whether the real polynomial

$$f(z)=z^4+5z^3+10z^2+10z+4 \qquad (7.4.39)$$

is stable. The test fraction of Corollary 7.33 is $t=g/f$, where

$$g(z)=5z^3+10z. \qquad (7.4.40)$$

Division of f by g yields

$$t(z)=\frac{1}{f(z)/g(z)}=\frac{1}{1+h_1(z)},$$

where

$$h_1(z)=\tfrac{1}{5}z+\frac{8z^2+4}{5z^3+10z}.$$

Thus

$$h_1(z) = \tfrac{1}{5}z + \frac{1}{h_2(z)},$$

where, upon division of $5z^3 + 10z$ by $8z^2 + 4$, we obtain

$$h_2(z) = \tfrac{5}{8}z + \frac{\tfrac{15}{2}z}{8z^2 + 4}.$$

Therefore

$$h_2(z) = \tfrac{5}{8}z + \frac{1}{h_3(z)},$$

where, upon division of $8z^2 + 4$ by $\tfrac{15}{2}z$, we arrive at

$$h_3(z) = \tfrac{16}{15}z + \frac{1}{\tfrac{15}{8}z}.$$

Combining the preceding results, we obtain the continued-fraction expression

$$t(z) = \frac{1}{1 + \tfrac{1}{5}z} + \frac{1}{\tfrac{5}{8}z} + \frac{1}{\tfrac{16}{15}z} + \frac{1}{\tfrac{15}{8}z}. \tag{7.4.41}$$

It follows from Corollary 7.33 that $f(z)$ is stable.

Example 3.   For what real values of the constant c will the polynomial

$$f(z) = z^4 + 5z^3 + 10z^2 + 10z + c \tag{7.4.42}$$

be stable? Following the procedure used in Example 2, we obtain for the test function

$$t(z) = \frac{1}{1 + \tfrac{1}{5}z} + \frac{1}{\tfrac{5}{8}z} + \frac{1}{d_3 z} + \frac{1}{d_4 z}, \tag{7.4.43a}$$

where

$$d_3 = \frac{8}{10 - \tfrac{5}{8}c} \quad \text{and} \quad d_4 = \frac{10 - \tfrac{5}{8}c}{c}. \tag{7.4.43b}$$

Therefore $d_3 > 0$ iff $c < 16$, and hence $d_4 > 0$ iff $c > 0$. Thus, by Corollary

7.33, f is a real stable polynomial iff

$$0 < c < 16.$$

Example 4. In Example 1 we considered the polynomial which is -1 multiplied by

$$f(z) = z^3 + z^2 - z - 1 = (z^2 - 1)(z + 1). \qquad (7.4.44)$$

The test function of Corollary 7.33 is $t = f/g$, where

$$g(z) = z^2 - 1.$$

Therefore

$$t(z) = \frac{1}{1 + z}. \qquad (7.4.45)$$

The right side is a continued fraction of the form

$$\frac{1}{1 + d_1 z} + \frac{1}{d_2 z} + \frac{1}{d_3 z} + \cdots + \frac{1}{d_m z}, \qquad d_j > 0, \qquad (7.4.46)$$

in Corollary 7.33. However, $m = 1$ instead $m = 3$ as is required in (7.4.38). Thus Corollary 7.33 implies that $f(z)$ is not stable. We have included this example to emphasize the fact that for f to be stable, it is necessary not only for $t(z)$ to be representable by a continued fraction of the form (7.4.46), but also for $m = n$, where n is the degree of the polynomial $f(z)$.

We conclude this section by stating two additional results on the location of zeros of a polynomial. For a proof of the *Hurwitz criterion for stability* (Theorem 7.34) see [Henrici, 1977, Theorem 12.7c]. Theorem 7.35 is due to Wall [1945] for real polynomials and was extended to complex polynomials by Frank [1946b]. A proof of that result can be found in [Wall, 1948, Theorem 48.1].

THEOREM 7.34. *Let*

$$f(z) = a_0 + a_1 z + a_2 z^2 + \cdots + a_n z^n, \qquad a_0 \neq 0, \qquad (7.4.47)$$

be a real polynomial of degree n. Let

$$D_1 = a_1, \qquad D_2 = \begin{vmatrix} a_1 & a_3 \\ a_0 & a_2 \end{vmatrix}, \qquad D_3 = \begin{vmatrix} a_1 & a_3 & a_5 \\ a_0 & a_2 & a_4 \\ 0 & a_1 & a_3 \end{vmatrix}, \ldots, \qquad (7.4.48)$$

where D_k is an n-by-n determinant, $a_m = 0$ for $m > n$, and $D_0 = 1$, $D_{-1} = 1/a_0$,

$D_{-2} = 1/a_0^2$. *Then f is stable iff the determinants (7.4.48) satisfy*

$$D_{2k} > 0, \qquad k = 1, 2, \ldots, [n/2],$$

and

$$\text{sign } D_{2k+1} = \text{sign } a_0, \qquad k = 0, 1, \ldots, \left[\!\left[\frac{n-1}{2}\right]\!\right].$$

We note that from the numerical point of view it is easier to determine the stability of a polynomial by the division method used in Example 2 than by computing the determinants D_k.

THEOREM 7.35. *Let*

$$f(z) = z^n + a_{n-1} z^{n-1} + a_{n-2} z^{n-2} + \cdots + a_1 z + a_0$$

be a polynomial with complex coefficients. Let the test fraction $t(z)$ of Theorem 7.32 be represented by a terminating continued fraction of the form (7.4.32a), where

$$\text{Re}(c_j) = 0, \qquad j = 1, 2, \ldots, n,$$

and where m of the d_j are positive and $n - m$ are negative. Then m of the zeros of $f(z)$ have negative real part, and $n - m$ have positive real part.

CHAPTER 8

Truncation-Error Analysis

8.1 Introduction

If a continued fraction $K(a_n/b_n)$ with nth approximant f_n converges to a finite limit f, then

$$f - f_n$$

is called the *truncation error of the nth approximant*. The problem we are concerned with here is the determination of realistic bounds for $|f - f_n|$. Such bounds have considerable importance for the computation of functions using continued-fraction representations. There is a substantial overlap between the theory discussed in Chapter 4 and truncation-error analysis, which we consider in this chapter.

The approach that will be used is the following: Suppose we are given a continued fraction $K = K(a_n/b_n)$ belonging to a family of continued fractions denoted by $\mathcal{K}$. Then for each $n = 1, 2, 3, \ldots$, we wish to find a region in $\mathbb{C}$ which contains all $(n+k)$th approximants of all continued fractions in $\mathcal{K}$ which have the elements $a_1, a_2, \ldots, a_n$ and $b_1, b_2, \ldots, b_n$ in common with K. Such a region is called an nth *inclusion region I_n of K with respect to the family* $\mathcal{K}$. Knowledge of I_n, if it is bounded, can be used to estimate the truncation error, since

$$|f - f_n| < \operatorname{diam} I_n.$$

The method, developed in Chapter 4, of estimating the radii R_n of nested circular disks $S_n(V_n)$ is one way of finding truncation error bounds [see Theorem 4.2 and note that $f_n = S_n(0)$]. For then

$$|f - f_n| < 2R_n = \operatorname{diam} S_n(V_n).$$

ENCYCLOPEDIA OF MATHEMATICS and Its Applications, Gian-Carlo Rota (ed.).
Vol. 11: William B. Jones and W. J. Thron, Continued Fractions.   ISBN 0-201-13510-8

In using it, however, the emphasis here is on restricting element regions sufficiently to get a rapidly decreasing sequence $\{R_n\}$. In Chapter 4 we were interested in determining the largest element regions for which $\lim R_n = 0$, even if the approach to zero is extremely slow. In addition, a number of convergence proofs are "existential" (that is, not "constructive,") in nature and thus yield no truncation-error estimates at all.

Since large-scale computations with continued fractions have become feasible only since the advent of electronic computers, it is not surprising that references to work in truncation-error analysis are all of recent origin. Important contributions have been made by [Baker, 1969], [Blanch, 1964], [Common, 1968], [Elliott, 1967], [Field, 1976, 1977, 1978a, b], [Field and Jones, 1972], [Gragg, 1968, 1970], [Hayden, 1965], [Henrici and Pfluger, 1966], [Jefferson, 1969a, b], [Jones and Snell, 1969], [Jones and Thron, 1971, 1976], [Lange, 1966], [Luke, 1958], [Merkes, 1966], [Reid, 1978], [Sweezy and Thron, 1967] and [Thron, 1958]. The papers by Elliott [1967] and Luke [1958] give asymptotic estimates of the truncation error and are not dealt with here.

Though conceptually distinct, the search for suitable numbers w_n such that $\{S_n(w_n)\}$ converges faster to $f = \lim S_n(0)$ than $\{S_n(0)\}$ does will also be discussed in this chapter, since it has practical implications. Such numbers w_n are sometimes called "converging factors" (see [Wynn, 1959]). One also speaks of accelerating convergence. Glaisher [1873/4] appears to have been the first to come upon an example where suitable choice of w_n accelerated the convergence of $\{S_n(w_n)\}$. Hamel [1918a, b] found other reasons suggesting the study of $\{S_n(w_n)\}$. Wynn [1959], Hayden [1965], Gill [1975, 1978a], and Thron and Waadeland [1980] were again concerned with the acceleration of convergence.

8.2 General Theory of Inclusion Regions and Truncation Errors

Jones and Thron [1976] developed general machinery for truncation-error analysis which applies not only to continued fractions but also to other infinite processes such as approximate integration and iterative methods for solving non-linear equations. Our treatment here is strongly influenced by that analysis. However, we restrict ourselves to that part of the theory which is directly applicable to continued fractions.

Let $\mathcal{K}$ be a family of continued fractions, and denote its elements by $K, K^*, K^\dagger, \cdots$. By $f_n(K)$ we shall mean the nth approximant of the continued fraction K, and by $a_n(K)$ and $b_n(K)$ the nth partial numerator and denominator, respectively. The symbol $\sim_n$ is defined to mean

$$K \sim_n K^* \quad \text{iff} \quad a_m(K) = a_m(K^*) \text{ and } b_m(K) = b_m(K^*), \quad m = 1, 2, \ldots, n.$$
$$(8.2.1)$$

For $K \in \mathcal{K}$ we let $\psi_n(\mathcal{K}, K)$ denote the subset of $\hat{C}$ defined by

$$\Psi_n(\mathcal{K}, K) = \left[\, f_m(K^*) : m \geqslant n, \; K^* \sim_n K, \; K^* \in \mathcal{K} \,\right] \qquad (8.2.2)$$

Every closed subset of $\hat{C}$ containing $\Psi_n(\mathcal{K}, K)$ will be called an *nth inclusion region for K with respect to* $\mathcal{K}$. Since $c(\Psi_n(\mathcal{K}, K))$ is the smallest such inclusion region, we call it the *best nth inclusion region of K with respect to* $\mathcal{K}$.

If a continued fraction K converges to a finite value, denoted by $f(K)$, and if $\Psi_n(\mathcal{K}, K)$ is bounded, then the truncation error $f(K) - f_n(K)$ of the nth approximant satisfies

$$| f(K) - f_n(K)| \leqslant \operatorname{diam} c(\Psi_n(\mathcal{K}, K)). \qquad (8.2.3)$$

Clearly $f(K)$ need not be known and in most practical cases is not known. The point of truncation-error analysis is to determine how far the unknown $f(K)$ may be from the known (computed) $f_n(K)$. In all of this the concept of *available information* both as to the class $\mathcal{K}$ and the number n of pairs of elements $\langle a_1, b_1 \rangle, \cdots, \langle a_n, b_n \rangle$ plays a key role. The available information may have some unsurmountable limits (beyond our control) or, more likely, may be practically limited by considerations of effort and/or cost. Whatever the reasons, it is clear that one can make meaningful statements about truncation-error bounds and inclusion regions only in terms of precisely stated assumptions about available information. The following result is an immediate consequence of the definition of Ψ_n.

THEOREM 8.1.

(A) *Let* $\mathcal{K}$ *and* $\mathcal{K}'$ *be families of continued fractions, and let* $\Psi_n(\mathcal{K}, K)$ *be the set defined in* (8.2.2). *Then*

$$\Psi_{n+1}(\mathcal{K}, K) \subseteq \Psi_n(\mathcal{K}, K), \qquad n = 1, 2, 3, \ldots, \qquad (8.2.4)$$

and

$$\Psi_n(\mathcal{K}, K) \subseteq \Psi_n(\mathcal{K}', K) \qquad if \;\; \mathcal{K} \subseteq \mathcal{K}'. \qquad (8.2.5)$$

(B) *Let* $\mathcal{K}_1, \mathcal{K}_2, \ldots, \mathcal{K}_m$ *be families of continued fractions, and assume that*

$$K \in \bigcap_{\nu=1}^{m} \mathcal{K}_\nu \neq \varnothing.$$

Then

$$\Psi_n\left(\bigcap_{\nu=1}^{m} \mathcal{K}_\nu, K \right) = \bigcap_{\nu=1}^{m} \Psi_n(\mathcal{K}_\nu, K). \qquad (8.2.6)$$

Before proceeding to give examples of families $\mathcal{K}$, we note that the knowledge of $\Psi_n(\mathcal{K}, K)$ explicitly gives much more information than just knowing $\operatorname{diam} c(\Psi_n(\mathcal{K}, K))$. We also observe that as far as inclusion regions are concerned convergence of the continued fraction K plays no role, but for the formulation of truncation error one must know that $f(K)$ exists and is finite. The following elementary examples illustrate the basic concepts of inclusion regions and families of continued fractions. Many other examples of interest and importance are given in the following section.

Example 1. For $a \in \mathbb{C}$ let $\mathcal{L}_a$ be the family of continued fractions defined by

$$\mathcal{L}_a = \left[K : K = \mathop{K}_{n=1}^{\infty} (a_n/1), \lim a_n = a \right].$$

Thus $\mathcal{L}_a$ consists of all limit periodic continued fractions $\mathop{K}(a_n/1)$ with $\lim a_n = a$. We consider this class mainly to illustrate the fact that Ψ_n can be the whole complex plane. Clearly we have, for all $K \in \mathcal{L}_a$ and $n = 1, 2, 3, \ldots,$

$$\Psi_n(\mathcal{L}_a, K) = \hat{\mathbb{C}}.$$

Example 2. For each positive number p, let Σ_p denote the family of continued fractions

$$\Sigma_p = \left[K : K = \mathop{K}_{n=1}^{\infty} (a_n/1), \sum_{n=1}^{\infty} |a_n| < p \right].$$

The interesting thing about this family is that the restrictions on each a_n depend on the choice of all preceding a_m, $m = 1, 2, \ldots, n - 1$.

It becomes desirable here to modify the notation for the linear fractional transformations (2.1.1) slightly to bring out the dependence on the continued fraction K. Hence we write

$$s_n(K, w) = \frac{a_n(K)}{b_n(K) + w}, \qquad n = 1, 2, 3, \ldots, \tag{8.2.7a}$$

and

$$S_1(K, w) = s_1(K, w), \tag{8.2.7b}$$

$$S_n(K, w) = S_{n-1}(K, s_n(K, w)), \qquad n = 2, 3, 4, \ldots. \tag{8.2.7c}$$

By $s_n^{-1}(K, w)$ and $S_n^{-1}(K, w)$ we mean the inverse with respect to w.

By imposing restrictions on $\langle a_n, b_n \rangle$, as discussed in Section 4.2, we arrive at a very important collection of families $\mathcal{K}$.

Example 3.   In the terminology of Section 4.2, let $\{\Omega_n\}$ be a sequence of element regions corresponding to a sequence of value regions $\{V_n\}$ (see Theorems 4.1 and 4.2). Let $\mathcal{K}(\{\Omega_n\})$ denote the family of continued fractions defined by

$$\mathcal{K}(\{\Omega_n\}) = [\, K : \langle a_n(K), b_n(K) \rangle \in \Omega_n, \, n = 1, 2, 3, \ldots \,]. \qquad (8.2.8)$$

Then it can be seen that

$$\Psi_n(\mathcal{K}(\{\Omega_n\}), K) \subseteq S_n(K, V_n), \qquad n = 1, 2, 3, \ldots. \qquad (8.2.9)$$

Equality holds in (8.2.9) if $V_n = W_n$, where W_n is defined by (4.2.5).

The phenomenon, shown in the above example, that inclusion regions $I_n(\mathcal{K}, K)$ can be obtained as $S_n(K, T_n(\mathcal{K}, K))$, is not restricted to families $\mathcal{K}(\{\Omega_n\})$. Whenever there are inclusion regions $I_n(\mathcal{K}, K)$, then

$$T_n(\mathcal{K}, K) = S_n^{-1}(K, I_n(\mathcal{K}, K))$$

will satisfy

$$I_n(\mathcal{K}, K) = S_n(K, T_n(\mathcal{K}, K)). \qquad (8.2.10)$$

Thus all families $\mathcal{K}$ can be classified by the behavior of their sequences $\{T_n(\mathcal{K}, K)\}$.

In Example 1 we had $T_n(\mathcal{L}_a, K) = \hat{\mathbb{C}}$ for all $K \in \mathcal{L}_a$, $n \geqslant 1$, $a \in \mathbb{C}$. In Example 2, $T_n(\Sigma_p, K)$ is fairly difficult to describe but it clearly depends upon K. In Example 3 the T_n are independent of K but do depend upon n. Finally, if one considers $\Omega_n = \Omega$, $n = 1, 2, 3, \ldots$, then one arrives at a set $T_n = T$, where T depends only on $\mathcal{K}$.

The formula (4.2.2b), that is,

$$\frac{a_n}{b_n + V_n} \subseteq V_{n-1} \qquad \text{for} \quad \langle a_n, b_n \rangle \in \Omega_n,$$

has a general analogue as follows:

$$\frac{a_n(K)}{b_n(K) + T_n(\mathcal{K}, K)} \subseteq T_{n-1}(\mathcal{K}, K).$$

Writing Ψ_n as a function of T_n suggests that, if the s_n are contraction maps (i.e. $|s_n'(K, w)| < 1$), then the Ψ_n will be small. This observation has been used by Hayden [1965], who credits Merkes with the idea. Hayden was able to get extremely good results by applying this approach to continued-fraction expansions of $\tan z$ and e^{-z}. Unfortunately, the applicability of the idea seems limited. However, fortunately, diam $S_n(K, T_n)$ may be substantially smaller than diam T_n even though $|S'_n(K, w)| < 1$ does not hold for all $w \in T_n$.

We conclude this section by observing that the following classes are of great importance:

$$S(z) = \left[K : K = \mathbf{K}(a_n z/1), a_n > 0, n = 1,2,3,\ldots \right],$$

$$\mathcal{T}(z) = \left[K : K = \mathbf{K}(F_n z/(1 + G_n z)), F_n > 0, G_n > 0, n = 1,2,3,\ldots \right],$$

$$\mathcal{P}(\{\zeta_n\}) = \left[K : K = \cfrac{1}{\beta_1 + \zeta_1} + \cfrac{-\alpha_1^2}{\beta_2 + \zeta_2} + \cfrac{-\alpha_2^2}{\beta_3 + \zeta_3} + \cdots, \alpha_n \neq 0, \right.$$

$$\operatorname{Im}(\beta_n) \geqslant 0, \left[\operatorname{Im}(\alpha_n)\right]^2 = \operatorname{Im}(\beta_{n+1}) \operatorname{Im}(\beta_n)(1 - h_{n-1}) h_n,$$

$$\left. 0 < h_n < 1 \right].$$

These are, respectively, the S-fractions, positive T-fractions, and positive definite continued fractions. Inclusion regions and truncation-error bounds for all three of these families are given in Section 8.3. We shall also return to these families in Chapter 9 in connections with the theory of moments.

8.3 Explicit Results on Inclusion Regions and Truncation Error Bounds

By restricting the element regions considered in Chapter 4 so as to get smaller radii R_n for the circular disks $S_n(V_n)$ derived there (Theorem 4.2), one can obtain a good deal of information on truncation-error bounds. We give three typical results. The first two are due to Field and Jones [1972]. Sweezy and Thron [1967] had the special case of Theorem 8.2 where c is real. For $\delta = 0$ one still has convergence (see Theorems 4.33 or Corollary 4.34), but it may be very slow.

THEOREM 8.2. *Let* $0 < \delta < 1$, $r > 0$ *and* c *a complex number be given. Assume that they are related by*

$$r^2 - |c|^2 = 1 - \delta. \tag{8.3.1}$$

Let $\mathbf{K}(1/b_n)$ *be a continued fraction with elements* b_n *satisfying*

$$|b_{2n-1} + \Gamma| \geqslant \rho, \quad |b_{2n} + \bar{\Gamma}| \geqslant \rho, \quad n = 1,2,3,\ldots, \tag{8.3.2a}$$

where

$$\Gamma = \frac{c(2-\delta)}{1-\delta}, \qquad \rho = \frac{r(2-\delta)}{1-\delta}. \qquad (8.3.2b)$$

Then $K(1/b_n)$ *converges to a finite value* f, *and if* f_n *denotes the* n*th approximant, then*

$$|f - f_n| < 2r\left(1 - \frac{\delta(2-\delta)}{1+(r+|c|)^2}\right)^{n-1}, \qquad n = 1, 2, 3, \ldots \qquad (8.3.3a)$$

and

$$0 < \frac{\delta(2-\delta)}{1+(r+|c|)^2} < 1. \qquad (8.3.3b)$$

Although the statement of the following theorem is somewhat more complicated, it provides much sharper error bounds.

THEOREM 8.3. *Let* $\{c_n\}$ *be a sequence of complex numbers, and let* $\{r_n\}$ *and* $\{\delta_n\}$ *be sequences of positive real numbers such that*

$$0 \leqslant |c_n| < r_n, \qquad n = 0, 1, 2, \ldots, \qquad (8.3.4a)$$

$$\delta_1 = 1; \qquad 0 < \delta_n \leqslant 1, \quad n = 2, 3, 4, \ldots. \qquad (8.3.4b)$$

Let $K(1/b_n)$ *be a continued fraction with elements* b_n *satisfying the conditions*

$$|b_n + c_n + \kappa_{n-1}| \geqslant r_n + \frac{\rho_{n-1}}{\delta_n}, \qquad n = 1, 2, 3, \ldots, \qquad (8.3.5a)$$

where

$$\kappa_{n-1} = \frac{\bar{c}_{n-1}}{r_{n-1}^2 - |c_{n-1}|^2}, \qquad \rho_{n-1} = \frac{r_{n-1}}{r_{n-1}^2 - |c_{n-1}|^2}, \qquad n = 1, 2, 3, \ldots. \qquad (8.3.5b)$$

Let B_n *and* f_n *denote the* n*th denominator and approximant of* $K(1/b_n)$, *respectively. Then, for* $n \geqslant 2$ *and* $p \geqslant 0$,

$$|f_{n+p} - f_n| < 2r_0 \prod_{j=2}^{n} g_j(\gamma_{j-1}, \delta_j) < 2r_0 \prod_{j=2}^{n} M_j(\delta_j) < 2r_0 \prod_{j=2}^{n} \delta_j, \qquad (8.3.6a)$$

where

$$g_j(\gamma_{j-1}, \delta_j) = \frac{\lambda_j(1 - \gamma_{j-1}^2)}{2(1/\delta_j + \lambda_j - \gamma_{j-1})(1/\delta_j - \gamma_{j-1})},$$ (8.3.6b)

$$0 \leqslant \gamma_{j-1} = \frac{|D_{j-1} - \kappa_{j-1}|}{\rho_{j-1}} < 1,$$ (8.3.6c)

$$D_{j-1} = \frac{B_{j-2}}{B_{j-1}}, \qquad \lambda_j = \frac{2r_j}{\rho_{j-1}}$$ (8.3.6d)

$$0 < M_j(\delta_j) = \frac{1 + \lambda_j \delta_j - \delta_j^2 - \sqrt{(1 - \delta_j^2)((1 + \lambda_j \delta_j)^2 - \delta_j^2)}}{\lambda_j \delta_j^2} \leqslant \delta_j.$$

(8.3.6e)

Remarks. If the continued fraction $K(1/b_n)$ converges to a finite value f, then Theorem 8.3 gives truncation-error bounds in (8.3.6a) if we replace f_{n+p} by f. The first inequality in (8.3.6a) was obtained by an analysis similar to that used in the proof of Theorem 4.40 (uniform parabola theorem). The second inequality in (8.3.6a) was determined by

$$M_j(\delta_i) = \max_{0 < \gamma_{j-1} < 1} g_j(\gamma_{j-1}, \delta_j).$$ (8.3.7)

The last inequality in (8.3.6a) can be proved directly from the definition of $M_j(\delta_j)$. From (8.3.5) one can see that the element region for b_n is made smaller if δ_n is decreased, and from (8.3.6) one can see that this decreases the truncation error bound. In practical applications one chooses the δ_n as small as possible. Some aids in choosing the δ_n were discussed in [Field and Jones, 1972, Lemma 3.1].

Example 1. Bessel functions of the first kind. A continued fraction representation of ratios of Bessel functions was given in Theorem 5.16. Setting $\nu = 0$ and making an equivalence transformation of (5.4.32), we obtain for all $m = 1, 2, 3, \ldots$,

$$G_m(z) = \frac{J_m(z)}{J_{m-1}(z)} = \frac{1}{\dfrac{2m}{z}} - \frac{1}{\dfrac{2(m+1)}{z}} - \frac{1}{\dfrac{2(m+2)}{z}} \cdots,$$ (8.3.8)

valid for all $z \in \mathbb{C}$. Let $f_{m,n}(z)$ denote the nth approximant of the continued fraction (8.3.8). Let r and s be real numbers such that

$$|s| < r.$$ (8.3.9)

Table 8.3.1.   Truncation-error Bounds for
Bessel Functions of the First Kind,
$|J_m(z)/J_{m-1}(z) - f_{m,n}(z)| < 2\Pi_{j=2}^n M_j^{(m)}$, $|z| < 1, m > 1, n > 2$

n	2	3	4	5	6	7	8
$2\Pi_2^n M_j^{(m)}$	1.4(−1)	4.0(−3)	6.5(−5)	6.6(−7)	4.6(−9)	2.4(−11)	9.4(−14)

Then from Theorem 8.3 it can be shown that for all $z \neq 0$ in the disk

$$D(r,s) = \left[z : \left| z + \frac{2msi}{1+r^2-s^2} \right| < \frac{2mr}{1+r^2-s^2} \right], \qquad (8.3.10)$$

we have

$$|G_m(z) - f_{m,n}(z)| < 2r \prod_{j=2}^{n} M_j^{(m)}, \qquad m = 1,2,3,\ldots, \quad n = 2,3,4,\ldots,$$

$$(8.3.11a)$$

where

$$M_j^{(m)} = \frac{1 + \lambda \delta_j^{(m)} - (\delta_j^{(m)})^2 - \sqrt{\left[1 - (\delta_j^{(m)})^2\right]\left[(1+\lambda\delta_j^{(m)})^2 - (\delta_j^{(m)})^2\right]}}{\lambda(\delta_j^{(m)})^2},$$

$$(8.3.11b)$$

$$\delta_1^{(m)} = 1; \qquad \delta_j^{(m)} = \frac{mr}{mr + (j-1)(r-s)(1+r^2-s^2)} < 1, \quad j > 2,$$

$$(8.3.11c)$$

$$\lambda = 2(r^2 - s^2). \qquad (8.3.11d)$$

Taking $s = 0$ and $r = 1$, we obtain truncation-error bounds (8.3.11a) for $|z| < 1$ given in Table 8.3.1. The entries in the table should be read as the first two-digit number multiplied by 10 to the power in parentheses. Large disks $D(r,s)$ for z in the upper or lower half plane can be obtained by choosing $s \neq 0$. Other truncation error bounds for $K(1/b_n)$ can be found in [Field, 1977, 1978a,b].

For continued fractions $K(a_n/1)$ we give the following result due to Jones and Snell [1969], which extends the results of Thron [1958].

THEOREM 8.4. *Let $\{P_n\}$ be a sequence of complex numbers $P_n = p_n e^{i\psi_n}$ such that*

$$|P_n - \tfrac{1}{2}| < \tfrac{1}{2} - \varepsilon, \qquad 0 < \varepsilon < \tfrac{1}{2}, \quad n = 0,1,2,\ldots. \qquad (8.3.12)$$

Let $\{E_n\}$ be the sequence of parabolic regions defined by

$$E_n = \left[z : |z| - \text{Re}(ze^{-i(\psi_n + \psi_{n-1})}) \leqslant 2kp_{n-1}(\cos \psi_n - p_n) \right], \quad (8.3.13)$$

where

$$0 < k < 1.$$

If $K(a_n/1)$ is a continued fraction with elements satisfying

$$a_n \in E_n, \quad 0 < |a_n| < M, \quad n = 1, 2, 3, \ldots, \quad (8.3.14)$$

for some $M > 0$, then $K(a_n/1)$ converges to a finite value f and

$$|f - f_n| < \frac{|a_1|(\cos \psi_1 - p_1)}{\left(1 + \dfrac{\varepsilon^2(1-k)}{M}\right)^{n-1}}, \quad n = 2, 3, 4, \ldots \quad (8.3.15)$$

For a description of the parabolic boundary of E_n see Figure 4.2.1(b). The parameter k permits the shrinkage of the regions. For $k = 1$ the continued fraction still converges (see Theorem 4.43) but no truncation-error bounds can be obtained from Theorem 8.4. More flexibility could be obtained by allowing k to vary with n, but this will not be done here.

In Theorems 8.2 and 8.4 the bounds on the truncation error are of the form d^n, $0 < d < 1$. Here d is independent of the particular continued fraction considered, and depends only on the class $\mathcal{K}$. One can thus determine *a priori* what n to use in order to insure that the approximant f_n provide the desired degree of approximation.

There are also the so called *a posteriori* bounds which depend upon $f_n - f_{n-1}$. One way of getting such bounds is as follows. Assume that $K(a_n/b_n)$ converges to a finite value f. Let f_n denote the nth approximant, and let S_n denote the linear fractional transformation (2.1.1b), so that $S_n(0) = f_n$. Then

$$f = S_n(f^{(n)}), \quad n = 1, 2, 3, \ldots, \quad (8.3.16a)$$

where $f^{(n)}$ is the nth tail of the continued fraction

$$f^{(n)} = \frac{a_{n+1}}{b_{n+1}} + \frac{a_{n+2}}{b_{n+2}} + \frac{a_{n+3}}{b_{n+3}} + \cdots . \quad (8.3.16b)$$

Now by (2.1.7)

$$f - S_n(w_n) = S_n(f^{(n)}) - S_n(w_n)$$

$$= \frac{A_n + f^{(n)}A_{n-1}}{B_n + f^{(n)}B_{n-1}} - \frac{A_n + w_n A_{n-1}}{B_n + w_n B_{n-1}}$$

$$= \frac{(A_{n-1}B_n - A_n B_{n-1})(f^{(n)} - w_n)}{(B_n + f^{(n)}B_{n-1})(B_n + w_n B_{n-1})}.$$

Setting

$$h_n = \frac{B_n}{B_{n-1}} = b_n + \frac{a_n}{b_{n-1} +} \cdots + \frac{a_2}{b_1} \qquad (8.3.17)$$

(Theorem 4.5), we have

$$f - S_n(w_n) = \frac{(f_{n-1} - f_n) h_n (f^{(n)} - w_n)}{(h_n + f^{(n)})(h_n + w_n)}, \qquad n = 2, 3, 4, \ldots, \qquad (8.3.18)$$

and setting $w_n = 0$ so that $S_n(0) = f_n$, we obtain

$$f - f_n = \frac{(f_{n-1} - f_n) f^{(n)}}{h_n + f^{(n)}}, \qquad n = 2, 3, 4, \ldots . \qquad (8.3.19)$$

We shall return to (8.3.18) in Section 8.4. Hayden [1965] used (8.3.19) as the basis of his investigations. In a slightly different form it was also used by Roach [1977] (see proof of Theorem 4.25). Hayden called the h_n critical points, and his results depend on imposing restrictions on $K(a_n/b_n)$ which allow conclusions as to the location of h_n, in particular with respect to $f^{(n)}$. If $K(a_n/b_n) \in \mathcal{K}(\Omega)$ (Example 3 of Section 8.2 with $\Omega_n = \Omega$ for all n), if a value region V is known for this Ω, and if $V/(b_n + 2V)$ is bounded, then (8.3.19) yields usable truncation-error bounds in terms of $f_{n-1} - f_n$. Here we have used the fact [see (8.3.17)] that $h_n \in (b_n + V)$.

Merkes [1966] used a different approach to obtain estimates

$$|f - f_n| \leqslant \begin{cases} \dfrac{|1 + \rho_{n-1}|(1 - M_{n-1})}{1 - |1 + \rho_{n-1}|(1 - M_{n-1})} |f_n - f_{n-1}|, \\[3mm] \dfrac{1 - M_{n-1}}{M_{n-1} - m_{n-1}} |f_n - f_{n-1}|. \end{cases} \qquad (8.3.20)$$

Here

$$\rho_n = -a_{n+1} \frac{B_{n-1}}{B_{n+1}}, \qquad (8.3.21)$$

and the M_n and m_n are minimal and maximal parameters of a chain sequence $\{c_n\}$ (see remarks at end of Section 4.4.3) related to the elements of $K(a_n/b_n)$ by

$$\left| \frac{a_{n+1}}{b_n b_{n+1}} \right| \leqslant c_n. \qquad (8.3.22)$$

Applying value-region considerations to (8.3.19), one can prove the following two results of Blanch [1964].

THEOREM 8.5. *In* $K(1/b_n)$ *let*

$$|b_n| > 2(1+\varepsilon), \qquad \varepsilon > 0, \quad n = 1, 2, 3, \ldots, \tag{8.3.23}$$

and let f_n *denote the* n *th approximant. Then the continued fraction converges to a finite limit* f *and*

$$|f - f_n| < \left(\sqrt{\frac{1}{4} + \frac{1}{2\varepsilon}} - \frac{1}{2} \right) |f_n - f_{n-1}|, \qquad n = 2, 3, 4, \ldots. \tag{8.3.24}$$

THEOREM 8.6. *In* $K(a_n/1)$ *let*

$$|a_n| < \tfrac{1}{4} - \varepsilon, \qquad \varepsilon > 0, \quad n = 1, 2, 3, \ldots, \tag{8.3.25}$$

and let f_n *denote the* n *th approximant. Then the continued fraction converges to a finite limit* f *and*

$$|f - f_n| < \frac{1 - 2\sqrt{\varepsilon}}{2\sqrt{\varepsilon}} |f_n - f_{n-1}|, \qquad n = 2, 3, 4, \ldots. \tag{8.3.26}$$

Simple sequences were introduced by Jones and Thron [1971]. A sequence of complex numbers $\{v_n\}$ is called a *simple sequence* if there exists a positive constant c, independent of n, such that

$$|v_{n+m} - v_n| \leqslant c |v_n - v_{n-1}|, \qquad n, m = 1, 2, 3, \ldots. \tag{8.3.27}$$

Simple sequences need not converge and convergent sequences need not be simple.

In the following we shall describe a number of families of continued fractions whose sequences of approximants are simple sequences. However, it will be convenient to deduce the results from a theorem on inclusion regions which we shall now give.

For that purpose we consider a family of continued fractions denoted by $\mathcal{B}(\theta, \gamma_0)$. Let γ_0 and θ be real numbers with $0 < |\theta| < \pi$. Then $\mathcal{B}(\theta, \gamma_0)$ is defined to be the family of all continued fractions $K(a_n/b_n)$ such that $\{a_n\}$ is an arbitrary sequence of non-zero complex numbers and the elements b_n satisfy the conditions

$$0 \leqslant \arg b_n - \gamma_n \leqslant \theta, \quad n = 1, 2, 3, \ldots, \qquad \text{if} \quad 0 < \theta < \pi, \tag{8.3.28a}$$

$$\theta \leqslant \arg b_n - \gamma_n \leqslant 0, \quad n = 1, 2, 3, \ldots, \qquad \text{if} \quad -\pi < \theta < 0, \tag{8.3.28b}$$

where the γ_n are defined inductively by

$$\gamma_n = \arg a_n - \gamma_{n-1} - \theta, \qquad n = 1, 2, 3, \ldots. \tag{8.3.28c}$$

It is tacitly assumed that $b_n \neq 0$ for all n. In the following, for each $K = \mathbf{K}(a_n/b_n)$ in $\mathcal{B}(\theta, \gamma_0)$, the linear fractional transformations $s_n(K, w)$ and $S_n(K, w)$ are defined as in (8.2.7).

THEOREM 8.7. Let γ_0, θ be real numbers with $0 < |\theta| < \pi$, and let $K = \mathbf{K}(a_n/b_n) \in \mathcal{B}(\theta, \gamma_0)$. Let λ_n, μ_n, ν_n be subsets of $\hat{\mathbf{C}}$ defined by

$$\lambda_n = \left[z : z = s_n^{-1}(K, te^{i\gamma_{n-1}}), \ -\infty < t \leqslant \infty \right], \qquad n = 1, 2, 3, \ldots, \qquad (8.3.29a)$$

$$\mu_n = \left[z : z = te^{i\gamma_n}, \ -\infty < t \leqslant \infty \right], \qquad n = 0, 1, 2, \ldots, \qquad (8.3.29b)$$

$$\nu_n = \left[z : z = s_{n+1}(K, te^{i\gamma_{n+1}}), \ -\infty < t \leqslant \infty \right], \qquad n = 0, 1, 2, \ldots. \qquad (8.3.29c)$$

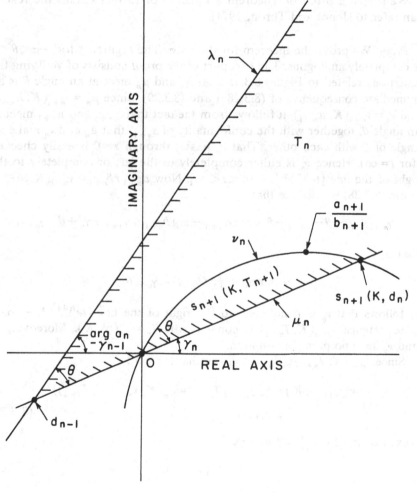

Figure 8.3.1. Angular opening $T_n(\mathcal{B}, K)$.

Finally, let $T_n(\mathcal{B}, K)$ *be the angular opening closed in* $\hat{C}$ *(see Figure 8.3.1) which is bounded by that ray of* μ_n *that contains* $z=0$ *and by that ray of* λ_n *that makes an angle of* θ *with the ray on* μ_n. *Then:*

(A) *For each* $n = 1, 2, 3, \ldots,$ *the set* $\Theta_n(\mathcal{B}, K)$ *defined by*

$$\Theta_n(\mathcal{B}, K) = S_n(K, T_n(\mathcal{B}, K)) \tag{8.3.30}$$

is an n*th inclusion region for the continued fraction* K *with respect to the family* $\mathcal{B}(\theta, \gamma_0)$.

(B) *For each* $n = 1, 2, 3, \ldots,$ $\Theta_n(\mathcal{B}, K)$ *is a convex lens-shaped region and*

$$\Theta_n(\mathcal{B}, K) \supseteq \Theta_{n+1}(\mathcal{B}, K). \tag{8.3.31}$$

We sketch a proof of Theorem 8.7 here. For further details the reader can refer to [Jones and Thron, 1971].

Proof. We prove the theorem for $0 < \theta < \pi$. The argument for $-\pi < \theta < 0$ is completely analogous. The first part of the proof consists of verifying the assertions related to Figure 8.3.1. That λ_n and μ_n meet at an angle θ is an immediate consequence of (8.3.28c) and (8.3.29). Since $\mu_n = s_{n+1}(K, \lambda_{n+1})$ and $\nu_n = s_{n+1}(K, \mu_{n+1})$, it follows from the fact that λ_{n+1} and μ_{n+1} meet at an angle θ, together with the conformality of s_{n+1}, that μ_n and ν_n make an angle of θ with each other. That ν_n passes through $z=0$ is easily checked (for $t = \infty$). Hence ν_n is either completely to the left or completely to the right of the line $[te^{i(\theta+\gamma_n)}: -\infty < t < \infty]$. Now $a_{n+1}/b_{n+1} = s_{n+1}(K, 0) \in \nu_n$. From (8.3.28) we deduce that

$$\gamma_{n+1} + \gamma_n + \theta - \gamma_{n+1} - \theta \leqslant \arg a_{n+1} - \arg b_{n+1} \leqslant \gamma_{n+1} + \gamma_n + \theta - \gamma_{n+1},$$

that is,

$$\gamma_n \leqslant \arg(a_{n+1}/b_{n+1}) \leqslant \gamma_n + \theta.$$

It follows that ν_n is completely to the right of the line $[te^{i(\theta+\gamma_n)}: -\infty < t < \infty]$. Hence $s_{n+1}(K, T_{n+1})$ is convex as is $T_n = T_n(\mathcal{B}, K)$. Moreover, λ_n and ν_n have no point in common.

Since $s_{n+1}(K, T_{n+1}) \subseteq T_n$, it follows that for $n \geqslant 1$

$$\Theta_{n+1}(\mathcal{B}, K) = S_{n+1}(K, T_{n+1}) = S_n(K, s_{n+1}(K, T_{n+1}))$$
$$\subseteq S_n(K, T_n) = \Theta_n(\mathcal{B}, K). \tag{8.3.32}$$

Next we introduce Γ_n defined by

$$\Gamma_n = S_n(K, \mu_n).$$

Then

$$\Gamma_{n+1} = S_{n+1}(K, \mu_{n+1}) = S_n(K, s_{n+1}(K, \mu_{n+1})) = S_n(K, \nu_n)$$

and

$$\Gamma_{n-1} = S_{n-1}(K, \mu_{n-1}) = S_{n-1}(K, s_n(\lambda_n)) = S_n(K, \lambda_n),$$

since $\nu_n = s_{n+1}(K, \mu_{n+1})$ and $\lambda_n = s_n^{-1}(K, \mu_{n-1})$. Thus $\Theta_n = \Theta_n(\mathcal{B}, K)$ is bounded by arcs of Γ_n and Γ_{n-1} (see Figure 8.3.2). The corners of Θ_n are $S_n(K, d_{n-1}) = \zeta_{n-1}$ and $S_n(K, \infty) = S_{n-1}(K, 0) = f_{n-1}$. Moreover, $S_n(K, 0) = f_n$ is on the boundary of Θ_n. Therefore we can conclude that all approximants f_{n+k}, $k \geqslant -1$, of $K = K(a_n/b_n)$ lie in Θ_n, so that Θ_n is an nth inclusion region for $K(a_n/b_n)$ with respect to the family $\mathcal{B}(\theta, \gamma_0)$.

It remains to prove that the lens-shaped regions Θ_n are convex. That $\Theta_1 = S_1(K, T_1) = s_1(K, T_1)$ is convex follows from the arguments used in proving the assertions related to Figure 8.3.1. Now we assume that Θ_n is convex. Then the arc of Γ_{n+1}, which bounds Θ_{n+1} and which makes an angle of θ with Γ_n at f_n, intersects Γ_n also at a point ζ_n. Moreover ζ_n lies on the arc of Γ_n between f_n and f_{n-1} which does not contain ζ_{n-1} (see Figure 8.3.2). This is true because

$$\zeta_{n-1} = S_n(K, d_{n-1}), \qquad f_n = S_n(K, 0),$$
$$\zeta_n = S_n(K, s_{n+1}(K, d_n)), \qquad f_{n-1} = S_n(K, \infty)$$

and the order of the points d_{n-1}, 0, $s_{n+1}(K, d_n)$, ∞ on μ_n (Figure 8.3.1) is

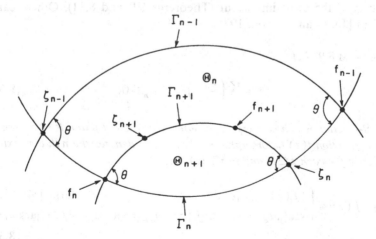

Figure 8.3.2. nth inclusion region Θ_n for the continued fraction $K(a_n/b_n)$ with respect to the family $\mathcal{B}(\theta, \gamma_0)$.

preserved on Γ_n under the linear fractional transformation S_n. Therefore since $\Theta_{n+1} \subseteq \Theta_n$, it follows that Θ_{n+1} must be convex. ∎

We note that for the radii r_n of the circles Γ_n in the proof of Theorem 8.7 (Figure 8.3.2), we have

$$r_1 > r_3 > \cdots > r_{2n-1} > \cdots \quad \text{and} \quad r_2 > r_4 > \cdots > r_{2n} > \cdots.$$

This follows from the fact that $\Gamma_{n-1} \cap \Gamma_{n+1} = \varnothing$.

From the inclusion regions established in Theorem 8.7 we now derive truncation-error bounds.

THEOREM 8.8. *Let* γ_0, θ *be real numbers with* $0 < |\theta| < \pi$. *If* $K(a_n/b_n)$ *belongs to* $\mathfrak{B}(\theta, \gamma_0)$ *and converges to a finite value* f, *then its* nth *approximants satisfy*

$$|f - f_n| \leqslant \begin{cases} |f_n - f_{n-1}| & \text{if } 0 < |\theta| \leqslant \pi/2, \\ \sec(|\theta| - \pi/2)|f_n - f_{n-1}| & \text{if } \pi/2 < |\theta| < \pi. \end{cases} \quad (8.3.33)$$

Proof. For $0 < |\theta| < \pi/2$, the proof follows from the simple observation that in that case

$$\text{diam } \Theta_{n+1} = |f_n - \zeta_n| \leqslant |f_n - f_{n-1}|$$

[see Figure 8.3.2]. For $\pi/2 < |\theta| < \pi$, the diameter of Θ_{n+1} will be the smaller of the diameters of Γ_n and Γ_{n+1}. For these, one easily obtains the bound $\sec(|\theta| - \pi/2)|f_n - \zeta_n|$, from which (8.3.33) follows. ∎

Theorems 8.7 and 8.8 have a number of important applications. We give here two of the most important (Theorems 8.9 and 8.11). Others can be found in [Jones and Thron, 1971].

THEOREM 8.9. *Let*

$$K = K\left(\frac{a_n z}{1}\right), \qquad a_n > 0, \quad (8.3.34)$$

be an S-*fraction which converges, for each* z *in the cut plane* $R = [z : |\arg z| < \pi]$, *to a function* $f(z)$ *holomorphic in* R. *If* $f_n(z)$ *denotes the* nth *approximant of* K, *then for each* $z \in R$ *and* $n = 2, 3, 4, \ldots$,

$$|f(z) - f_n(z)| \leqslant \begin{cases} |f_n(z) - f_{n-1}(z)| & \text{if } |\arg z| \leqslant \pi/2, \\ \sec(|\arg z| - \pi/2)|f_n(z) - f_{n-1}(z)| & \text{if } \pi/2 < |\arg z| < \pi. \end{cases}$$
$$(8.3.35)$$

Proof. We recall first that necessary and sufficient conditions for convergence of (8.3.34) are given by Theorem 4.58 (Stieltjes), and that if the

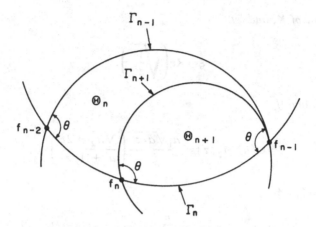

Figure 8.3.3. nth inclusion region Θ_n for S-fraction $\mathbf{K}(a_n z/1)$ with respect to the family $\mathfrak{S}(z)=\mathfrak{B}(\theta,0)$.

S-fraction K converges at a single point in R, then it converges at every point in R to a holomorphic function $f(z)$. If arg $z=0$, then the assertion of the theorem follows directly from Theorem 4.28 (Seidel-Stern). For $0<|\arg z|<\pi$, the Θ_n as defined in Theorem 8.7, with $\theta=\arg z$, $\gamma_n=0$ (for $n>0$), $d_n=-1$ and hence

$$\zeta_n=S_{n+1}(K,d_n)=S_n(K,\infty)=S_{n-1}(K,0)=f_{n-1},$$

will be the nth inclusion regions for K with respect to the family $\mathfrak{S}(z)=\mathfrak{B}(\theta,0)$ of the S-fractions $\mathbf{K}(a_n z/1)$ (see Figure 8.3.3). Thus (8.3.35) follows from (8.3.33). ∎

   This result was originally proved by Henrici and Pfluger [1966]. By using integral representations of S-fractions [see Theorem 9.4(A)], they were able to show that the Θ_n in this case are the best nth inclusion regions. They also obtained *a priori* bounds for $|f_n(z)-f_{n-1}(z)|$ as follows:

THEOREM 8.10. *Let*

$$K=\mathbf{K}\!\left(\frac{a_n z}{1}\right),\qquad a_n>0,$$

be an S-fraction that corresponds at $z=0$ to the fps

$$L=c_0 z+c_1 z^2+c_2 z^3+\cdots. \tag{8.3.36}$$

Further let K converge for each z in the cut plane $R=[z:|\arg z|<\pi]$ to a function $f(z)$ holomorphic in R. For each $z\in R$, let $f_n(z)$ denote the nth *approximant of K, and let*

$$\xi = \mathrm{Re}\left(\sqrt{\frac{1}{z}}\right).$$

Then $\xi > 0$, and if

$$C = C(a_1, a_2, z) = \frac{a_1\sqrt{a_2 z}}{\xi}\left(\frac{\sqrt{a_2} + 2\xi}{a_2 + |1/z|}\right)^{1/2}, \qquad (8.3.37)$$

then for $n = 2, 3, 4, \ldots$,

$$|f_n(z) - f_{n-1}(z)| \leqslant C \prod_{k=2}^{n}\left(1 + 2\xi a_k^{-1/2}\right)^{-1/2}, \qquad (8.3.38\mathrm{a})$$

$$f_{n+1}(z) - f_n(z)| \leqslant C\left(1 + 2\xi\left|\frac{c_n}{c_0}\right|^{1/2n}\right)^{-n/2}, \qquad (8.3.38\mathrm{b})$$

$$|f_{n+1}(z) - f_n(z)| \leqslant C\left(1 + \frac{2\xi}{e}\sum_{k=1}^{n}\left|\frac{c_0}{c_k}\right|^{1/2k}\right)^{-1/2}. \qquad (8.3.38\mathrm{c})$$

Using each of the three inequalities in (8.3.38) with (8.3.35) gives *a priori* truncation-error bounds for the S-fraction $K(a_n z/1)$. The last two inequalities are expressed in terms of the coefficients c_k of the corresponding power series L. Moreover, (8.3.38c) can be thought of as a quantitative form of Carleman's sufficient condition for convergence of an S-fraction [see remarks following Theorem 9.8(A)]. An application of the *a priori* bounds in Theorem 8.10 will be given in Chapter 11. We consider now some applications of Theorem 8.9.

Example 2. Natural logarithm. After an equivalence transformation, (6.1.17) can be written as an S-fraction

$$\log(1 + z) = \frac{z}{1} + \frac{\dfrac{1^2}{1 \times 2}z}{1} + \frac{\dfrac{1^2}{2 \times 3}z}{1} + \frac{\dfrac{2^2}{3 \times 4}z}{1} + \frac{\dfrac{2^2}{4 \times 5}z}{1} + \cdots, \qquad (8.3.39)$$

valid for all z in the cut plane with cut along the negative real axis from -1 to $-\infty$. Letting $f_n(z)$ denote the nth approximant of the S-fraction and setting $z = 1$, we obtain from Theorem 8.9 the truncation-error bounds $|f_n(1) - f_{n-1}(1)|$ given in Table 8.3.2. Also given are actual truncation errors

Table 8.3.2. Truncation-Error Bounds for $\log(1+z)$

n	2	3	4	5	6
$\|f_n(1)-f_{n-1}(1)\|$	3.3×10^{-1}	3.3×10^{-2}	7.7×10^{-3}	1.0×10^{-3}	2.1×10^{-4}
$\|\ln 2-f_n(1)\|$	2.6×10^{-2}	6.9×10^{-3}	8.4×10^{-4}	1.9×10^{-4}	2.5×10^{-5}

$|\log 2 - f_n(1)|$, from which one can see that the error bounds have about the right order of magnitude.

Example 3. Tangent function. After an equivalence transformation, (6.1.54) can be expressed as

$$f(z) = -z\tan z = \cfrac{-z^2}{1} \; \cfrac{-\dfrac{1}{1\times3}z^2}{+ \quad 1} \; \cfrac{-\dfrac{1}{3\times5}z^2}{+ \quad 1} \; \cfrac{-\dfrac{1}{5\times7}z^2}{+ \quad 1} \; + \cdots,$$

$$(8.3.40)$$

which is valid for all $z\in\mathbb{C}$. Letting $z=e^{i\pi/4}$, we obtain from Theorem 8.9 the truncation-error bounds $|f_n(e^{i\pi/4})-f_{n-1}(e^{i\pi/4})|$ given in Table 8.3.3. Here $f_n(z)$ denotes the nth approximant of (8.3.40). Also shown is the actual truncation error $|f(e^{i\pi/4})-f_n(e^{i\pi/4})|$, and again we see that the error bound has about the right order of magnitude.

In applications of Theorems 8.9 and 8.10 it is useful to note some equivalent forms of S-fractions. For z in the cut plane $R=[z:|\arg z|<\pi]$ we let $\zeta=1/z$ and $w=1/\sqrt{z}$. Then *the following continued fractions are equivalent*:

$$\frac{a_1 z}{1} + \frac{a_2 z}{1} + \frac{a_3 z}{1} + \frac{a_4 z}{1} + \cdots, \qquad a_n>0, \qquad (8.3.41a)$$

$$\frac{a_1}{\zeta} + \frac{a_2}{1} + \frac{a_3}{\zeta} + \frac{a_4}{1} + \cdots, \qquad a_n>0, \qquad (8.3.41b)$$

$$\frac{1}{\zeta}\left(\frac{a_1}{1} + \frac{a_2}{\zeta} + \frac{a_3}{1} + \frac{a_4}{\zeta} + \cdots\right), \qquad a_n>0, \qquad (8.3.41c)$$

$$\frac{1}{w}\left(\frac{a_1}{w} + \frac{a_2}{w} + \frac{a_3}{w} + \frac{a_4}{w} + \cdots\right), \qquad a_n>0. \qquad (8.3.41d)$$

Here $|\arg\zeta|<\pi$ and $|\arg w|<\pi/2$ [or equivalently $\mathrm{Re}(w)>0$].

Table 8.3.3. Truncation-Error Bounds for $f(z)=-z\tan z$

n	2	3	4	5	6
$\|f_n(e^{i\pi/4})-f_{n-1}(e^{i\pi/4})\|$	3.2×10^{-1}	2.5×10^{-2}	5.5×10^{-3}	8.6×10^{-6}	8.7×10^{-8}
$\|f(e^{i\pi/4})-f_n(e^{i\pi/4})\|$	2.0×10^{-2}	5.5×10^{-4}	8.6×10^{-6}	8.7×10^{-8}	8.0×10^{-10}

Example 4. Complementary error function. By (6.2.25) and (6.2.19) the function

$$f(w) = \frac{\sqrt{\pi}\, e^{w^2}}{w} \operatorname{erfc}(w) = \frac{2e^{w^2}}{w} \int_w^\infty e^{-t^2}\, dt \qquad (8.3.42)$$

is represented, for all w such that $\operatorname{Re}(w) > 0$, by the continued fraction

$$f(w) = \frac{1}{w}\left(\frac{1}{w} + \frac{\frac{1}{2}}{w} + \frac{1}{w} + \frac{\frac{3}{2}}{w} + \frac{2}{w} + \frac{\frac{5}{2}}{w} + \frac{3}{w} + \cdots\right). \qquad (8.3.43)$$

Since (8.3.43) has the same form as (8.3.41d), it is equivalent to an S-fraction (8.3.41a), and hence Theorem 8.9 can be applied to estimate truncation errors. Let $f_n(w)$ denote the nth approximant of (8.3.43). Then setting $w = e^{i\pi/4}$ and $f_n = f_n(e^{i\pi/4})$, we give the truncation-error bounds $|f_n - f_{n-1}|$ in Table 8.3.4. Also given are the values of the approximants f_n. The inclusion regions Θ_n for this continued fraction are shown in Figure 8.3.4. From this it can be seen that the truncation-error bounds have about the right order of magnitude.

Table 8.3.4. Truncation-Error bounds $|f_n - f_{n-1}|$ for $f(w) = (\sqrt{\pi}\, e^{w^2}/w) \operatorname{erfc}(w)$ at $w = e^{i\pi/4}$ and nth approximants f_n

n	$\lvert f_n - f_{n-1}\rvert$	f_n
1		$-i$
2	4.5×10^{-1}	$\dfrac{2-4i}{5}$
3	2.5×10^{-1}	$\dfrac{2-10i}{13}$
4	1.4×10^{-1}	$\dfrac{38-124i}{145}$
5	8.7×10^{-2}	$\dfrac{118-404i}{521}$
6	5.6×10^{-2}	$\dfrac{1982-7278i}{8749}$
7		$\dfrac{9950-32860i}{41245}$

The next theorem that we deduce from Theorems 8.7 and 8.8 generalizes a result of Jefferson [1969b] to general T-fractions.

THEOREM 8.11. *Let*

$$\overset{\infty}{\underset{n=1}{\mathrm{K}}}\left(\frac{z}{e_n + d_n z}\right) \qquad (8.3.44)$$

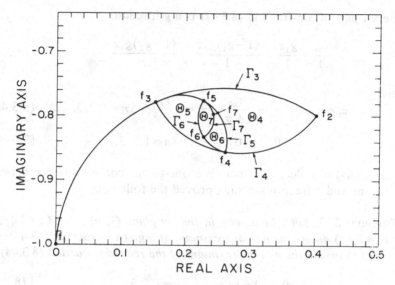

Figure 8.3.4.   Inclusion regions for $f(w) = (\sqrt{\pi}\ e^{w^2}/w)\,\mathrm{erfc}(w)$ at $w = e^{i\pi/4}$.

be a general T-fraction with

$$e_n > 0, \quad d_n > 0, \qquad n = 1, 2, 3, \ldots. \tag{8.3.45}$$

Assume that (8.3.44) *converges for all z in the cut plane* $R = [z : |\arg z| < \pi]$ *to a function* $f(z)$ *holomorphic in R, and let* $f_n(z)$ *denote the n th approximant of* (8.3.44). *Then for each* $z \in R$ *and* $n = 2, 3, 4, \ldots,$

$$|f(z) - f_n(z)| \le \begin{cases} |f_n(z) - f_{n-1}(z)| & \text{if } |\arg z| \le \pi/2, \\ \sec(|\arg z| - \pi/2)|f_n(z) - f_{n-1}(z)| & \text{if } \pi/2 < |\arg z| < \pi. \end{cases} \tag{8.3.46}$$

Proof. The proof is similar to that of Theorem 8.9. In this case the set Θ_n, as defined in Theorem 8.7, will be an nth inclusion region of (8.3.44) with respect to the family $\tau(z)$ if we choose $\gamma_n = 0$ (for $n \ge 0$), $\theta = \arg z$, and $\zeta_n = S_{n+1}(e_n)$ (see Figure 8.3.2). ∎

A result for g-fractions (see Sections 4.5.1 and 7.1.3) which is not subsumed under Theorem 8.7 was given by Gragg [1968] in terms of π-fractions

$$\frac{\pi_0}{1+z} - \frac{z}{1} + \frac{\pi_1}{1+z} - \frac{z}{1} + \frac{\pi_2}{1+z} - \frac{z}{1} + \cdots, \qquad \pi_n > 0, \quad |\arg(1+z)| < \pi. \tag{8.3.47}$$

The even part of a π-fraction (8.3.47) is a g-fraction

$$\frac{s_0}{1} + \frac{g_1 z}{1} + \frac{(1-g_1)g_2 z}{1} + \frac{(1-g_2)g_3 z}{1} + \cdots, \qquad (8.3.48)$$

where

$$\pi_0 = s_0; \qquad g_n = \frac{\pi_n}{1+\pi_n}, \qquad \pi_n = \frac{g_n}{1-g_n}, \qquad n=1,2,3,\ldots, \qquad (8.3.49a)$$

$$s_0 > 0; \qquad 0 < g_n < 1, \qquad n=1,2,3,\ldots. \qquad (8.3.49b)$$

From (8.3.49) it is clear that there is a one-to-one correspondence between π-fractions and g-fractions. Gragg proved the following:

THEOREM 8.12. *Let z be a point in the cut plane $Q_0 = [z : |\arg(z+1)| < \pi]$.*
Let $P_n = P_n(z)$ denote the nth approximant of a given π-fraction (8.3.47), and
let $f_n = f_n(z)$ denote the nth approximant of the related g-fraction (8.3.48), so
that

$$P_{2n}(z) = f_n(z), \qquad n=1,2,3,\ldots. \qquad (8.3.50)$$

Let $\Gamma_n(z)$ $[\Gamma'_n(z)]$ denote the circle determined by the three points P_{2n-2},
P_{2n-1}, P_{2n} $[P_{2n-3}, P_{2n-1}, P_{2n}]$, and let $\Delta_n(z)[\Delta'_n(z)]$ denote the closed disk
consisting of Γ_n $[\Gamma'_n]$ and its interior. Let

$$\Theta_n(z) = \Delta_n(z) \cap \Delta'_n(z), \qquad n=2,3,4,\ldots. \qquad (8.3.51)$$

Then for each $n=2,3,4,\ldots$:

(A) *$\Theta_n(z)$ is a convex nth inclusion region for the g-fraction (8.3.48).*
(B)

$$\Theta_{n+1}(z) \subseteq \Theta_n(z). \qquad (8.3.52)$$

(C) *If $f(z)$ denotes the holomorphic function to which the g-fraction (8.3.48)*
converges in Q_0, then

$$|f(z) - f_n(z)| < \operatorname{diam} \Theta_n(z) \leq K(z) \left| \frac{1 - \sqrt{1+z}}{1 + \sqrt{1+z}} \right|^{n-1}, \qquad (8.3.53a)$$

where

$$K(z) = \max\left[1, \tan\left| \frac{\arg z}{2} \right| \right] \frac{\pi_0}{\operatorname{Re}(\sqrt{1+z})} \left| \sqrt{1+z} - \frac{1}{\sqrt{1+z}} \right|. \qquad (8.3.53b)$$

(D) *The circles Γ_n and Γ'_n intersect at the angle $\arg z$.*

We note that (8.3.53) gives an *a priori* bound for the truncation error. An *a posteriori* bound can be obtained from the fact that

$$\operatorname{diam}\Theta_n(z)=\max\left[1,\tan\left|\frac{\arg z}{2}\right|\right]|P_{2n}-P_{2n-1}|.$$

For a schematic diagram of the inclusion regions $\Theta_n(z)$ described in Theorem 8.12, see Figure 8.3.5. The following example is also due to Gragg [1968].

Example 5. Natural logarithm.   The function

$$f(z)=\frac{\log(1+z)}{z}=\int_0^1\frac{dt}{1+zt} \tag{8.3.54}$$

belongs to the class $\mathcal{W}$ (see Section 7.1.3). For z in the disk $|z|<1$, $f(z)$ is represented by the convergent power series

$$f(z)=1-\tfrac{1}{2}z+\tfrac{1}{3}z^2-\tfrac{1}{4}z^3+\cdots, \tag{8.3.55}$$

and for z in the cut plane $Q_0=[z:|\arg(1+z)|<\pi]$ by the π-fraction

$$\frac{1}{1+z}-\frac{z}{1}+\frac{\tfrac{1}{2}}{1+z}-\frac{z}{1}+\frac{1}{1+z}-\frac{z}{1}+\frac{\tfrac{2}{3}}{1+z}-\frac{z}{1}+\cdots. \tag{8.3.56}$$

Here $\pi_0=\pi_{2n+1}=1$ for $n\geqslant0$ and $\pi_{2n}=n/(n+1)$ for $n\geqslant1$.

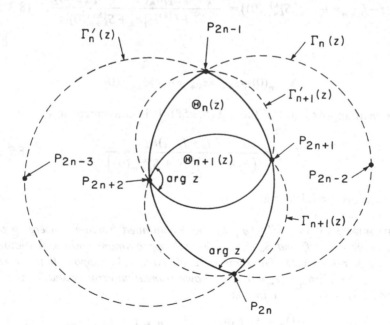

Figure 8.3.5. nth inclusion region $\Theta_n(z)$ for a g-fraction.

An extension of Theorem 8.12 was given by Gragg [1970] for functions which can be expressed in the form

$$f(z) = \int_a^b \frac{d\psi(t)}{z-t} \qquad \text{for } z \in \mathbb{C} \sim [z: a \leqslant z \leqslant b], \qquad (8.3.57)$$

where $-\infty < a < b < +\infty$ and $\psi(t)$ is a bounded non-decreasing function with infinitely many points of increase on $[a, b]$. Baker [1969] obtained inclusion regions for S-fractions which converge in a neighborhood of the origin, similar to the results by Henrici and Pfluger [1966] and Gragg [1968, 1970]. The connections between the results of Baker [1969] and Gragg [1968] are brought out by Field [1976]. Common in [1968] obtained inclusion regions for S-fractions for the case in which z is real.

Truncation-error estimates, such as those given in Theorems 8.2, 8.3, 8.4, 8.5, and 8.6, may apply only to the "tail"

$$f^{(k)} = \frac{a_{k+1}}{b_{k+1}} + \frac{a_{k+2}}{b_{k+2}} + \frac{a_{k+3}}{b_{k+3}} + \cdots \qquad (8.3.58)$$

of a continued fraction, since the required conditions may hold only for $\langle a_n, b_n \rangle$ with $n > k$. We now investigate what can be said about truncation errors for the whole continued fraction. Using the formula (8.3.18), one obtains

$$f - f_{k+m} = f - S_k\big(S_{k+m}^{(k)}(0)\big) = \frac{(f_k - f_{k-1})h_k\big(f^{(k)} - S_{k+m}^{(k)}(0)\big)}{|h_k + f^{(k)}| \cdot |h_k + S_{k+m}^{(k)}(0)|}. \qquad (8.3.59)$$

Here

$$S_{k+m}^{(k)}(0) = s_{k+1} \circ s_{k+2} \circ \cdots \circ s_{k+m}(0)$$

is the mth approximant of the tail (8.3.58). It is convenient to define

$$D_{k,m} = \frac{(f_k - f_{k-1})h_k}{(h_k + f^{(k)})\big[h_k + S_{k+m}^{(k)}(0)\big]}. \qquad (8.3.60)$$

Our result is the following

THEOREM 8.13. *Let* $\mathbf{K}_{n=1}^{\infty}(a_n/b_n)$ *be a continued fraction converging to a finite value* f. *Let* f_n *and* B_n *denote the* n*th approximant and denominator of* $\mathbf{K}(a_n/b_n)$, *respectively. For some natural number* k, *suppose that for each* $m \geqslant 1$, ε_m, *with* $\lim_{m \to \infty} \varepsilon_m = 0$, *is a known truncation-error bound for the* k*th tail* (8.3.58) *of* $\mathbf{K}(a_n/b_n)$, *that is,*

$$|f^{(k)} - S_{k+m}^{(k)}(0)| < \varepsilon_m, \qquad m = 1, 2, 3, \ldots. \qquad (8.3.61)$$

Here $S_{k+m}^{(k)}(0)$ denotes the m th approximant of $f^{(k)}$. Further suppose that there is a value region V such that

$$S_{k+m}^{(k)}(0) \in V, \qquad m = 1, 2, 3, \ldots. \qquad (8.3.62)$$

If there exists a $\delta_k > 0$ such that

$$|h_k + V| > \delta_k, \qquad where \quad h_k = \frac{B_k}{B_{k-1}}, \qquad (8.3.63)$$

then

$$|f - f_{k+m}| < \frac{|f_k - f_{k-1}||h_k|}{\delta_k^2} \varepsilon_m, \qquad m = 1, 2, 3, \ldots. \qquad (8.3.64)$$

Since h_k can be computed explicitly and V is assumed known, then a δ_k satisfying (8.3.63) can be computed explicitly. Hence

$$\frac{|f_k - f_{k-1}||h_k|}{\delta_k^2} \qquad (8.3.65)$$

is computable, and it provides a computable bound for $D_{k,m}$. The number $D_{k,m}$ is approximately $S_k'(f^{(k)})$, or $S_k'(S_{k+m}^{(k)}(0))$, for m sufficiently large.

We now study continued fractions of the form

$$\frac{(a+1)z}{1} + \frac{(b+1)z}{1} + \frac{(a+2)z}{1} + \frac{(b+2)z}{1} + \cdots, \qquad (8.3.66)$$

where a and b are non-real constants, which arise in connection with expansions of ratios of certain confluent hypergeometric series (see Section 6.1.4). In Theorem 6.6 we showed that these continued fractions converge to meromorphic functions in the cut plane $R = [z : |\arg z| < \pi]$. We now consider the speed of convergence of (8.3.66) for fixed a, b at a fixed value of $z \in R$.

Let $P^{(z)}$ be the parabolic region defined by

$$P^{(z)} = \left[w : |w| - \text{Re}(we^{-i\arg z}) < \tfrac{1}{4}(1 + \cos \arg z) \right] \qquad (8.3.67)$$

(see Figure 8.3.6, with $k^* = 2$). Then we know that a continued fraction $\mathbf{K}_{n=1}^{\infty}(a_n/1)$ converges if $a_n \in P^{(z)}$ for all $n \geq 1$, provided a_n does not tend to infinity too fast (see Theorem 4.42 with $2\alpha = \arg z$).

Now set

$$a_{2n-1} = (a+n-1+k^*), \qquad a_{2n} = (b+n+k^*), \qquad n \geq 1. \quad (8.3.68)$$

Here $k^* \geq 0$ is so chosen that

$$a_n \in P^{(z)}, \qquad n \geq 1.$$

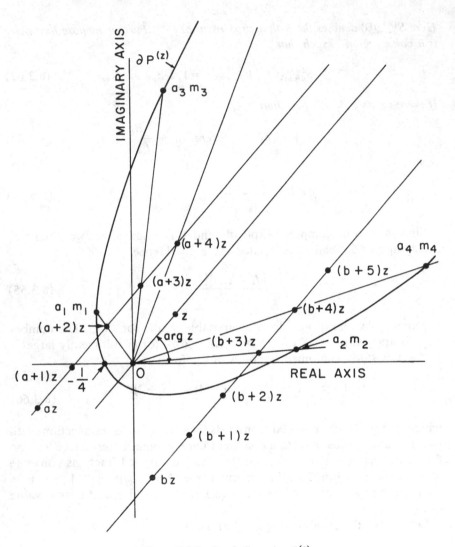

Figure 8.3.6. Parabolic region $P^{(z)}$.

Next, select $m_n \geqslant 1$ so that $a_n m_n$ is on the boundary of $P(z)$ for all $n \geqslant 1$. Finally let

$$\frac{8a_n m_n e^{-i \arg z}}{1 + \cos \arg z} = u_n + i v_n, \qquad n \geqslant 1. \qquad (8.3.69)$$

Then $v_n^2 = 4u_n + 4$. Moreover, the conditions of [Thron, 1958] and [Jones and Snell, 1969] are satisfied with

$$g_n = \tfrac{1}{2}, \qquad 2\alpha = \arg z, \qquad k_n = \frac{1}{m_n}$$

and

$$x_n = \mathrm{Re}\left(\frac{2h_n e^{-i(\arg z)/2}}{\cos(\arg z)/2}\right) \ge 1 + \frac{1}{n}. \tag{8.3.70}$$

In those articles one obtains the estimate

$$\frac{R_n}{R_{n-1}} \le \frac{k_n(u_n^2 + v_n^2)^{1/2}(x_{n-1} - 1)}{(x_{n-1} - k_n)(x_{n-1} + k_n + u_n k_n)}, \tag{8.3.71}$$

where R_n denotes the radius of the circular disk $S_n(V)$ and V is the half plane

$$V = \left[w : \mathrm{Re}(we^{-i(\arg z)/2}) > -\tfrac{1}{2}\cos(\arg z)/2\right].$$

Using the fact that

$$k_n(u_n^2 + v_n^2)^{1/2} = (u_n + 2)k_n = \frac{8|c_n + n + k^*|\cdot|z|}{1 + \cos \arg z} = d_n, \tag{8.3.72}$$

one arrives at

$$\frac{R_n}{R_{n-1}} \le \frac{d_n[x_{n-1} - k_n - (1 - k_n)]}{(x_{n-1} - k_n)(x_{n-1} - k_n + d_n)}. \tag{8.3.73}$$

Substituting $\delta = 1/d_n$, $t = x_{n-1} - k_n$ and $r = 1 - k_n$ in (8.3.73) leads to

$$\frac{R_n}{R_{n-1}} \le Q(t) = \frac{t - r}{\delta t^2 + t}.$$

An elementary calculation shows that $\max_t Q(t)$ is assumed for

$$t = r + r\sqrt{1 + (1/r\delta)}$$

and is

$$\max_t Q(t) = \frac{1}{(\sqrt{r\delta} + \sqrt{1 + r\delta})^2} \le \frac{1}{1 + 3r\delta}. \tag{8.3.74}$$

Since $k_n = O(1/n)$, we have, for some n^*,

$$r\delta = \frac{1 - k_n}{d_n} = \frac{(1 - O(1/n))(1 + \cos \arg z)}{8|z||a_n|}$$

$$> \frac{K(z)}{n} \quad \text{for} \quad n > n^*. \tag{8.3.75}$$

Here

$$K(z) = \frac{1 + \cos \arg z}{5|z|}.\qquad\qquad(8.3.76)$$

For $n > n^*$ we thus have

$$\frac{R_n}{R_{n-1}} \leqslant \frac{1}{1 + 3K(z)/n}\qquad\qquad(8.3.77)$$

and hence

$$R_n \leqslant R_{n^*} \prod_{m=n^*+1}^{n} \frac{1}{1 + 3K(z)/m}, \quad n > n^*.$$

Since

$$\underset{m=1}{\overset{n+p}{\mathbf{K}}}\left(\frac{a_m}{1}\right) \in S_n(V), \quad p > 0,$$

it follows that

$$\left|\underset{m=1}{\overset{n+p}{\mathbf{K}}}\left(\frac{a_m}{1}\right) - \underset{m=1}{\overset{n}{\mathbf{K}}}\left(\frac{a_m}{1}\right)\right| < 2R_n \leqslant 2R_{n^*} \prod_{m=n^*+1}^{n} \frac{1}{1 + 3K(z)/m}.$$
$$\qquad\qquad(8.3.78)$$

Now it is known (see, for example, Lange's thesis [1960]) that

$$\prod_{m=k}^{n} \left(1 + \frac{c}{m}\right) > \left(\frac{n+c}{k+c}\right)^c, \quad c > 0.\qquad\qquad(8.3.79)$$

Substituting (8.3.79) into (8.3.78), one finally arrives at the following estimates for the speed of convergence:

$$\left|\underset{m=1}{\overset{\infty}{\mathbf{K}}}\left(\frac{a_m}{1}\right) - \underset{m=1}{\overset{n}{\mathbf{K}}}\left(\frac{a_m}{1}\right)\right| < 2R_{n^*}\left(\frac{n^*+1+3K(z)}{n+3K(z)}\right)^{-3K(z)}$$
$$< A(z)n^{-3K(z)}.\qquad\qquad(8.3.80)$$

In view of the fact that $\lim a_n = \infty$, one had to expect extremely slow convergence for the continued fraction (8.3.66). That the convergence is somewhat better than this is due to the fact that the elements a_n do not lie on the boundary of the parabolic region $P^{(z)}$. It is also of interest to note how the speed of convergence depends on z, being quite good for small $|z|$ and $|\arg z| < \pi - \varepsilon$, $\varepsilon > 0$.

For the family $\mathcal{P}(\{\zeta_n\})$ of positive definite continued fractions (for definition see end of Section 8.2), Jones and Thron [1976, Theorem 9] were able to prove the following result giving truncation error bounds and best inclusion regions.

THEOREM 8.14. *Let* $\{\zeta_n\}$ *be a fixed sequence of complex numbers satisfying*

$$\text{Im}(\zeta_1)>0, \quad \text{Im}(\zeta_n)\geqslant 0, \quad n=2,3,4,\ldots \qquad (8.3.81)$$

and let $\mathcal{P}(\{\zeta_n\})$ *be the family of all positive definite continued fractions*

$$K=\cfrac{1}{\beta_1+\zeta_1} \+ \cfrac{-\alpha_1^2}{\beta_2+\zeta_2} \+ \cfrac{-\alpha_2^2}{\beta_3+\zeta_3} \+ \cdots, \qquad \alpha_n\neq 0. \qquad (8.3.82)$$

For each $K\in\mathcal{P}(\{\zeta_n\})$ *of the form* (8.3.82) *let*

$$a_1(K)=1, \quad a_{n+1}(K)=-\alpha_n^2, \ d_n(K)=\beta_n, \quad n=1,2,3,\ldots,$$

and let $\{M_n(K)\}$ *be a sequence of real numbers and* $\{U_n(K)\}$ *be a sequence of half-plane regions closed in* $\hat{\mathbb{C}}$ *defined, for* $n=1,2,3,\ldots,$ *by*

$$M_0(K)=0, \qquad (8.3.83a)$$

$$M_n(K)=\begin{cases} \dfrac{|a_{n+1}(K)|+\text{Re}(a_{n+1}(K))}{2\,\text{Im}(d_n(K))\,\text{Im}(d_{n+1}(K))\left[1-M_{n-1}(K)\right]} \\ \quad \text{if } \text{Im}(d_n(K))\,\text{Im}(d_{n+1}(K))\left[1-M_{n-1}(K)\right]\neq 0, \quad (8.3.83b) \\ 0 \quad \text{otherwise}, \end{cases}$$

$$U_n(K)=\left[\,u: u\in\hat{\mathbb{C}}, \ \text{Im}(u)\geqslant -\text{Im}(d_n(K))\left[1-M_{n-1}(K)\right]\right]. \tag*{(8.3.84)}$$

Then for each $K\in\mathcal{P}(\{\zeta_n\})$ *the following hold*:

(A) *For each* $n=1,2,3,\ldots,$

$$0\leqslant M_{n-1}(K)\leqslant 1, \qquad (8.3.85)$$

and

$$c(\psi_n(\mathcal{P}(\{\zeta_n\}),K))=S_n(K,U_n(K)), \qquad (8.3.86)$$

that is, $S_n(K,U_n(K))$ *is the best* n *th inclusion region for* K *with respect to* $\mathcal{P}(\{\zeta_n\})$. *Here* $S_n(K,w)$ *is defined as in* (8.2.7).
(B) $\{S_n(K,w)\}$ *is a nested sequences of circular disks.*
(C) *For each* $n=1,2,3,\ldots,$ *if* $c_n(K)$ *and* $\rho_n(K)$ *denote the center and radius, respectively, of* $S_n(K,U_n(K))$, *and if* K *converges to a finite value* f,

then

$$|f-f_n(K)| \leqslant |f_n(K)-c_n(K)|+\rho_n(K). \qquad (8.3.87)$$

Here $f_n(K)$ denotes the nth approximant of K.

We note that

$$|a_{n+1}(K)|+\mathrm{Re}(a_{n+1}(K))=|\alpha_n^2|-\mathrm{Re}(\alpha_n^2)=2[\mathrm{Im}(\alpha_n)]^2. \qquad (8.3.88)$$

If one sets $\zeta_n=\zeta$, for all $n=1,2,3,\ldots$, then the positive definite continued fraction (8.3.82) becomes a positive definite J-fraction (see Section 4.5.1 and Theorem 4.61). If, in addition, all α_n and β_n are real, then (8.3.82) is a real J-fraction. The family of real J-fractions will be denoted by $\mathfrak{R}(\zeta)$. For this important special case we have the following Corollary of Theorem 8.14:

COROLLARY 8.15. *Let ζ be a fixed complex number such that $\mathrm{Im}(\zeta)\neq 0$, and let $K\in\mathfrak{R}(\zeta)$, that is, K is a continued fraction of the form*

$$K=\cfrac{1}{\beta_1+\zeta} \; \cfrac{-\alpha_1^2}{+ \; \beta_2+\zeta} \; \cfrac{-\alpha_2^2}{+ \; \beta_3+\zeta} \; + \; \cdots, \qquad \alpha_n\neq 0, \qquad (8.3.89)$$

where the α_n and β_n are real constants. Let U be the half-plane region closed in $\hat{\mathbb{C}}$ defined by

$$U=\begin{cases} [u:\mathrm{Im}(u)\geqslant 0] & \text{if} \quad \mathrm{Im}(\zeta)>0, \\ [u:\mathrm{Im}(u)\leqslant 0] & \text{if} \quad \mathrm{Im}(\zeta)<0. \end{cases} \qquad (8.3.90)$$

Let $S_n(K,w)$ be defined as in (8.3.7), and let $f_n(K)$ denote the nth approximant of K. Then the following statements hold:

(A) *For each $n=1,2,3,\ldots$,*

$$c(\psi_n(\mathfrak{R}(\zeta),K))=S_n(K,U),$$

that is, $S_n(K,U)$ is the best nth inclusion region for K with respect to $\mathfrak{R}(\zeta)$.

(B) *$\{S_n(K,U)\}$ is a nested sequence of closed circular disks, and $\partial S_n(K,U)$ and $\partial S_{n+1}(K,U)$ are tangent to each other at $f_n(K)=S_n(K,0)$ (see Figure 8.3.7).*

(C) *If K converges to a finite value f, then*

$$|f-f_n(K)| \leqslant 2\rho_n(K), \quad n=1,2,3,\ldots, \qquad (8.3.91)$$

where $\rho_n(K)$ is the radius of $S_n(K,U)$.

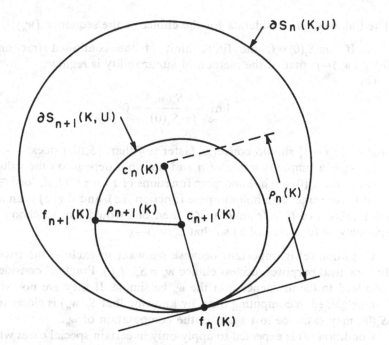

Figure 8.3.7. $S_n(K, U)$ is the best nth inclusion region for the real J-fraction K with respect to $\Re(\zeta)$.

(D) *If $c_n(K)$ denotes the center of $S_n(K, U)$, then the $\rho_n(K)$ and $c_n(K)$ can be computed successively by*

$$\rho_1(K) = \frac{1}{2|\operatorname{Im}(\zeta)|}, \qquad c_1(K) = -\frac{i}{2\operatorname{Im}(\zeta)}, \qquad (8.3.92a)$$

$$\rho_{n+1}(K) = \frac{|f_{n+1}(K) - f_n(K)|^2}{2\rho_n(K)\operatorname{Re}\left(\dfrac{f_{n+1}(K) - f_n(K)}{c_n(K) - f_n(K)}\right)}, \qquad n = 1, 2, 3, \ldots,$$

$$(8.3.92b)$$

$$c_{n+1}(K) = f_n(K) + \frac{\rho_{n+1}(K)}{\rho_n(K)}[c_n(K) - f_n(K)], \qquad n = 1, 2, 3, \ldots.$$

$$(8.3.92c)$$

8.4 Accelerating Convergence

In their most general setting the results to be presented here are methods of summability. In the notation of (2.1.1), we replace the sequence of approximants $\{S_n(0)\}$ of a continued fraction by a sequence $\{S_n(w_n)\}$.

The following are desiderata for the choice of the sequence $\{w_n\}$:

(a) If $\lim S_n(0) = f$, the finite limit of the continued fraction, then $\lim S_n(w_n) = f$; that is, the method of summability is *regular*.
(b)

$$\lim_{n\to\infty} \frac{f - S_n(w_n)}{f - S_n(0)} = 0;$$

that is, $\{S_n(w_n)\}$ should converge faster to f than $\{S_n(0)\}$ does.

(c) w_n is a simple function of n and does not depend on the value f.

(d) If the $S_n(0)$ are holomorphic functions of z for $z \in Q$, if, for $z \in D \subseteq Q$, $\{S_n(0)\}$ converges to a holomorphic function $f(z)$, and if $g(z)$ is an analytic continuation of $f(z)$ from D to Q, then one would like to choose the w_n (possibly as functions of z) so that $S_n(w_n) \to g$.

Condition (c) is important because we want to exclude the trivial and, for practical purposes, useless choice $w_n = S_n^{-1}(f)$. Practical considerations also lead to the insistence that the w_n be simple. If they are not, whatever can be gained in computing time, by knowing that $S_n(w_n)$ is closer to f than $S_n(0)$, may easily be lost again in the computation of w_n.

Condition (d) is expected to apply only in certain special cases when one is dealing with sequences of functions. It appears to be difficult to state general theorems for such cases; however, we shall give an example of this phenomenon in Chapter 12.

We now concentrate on the case in which the continued fraction under consideration converges to a finite value f. We recall the notation $f_n = S_n(0)$,

$$f^{(k)} = \mathop{\mathbf{K}}_{n=k+1}^{\infty} \left(\frac{a_n}{b_n}\right) \quad \text{and} \quad h_n = \frac{B_n}{B_{n-1}} = b_n + \frac{a_n}{b_{n-1} +} \cdots + \frac{a_2}{b_1}.$$

From (8.3.18) and (8.3.19) one then obtains

$$\frac{f - S_n(w_n)}{f - f_n} = \frac{h_n}{h_n + w_n}\left(1 - \frac{w_n}{f^{(n)}}\right). \tag{8.4.1}$$

There might be unusual conditions under which one could try to have $h_n/(h_n + w_n) \to 0$; however, in general it seems more sensible to aim for a bound on $h_n/(h_n + w_n)$ and for $1 - (w_n/f^{(n)})$ to tend to zero. The latter involves knowing something about $f^{(n)}$. Wynn [1959], in generalizing Glaisher's [1873/4] example, used recurrence relations to estimate $f^{(n)}$ and then set w_n equal to a first approximation of $f^{(n)}$. Hayden [1965] improved Wynn's method of estimating $f^{(n)}$ in special cases.

In general $f^{(n)}$ does not converge and hence a choice of w_n would be difficult at best. We restrict ourselves therefore to studying continued fractions where the $f^{(n)}$ converge. There is not much loss in generality if we consider only continued fractions of the form $K(a_n/1)$. Then, from

$$f^{(n)} = \frac{a_n}{1 + f^{(n+1)}},$$

we conclude that $\lim a_n$ must exist. Thus we are dealing either with a limit periodic continued fraction or with one for which $\lim a_n = \infty$. Some of the latter type are considered by Wynn; however, a general theory appears to be difficult.

Beginning in 1973, Gill [1973, 1975, 1978a, b] considered among other things the subject of accelerating convergence of limit periodic continued fractions by replacing $\{S_n(0)\}$ with $\{S_n(-\frac{1}{2} + \sqrt{\frac{1}{4} + a})\}$. His results in this area are essentially subsumed under the following theorem of Thron and Waadeland [1980].

THEOREM 8.16. *Let* $K(a_n/1)$ *be a periodic continued fraction with*

$$\lim_{n \to \infty} a_n = a,$$

where $|\arg(a + \frac{1}{4})| < \pi$ *and* $a \neq -\frac{1}{4}, 0, \infty$. *Further assume that for all* $n \geqslant 1$

$$|a_n - a| \leqslant \min\left[\frac{1}{2}\left(|\frac{1}{4} + a| + \frac{1}{4} - |a|\right), \frac{1}{2}|a|\right].$$

Now set

$$d_n = \max_{m > n} |a_m - a|.$$

Then

$$\left|\frac{f - S_n\left(-\frac{1}{2} + \sqrt{\frac{1}{4} + a}\right)}{f - S_n(0)}\right| \leqslant 2d_n \frac{|a| + |\frac{1}{2} + a\sqrt{\frac{1}{4} + a}|}{|a|\left(\frac{1}{4} + |\frac{1}{4} + a| - |a|\right)}.$$

Here it is understood that Re $(\sqrt{\frac{1}{4} + a}) > 0$.

Thron and Waadeland also have a result for the case $a = -\frac{1}{4}$ in the preceding reference.

The proof of Theorem 8.16 is based on value-theoretic considerations. This theorem, together with Theorem 8.13, allows conclusions as to acceleration of convergence of $\{S_n(f^*)\}$ over $\{S_n(0)\}$. Thus, for all convergent limit periodic continued fractions $K(a_n/1)$, the quantity $S_n(f^*)$ is *eventually* a better approximation to f than $S_n(0)$ is. However, this may be true only for large n, while for smaller n the approximation might actually be worse. There is room for results giving, under suitable restrictions on a_n, more precise statements about $1 - (f^*/f^{(n)})$.

CHAPTER 9

Asymptotic Expansions and Moment Problems

9.1 Introduction

Let

$$c_0 + c_1 z^{-1} + c_2 z^{-2} + \cdots \tag{9.1.1}$$

be a formal power series (fps) at $z = \infty$. Let $f(z)$ be a function holomorphic in a region S with $\infty \in c(S)$. Then we say that (9.1.1) is an *asymptotic expansion of $f(z)$ at $z = \infty$, with respect to S,* if there exist sequences of positive numbers $\{\eta_n\}$ and $\{\rho_n\}$ such that, for each $n = 0, 1, 2, \ldots$,

$$\left| f(z) - \sum_{k=0}^{n} c_k z^{-k} \right| \leqslant \eta_n |z|^{-n-1} \quad \text{for} \quad |z| > \rho_n, \ z \in S. \tag{9.1.2}$$

The reader is referred to [Henrici, 1977, Chapter 11] for a fuller discussion of asymptotic expansions. Here we only state a few of the important results. If $f(z)$ and S are given, then there is at most one asymptotic expansion (there may not be any). If a series (9.1.1) is given, then there always exists at least one function $f(z)$ and one region S (an angular opening at ∞) such that $f(z)$ is holomorphic in S and the series is the asymptotic expansion of $f(z)$ at $z = \infty$, with respect to S. This result is due to Ritt [1916] (see [Henrici, 1977, Theorem 11.3b]). That, for a given series (9.1.1), there cannot be a unique function $f(z)$ which is its asymptotic expansion, follows from the fact that e^{-z} has the asymptotic expansion

$$0 + 0 \cdot z^{-1} + 0 \cdot z^{-2} + \cdots, \quad \text{for} \quad |\arg z| < \alpha, \ \alpha > 0.$$

ENCYCLOPEDIA OF MATHEMATICS and Its Applications, Gian-Carlo Rota (ed.).
Vol. 11: William B. Jones and W. J. Thron, Continued Fractions. ISBN 0-201-13510-8

If $S \supset [z : |z| > \rho, \ \rho > 0]$, then a function $f(z)$ holomorphic in S has an asymptotic expansion at ∞, with respect to S, which is its Laurent expansion in S, and hence the expansion converges in S.

If the series (9.1.1) diverges, then it can be the asymptotic expansion of a function $f(z)$ only with respect to a region S which does not contain a deleted neighborhood of ∞.

The following important problem comes up naturally in the study of continued fractions. Let a series (9.1.1) be given, and let $K(a_n(z)/b_n(z))$ be a continued fraction that corresponds to (9.1.1) at $z = \infty$ and that converges to a holomorphic function $f(z)$ in a region D with $z = \infty$ on its boundary. Is the series (9.1.1) the asymptotic expansion of $f(z)$ at $z = \infty$, with respect to D?

Surprisingly little is known about this question. In some special cases one might hope to answer it by using minimal solutions of appropriate three-term recurrence relations (see Sections 5.3 and 6.2).

The only known general answer is provided by the theory of moments developed by Stieltjes [1894] for S-fractions, by Grommer [1914] and Hamburger [1920/21] for real J-fractions and H-fractions, and by Jones, Thron and Waadeland [1980] for positive T-fractions. We shall summarize the results here. Proofs and further details can be found in Perron [1957a], Wall [1948] and [Jones, Thron and Waadeland, 1980]. Some applications of moment theory to the representation of analytic functions by continued fractions are given in Section 9.6. We begin with a statement of three moment problems and a group of theorems indicating the connection with asymptotic expansions.

9.2 Moment Problems

For each pair (a, b), such that $-\infty \leqslant a < b \leqslant +\infty$, we shall let $\Phi(a, b)$ denote the family of all real-valued, bounded, monotone non-decreasing functions $\psi(t)$ with infinitely many points of increase on $a \leqslant t \leqslant b$. The classical moment problem is that of Stieltjes and is the following:

Stieltjes Moment Problem. For a given sequence $\{c_n\}_{n=0}^{\infty}$ of real numbers, find conditions on the sequence to insure the existence of a $\psi \in \Phi(0, \infty)$ such that

$$c_n = \int_0^{\infty} (-t)^n \, d\psi(t), \qquad n = 0, 1, 2, \ldots . \qquad (9.2.1)$$

Such a function $\psi(t)$ is called a *solution* of the moment problem. The *Hamburger moment problem* asks for a $\psi \in \Phi(-\infty, \infty)$ such that

$$c_n = \int_{-\infty}^{\infty} (-t)^n \, d\psi(t), \qquad n = 0, 1, 2, \ldots . \qquad (9.2.2)$$

Clearly, any solution of the Stieltjes moment problem also solves the Hamburger moment problem (one simply defines $\psi(t)=0$ for $t<0$).

Another moment problem is concerned with a double sequence

$$\ldots, c_{-2}, c_{-1}, c_0, c_1, c_2, \ldots \tag{9.2.3}$$

of real numbers and seeks a solution $\psi \in \Phi(0, \infty)$ such that

$$c_n = \int_0^\infty (-t)^n \, d\psi(t), \qquad n=0, \pm1, \pm2, \ldots. \tag{9.2.4}$$

We call this a *strong Stieltjes moment problem*. A solution of this problem for a double sequence $\{c_n\}_{-\infty}^\infty$ is always a solution of the classical Stieltjes moment problem for the sequence $\{c_n\}_0^\infty$. The following results hold.

THEOREM 9.1A. *Let* $\psi \in \Phi(0, \infty)$ *be a solution of the Stieltjes moment problem for a sequence* $\{c_n\}_0^\infty$. *Then the integral*

$$\int_0^\infty \frac{z \, d\psi(t)}{z+t} \tag{9.2.5}$$

is a holomorphic function $F(z)$ *for* z *in the cut plane* $R=[z:|\arg z|<\pi]$, *and the series*

$$c_0 + c_1 z^{-1} + c_2 z^{-2} + \cdots \tag{9.2.6}$$

is the asymptotic expansion of $F(z)$ *at* $z=\infty$, *with respect to* $R_\alpha = [z:|\arg z|<\alpha], 0<\alpha<\pi$.

THEOREM 9.1B. *Let* $\psi \in \Phi(-\infty, \infty)$ *be a solution of the Hamburger moment problem for a sequence* $\{c_n\}_0^\infty$. *Let* R^+ *and* R^- *denote the half-plane regions*

$$R^+ = [z:\operatorname{Im}(z)>0], \qquad R^- = [z:\operatorname{Im}(z)<0].$$

Then the integral

$$\int_{-\infty}^\infty \frac{z \, d\psi(t)}{z+t} \tag{9.2.7}$$

is a holomorphic function $F^+(z)$, *for* $z \in R^+$, *and a holomorphic function* $F^-(z)$, *for* $z \in R^-$, *and the series*

$$c_0 + c_1 z^{-1} + c_2 z^{-2} + \cdots \tag{9.2.8}$$

is the asymptotic expansion of $F^+(z)$ $[F^-(z)]$ *at* $z=\infty$, *with respect to* R^+ $[R^-]$.

By analogy with the definition at $z = \infty$, we defined a series

$$d_0 + d_1 z + d_2 z^2 + \cdots \qquad (9.2.9)$$

to be *an asymptotic expansion of a function* $f(z)$ *at* $z = 0$, *with respect to a region* S, *with* $0 \in c(S)$, if there exist sequences of positive numbers $\{\eta_n\}$ and $\{\rho_n\}$ such that for each $n = 0, 1, 2, \ldots$,

$$\left| f(z) - \sum_{k=0}^{n} d_k z^k \right| \leqslant \eta_n |z|^{n+1} \qquad \text{for} \quad |z| < \rho_n, \ z \in S. \qquad (9.2.10)$$

We can now state our third theorem.

THEOREM 9.1C. *Let* $\psi \in \Phi(0, \infty)$ *be a solution of the strong Stieltjes moment problem for a double sequence* $\{c_n\}_{n=-\infty}^{\infty}$. *Then the integral*

$$\int_0^\infty \frac{z \, d\psi(t)}{z + t} \qquad (9.2.11)$$

is a holomorphic function $G(z)$ *for* z *in the cut plane* $R = [z : |\arg z| < \pi]$. *The series*

$$c_0 + c_1 z^{-1} + c_2 z^{-2} + \cdots \qquad (9.2.12)$$

is the asymptotic expansion of $G(z)$ *at* $z = \infty$ *with respect to* R. *The series*

$$-c_{-1} z - c_{-2} z^2 - c_{-3} z^3 - \cdots \qquad (9.2.13)$$

is the asymptotic expansion of $G(z)$ *at* $z = 0$ *with respect to* R.

It turns out that the three moment problems mentioned here can be solved by means of continued fractions. Non-continued-fraction solutions, as well as solutions to other moment problems, are discussed in the monographs of Shohat and Tamarkin [1943], and Akhieser [1965]. Other good references are [Widder, 1946] and [Henrici, 1977].

The mean by which continued fractions enter moment problems is that certain continued fractions with variable elements can be represented by Stieltjes integrals. This is the subject of the following section.

9.3 Integral Representations of Continued Fractions

We begin by stating conditions to insure that continued fractions of certain types have real and/or positive coefficients. For regular C-fractions and associated continued fractions we replace z by $1/z$ to have corresponding power series at $z = \infty$. A continued fraction of the form

$$\frac{a_1}{1} + \frac{a_2}{z} + \frac{a_3}{1} + \frac{a_4}{z} + \cdots, \qquad a_n \neq 0 \qquad (9.3.1)$$

will be called a *modified regular C-fraction* [see (8.3.41)]. It is called a *modified S-fraction* if $a_n > 0$ for all n. We recall the definition of the Hankel determinant $H_k^{(n)}$ (of dimension k) associated with a sequence $\{c_n\}$:

$$H_0^{(n)} = 1; \qquad H_k^{(n)} = \begin{vmatrix} c_n & c_{n+1} & \cdots & c_{n+k-1} \\ c_{n+1} & c_{n+2} & \cdots & c_{n+k} \\ \vdots & \vdots & & \vdots \\ c_{n+k-1} & c_{n+k} & \cdots & c_{n+2k-2} \end{vmatrix}, \quad k=1,2,3,\ldots.$$

(9.3.2)

The next three results follow from Theorems 7.2, 7.14 and 7.18, respectively.

THEOREM 9.2A. *Let $\{c_n\}_{n=0}^{\infty}$ be a sequence of real numbers, and let*

$$\frac{a_1}{1} + \frac{a_2}{z} + \frac{a_3}{1} + \frac{a_4}{z} + \cdots, \qquad a_n \neq 0, \tag{9.3.3}$$

be a modified regular C-fraction which corresponds to the fLs

$$c_0 + c_1 z^{-1} + c_2 z^{-2} + \cdots \tag{9.3.4}$$

at $z = \infty$. Then a necessary and sufficient condition for

$$a_n > 0, \qquad n = 1,2,3,\ldots \tag{9.3.5}$$

is that

$$H_n^{(0)} > 0 \text{ and } (-1)^n H_n^{(1)} > 0 \qquad \text{for} \quad n = 1,2,3,\ldots. \tag{9.3.6}$$

THEOREM 9.2B. *Let $\{c_n\}_{n=0}^{\infty}$ be a sequence of real numbers, and let the continued fraction*

$$\frac{1}{b_1} + \frac{1}{b_2 z} + \frac{1}{b_3} + \frac{1}{b_4 z} + \cdots, \tag{9.3.7}$$

which is equivalent to (9.3.3), correspond to the fLs

$$c_0 + c_1 z^{-1} + c_2 z^{-2} + \cdots \tag{9.3.8}$$

at $z = \infty$. Then a necessary and sufficient condition for

$$b_n \text{ real}, \quad b_{2n-1} > 0 \text{ and } b_{2n} \neq 0, \qquad n = 1,2,3,\ldots, \tag{9.3.9}$$

is that

$$H_n^{(0)} > 0 \qquad \text{for} \quad n = 1,2,3,\ldots. \tag{9.3.10}$$

The same condition (9.3.10) *insures that in the J-fraction*

$$\frac{k_1 z}{l_1 + z} - \frac{k_2}{l_2 + z} - \frac{k_3}{l_3 + z} - \cdots, \tag{9.3.11}$$

corresponding to (9.3.4) *at* $z = \infty$, *one has*

$$k_n > 0 \text{ and } l_n \text{ real} \quad \text{for} \quad n = 1, 2, 3, \ldots. \tag{9.3.12}$$

A continued fraction (9.3.7) satisfying (9.3.9) is called an *H-fraction*, and a continued fraction (9.3.11) satisfying (9.3.12) is called a *real J-fraction* (or, by Perron [1957a], a *G-fraction*; see also Section 4.5.1).

THEOREM 9.2C. *Let* $\{c_n\}_{n=-\infty}^{\infty}$ *be a double sequence of real numbers, and let*

$$\frac{F_1 z}{1 + G_1 z} + \frac{F_2 z}{1 + G_2 z} + \frac{F_3 z}{1 + G_3 z} + \cdots \tag{9.3.13}$$

be a general T-fraction corresponding to

$$c_0 + c_1 z^{-1} + c_2 z^{-2} + \cdots \tag{9.3.14}$$

at $z = \infty$ *and to*

$$-c_{-1} z - c_{-2} z^2 - c_{-3} z^3 - \cdots \tag{9.3.15}$$

at $z = 0$. *Then a necessary and sufficient condition for*

$$F_n > 0, \quad G_n > 0, \quad n = 1, 2, 3, \ldots, \tag{9.3.16}$$

[*that is, for* (9.3.13) *to be a positive T-fraction*] *is that*

$$H_{2n+1}^{(-2n)} > 0, \quad H_{2n}^{(-2n+1)} > 0, \quad H_{2n}^{(-2n)} > 0, \quad H_{2n-1}^{(-2n+1)} < 0,$$
$$\text{for} \quad n = 1, 2, 3, \ldots. \tag{9.3.17}$$

Here we note that in (9.3.2) the definition of $H_k^{(n)}$ is extended to all integer values of n.

The next step in arriving at integral representations is to study the approximants of the continued fractions under consideration. It is here that the property of the coefficients being real and/or positive plays a key role.

THEOREM 9.3A. *For a modified S-fraction* (9.3.3) *the zeros* $(-t_k^{(n)})$ *of the* nth *denominator* $B_n(z)$ *are all distinct and negative. In addition, if* $A_n(z)$ *is* *the n th numerator, then A_n/B_n has a partial-fraction decomposition*

$$\frac{A_n(z)}{B_n(z)} = z \sum_{k=1}^{\lambda_n} \frac{M_k^{(n)}}{z + t_k^{(n)}}, \qquad \lambda_n = \frac{n}{2} \text{ or } \frac{n+1}{2}, \qquad (9.3.18)$$

such that

$$\sum_{k=1}^{\lambda_n} M_k^{(n)} = a_1 \quad and \quad M_k^{(n)} > 0 \quad for \quad k = 1, 2, \ldots, n. \qquad (9.3.19)$$

THEOREM 9.3B. *For an H-fraction (9.3.7) the zeros $(-t_k^{(n)})$ of the n th denominator $B_n(z)$ are all distinct and real. In addition, if $A_n(z)$ is the n th numerator, then A_n/B_n has a partial-fraction decomposition*

$$\frac{A_n(z)}{B_n(z)} = z \sum_{k=1}^{\lambda_n} \frac{M_k^{(n)}}{z + t_k^{(n)}}, \qquad \lambda_n = \frac{n}{2} \text{ or } \frac{n+1}{2}, \qquad (9.3.20)$$

such that

$$\sum_{k=1}^{\lambda_n} M_k^{(n)} = \frac{1}{b_1} \quad and \quad M_k^{(n)} > 0, \quad n = 1, 2, \ldots, \lambda_n. \qquad (9.3.21)$$

For a real J-fraction (9.3.11) one considers the generalized approximants

$$\frac{P_n(z, \tau)}{Q_n(z, \tau)}, \qquad (9.3.22)$$

where

$$P_n(z, \tau) = K_n(z) + \tau K_{n-1}(z), \qquad (9.3.23a)$$

$$Q_n(z, \tau) = L_n(z) + \tau L_{n-1}(z), \qquad (9.3.23b)$$

$$-\infty < \tau < \infty, \qquad (9.3.23c)$$

and where $K_n(z)$ and $L_n(z)$ denote the n th numerator and denominator, respectively, of (9.3.11). The zeros $[-t_k^{(n)}(\tau)]$ of $Q_n(z, \tau)$ are real and distinct, and P_n/Q_n has a partial-fraction decomposition

$$\frac{P_n(z, \tau)}{Q_n(z, \tau)} = z \sum_{k=1}^{\lambda_n} \frac{M_k^{(n)}(\tau)}{z + t_k^{(n)}(\tau)}, \qquad \lambda_n = n, \qquad (9.3.24)$$

such that

$$\sum_{k=1}^{\lambda_n} M_k^{(n)}(\tau) = k_1 \quad and \quad M_k^{(n)}(\tau) > 0, \quad k = 1, 2, \ldots, \lambda_n. \qquad (9.3.25)$$

THEOREM 9.3C. *For a positive T-fraction* (9.3.13), *the zeros* $(-t_k^{(n)})$ *of the* n *th denominator* $B_n(z)$ *are all distinct and negative. In addition, if* $A_n(z)$ *is the* n *th numerator, then* A_n/B_n *has a partial-fraction decomposition*

$$\frac{A_n(z)}{B_n(z)} = z \sum_{k=1}^{\lambda_n} \frac{M_k^{(n)}}{z + t_k^{(n)}}, \qquad \lambda_n = n, \qquad (9.3.26)$$

such that

$$\sum_{k=1}^{\lambda_n} M_k^{(n)} = \frac{F_1}{G_1} \quad and \quad M_k^{(n)} > 0, \quad k = 1, 2, \dots, \lambda_n. \qquad (9.3.27)$$

In all of the cases considered in the previous three theorems one can order the $t_k^{(n)}$ according to size and define step functions $\psi_n(t)$ by

$$\psi_n(t) = \begin{cases} 0 & \text{if} \quad -\infty < t \leqslant t_1^{(n)}, \\ \sum_{k=1}^{m} M_k^{(n)} & \text{if} \quad t_m^{(n)} < t \leqslant t_{m+1}^{(n)}, \ m = 1, 2, \dots, \lambda_n, \\ \beta & \text{if} \quad t_{\lambda_n}^{(n)} < t \leqslant \infty, \end{cases} \qquad (9.3.28)$$

where β is a_1, $1/b_1$, k_1, or F_1/G_1, respectively. One thus obtains in each case a Stieltjes-integral representation of the nth approximant A_n/B_n of the form

$$\frac{A_n(z)}{B_n(z)} = \int_{-\infty}^{\infty} \frac{z \, d\psi_n(t)}{z + t}. \qquad (9.3.29)$$

(For the real *J*-fraction the nth approximant was replaced by the generalized approximant P_n/Q_n in Theorem 9.3B.) From this, using the Grommer selection principle, one derives the following three theorems.

THEOREM 9.4A. *For every sequence of approximants of a modified S-fraction* (9.3.3), *there exists a subsequence* $\{n_k\}$ *and a function* $\psi \in \Phi(0, \infty)$ *such that*

$$\lim_{k \to \infty} \frac{A_{n_k}(z)}{B_{n_k}(z)} = \int_0^{\infty} \frac{z \, d\psi(t)}{z + t}. \qquad (9.3.30)$$

The convergence is uniform on every compact subset of $R = [z : |\arg z| < \pi]$.

THEOREM 9.4B. *For every sequence of approximants of an H-fraction* (9.3.7), *there exists a subsequence* $\{n_k\}$ *and a function* $\psi \in \Phi(-\infty, \infty)$ *such that*

$$\lim_{k \to \infty} \frac{A_{n_k}(z)}{B_{n_k}(z)} = \int_{-\infty}^{\infty} \frac{z \, d\psi(t)}{z + t}. \tag{9.3.31}$$

The convergence is uniform on every compact subset of R^+ or R^-.

  For every sequence of generalized approximants (with varying τ_n) of a real J-fraction (9.3.11), there exists a subsequence $\{n_k\}$ and a function $\psi \in \Phi(-\infty, \infty)$ such that

$$\lim_{k \to \infty} \frac{P_{n_k}(z, \tau_{n_k})}{Q_{n_k}(z, \tau_{n_k})} = \int_{-\infty}^{\infty} \frac{z \, d\psi(t)}{z + t}. \tag{9.3.32}$$

The convergence is uniform on every compact subset of R^+ or R^-.

THEOREM 9.4C. *For every sequence of approximants of a positive T-fraction (9.3.13), there exists a subsequence $\{n_k\}$ and a function $\psi \in \Phi(0, \infty)$ such that*

$$\lim_{k \to \infty} \frac{A_{n_k}(z)}{B_{n_k}(z)} = \int_{0}^{\infty} \frac{z \, d\psi(t)}{z + t}. \tag{9.3.33}$$

The convergence is uniform on every compact subset of $R = [z : |\arg z| < \pi]$.

9.4 Asymptotic Expansions for Continued Fractions

  Using the known convergence behavior of the continued fractions involved (Theorem 4.58 for S-fractions, Theorem 4.32 for H-fractions and Corollary 4.65 for positive T-fractions), we obtain some of the results below. For real J-fractions one is concerned with complete convergence, that is, the convergence of all possible sequences of generalized approximants. We do have convergence results for real J-fractions in Theorem 4.61 and (even more to the point) Corollary 8.15. They do not, however, directly apply to the situation considered here.

  THEOREM 9.5A. *For every modified S-fraction (9.3.3), there exist two functions $\psi_0, \psi_1 \in \Phi(0, \infty)$ such that*

$$\lim_{n \to \infty} \frac{A_{2n+\sigma}(z)}{B_{2n+\sigma}(z)} = \int_{0}^{\infty} \frac{z \, d\psi_\sigma(t)}{z + t}, \qquad \sigma = 0, 1. \tag{9.4.1}$$

The convergence is uniform on every compact subset of $R = [z : |\arg z| < \pi]$. If at least one of the series (4.5.21) diverges, then there exists a function

$\psi \in \Phi(0, \infty)$ *such that*

$$\lim_{n\to\infty} \frac{A_n(z)}{B_n(z)} = \int_0^\infty \frac{z\,d\psi(t)}{z+t}, \tag{9.4.2}$$

and the convergence is uniform on every compact subset of R.

THEOREM 9.5B. *For every H-fraction (9.3.7) with*

$$\sum_{k=1}^\infty |b_k| < \infty, \tag{9.4.3}$$

there exist two distinct functions $\psi_0, \psi_1 \in \Phi(-\infty, \infty)$ *such that*

$$\lim_{n\to\infty} \frac{A_{2n+\sigma}(z)}{B_{2n+\sigma}(z)} = \int_{-\infty}^\infty \frac{z\,d\psi_\sigma(t)}{z+t}, \qquad \sigma = 0, 1. \tag{9.4.4}$$

The convergence is uniform on every compact subset of R^+ *or of* R^-. *If*

$$\sum_{k=1}^\infty |b_k| = \infty \tag{9.4.5}$$

and at least one of conditions (A), (B), (C) *of Theorem 4.32 is satisfied, then there exists a function* $\psi \in \Phi(-\infty, \infty)$ *such that*

$$\lim_{n\to\infty} \frac{A_n(z)}{B_n(z)} = \int_{-\infty}^\infty \frac{z\,d\psi(t)}{z+t} \tag{9.4.6}$$

and the convergence is uniform on every compact subset of R^+ *or of* R^-.

For every real J-fraction (9.3.11), with n th numerator $K_n(z)$ *and denominator* $L_n(z)$, *which is such that at least one of the series*

$$\sum_{n=1}^\infty \frac{L_n(0)}{k_1 k_2 \cdots k_{n+1}}, \qquad \sum_{n=1}^\infty \frac{K_n(0)}{k_1 k_2 \cdots k_{n+1}} \tag{9.4.7}$$

diverges, there exists a function $\psi \in \Phi(-\infty, \infty)$ *such that*

$$\lim_{n\to\infty} \frac{P_n(z, \tau_n)}{Q_n(z, \tau_n)} = \int_{-\infty}^\infty \frac{z\,d\psi(t)}{z+t}. \tag{9.4.8}$$

Here $\{\tau_n\}$ *can be any sequence of real numbers. The convergence is uniform for z in every compact subset of* R^+ *or of* R^-.

As Perron [1957a, p. 217] points out, the cases for H-fractions considered above do not exhaust all possibilities. Other divergence phenomena can occur.

THEOREM 9.5C. *For every positive T-fraction (9.3.13), there exist two functions $\psi_0, \psi_1 \in \Phi(0, \infty)$ such that*

$$\lim_{n \to \infty} \frac{A_{2n+\sigma}(z)}{B_{2n+\sigma}(z)} = \int_0^\infty \frac{z \, d\psi_\sigma(t)}{z+t}, \qquad \sigma = 0, 1. \qquad (9.4.9)$$

The convergence is uniform on every compact subset of $R = [z : |\arg z| < \pi]$. Let $\{e_n\}$ and $\{d_n\}$ be defined by

$$e_1 = \frac{1}{F_1}, \qquad\qquad d_1 = \frac{G_1}{F_1}, \qquad\qquad (9.4.10a)$$

$$e_{2n-1} = \frac{\displaystyle\prod_{k=1}^{n-1} F_{2k}}{\displaystyle\prod_{k=1}^{n} F_{2k-1}}, \qquad d_{2n-1} = \frac{G_{2n-1} \displaystyle\prod_{k=1}^{n-1} F_{2k}}{\displaystyle\prod_{k=1}^{n} F_{2k-1}}, \qquad n = 2, 3, 4, \ldots,$$

$$(9.4.10b)$$

$$e_{2n} = \frac{\displaystyle\prod_{k=1}^{n} F_{2k-1}}{\displaystyle\prod_{k=1}^{n} F_{2k}}, \qquad d_{2n} = \frac{G_{2n} \displaystyle\prod_{k=1}^{n} F_{2k-1}}{\displaystyle\prod_{k=1}^{n} F_{2k}}, \qquad n = 1, 2, 3, \ldots,$$

$$(9.4.10c)$$

so that the positive T-fraction (9.3.13) is equivalent to

$$\frac{z}{e_1 + d_1 z} + \frac{z}{e_2 + d_2 z} + \frac{z}{e_3 + d_3 z} + \cdots \qquad (e_n > 0, \quad d_n > 0). \quad (9.4.11)$$

If

$$\sum_{n=1}^{\infty} e_n = \infty \quad or \quad \sum_{n=1}^{\infty} d_n = \infty, \qquad (9.4.12)$$

then there exists a function $\psi \in \Phi(0, \infty)$ such that

$$\lim_{n \to \infty} \frac{A_n(z)}{B_n(z)} = \int_0^\infty \frac{z \, d\psi(t)}{z+t}. \qquad (9.4.13)$$

The convergence is uniform on every compact subset of R.

The next group of theorems asserts that the functions ψ_σ are solutions of the moment problem.

THEOREM 9.6A. *Let a modified S-fraction* (9.3.3) *corresponding to*

$$c_0 + c_1 z^{-1} + c_2 z^{-2} + \cdots \qquad (9.4.14)$$

at $z = \infty$ *be given. For* $\sigma = 0, 1,$ *let*

$$\int_0^\infty \frac{z\, d\psi_\sigma(t)}{z+t} \qquad (9.4.15)$$

denote the function to which $A_{2n+\sigma}(z)/B_{2n+\sigma}(z)$ *converges on* R, *where* $\psi_\sigma \in \Phi(0, \infty)$. *Then*

$$c_k = \int_0^\infty (-t)^k\, d\psi_\sigma(t), \qquad k = 0, 1, 2, \ldots . \qquad (9.4.16)$$

THEOREM 9.6B. *Let an H-fraction* (9.3.7) *[real J-fraction* (9.3.11)*] corresponding to*

$$c_0 + c_1 z^{-1} + c_2 z^{-2} + \cdots \qquad (9.4.17)$$

at $z = \infty$ *be given. Let*

$$\int_{-\infty}^\infty \frac{z\, d\psi(t)}{z+t} \qquad (9.4.18)$$

denote the function to which $A_{n_k}(z)/B_{n_k}(z)$ *[$P_{n_k}(z, \tau_k)/Q_{n_k}(z, \tau_k)$] converges on* R^+ *or* R^-, *where* $\psi \in \Phi(-\infty, \infty)$. *Then*

$$c_k = \int_{-\infty}^\infty (-t)^k\, d\psi(t), \qquad k = 0, 1, 2, \ldots . \qquad (9.4.19)$$

THEOREM 9.6C. *Let a positive T-fraction* (9.3.13) *corresponding to*

$$c_0 + c_1 z^{-1} + c_2 z^{-2} + \cdots$$

at $z = \infty$ *and to*

$$-c_{-1} z - c_{-2} z^2 - c_{-3} z^3 - \cdots$$

at $z = 0$ *be given. For* $\sigma = 0, 1,$ *let*

$$\int_0^\infty \frac{z\, d\psi_\sigma(t)}{z+t}$$

denote the function to which $A_{2n+\sigma}(z)/B_{2n+\sigma}(z)$ converges on R, where $\psi_\sigma \in \Phi(0, \infty)$. Then

$$c_k = \int_0^\infty (-t)^k \, d\psi_\sigma(t), \qquad k=0, \pm 1, \pm 2, \ldots .$$

Pulling all of these results together, we obtain the following theorems concerning subsequences of approximants of continued fractions converging to functions and the asymptotic expansions of these functions.

THEOREM 9.7A. *Let $F_1(z)$ and $F_2(z)$ denote the functions to which the odd and even parts of a modified S-fraction converge, respectively. Let $\sum_0^\infty c_n z^{-n}$ denote the series to which the modified S-fraction corresponds at $z = \infty$. Then $\sum_0^\infty c_n z^{-n}$ is the asymptotic expansion of $F_1(z)$ and $F_2(z)$ at $z = \infty$ with respect to $R_\alpha = [z : |\arg z| < \alpha], 0 < \alpha < \pi$.*

THEOREM 9.7B. *Let an H-fraction (real J-fraction) be given. For an arbitrary sequence of approximants (generalized approximants), there exists a subsequence which converges to a function. Let $G(z)$ denote such a function. Let $\sum_0^\infty c_n z^{-n}$ denote the series to which the continued fraction corresponds at $z = \infty$. Then $\sum_0^\infty c_n z^{-n}$ is the asymptotic expansion of $G(z)$ at $z = \infty$, with respect to $I_\varepsilon = [z : |\mathrm{Im}\, z| > \varepsilon], \varepsilon > 0$.*

THEOREM 9.7C. *Let $F_1(z)$ and $F_2(z)$ denote the functions to which the odd and even parts of a positive T-fraction (9.3.13) converge. Let $L^* = \sum_0^\infty c_n z^{-n}$ ($L = -\sum_1^\infty c_{-n} z^n$) denote the series to which the general T-fraction corresponds at $z = \infty$ ($z = 0$). Then $L^*(L)$ is the asymptotic expansion of $F_1(z)$ and of $F_2(z)$ at $z = \infty$ ($z = 0$), with respect to $R_\alpha = [z : |\arg z| < \alpha], 0 < \alpha < \pi$.*

9.5 Solutions of the Moment Problems

The existence of solutions of the various moment problems stated in Section 9.2 is insured by the results of the preceding two sections. Uniqueness proofs require additional arguments which can be found in [Perron, 1957a] and [Jones, Thron and Waadeland, 1980]. A solution ψ to one of the moment problems is said to be *unique* if any two solutions differ only in their at most denumerable points of discontinuity.

THEOREM 9.8A. *The Stieltjes moment problem for a sequence $\{c_n\}$ has a solution iff*

$$H_n^{(0)} > 0 \quad \text{and} \quad (-1)^n H_n^{(1)} > 0, \qquad n = 1, 2, 3, \ldots . \tag{9.5.1}$$

The solution is unique iff at least one of the series (4.5.21) diverges. Here the

a_n *are defined by* (9.3.3) [*see also* (7.1.14)], *and the Hankel determinants* $H_k^{(n)}$
are defined by (9.3.2)

Carleman has shown that

$$\sum_{n=1}^{\infty} c_n^{-1/2n} = \infty \tag{9.5.2}$$

is sufficient for the uniqueness of a Stieltjes moment problem for a
sequence $\{c_n\}$ which is known to have at least one solution (see, for
example, [Shohat and Tamarkin, 1943, p. 20]). A by-product is that this
condition is sufficient for the convergence of the modified S-fraction
(9.3.3) which corresponds to the fLs (9.3.4). The condition (9.5.2) is known
as *Carleman's criterion*. An application of this criterion is given in Section
9.6 for a continued fraction related to the gamma function.

THEOREM 9.8B. *The Hamburger moment problem for a sequence* $\{c_n\}$ *has a
solution iff*

$$H_n^{(0)} > 0, \qquad n = 1, 2, 3, \ldots . \tag{9.5.3}$$

The solution is unique iff the real J-fraction (9.3.11), *which corresponds to*

$$c_0 + c_1 z^{-1} + c_2 z^{-2} + \cdots$$

at $z = \infty$, *has n th numerator* $K_n(z)$ *and denominator* $L_n(z)$ *such that at least
one of the series*

$$\sum_{n=1}^{\infty} \frac{L_n(0)}{k_1 k_2 \cdots k_{n+1}}, \quad \sum_{n=1}^{\infty} \frac{K_n(0)}{k_1 k_2 \cdots k_{n+1}} \tag{9.5.4}$$

diverges.

THEOREM 9.8C. *The strong Stieltjes moment problem for a double sequence*
$\{c_n\}_{n=-\infty}^{\infty}$ *has a solution iff*

$$H_{2n+1}^{(-2n)} > 0, \quad H_{2n}^{(-2n+1)} > 0, \quad H_{2n}^{(-2n)} > 0, \quad H_{2n-1}^{(-2n+1)} < 0, \qquad n = 1, 2, 3, \ldots .$$

The solution is unique iff at least one of the series Σe_n *or* Σd_n, *as defined in
Theorem* 9.5C, *diverges.*

We conclude this section with the following result of Markoff [1895], a
proof of which can be found in [Perron, 1957a] (see also [Szegö, 1968] and
the discussion at the end of Section 7.2).

THEOREM 9.9 (Markoff). *Let a and b be real numbers, $a < b$, and let $\psi \in \Phi(a, b)$. Then the real J-fraction (9.3.11), which corresponds to the series*

$$\sum_{n=0}^{\infty} \left(\int_a^b (-t)^n \, d\psi(t) \right) z^{-n} \tag{9.5.5}$$

at $z = \infty$, converges to the function

$$\int_a^b \frac{z \, d\psi(t)}{z + t} \tag{9.5.6}$$

for all $z \in \mathbb{C}$ such that $z \notin [w : -b \leqslant w \leqslant -a]$.

If $\psi \in \Phi(0, a)$, $a > 0$, then the modified S-fraction (9.3.3), which corresponds to the series

$$\sum_{n=0}^{\infty} \left(\int_0^a (-t)^n \, d\psi(t) \right) z^{-n} \tag{9.5.7}$$

at $z = \infty$, converges to the function

$$\int_0^a \frac{z \, d\psi(t)}{z + t} \tag{9.5.8}$$

for all $z \in \mathbb{C}$ such that $z \notin [w : -a \leqslant w \leqslant 0]$.

9.6 Representations of Analytic Functions

For each positive real number a, the function $F(z)$ defined by

$$F(z) = \frac{z}{\Gamma(a)} \int_0^\infty \frac{e^{-t} t^{a-1}}{z + t} \, dt = \int_0^\infty \frac{z \, d\psi(t)}{z + t}, \tag{9.6.1a}$$

where

$$\psi(t) = \frac{1}{\Gamma(a)} \int_0^t e^{-\tau} \tau^{a-1} \, d\tau, \tag{9.6.1b}$$

is holomorphic for $z \in R = [z : |\arg z| < \pi]$. We shall show that the theory developed in the present chapter can be used to obtain a continued-fraction representation of $F(z)$. First we write $F(z)$ in the form

$$F(z) = \frac{1}{\Gamma(a)} \int_0^\infty \frac{1}{1 + t/z} e^{-t} t^{a-1} \, dt.$$

Then expanding $1/(1+t/z)$ in a binomial series and integrating term by term, we obtain the divergent formal Laurent series

$$\Omega\left(a,1;-\frac{1}{z}\right) = \sum_{k=0}^{\infty} c_k z^{-k}, \qquad (9.6.2a)$$

where

$$c_k = \int_0^{\infty} (-t)^k \, d\psi(t)$$

$$= \frac{(-1)^k}{\Gamma(a)} \int_0^{\infty} e^{-t} t^{a+k-1} \, dt = (-1)^k \frac{\Gamma(a+k)}{\Gamma(a)}$$

$$= \begin{cases} 1 & \text{if } k=0 \\ (-1)^k a(a+1)(a+2)\cdots(a+k-1) & \text{if } k \geqslant 1. \end{cases} \qquad (9.6.2b)$$

From Theorem 6.5 we see that the continued fraction

$$\frac{1}{1+} \frac{a(1/z)}{1} + \frac{1(1/z)}{1} + \frac{(a+1)(1/z)}{1} + \frac{2(1/z)}{1} + \frac{(a+2)(1/z)}{1} + \cdots$$

$$(9.6.3)$$

corresponds at $z=\infty$ to the fLs $\Omega(a,1;1/z)$. We are therefore led to the following.

THEOREM 9.10. *Let a be a positive real number. Then:*

(A) *the modified S-fraction*

$$\frac{1}{1+} \frac{a}{z+} \frac{1}{1+} \frac{a+1}{z} + \frac{2}{1+} \frac{a+2}{z} + \frac{3}{1+} \cdots \qquad (9.6.4)$$

converges for all z in the cut plane

$$R = [z : |\arg z| < \pi] \qquad (9.6.5)$$

to the function $F(z)$ defined by (9.6.1), which is holomorphic in R. The convergence is uniform on every compact subset of R.

(B) *The continued fraction (9.6.4) corresponds to the fLs $\Omega(a,1;-1/z)$ at $z=\infty$.*

(C) *The divergent series $\Omega(a,1;-1/z)$ is the asymptotic expansion of $F(z)$ with respect to R.*

Proof. Since the c_k defined by (9.6.2b) satisfy

$$c_k = \int_0^\infty (-t)^k \, d\psi(t), \qquad k=0,1,2,\ldots, \tag{9.6.6}$$

where $\psi(t)$ is given by (9.6.1b), the function $\psi(t)$ is a solution of the Stieltjes moment problem defined by $\{c_k\}$. It follows from Theorem 9.1A that $F(z)$ is holomorphic in R, and the divergent series $\Omega(a,1;-1/z)$ is the asymptotic expansion of $F(z)$ with respect to R. This proves (C). The modified S-fraction (9.6.4) is equivalent to (9.6.3), and hence it corresponds to $\Omega(a,1;-1/z)$ at $z=\infty$, as asserted in (B). It follows from Theorems 4.58 and 6.6 that the modified S-fraction (9.6.4) converges to a function $g(z)$ holomorphic in R and that the convergence is uniform on every compact subset of R. Hence, also by Theorem 4.58 (Stieltjes), at least one of the series (4.5.21) diverges. Here

$$a_1 = 1; \qquad a_{2n} = a+n-1, \; a_{2n+1} = n, \quad n=1,2,3,\ldots.$$

Therefore by Theorem 9.5A there exists a function $\varphi(t) \in \Phi(0,\infty)$ such that

$$g(z) = \int_0^\infty \frac{z \, d\varphi(t)}{z+t}, \qquad \text{for all} \quad z \in R. \tag{9.6.7}$$

Since the modified S-fraction (7.6.4) corresponds at $z=\infty$ to $\Omega(a,1;-1/z)$ $=\Sigma_0^\infty(c_k/z^k)$, it follows from Theorem 9.6A that

$$c_k = \int_0^\infty (-t)^k \, d\varphi(t), \qquad k=0,1,2,\ldots. \tag{9.6.8}$$

Thus $\varphi(t)$ solves the Stieltjes moment problem for $\{c_k\}$. Theorem 9.8A implies that the Stieltjes moment problem for $\{c_k\}$ has a unique solution. Thus $\varphi(t) = \psi(t)$, except possibly at the points of discontinuity, and hence $g(z) = F(z)$ for all $z \in R$.   ∎

A number of extensions of Theorem 9.10 are given by Wall [1948, Section 92]. For example, he shows that the modified S-fraction (9.6.4) converges to the function $F(z)$ defined by (9.6.1), holomorphic for $z \in R$, even if a is complex with $\text{Re}(a) > 0$. Moreover, it is shown that if a and b are complex numbers with $\text{Re}(a) > 0$, then the continued fraction

$$\frac{1}{1} + \frac{az}{1} + \frac{(b+1)z}{1} + \frac{(a+1)z}{1} + \frac{(b+2)z}{1} + \frac{(a+2)z}{1} + \cdots \tag{9.6.9}$$

converges for $z \in R$ to the function

$$\frac{\int_0^\infty \dfrac{e^{-t}t^{a-1}\,dt}{(1+zt)^{b+1}}}{\int_0^\infty \dfrac{e^{-t}t^{a-1}\,dt}{(1+zt)^b}}. \qquad (9.6.10)$$

By use of the identity

$$\frac{z^a}{\Gamma(b)}\int_0^\infty \frac{e^{-t}t^{b-1}}{(z+t)^a}\,dt = \frac{z^b}{\Gamma(a)}\int_0^\infty \frac{e^{-t}t^{a-1}}{(z+t)^b}\,dt \qquad (9.6.11)$$

(see [Wall, 1948, p. 355]), taking $b=1$, we see that the function $F(z)$ of (9.6.1) can be expressed by

$$F(z) = \int_0^\infty \frac{z^a e^{-t}}{(z+t)^a}\,dt. \qquad (9.6.12)$$

For each $a \in C$, the function $F(z)$ in (9.6.12) is holomorphic for $z \in R$, and it can be shown that it is represented for $z \in R$ by the convergent continued fraction (9.6.4), provided $a \notin [0, -1, -2, \ldots]$.

In (9.6.12), if we let $z = x$ be real and positive and replace $x + t$ by τ, then we obtain

$$F(x) = e^x x^a \int_x^\infty e^{-\tau}\tau^{-a}\,d\tau. \qquad (9.6.13)$$

Finally, replacing a by $1-a$ in (9.6.13), we obtain the function

$$e^x x^{1-a}\int_x^\infty e^{-\tau}\tau^{a-1}\,d\tau = e^x x^{1-a}\Gamma(a, x), \qquad (9.6.14)$$

where

$$\Gamma(a, x) = \int_x^\infty e^{-\tau}\tau^{a-1}\,d\tau \qquad (9.6.15)$$

is an *incomplete gamma function* (see [Abramowitz and Stegun, 1964, Section 6.5]). After making the same replacements in the continued fraction (9.6.4) and an equivalence transformation, we arrive at the equation

$$\Gamma(a, x) = e^{-x}x^a\left(\frac{1}{x+}\ \frac{1-a}{1}\ +\frac{1}{x+}\ \frac{2-a}{1}\ +\frac{2}{x+}\ \frac{3-a}{1}\ +\frac{3}{x+}\ \cdots\right),$$

$$(9.6.16)$$

which is valid for all $x>0$ and $a \in C$ with $a \notin [1,2,3,\dots]$. The continued fraction in parentheses on the right side of (9.6.16) converges to a function of x holomorphic in the cut plane R. In defining $\Gamma(a,x)$ for complex values of x, the path of integration in (9.6.15) must lie in R. Thus we have

Example 1. Incomplete gamma function.

$$\Gamma(a,z)=e^{-z}z^{a}\left(\frac{1}{z} + \frac{1-a}{1} + \frac{1}{z} + \frac{2-a}{1} + \frac{2}{z} + \frac{3-a}{1} + \frac{3}{z} + \cdots\right)$$

(9.6.17)

is valid for all z in the cut plane

$$R=[z:|\arg z|<\pi]$$

and all $a \in C$ such that $a \notin [1,2,3,\dots]$. By an equivalence transformation, (9.6.17) can also be written in the familiar form

$$\Gamma(a,z)=e^{-z}z^{a-1}\left(\frac{1}{1} + \frac{1-a}{z} + \frac{1}{1} + \frac{2-a}{z} + \frac{2}{1} + \frac{3-a}{z} + \frac{3}{1} + \cdots\right).$$

(9.6.18)

Example 2. Complementary error function erfc(z) [see (6.2.19)].

$$\begin{aligned}
\mathrm{erfc}(z) &= \frac{1}{\sqrt{\pi}}\Gamma\left(\tfrac{1}{2},z^{2}\right) \\
&= \frac{e^{-z^{2}}}{\sqrt{\pi}\,z}\left(\frac{1}{1} + \frac{\frac{1}{2}}{z^{2}} + \frac{1}{1} + \frac{\frac{3}{2}}{z^{2}} + \frac{2}{1} + \frac{\frac{5}{2}}{z^{2}} + \frac{3}{1} + \cdots\right) \\
&= \frac{e^{-z^{2}}}{\sqrt{\pi}}\left(\frac{1}{z} + \frac{\frac{1}{2}}{z} + \frac{1}{z} + \frac{\frac{3}{2}}{z} + \frac{2}{z} + \frac{\frac{5}{2}}{z} + \frac{3}{z} + \cdots\right) \\
&= \frac{ze^{-z^{2}}}{\sqrt{\pi}}\left(\frac{1}{z^{2}} + \frac{\frac{1}{2}}{1} + \frac{1}{z^{2}} + \frac{\frac{3}{2}}{1} + \frac{2}{z^{2}} + \frac{\frac{5}{2}}{1} + \frac{3}{z^{2}} + \cdots\right)
\end{aligned}$$

(9.6.19)

are all valid for $\mathrm{Re}(z)>0$ [see (6.2.25)]. The second and third continued fractions are obtained from the first by equivalence transformations.

Example 3. Gamma Function $\Gamma(z)$. We consider the logarithm of the gamma function,

$$\log\Gamma(z)=\left(z-\tfrac{1}{2}\right)\mathrm{Log}\,z-z+\tfrac{1}{2}\mathrm{Log}\,2\pi+J(z).$$

(9.6.20)

Here $z \in R=[z:|\arg z|<\pi]$, and Log z denotes the principal branch of the logarithm, which is real when z is real and positive. Binet's function $J(z)$ is holomorphic in R and tends to zero as $z \to \infty$ in such a way that the distance from z to the negative real axis grows without bound. The function $J(z)$ can be written as

$$J(z) = \frac{1}{\pi} \int_0^\infty \frac{z}{z^2+t^2} \operatorname{Log} \frac{1}{1-e^{2\pi t}} dt. \tag{9.6.21}$$

If $F(z)$ is defined by

$$F(z) = zJ(z^{1/2}), \tag{9.6.22}$$

then it can be shown that

$$F(z) = \int_0^\infty \frac{z \, d\psi(t)}{z+t}, \tag{9.6.23}$$

where

$$\psi(t) = \frac{1}{2\pi} \int_0^t \frac{1}{\sqrt{\tau}} \operatorname{Log} \frac{1}{1-e^{-2\pi\sqrt{\tau}}} d\tau, \tag{9.6.24}$$

and $\psi \in \Phi(0, \infty)$.

It can be further shown that

$$c_n = \int_0^\infty (-t)^n \, d\psi(t) = \frac{\beta_{2n+2}}{(2n+1)(2n+2)}, \qquad n = 0, 1, 2, \ldots, \tag{9.6.25}$$

where β_m denotes the mth Bernoulli number (see, for example, [Henrici, 1977, Section 11.1]). It follows from (9.6.25) and Theorem 9.1A that the series

$$c_0 + c_1 z^{-1} + c_2 z^{-2} + \cdots \tag{9.6.26}$$

is the asymptotic expansion for

$$F(z) = \int_0^\infty \frac{z \, d\psi(t)}{z+t} \tag{9.6.27}$$

at $z = \infty$, with respect to R. Moreover, ψ is a solution to the Stieltjes moment problem for the sequence $\{c_n\}$. Since

$$\beta_{2n} \sim \frac{(-1)^n 2(2n)!}{(2\pi)^{2n}} \qquad (n \to \infty), \tag{9.6.28}$$

it follows that Carleman's criterion (9.5.2) is satisfied, and hence the solution to the moment problem for $\{c_n\}$ is unique. Thus, by Theorems 7.2, 9.2A and 9.8A, there exists a modified S-fraction

$$\frac{a_1}{1} + \frac{a_2}{z} + \frac{a_3}{1} + \frac{a_4}{z} + \cdots, \qquad a_n > 0 \qquad (9.6.29)$$

which corresponds at $z = \infty$ to the series (9.6.26). By Theorem 4.58 (Stieltjes), the continued fraction converges to a function $G(z)$ holomorphic in $R = [z : |\arg z| < \pi]$. Theorem 9.5A implies that there exists a function $\varphi \in \Phi(0, \infty)$ such that

$$G(z) = \int_0^\infty \frac{z \, d\varphi(t)}{z+t} \qquad (9.6.30)$$

for all $z \in R$. By Theorem 9.6A

$$c_k = \int_0^\infty (-t)^k \, d\varphi(t), \qquad k = 0, 1, 2, \ldots, \qquad (9.6.31)$$

so that $\varphi(t)$ is a solution to the Stieltjes moment problem for $\{c_n\}$. Since the moment problem has a unique solution, we must have $\varphi = \psi$. Thus we have proved that there exists a modified S-fraction (9.6.29) which corresponds to the series (9.6.26) at $z = \infty$, and that this continued fraction converges to the function $F(z)$ for all $z \in R$. Therefore we obtain

$$J(\sqrt{z}\,) = \frac{1}{\sqrt{z}} \left(\frac{a_1}{1} + \frac{a_2}{z} + \frac{a_3}{1} + \frac{a_4}{z} + \cdots \right) \qquad (9.6.32)$$

for all $z \in R$. Replacing $\sqrt{z}$ by z in (9.6.32) and making an equivalence transformation, we obtain

$$J(z) = \frac{a_1}{z} + \frac{a_2}{z} + \frac{a_3}{z} + \frac{a_4}{z} + \cdots, \qquad a_n > 0, \qquad (9.6.33)$$

valid for all z such that $\mathrm{Re}(z) > 0$. By using the quotient-difference algorithm (Section 7.1.2), one can compute the coefficients a_n in (9.6.34). The first few a_n are given by

$$J(z) = \frac{\frac{1}{12}}{z} + \frac{\frac{1}{30}}{z} + \frac{\frac{53}{210}}{z} + \frac{\frac{195}{371}}{z} + \frac{\frac{22999}{22737}}{z} + \frac{\frac{29944523}{19733142}}{z} + \frac{\frac{109535241009}{48264275462}}{z} + \cdots.$$
$$(9.6.34)$$

Numerical experiments by Henrici and Pfluger [1966] indicate that the convergence of the continued fraction (9.6.34) is very slow. However, they point out that the first few approximants, up to $n = 7$, give relatively good approximations, due in part to the fact that the first few Bernoulli numbers are small.

CHAPTER 10

Numerical Stability in Evaluating Continued Fractions

It has been seen that continued-fraction expansions provide a useful means for representing and computing values of analytic functions. Some algorithms for computing the approximants of a continued fraction were described in Section 2.1.4. In the forward recurrence algorithm (FR algorithm) one uses the difference equations (2.1.6) to compute successively the nth numerators A_n and denominators B_n and then the nth approximant $f_n = A_n/B_n$. One difficulty with the FR algorithm is that, although the sequence $\{f_n\}$ may converge to a finite limit, the A_n and B_n may both tend to infinity or to zero, thus making it necessary to rescale from time to time, to prevent machine overflow or underflow.

Blanch in [1964] gave an analysis of rounding errors obtained from the computation of continued-fraction approximants which seems to indicate that the backward recurrence algorithm (BR algorithm) is numerically more stable than the FR algorithm. The problem had also been studied by Macon and Baskervill [1956]. Explicit upper bounds for the rounding error produced by the BR algorithm were given by Jones and Thron in [1974a, b]. Their results, which evolved from work included in [Blanch, 1964], are described in the present chapter.

The basic results on roundoff error in the BR algorithm are contained in Theorems 10.1 and 10.4. The main assumption about continued fractions $K(a_n/b_n)$ in Theorem 10.4 is that there exists a sequence $\{V_n\}$ of non-empty subsets of $\hat{C}$ such that, for all $n \geqslant 1$,

$$0 \in V_n$$

ENCYCLOPEDIA OF MATHEMATICS and Its Applications, Gian-Carlo Rota (ed.). Vol. 11: William B. Jones and W. J. Thron, Continued Fractions.   ISBN 0-201-13510-8

and

$$\frac{a_n}{b_n + V_n} \subseteq V_{n-1}.$$

This property was seen in Chapter 4 to play a fundamental role in the development of much of the known convergence theory of continued fractions. In Chapter 8 it was seen that it is also useful in determining truncation-error bounds. Therefore its occurrence here seems quite natural. In Corollary 10.2 the roundoff-error bounds are stated in terms of the number of decimal digits carried in the machine arithmetic operations.

The principal application given here is to the important class of S-fractions (Section 10.3). Several specific examples are included. From these results and examples, the reader should be able to apply the general theory to other classes of continued fractions.

10.1 General Estimates of Relative Roundoff Error

The *backward recurrence algorithm* (BR algorithm) for computing the nth approximant

$$f_n = \frac{a_1}{b_1} + \frac{a_2}{b_2} + \cdots + \frac{a_n}{b_n}, \qquad a_k, b_k \neq 0, \tag{10.1.1}$$

of a continued fraction $K(a_n / b_n)$ consists of setting

$$G_{n+1}^{(n)} = 0 \tag{10.1.2a}$$

and computing successively, from "tail to head,"

$$G_k^{(n)} = a_k + \left(b_k + G_{k+1}^{(n)} \right), \qquad k = n, n-1, \ldots, 1. \tag{10.1.2b}$$

We then have

$$f_n = G_1^{(n)} \tag{10.1.2c}$$

(see Section 2.1.4). In the following we let α_k and β_k denote the relative errors in the machine (rounded) values $\hat{a}_k$ and $\hat{b}_k$ of the elements a_k and b_k, respectively, so that

$$\hat{a}_k = a_k(1 + \alpha_k), \qquad \hat{b}_k = b_k(1 + \beta_k). \tag{10.1.3}$$

Further, let $\hat{G}_k^{(n)}$ denote the computed value of $G_k^{(n)}$, using machine numbers $\hat{a}_k, \hat{b}_k$ and machine operations for division and addition (denoted by $\oplus$ and $\oplus$, respectively). Then

$$\hat{G}_{n+1}^{(n)} = 0,$$
(10.1.4a)

and

$$\hat{G}_k^{(n)} = \hat{a}_k \oplus (\hat{b}_k \oplus \hat{G}_{k+1}^{(n)}), \qquad k = n, n-1, \ldots, 1.$$
(10.1.4b)

The number

$$\hat{f}_n = \hat{G}_1^{(n)}$$
(10.1.4c)

is the computed (approximate) value of $f_n = G_1^{(n)}$. If $\varepsilon_k^{(n)}$ denotes the relative error in $\hat{G}_k^{(n)}$, then $\varepsilon_{n+1}^{(n)} = 0$ and

$$\hat{G}_k^{(n)} = G_k^{(n)}(1 + \varepsilon_k^{(n)}), \qquad k = n, n-1, \ldots, 1.$$
(10.1.5)

It is also convenient to introduce the relative error $\gamma_k^{(n)}$ defined by

$$\hat{G}_k^{(n)} = \left[\hat{a}_k \div \left(\hat{b}_k + \hat{G}_{k+1}^{(n)} \right) \right](1 + \gamma_k^{(n)}), \qquad k = 1, 2, \ldots, n.$$
(10.1.6)

The expression on the right side of (10.1.6) involves machine numbers but exact arithmetic operations (no rounding). Combining (10.1.3), (10.1.5) and (10.1.6) with the definition

$$g_k^{(n)} = \frac{G_{k+1}^{(n)}}{b_k + G_{k+1}^{(n)}}, \qquad k = 1, 2, \ldots, n,$$
(10.1.7)

one obtains

$$\varepsilon_k^{(n)} = \frac{(1 + \alpha_k)(1 + \gamma_k^{(n)})}{1 + \beta_k + g_k^{(n)}(\varepsilon_{k+1}^{(n)} - \beta_k)} - 1,$$

or

$$\varepsilon_k^{(n)} = \frac{\alpha_k - \beta_k + \gamma_k^{(n)} + \alpha_k \gamma_k^{(n)} - g_k^{(n)}(\varepsilon_{k+1}^{(n)} - \beta_k)}{1 + \beta_k + g_k^{(n)}(\varepsilon_{k+1}^{(n)} - \beta_k)}, \qquad k = 1, 2, \ldots, n$$

(10.1.8)

We are interested in estimating the numbers $\varepsilon_k^{(n)}$, and particularly $\varepsilon_1^{(n)}$, which is the relative error in the machine approximation $\hat{f}_n = \hat{G}_1^{(n)}$. Such estimates are provided by the following theorem, a proof of which can be found in [Jones and Thron, 1974a, Theorem 3.1].

THEOREM 10.1. *For each* $k=1,2,\ldots,n$ *let* $\varepsilon_k^{(n)}$ *satisfy* (10.1.8) *with* $g_n^{(n)}=\varepsilon_{n+1}^{(n)}=0$. *Let* α, β, γ, η *and* ω *be non-negative numbers such that for* $k=1,2,\ldots,n$,

$$|\alpha_k|\leqslant\alpha\omega,\quad |\beta_k|\leqslant\beta\omega,\quad |\gamma_k^{(n)}|\leqslant\gamma\omega,\quad |g_k^{(n)}|\leqslant\eta, \qquad (10.1.9a)$$

where

$$\alpha=0\quad or\quad \alpha\geqslant1, \qquad (10.1.9b)$$

$$\beta=0\quad or\quad \beta\geqslant1, \qquad (10.1.9c)$$

$$\gamma\geqslant1,\quad \alpha+\beta+\gamma\geqslant2\quad and\quad \eta>0. \qquad (10.1.9d)$$

Then, for $k=1,2,\ldots,n$,

$$|\varepsilon_k^{(n)}|\leqslant\omega(1+\alpha+\beta+\gamma+\beta\eta)\sum_{j=0}^{n-k}\eta^j, \qquad (10.1.10)$$

provided that

$$0\leqslant\omega<\frac{1}{16(\alpha+\beta+\gamma)^2}, \qquad (10.1.11a)$$

and

$$0\leqslant\omega<\frac{2}{\left(1+\beta+\beta\eta+\eta(1+\alpha+\beta+\gamma+\beta\eta)\sum_{j=0}^{n-k}\eta^j\right)^2}. \qquad (10.1.11b)$$

Remarks.

1. Typically one will have

$$\omega=\left(\tfrac{1}{2}\right)10^{-\nu+1}, \qquad (10.1.12)$$

where ν is the number of significant decimal digits carried in the machine computations. ν is sometimes called the *machine constant*. The parameter α (or β) will be zero if $\hat{a}_k=a_k$ (or $\hat{b}_k=b_k$) for $k=1,2,\ldots,n$. If $\hat{a}_k$ (or $\hat{b}_k$) is correctly rounded in the last machine decimal digit, then one can take $\alpha=1$ (or $\beta=1$). Normally one can take $\gamma=2$.

2. The principal difficulty in applying Theorem 10.1 is finding a suitable estimate η of the quantities $|g_k^{(n)}|$. Methods for obtaining such estimates are given in Sections 10.2 and 10.3.

The following corollary of Theorem 10.1 gives estimates of $|\varepsilon_1^{(n)}|$ for the important special case in which we have $\alpha=\beta=1$, $\gamma=2$ and $0<\eta\leqslant 1$. The estimates are expressed in terms of the parameter η and the machine constant ν.

COROLLARY 10.2. *In addition to the hypotheses of Theorem 10.1, we assume that*

$$\alpha=\beta=1, \qquad \gamma=2 \quad and \quad \omega=\left(\tfrac{1}{2}\right)10^{-\nu+1}. \tag{10.1.13}$$

Then, for all ν and n such that

$$\nu>4 \quad and \quad n\leqslant 10^{(\nu-2)/2}, \tag{10.1.14}$$

we have

$$|\varepsilon_1^{(n)}|<3\left(\frac{1-\eta^n}{1-\eta}\right)10^{-\nu+1}<\frac{3\times10^{-\nu+1}}{1-\eta} \qquad if \quad 0<\eta<1,$$
$$\tag{10.1.15a}$$

and

$$|\varepsilon_1^{(n)}|<3n\times10^{-\nu+1} \qquad if \quad \eta=1. \tag{10.1.15b}$$

We note that when the conditions of Corollary 10.2 are satisfied, we will have

$$\varepsilon_1^{(n)}=O(1) \qquad if \quad 0<\eta<1 \tag{10.1.16a}$$

and

$$\varepsilon_1^{(n)}=O(n) \qquad if \quad \eta=1. \tag{10.1.16b}$$

The first case insures numerical stability of the BR algorithm, and the second shows that it is almost stable. The conditions (10.1.14) insure (10.1.11). In most of the digital computers presently in use one has a machine constant $\nu>10$. Hence the restriction (10.1.14) will not normally impose a serious problem.

10.2 Methods for Estimating $g_k^{(n)}$

In Theorem 10.4 we shall give two methods for estimating the $g_k^{(n)}$. First, however, we state a result (Theorem 10.3) that the $g_k^{(n)}$ are invariant under equivalence transformations of the continued fraction. The significance of this property is that it shows that there is no need to search for an optimal form of a continued fraction for the purpose of minimizing the $g_k^{(n)}$.

THEOREM 10.3. *Let* $K(a_n/b_n)$ *and* $K(a_n^*/b_n^*)$ *be equivalent continued fractions, so that there exists a sequence of non-zero complex constants* $\{r_n\}$ *satisfying, for* $n = 1,2,3,\ldots,$

$$a_n = r_n r_{n-1} a_n^* \qquad (r_0 = 1), \tag{10.2.1a}$$

$$b_n = r_n b_n^*. \tag{10.2.1b}$$

For each $n = 1,2,3,\ldots$ *and* $k = 1,2,\ldots, n,$ *let*

$$g_k^{(n)} = \frac{G_{k+1}^{(n)}}{b_k + G_{k+1}^{(n)}} \quad and \quad g_k^{(n)*} = \frac{G_{k+1}^{(n)*}}{b_k^* + G_{k+1}^{(n)*}}, \tag{10.2.3}$$

where

$$G_{n+1}^{(n)} = 0, \qquad G_k^{(n)} = \frac{a_k}{b_k + G_{k+1}^{(n)}}, \qquad k = n, n-1, \ldots, 1, \tag{10.2.3a}$$

$$G_{n+1}^{(n)*} = 0, \qquad G_k^{(n)*} = \frac{a_k^*}{b_k^* + G_{k+1}^{(n)*}}, \qquad k = n, n-1, \ldots, 1. \tag{10.2.3b}$$

Then, for each $n = 1,2,3,\ldots$ *and* $k = 1,2,\ldots, n,$

$$G_k^{(n)} = r_{k-1} G_k^{(n)*} \tag{10.2.4a}$$

and

$$g_k^{(n)} = g_k^{(n)*}. \tag{10.2.4b}$$

In Chapter 4 it was shown that many of the convergence theorems for continued fractions $K(a_n/b_n)$ are based on properties of the form

$$s_n(V_n) \subseteq V_{n-1}, \tag{10.2.5}$$

where $s_n(w) = a_n/(b_n + w)$ and $\{V_n\}$ is a sequence of subsets of the extended plane $\hat{C}$. It will now be seen that (10.2.5) also plays a basic role in determining estimates of the $g_k^{(n)}$.

THEOREM 10.4. *Let* $n \geqslant 2,$ *let*

$$f_n = \frac{a_1}{b_1} + \frac{a_2}{b_2} + \cdots + \frac{a_n}{b_n} \tag{10.2.6}$$

be the n th approximant of a continued fraction $K(a_n/b_n)$, and let $V_1, V_2, \ldots, V_n$ be non-empty subsets of $\hat{C}$ such that

$$0 \in V_n \tag{10.2.7}$$

and

$$s_k(V_k) = \frac{a_k}{b_k + V_k} \subseteq V_{k-1}, \qquad k = 2, 3, \ldots, n. \tag{10.2.8}$$

Further, let

$$A^{(n)} = \max[|a_k| : k = 2, 3, \ldots, n], \tag{10.2.9a}$$

$$\delta^{(n)} = \min[d(-b_k, V_k) : k = 1, 2, \ldots, n], \tag{10.2.9b}$$

and

$$M^{(n)} = \max\left[|w| : w \in \frac{V_k}{b_k + V_k}, k = 1, 2, \ldots, n\right]. \tag{10.2.10}$$

Here $d(-b_k, V_k)$ denotes the distance from the point $-b_k$ to the set V_k. Then:

(A)

$$G_k^{(n)} \in V_{k-1}, \qquad k = 2, 3, \ldots, n, n+1, \tag{10.2.11}$$

and

$$|b_k + G_{k+1}^{(n)}| \geq d(-b_k, V_k). \tag{10.2.12}$$

Here $G_k^{(n)}$ is defined by (10.1.2b).

(B) If $g_k^{(n)}$ is defined by (10.1.7), then

$$|g_k^{(n)}| \leq \frac{A^{(n)}}{(\delta^{(n)})^2}, \qquad k = 1, 2, \ldots, n. \tag{10.2.13}$$

(C)

$$|g_k^{(n)}| \leq M^{(n)}, \qquad k = 1, 2, \ldots, n. \tag{10.2.14}$$

Proof. (A): The proof of (10.2.11) is by a backward induction on k, beginning with $k = n+1$. By (10.2.7) and (10.1.2a) we have $G_{n+1}^{(n)} = 0 \in V_n$. Now we assume that $G_{k+1}^{(n)} \in V_k$ for some k with $1 \leq k \leq n$. Then by (10.2.8), we have

$$G_k^{(n)} = s_k(G_{k+1}^{(n)}) \in s_k(V_k) \subseteq V_{k-1}. \tag{10.2.15}$$

This proves (10.2.11), which implies (10.2.12).
  (B): By (10.1.2) and (10.1.7) we obtain

$$|g_k^{(n)}| = \left| \frac{G_{k+1}^{(n)}}{b_k + G_{k+1}^{(n)}} \right| = \left| \frac{a_{k+1}}{(b_k + G_{k+1}^{(n)})(b_{k+1} + G_{k+2}^{(n)})} \right|, \qquad k = 1, 2, \ldots, n-1.$$
$$\tag{10.2.16}$$

Hence (10.2.9) and (10.2.12) imply (10.2.13).
  (C): is an immediate consequence of (10.2.10) and (10.2.11). ∎

10.3 Applications

  We shall illustrate the uses of Theorem 10.4 by applying it to S-fractions. Another application to the class of all continued fractions, dealt with in Theorem 4.43 (multiple parabola theorem), was given in [Jones and Thron, 1974a]. These examples should be helpful to the reader in making other applications of Theorem 10.4.

  THEOREM 10.5. *Let*

$$f_n = \frac{a_1 z}{1} + \frac{a_2 z}{1} + \cdots + \frac{a_n z}{1} \tag{10.3.1}$$

be the nth approximant of an S-fraction $K(a_n z / 1)$ *such that*

$$0 < a_k \leqslant A(n), \qquad k = 1, 2, \ldots, n, \tag{10.3.2a}$$

$$z = r e^{i\theta}, \qquad r > 0, \quad |\theta| < \pi. \tag{10.3.2b}$$

Also let

$$G_{n+1}^{(n)} = 0, \qquad G_k^{(n)} = \frac{a_k z}{1} + \cdots + \frac{a_n z}{1}, \quad k = 1, 2, \ldots, n, \tag{10.3.3a}$$

and

$$g_k^{(n)} = \frac{G_{k+1}^{(n)}}{1 + G_{k+1}^{(n)}}, \qquad k = 1, 2, \ldots, n. \tag{10.3.3b}$$

Then for $k = 1, 2, \ldots, n$:

(A)

$$|g_k^{(n)}| \leqslant \frac{rA(n)}{\left[1 + 2rA(n)\cos\theta + r^2 A(n)^2\right]^{1/2}} < 1 \quad \textit{if} \quad |\theta| < \frac{\pi}{2}.$$

$$(10.3.4)$$

(B) *If* $\pi/2 < |\theta| < \pi$, *then*

$$|g_k^{(n)}| \leqslant \frac{rA(n)}{1 + 2rA(n)\cos\theta + r^2 A(n)^2} \qquad (10.3.5a)$$

provided $rA(n) < \cos(\pi - \theta)$,

$$|g_k^{(n)}| \leqslant \frac{rA(n)}{\left[1 + 2rA(n)\cos\theta + r^2 A(n)^2\right]^{1/2}} \qquad (10.3.5b)$$

provided $\cos(\pi - \theta) \leqslant rA(n) \leqslant \sec(\pi - \theta)$, *and*

$$|g_k^{(n)}| \leqslant rA(n)\csc^2\theta \qquad (10.3.5c)$$

provided $\cos(\pi - \theta) \leqslant rA(n)$.

We shall prove Theorem 10.5 in order to illustrate the method involved. Our proof makes use of the following two lemmas.

LEMMA 10.6. *If* f_n *and* $G_k^{(n)}$ *are defined as in Theorem* 10.5, *then*

$$\frac{a_k z}{1 + V} \subseteq V, \qquad k = 1, 2, \ldots, n, \qquad (10.3.6)$$

where $V = V(A(n), r, \theta)$ *is the convex lens-shaped region* (Figure 10.3.1), *with interior angle* $|\theta|$, *bounded by the ray issuing from the origin in the direction* θ *and the circular arc starting at the origin, tangent to the real axis, and extending to the point* $rA(n)e^{i\theta}$. *Moreover*,

$$G_k^{(n)} \in V, \qquad k = 1, 2, \ldots, n. \qquad (10.3.7)$$

Proof. The region V has vertices at the two points 0 and $rA(n)e^{i\theta}$. Hence $1 + V$ has its vertices at 1 and $1 + rA(n)e^{i\theta}$, and $1/(1 + V)$ is a lens-shaped region with vertices at 1 and $1/[1 + rA(n)e^{i\theta}]$. The region $1/(1 + V)$ is contained in the lens-shaped region X which is bounded by the real axis and the circular arc passing through 0 (at an angle $-\theta$ with respect to the

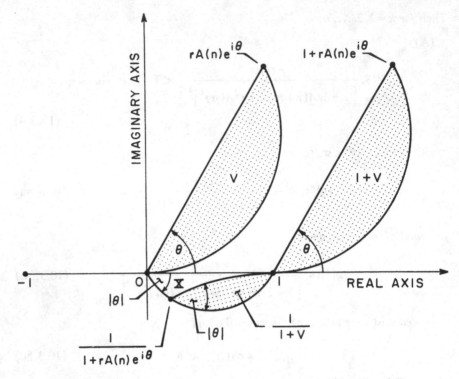

Figure 10.3.1. Schematic diagram of regions, V, $1+V$, $1/(1+V)$ and X.

real axis) and passing through 1. The point $1/[1+rA(n)e^{i\theta}]$ is located on this circular arc which bounds X (see Figure 10.3.1). Thus (10.3.2) implies that

$$\frac{a_k z}{1+V} \subseteq a_k z X \subseteq V \quad \text{for} \quad k = 1, 2, \ldots, n,$$

which proves (10.3.6). Since $0 \in V$ and (10.3.6) holds, (10.2.11) in Theorem 10.4 can be applied to prove (10.3.7). ∎

LEMMA 10.7. *Let* f_n, a_k, $z = re^{i\theta}$, $G_k^{(n)}$ *and* $g_k^{(n)}$ *be defined as they are in Theorem* 10.5. *Let*

$$W = \frac{V}{1+V}, \tag{10.3.8}$$

where V *is the lens-shaped region defined in Lemma* 10.6. *Then* W *is the convex lens-shaped region with the same interior angle* $|\theta|$ *as* V, *with vertices at* 0 *and at*

$$w_0 = \frac{rA(n)e^{i\theta}}{1+rA(n)e^{i\theta}}, \tag{10.3.9}$$

such that one of its bounding circular arcs is tangent to the real axis at 0 (*see Figure* 10.3.2). *Moreover*

$$g_k^{(n)} \in W, \qquad k=1,2,\ldots,n. \tag{10.3.10}$$

Proof. Since V is a convex lens-shaped region, $1+V$ is also. Clearly $1/(1+V)$ is a lens-shaped region with the same angular opening $|\theta|$ as V. That it is also convex can be seen from the fact that $1+V$ passes through 1 and that its bounding circular arc is tangent to the real axis at 1. It follows that

$$W = \frac{V}{1+V} = 1 - \frac{1}{1+V}$$

is also a convex lens-shaped region with the same interior angle $|\theta|$ as V.

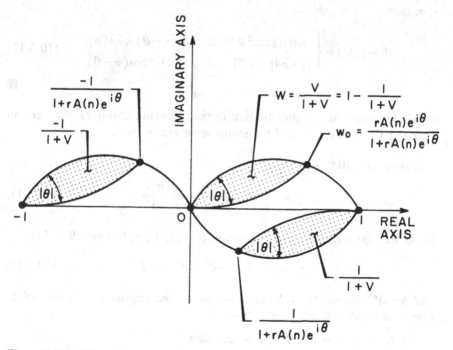

Figure 10.3.2. Schematic diagram of regions $1/(1+V)$, $-1/(1+V)$ and $W = V/(1+V)$.

That 0 and w_0 are the vertices of W can be shown by a simple calculation. The statement (10.3.10) follows from (10.3.7) and (10.3.8).   ∎

Proof of Theorem 10.5.   Statement (A) can be verified from the geometry of the region W described in Lemma 10.7 (see Figure 10.3.2), the fact that

$$|w_0| = \frac{rA(n)}{\sqrt{1 + 2rA(n)\cos\theta + r^2A(n)^2}},$$ (10.3.11)

and Theorem 10.4(C). A proof of (10.3.5b) can also be made from (10.3.10), Theorem 10.4(C) and the following argument: When the convex lens-shaped region W has an interior angle $|\theta|$ greater than $\pi/2$, then the distance $|w_0|$ between the vertices of W can be smaller than the diameter of W. This will, in fact, be the case precisely when one of the angles which the straight line passing through 0 and w_0 makes with one of the bounding arcs of W exceeds $\pi/2$. By a simple geometric argument, one can show that one of the angles discussed above will exceed $\pi/2$ iff either $rA(n) > \sec(\pi - \theta)$ or $rA(n) < \cos(\pi - \theta)$. This proves (10.3.5b). Proofs of (10.3.5a) and (10.3.5c) can be made using Theorem 10.4(B) and the following properties of the region V of Lemma 10.6 (see Figure 10.3.1): If $\pi/2 < |\theta| < \pi$, then

$$d(-1, V) = \begin{cases} rA(n)\csc^2\theta & \text{if } \cos(\pi - \theta) < rA(n), \\ |1 + rA(n)e^{i\theta}| & \text{if } rA(n) < \cos(\pi - \theta). \end{cases}$$ (10.3.12)

∎

We conclude this chapter by noting the following corollary of Theorem 10.5 and Corollary 10.2 and by giving some examples.

COROLLARY 10.8.   *Let*

$$f_n = \frac{a_1 z}{1} + \frac{a_2 z}{1} + \cdots + \frac{a_n z}{1}$$ (10.3.13)

be the n th approximant of an S-fraction $K(a_n z/1)$, $a_k > 0$ *for all n. Let*

$$z = re^{i\theta}, \qquad r > 0, \qquad |\theta| \leqslant \pi/2,$$ (10.3.14)

and let $\varepsilon_1^{(n)}$ *denote the relative error in* $\hat{f}_n$, *the approximate value of* f_n, *computed by the BR algorithm. Then:*

(A) *If* $\{a_k\}_1^\infty$ *is a bounded sequence, then*

$$\varepsilon_1^{(n)} = O(1).$$ (10.3.15)

(B) *If* $\{a_k\}_1^\infty$ *is an unbounded sequence, then*

$$\varepsilon_1^{(n)} = O(n). \tag{10.3.16}$$

The results of Corollary 10.8 and of the following examples are, of course, subject to the restrictions (10.1.14) of Corollary 10.2. Many examples of analytic functions represented by S-fractions have been given in previous chapters.

Example 1. Arctangent. From (6.1.14)

$$z \arctan z = \frac{z^2}{1} + \frac{1^2 z^2}{3} + \frac{2^2 z^2}{5} + \frac{3^2 z^2}{7} + \cdots \tag{10.3.17}$$

is valid for all z in the cut plane with cuts along the imaginary axis from $+i$ to $+i\infty$ and from $-i$ to $-i\infty$. It follows from Corollary 10.8 that the relative error $\varepsilon_1^{(n)}$ in computing the nth approximant of (10.3.17) by the BR algorithm is such that

$$\varepsilon_1^{(n)} = O(1) \qquad \text{for} \quad |\arg z| \leqslant \pi/4. \tag{10.3.18}$$

Thus the BR algorithm is stable for this range of z.

Example 2. Bessel functions. From (6.1.53) we have, for $m > -1$,

$$\frac{-z J_{m+1}(z)}{J_m(z)} = \frac{-z^2}{2(m+1)} + \frac{-z^2}{2(m+2)} + \frac{-z^2}{2(m+3)} + \cdots, \tag{10.3.19}$$

which is valid for all $z \in \mathbb{C}$. The continued fraction in (10.3.19) is equivalent to an S-fraction $\mathbf{K}(a_n w/1)$ where $w = -z^2$ and where $\{a_n\}$ is a bounded sequence of positive numbers. Hence by Corollary 10.8, the relative error $\varepsilon_1^{(n)}$, in computing the nth approximant of (10.3.19) by the BR algorithm, satisfies

$$\varepsilon_1^{(n)} = O(1) \qquad \text{for} \frac{\pi}{4} \leqslant \arg z \leqslant \frac{3\pi}{4}$$

$$\text{and for} \qquad -\frac{3\pi}{4} \leqslant \arg z \leqslant -\frac{\pi}{4}. \tag{10.3.20}$$

This insures the numerical stability of the BR algorithm.

Example 3. Complementary error function. By (6.2.25)

$$\sqrt{\pi}\, e^{z^2} \operatorname{erfc} z = \frac{1}{z} + \frac{\frac{1}{2}}{z} + \frac{1}{z} + \frac{\frac{3}{2}}{z} + \frac{2}{z} + \frac{\frac{5}{2}}{z} + \frac{3}{z} + \cdots \tag{10.3.21}$$

is valid for $\text{Re}(z) > 0$. By making an equivalence transformation we see that

$$\frac{\sqrt{\pi}\, e^{z^2}}{z} \operatorname{erfc} z = \frac{1(1/z^2)}{1} + \frac{\frac{1}{2}(1/z^2)}{1} + \frac{1(1/z^2)}{1}$$
$$+ \frac{\frac{3}{2}(1/z^2)}{1} + \frac{2(1/z^2)}{1} + \frac{\frac{5}{2}(1/z^2)}{1} + \cdots \quad (10.3.22)$$

is also valid for all z such that $\text{Re}(z) > 0$. With $w = 1/z^2$, (10.3.22) is an S-fraction $K(a_n w/1)$ for which $\{a_k\}_1^\infty$ is an unbounded sequence of positive numbers. Hence by Corollary 10.8, the relative error $\varepsilon_1^{(n)}$, in computing the nth approximant of (10.3.22) by the BR algorithm, satisfies

$$\varepsilon_1^{(n)} = O(n) \qquad \text{for} \quad |\arg z| < \pi/4. \qquad (10.3.23)$$

Thus for this case we are assured that the BR algorithm is almost stable.

CHAPTER 11

Application of Continued Fractions to Birth-Death Processes

We consider a birth-death process in which a population of size m at time $t=0$ is changing because of birth (or immigration) at a rate λ_r and death (or emigration) at a rate μ_r when the population has size r. Each individual is assumed to have a probability $\lambda_r \Delta t + O((\Delta t)^2)$ of producing a new individual and a probability $\mu_r \Delta t + O((\Delta t)^2)$ of dying during a short interval of time $(t, t+\Delta t)$. The rates λ_r and μ_r are positive functions of r, independent of t. For an extensive treatment of birth-death processes the reader can refer to [Karlin, 1969]. Since birth-death processes are known to be connected to the Stieltjes moment problem [Karlin and McGregor, 1957], it was not surprising that Murphy and O'Donohoe [1975] found that continued fractions are also closely related. In particular, they obtained a continued fraction representing the Laplace transform of $p_r(t)$, the probability of having a population of size r at time t. They also discussed computational procedures and numerical examples. Their investigation was extended by Jones and Magnus [1977], who employed the Henrici-Pfluger truncation-error bounds for S-fractions (Theorem 8.10) to establish convergence of the approximations to $p_r(t)$. The role of continued fractions in birth-death processes is described in Section 11.1. In Section 11.2 we give some procedures for computing the $p_r(t)$ based on the quotient-difference algorithm and discuss a numerical example.

11.1 Birth-Death Processes

In a birth-death process such as that described above, the differential-difference equations that govern the growth of the population are given by

$$p_0'(t) = -\lambda_0 p_0(t) + \mu_0 p_1(t), \tag{11.1.1a}$$

$$p_r'(t) = \lambda_{r-1} p_{r-1}(t) - (\lambda_r + \mu_r) p_r(t) + \mu_{r+1} p_{r+1}(t), \quad r = 1, 2, 3, \ldots . \tag{11.1.b}$$

ENCYCLOPEDIA OF MATHEMATICS and Its Applications, Gian-Carlo Rota (ed.). Vol. 11: William B. Jones and W. J. Thron, Continued Fractions.   ISBN 0-201-13510-8

To solve these equations it is useful to introduce the Laplace transform

$$P_r(s) = \mathcal{L}\{p_r(t)\} = \int_0^\infty e^{-st} p_r(t)\, dt \qquad (11.1.2)$$

and consider the transformed system of second-order linear difference equations

$$\mu_1 P_1(s) = (\lambda_0 + s)P_0(s) - \delta_{0,m}, \qquad (11.1.3a)$$

$$\mu_{r+1} P_{r+1}(s) = -\lambda_{r-1} P_{r-1}(s) + (\lambda_r + \mu_r + s)P_r(s) - \delta_{r,m}, \qquad r = 1,2,3,\ldots, \qquad (11.1.3b)$$

where $\delta_{r,m}$ denotes the Kronecker delta. Introducing the notation

$$L_r = \lambda_0 \lambda_1 \cdots \lambda_r, \qquad r = 0,1,2,\ldots, \qquad (11.1.4a)$$

$$M_r = \mu_1 \mu_2 \cdots \mu_r, \qquad r = 1,2,3,\ldots, \qquad (11.1.4b)$$

and normalizing the above equations by

$$f_0^{(m)}(s) = P_0(s), \qquad (11.1.5a)$$

$$f_r^{(m)}(s) = (-1)^r M_r P_r(s), \qquad r = 1,2,3,\ldots, \qquad (11.1.5b)$$

leads to the system of equations

$$f_1^{(m)}(s) = \delta_{0,m} - (\lambda_0 + s)f_0^{(m)}(s), \qquad (11.1.6a)$$

$$f_{r+1}^{(m)}(s) = -\lambda_{r-1}\mu_r f_{r-1}^{(m)}(s) - (\lambda_r + \mu_r + s)f_r^{(m)}(s) + (-1)^r M_r \delta_{r,m}, \qquad r = 1,2,3,\ldots. \qquad (11.1.6b)$$

We note that here (m) denotes a superscript and not a derivative. Setting $m=0$ and $f_r^{(0)} = f_r(s)$, we deduce that

$$f_0(s) = \cfrac{1}{\lambda_0 + s + f_1(s)/f_0(s)}$$

$$= \cfrac{1}{\lambda_0 + s} - \cfrac{\lambda_0 \mu_1}{\lambda_1 + \mu_1 + s} - \cdots - \cfrac{\lambda_{r-1}\mu_r}{\lambda_r + \mu_r + s + f_{r+1}(s)/f_r(s)},$$
$$r = 1,2,3,\ldots, \qquad (11.1.7)$$

which leads us to consider the real J-fraction

$$\cfrac{1}{\lambda_0 + s} - \cfrac{\lambda_0 \mu_1}{\lambda_1 + \mu_1 + s} - \cfrac{\lambda_1 \mu_2}{\lambda_2 + \mu_2 + s} - \cdots. \qquad (11.1.8)$$

We shall let $A_n(s)$ and $B_n(s)$ denote the nth numerator and denominator

of (11.1.8), respectively. It is easily seen (see Section 2.4.2) that the
J-fraction (11.1.8) is the even part of the continued fraction

$$\frac{1}{s} + \frac{\lambda_0}{1} + \frac{\mu_1}{s} + \frac{\lambda_1}{1} + \frac{\mu_2}{s} + \frac{\lambda_2}{1} + \cdots, \tag{11.1.9}$$

which is equivalent to the S-fraction

$$\frac{z}{1} + \frac{\lambda_0 z}{1} + \frac{\mu_1 z}{1} + \frac{\lambda_1 z}{1} + \frac{\mu_2 z}{1} + \frac{\lambda_2 z}{1} + \cdots, \tag{11.1.10}$$

where $z = 1/s$ [see (8.3.41)]. By Theorem 4.58 (Stieltjes), the even and odd
parts of (11.1.9) converge to functions holomorphic in the cut plane

$$R = [s : |\arg s| < \pi] \tag{11.1.11}$$

and the convergence is uniform on every compact subset of R. Moreover,
the continued fraction (11.1.9) itself converges to a function holomorphic
in R iff at least one of the two series

$$\sum_{r=1}^{\infty} \frac{\mu_1 \mu_2 \cdots \mu_r}{\lambda_1 \lambda_2 \cdots \lambda_r}, \quad \sum_{r=0}^{\infty} \frac{\lambda_0 \lambda_1 \cdots \lambda_r}{\mu_1 \mu_2 \cdots \mu_{r+1}} \tag{11.1.12}$$

diverges. It follows that the J-fraction (11.1.8) is convergent for all $s \in R$.
Subsequently we shall assume that at least one of the series (11.1.12)
diverges.

We denote the value of (11.1.8) by f_0 and define $f_1, f_2, f_3, \ldots$ recursively
by (11.1.6) with $m = 0$ and with $f_r^{(0)}$ replaced by f_r. We thus obtain a
solution of the system (11.1.6) with $m = 0$ and note that two consecutive f_r's
cannot equal zero. Hence, for all r, $f_{r+1}/f_r \in \hat{C}$. For each $r > 0$, the following two continued fractions are seen to converge to functions meromorphic
in R, namely,

$$\frac{f_{r+1}}{f_r} = -\frac{\lambda_r \mu_{r+1}}{\lambda_{r+1} + \mu_{r+1} + s} - \frac{\lambda_{r+1} \mu_{r+2}}{\lambda_{r+2} + \mu_{r+2} + s} - \cdots \tag{11.1.13}$$

and

$$f_r = \frac{(-1)^r L_{r-1} M_r / B_r}{B_{r+1}/B_r} - \frac{\lambda_r \mu_{r+1}}{\lambda_{r+1} + \mu_{r+1} + s} - \frac{\lambda_{r+1} \mu_{r+2}}{\lambda_{r+2} + \mu_{r+2} + s} - \cdots . \tag{11.1.14}$$

The latter continued fraction is equivalent to

$$\frac{(-1)^r L_{r-1} M_r}{B_{r+1}} - \frac{\lambda_r \mu_{r+1} B_r}{\lambda_{r+1} + \mu_{r+1} + s} - \frac{\lambda_{r+1} \mu_{r+2}}{\lambda_{r+2} + \mu_{r+2} + s} - \cdots, \tag{11.1.15}$$

which has an nth denominator $B_n^{(r)}$ given by

$$B_n^{(r)} = B_{r+n}.$$ (11.1.16)

If $m > 0$ and $r = 1, 2, \ldots, m$, then an induction on r shows that

$$B_r f_{r-1}^{(m)} + B_{r-1} f_r^{(m)} = 0$$ (11.1.17)

for all solutions of (11.1.6). In particular, for $r = m$, we have

$$B_m f_{m-1}^{(m)} - B_{m-1} f_m^{(m)}, \; = 0$$ (11.1.18)

which, when inserted in (11.1.6) with $r = m$, gives

$$f_{m+1}^{(m)} = (-1)^m M_m - \frac{B_{m+1}}{B_m} f_m^{(m)}.$$ (11.1.19)

Equations (11.1.19) and (11.1.6) with $r = m+1, m+2, \ldots$ together form a system of equations analogous to (11.1.6) with $m = 0$. This leads us to a convergent continued fraction similar to (11.1.8), whose possibly infinite value we denote by $f_m^{(m)}$, namely

$$f_m^{(m)} = \frac{(-1)^m M_m}{B_{m+1}/B_m} - \frac{\lambda_m \mu_{m+1}}{\lambda_{m+1} + \mu_{m+1} + s} - \frac{\lambda_{m+1} \mu_{m+2}}{\lambda_{m+2} + \mu_{m+2} + s} - \cdots .$$ (11.1.20)

From the theory of S-fractions (see Theorem 9.3A), the zeros of the B_r's are distinct and all lie on the negative real axis. Therefore (11.1.6) and (11.1.18) determine the other $f_r^{(m)}$ in terms of $f_m^{(m)}$, and this gives a solution to the system (11.1.6). By comparing (11.1.14) and (11.1.20), applying (11.1.18) repeatedly and employing (11.1.6), we find that

$$f_r^{(m)} = (-1)^{m-r} \frac{B_r}{L_{m-1}} f_m \qquad \text{if} \quad r = 0, 1, \ldots, m,$$ (11.1.21a)

and

$$f_r^{(m)} = \frac{B_m}{L_{m-1}} f_r \qquad \text{if} \quad r = m, m+1, m+2, \cdots .$$ (11.1.21b)

Equations (11.1.5), (11.1.14) and (11.1.21) lead to continued-fraction representations of the $P_r(s)$ given by the following.

THEOREM 11.1. *Let* $\{\lambda_r\}_0^\infty$ *and* $\{\mu_r\}_1^\infty$ *be sequences of positive numbers, and for each* $k = 0, 1, 2, \ldots,$ *let* $B_k(s)$ *denote the kth denominator of the real*

J-fraction (11.1.8). *Let* $\{P_r(s)\}$ *be defined as follows: for* $r = 0, 1, 2, \ldots, m$,

$$P_r(s) = \frac{M_m}{M_r} \frac{B_r}{B_m} \left(\frac{1}{B_{m+1}/B_m} - \frac{\lambda_m \mu_{m+1}}{\lambda_{m+1} + \mu_{m+1} + s} - \frac{\lambda_{m+1} \mu_{m+2}}{\lambda_{m+2} + \mu_{m+2} + s} - \cdots \right),$$

(11.1.22a)

and for $r = m, m+1, m+2, \ldots,$

$$P_r(s) = \frac{L_{r-1}}{L_{m-1}} \frac{B_m}{B_r} \left(\frac{1}{B_{r+1}/B_r} - \frac{\lambda_r \mu_{r+1}}{\lambda_{r+1} + \mu_{r+1} + s} - \frac{\lambda_{r+1} \mu_{r+2}}{\lambda_{r+2} + \mu_{r+2} + s} - \cdots \right).$$

(11.1.22b)

Here the L_r *and* M_r *are defined by* (11.1.4). *Then each continued fraction in* (11.1.22) *converges to a function* $P_r(s)$ *meromorphic in the cut plane* $R = [s : |\arg s| < \pi]$, *and* $\{P_r(s)\}$ *is a solution of the system of difference equations* (11.1.3).

Using the inverse Laplace transform $\mathcal{L}^{-1}$, we define $\{p_r(t)\}$ by

$$p_r(t) = \mathcal{L}^{-1}\{P_r(s)\} = \frac{1}{2\pi} \int_{-\infty}^{\infty} e^{st} P_r(s) \, d\omega, \qquad r = 0, 1, 2, \ldots, \quad (11.1.23)$$

where $s = c + i\omega$, $c > 0$, and where $\{P_r(s)\}$, defined by (11.1.22), is a solution of the difference equations (11.1.3). It follows that $\{p_r(t)\}$ is a solution of the system of differential-difference equations (11.1.1). Our main interest here is in using the continued fractions (11.1.22) to compute the $p_r(t)$. For this purpose we let $P_{r,n}(s)$ denote the nth approximant of the continued fraction in (11.1.22) and let

$$p_{r,n}(t) = \mathcal{L}^{-1}\{P_{r,n}(s)\} = \frac{1}{2\pi} \int_{-\infty}^{\infty} e^{st} P_{r,n}(s) \, d\omega, \qquad (11.1.24)$$

where $s = c + i\omega$, $c > 0$. By using the fact that the real J-fraction (11.1.8) is the even part of (11.1.9), which is equivalent to the S-fraction, and then employing the Henrici-Pfluger estimates of the truncation error of a convergent S-fraction (Theorem 8.10), Jones and Magnus [1977] were able to prove the following result.

THEOREM 11.2. *Let* $\{\lambda_r\}_0^\infty$ *and* $\{\mu_r\}_1^\infty$ *be sequences of positive integers such that at least one of the series* (11.1.12) *is divergent. For each* $r \geq 0$ *and* $n \geq 1$, *let* $P_{r,n}(s)$ *denote the* nth *approximant of the continued fraction for* $P_r(s)$ *in*

Theorem 11.1 and let $p_{r,n}(t)$ and $p_r(t)$ be defined by (11.1.23) and (11.1.24), respectively. Then:

(A) *For each* $r = m, m+1, m+2, \ldots$, *there exists a function* $K(t)$, *independent of* n, *such that*

$$|p_r(t) - p_{r,n}(t)|^2 < K(t) \prod_{k=1}^{n-1} \frac{\left(\sqrt{\lambda_{r+k}} + 2\sqrt{c}\right)^{1/x_n} \left(\sqrt{\mu_{r+k}} + 2\sqrt{c}\right)^{1/x_n}}{\left(1 + 2\sqrt{c}/\sqrt{\lambda_{r+k}}\right)\left(1 + 2\sqrt{c}/\sqrt{\mu_{r+k}}\right)},$$

(11.1.25)

where $x_n = r - m + \frac{7}{4} + n/2$. *A similar result holds for* $r = 0, 1, \ldots, m$.

(B) *If there exists a constant* α *such that*

$$0 < \lambda_r \leq \alpha r \text{ and } 0 < \mu_r \leq \alpha r \quad \text{for} \quad r = 1, 2, 3, \ldots,$$

(11.1.26)

then

$$\lim_{n \to \infty} p_{r,n}(t) = p_r(t).$$

(11.1.27)

Remarks.

(1) The condition (11.1.26) will hold if any one of the following is satisfied:

$\{\lambda_r\}_0^\infty$ and $\{\mu_r\}_1^\infty$ are both bounded.   (11.1.28)

$\{\mu_r\}_1^\infty$ is bounded and $0 < \lambda_r = \alpha r$, for $r \geq 1$.   (11.1.29)

$\{\lambda_r\}_0^\infty$ is bounded and $0 < \mu_r = \alpha r$, for $r \geq 1$.   (11.1.30)

(2) By further analysis along the lines used in proving Theorem 11.2, it is expected that computable estimates can be obtained for the speed of convergence of $\{p_{r,n}(t)\}_{n=1}^\infty$ to its limit $p_r(t)$.

11.2 Computational Procedures

In this section we describe a procedure for computing

$$p_{r,n}(t) = \mathcal{L}^{-1}\{P_{r,n}(s)\},$$

(11.2.1)

the nth approximation of the probability $p_r(t)$ that the population will have size r at time t, assuming that the size is m at time $t = 0$. Fortunately $P_{r,n}(s)$ has a partial-fraction representation that helps find its inverse Laplace transform (11.2.1).

Let $A_{r,n}(s)$, $B_{r,n}(s)$ and $P_{r,n}(s)$ denote the nth numerator, denominator and approximant, respectively, of the continued fraction representing $P_r(s)$ in Theorem 11.1. Then it can be seen that $A_{r,n}(s)$ and $B_{r,n}(s)$ are polynomials in s of degrees $\deg(A_{r,n})$, $\deg(B_{r,n})$ given by

$$\deg(A_{r,n}) \leq m+n-1, \tag{11.2.2a}$$

and

$$\deg(B_{r,n}) = \begin{cases} m+n & \text{if } r \leq m, \\ r+n & \text{if } r \geq m. \end{cases} \tag{11.2.2b}$$

Moreover

$$B_{r,n}(s) = \begin{cases} B_{m+n}(s) & \text{if } r \leq m, \\ B_{r+n}(s) & \text{if } r \geq m. \end{cases} \tag{11.2.3}$$

For convenience we define $K = K(r, m, n)$ by

$$K = K(r, m, n) = \begin{cases} m+n & \text{if } r \leq m, \\ r+n & \text{if } r \geq m, \end{cases} \tag{11.2.4}$$

and denote the zeros of $B_{r,n}(s)$ by $s_1^{(r,m,n)}, s_2^{(r,m,n)}, \ldots, s_K^{(r,m,n)}$. It is well known that these zeros are real, negative and distinct (see Theorem 9.3A). It follows that $P_{r,n}(s)$ has a partial-fraction representation of the form

$$P_{r,n}(s) = \frac{A_{r,n}(s)}{B_{r,n}(s)} = \sum_{j=1}^{K} \frac{H_j^{(r,m,n)}}{s - s_j^{(r,m,n)}},$$
$$r = 0, 1, 2, \ldots, \quad n = 1, 2, 3, \ldots. \tag{11.2.5}$$

Hence the inverse Laplace transform $p_{r,n}(t)$ is given by

$$p_{r,n}(t) = \mathcal{L}^{-1}\{P_{r,n}(s)\} = \sum_{j=1}^{K} H_j^{(r,m,n)} e^{s_j^{(r,m,n)} t}. \tag{11.2.6}$$

Here $H_j^{(r,m,n)}$ is the constant which can be computed by

$$H_j^{(r,m,n)} = \frac{B_r\left(s_j^{(r,m,n)}\right) B_m\left(s_j^{(r,m,n)}\right) A_K\left(s_j^{(r,m,n)}\right)}{L_{m-1} M_r C_K\left(s_j^{(r,m,n)}\right)}, \quad j = 1, 2, \ldots, K, \tag{11.2.7a}$$

where

$$C_K\left(s_j^{(r,m,n)}\right) = \prod_{\substack{i=1 \\ i \neq j}}^{K} \left(s_j^{(r,m,n)} - s_i^{(r,m,n)}\right), \quad j = 1, 2, \ldots, K. \tag{11.2.7b}$$

The values of $B_r(s)$, $B_m(s)$ and $A_K(s)$ in (11.2.5a) can be computed by means of the difference equations

$$A_1(s)=1, \qquad A_2(s)=(\lambda_1+\mu_1+s), \tag{11.2.8a}$$

$$A_j(s)=(\lambda_{j-1}+\mu_{j-1}+s)A_{j-1}(s)-\lambda_{j-2}\mu_{j-1}A_{j-2}(s), \qquad j=3,4,\dots,K, \tag{11.2.8b}$$

and

$$B_1(s)=\lambda_0+s, \qquad B_2(s)=(\lambda_1+\mu_1+s)B_1(s)-\lambda_0\mu_1, \tag{11.2.9a}$$

$$B_j(s)=(\lambda_{j-1}+\mu_{j-1}+s)B_j(s)-\lambda_{j-2}\mu_{j-1}B_{j-2}(s), \qquad j=3,4,\dots,r. \tag{11.2.9b}$$

Finally, the zeros $s_j^{(r,m,n)}$ can be approximated by means of the quotient-difference algorithm (Section 7.1.2) as follows: Suppose we are given $L, K, \lambda_0, \lambda_1,\dots,\lambda_{K-1}$, and $\mu_1, \mu_2,\dots,\mu_{K-1}$. We set

$$q_j^{(1)}=-\lambda_{j-1}, \qquad j=1,2,\dots,K, \tag{11.2.10a}$$

$$e_j^{(1)}=-\mu_j, \qquad j=1,2,\dots,K-1, \tag{11.2.10b}$$

and compute:
for $i=1,2,\dots,L-1$,

$$
\left|
\begin{aligned}
&q_1^{(i+1)}=e_1^{(i)}+q_1^{(i)}, \\[6pt]
&e_1^{(i+1)}=\frac{q_2^{(i)}}{q_1^{(i)}}e_1^{(i)}, \\[6pt]
&\qquad j=2,3,\dots,K-1, \\[4pt]
&\left|
\begin{aligned}
&q_j^{(i+1)}=e_j^{(i)}-e_{j-1}^{(i+1)}+q_j^{(i)}, \\[6pt]
&e_j^{(i+1)}=\frac{q_{j+1}^{(i)}}{q_j^{(i+1)}}e_j^{(i)},
\end{aligned}
\right. \\[6pt]
&q_K^{(i+1)}=e_K^{(i)}-e_{K-1}^{(i+1)}+q_K^{(i)}.
\end{aligned}
\right.
\tag{11.2.10c}
$$

It follows from Theorems 7.8, 7.10, and 9.3A and Equation (9.3.29) that

$$s_j^{(r,m,n)}= \lim_{L\to\infty} q_j^{(L)}. \tag{11.2.11}$$

Thus, for L sufficiently large, $q_j^{(L)}$ can be used to approximate the zero $s_j^{(r,m,n)}$. The reader should be cautioned again, however, concerning the numerical instability of the quotient-difference algorithm mentioned in

Table 11.2.1a. Values of $q_j^{(i)}$ from Quotient-Difference Algorithm; $m=r=0, n=5$

The numbers in this table are in floating-point form. The first 13-digit number should be multiplied by 10 to the exponent given by the last two digits and sign. The letter D indicates that the computations were made in double precision arithmetic.

i	$q_1^{(i)}$			$q_2^{(i)}$			$q_3^{(i)}$		
1	-2.0000	00000	000D-01	-2.0000	00000	000D-01	-2.0000	00000	000D-01
10	-1.7329	35919	966D+00	-1.6489	93209	611D+00	-9.8419	28865	211D-01
20	-2.2632	08117	753D+00	-1.4581	66006	518D+00	-8.6543	60873	559D-01
30	-2.2758	38542	422D+00	-1.4511	98299	924D+00	-8.6477	20514	521D-01
40	-2.2759	80783	306D+00	-1.4511	13944	886D+00	-8.6476	82770	040D-01
50	-2.2759	82362	446D+00	-1.4511	12973	876D+00	-8.6476	82556	635D-01
60	-2.2759	82379	974D+00	-1.4511	12962	903D+00	-8.6476	82555	430D-01
70	-2.2759	82380	169D+00	-1.4511	12962	780D+00	-8.6476	82555	423D-01
80	-2.2759	82380	171D+00	-1.4511	12962	779D+00	-8.6476	82555	423D-01
90	-2.2759	82380	171D+00	-1.4511	12962	779D+00	-8.6476	82555	423D-01

Section 7.1.2. For very large values of K the procedure may break down because of the difficulty of computing the zeros $s_j^{(r, m, n)}$. We now describe some numerical results obtained for this problem using the above procedures.

Example. We consider an immigration-death process in which the birth (or immigration) rates λ_r and death rates μ_r are given by

$$\lambda_r = 0.2, \qquad r = 0, 1, 2, \ldots,$$
$$\mu_r = 0.4r, \qquad r = 1, 2, 3, \ldots.$$

Table 11.2.1b. Values of $q_j^{(i)}$ from Quotient-Difference Algorithm; $m=r=0, n=5$

The numbers in this table are in floating-point form. The first 13-digit number should be multiplied by 10 to the exponent given by the last two digits and sign. The letter D indicates that the computations were made in double-precision arithmetic.

i	$q_4^{(i)}$			$q_5^{(i)}$		
1	-2.0000	00000	000D-01	-2.0000	00000	000D-01
10	-4.1419	15020	738D-01	-2.7470	54029	303D-04
20	-4.0786	84560	557D-01	-2.7470	54029	303D-04
30	-4.0786	16978	401D-01	-2.7470	54029	303D-04
40	-4.0786	16961	059D-01	-2.7470	54029	303D-04
50	-4.0786	16961	050D-01	-2.7470	54029	303D-04
60	-4.0786	16961	050D-01	-2.7470	54029	303D-04
70	-4.0786	16961	050D-01	-2.7470	54029	303D-04
80	-4.0786	16961	050D-01	-2.7470	54029	303D-04
90	-4.0786	16961	050D-01	-2.7470	54029	303D-04

This is sometimes called a queuing problem with infinitely many servers. Some results from applying the quotient-difference algorithm (11.2.10), with $m=r=0$ and $n=5$, are given in Table 11.2.1. From these figures one can see the convergence of each $q_j^{(i)}$ to $s_j^{(0,0,5)}$, $j=1,2,3,4,5$. The resulting approximations of $s_j^{(0,0,5)}$ are given in Table 11.2.2, together with the values of $H_j^{(0,0,5)}$. Also given in Table 11.2.2 are values of $s_j^{(0,0,n)}$ and $H_j^{(0,0,n)}$ for $n=3,5$ and 10. Using values of $s_j^{(r,m,n)}$ and $H_j^{(r,m,n)}$ such as those given in Table 11.2.2, we have computed values of $p_{r,n}(t)$ for $m=r=0$, $n=3,4,\dots,10$ and $t=1,5,10,20$, given in Table 11.2.3. Those numbers illustrate the rate of convergence of $p_{0,n}(t)$ to $p_0(t)$ as n increases. Finally, in Tables 11.2.4, we give rounded values of $p_{r,10}(t)$ for $m=0$, $r=0,1,2,3,4,5$ and for various

Table 11.2.2.   Values of $s_j^{(r,m,n)}$ and $H_j^{(r,m,n)}$; $m=r=0$

The numbers in this table are in floating-point form. The first 13-digit number should be multiplied by 10 to the exponent given by the last two digits and sign. The letter D indicates that the computations were made in double-precision arithmetic; E, that it was made in single-precision arithmetic.

		$n=K=3$				
j	$s_j^{(0,0,3)}$			$H_j^{(0,0,3)}$		
1	−1.2691	92659	142D+00	2.1348	89373	673E−02
2	−5.1865	43144	812D−01	3.1788	77761	259E−01
3	−1.2153	02637	679D−02	6.6076	33301	373E−01

		$n=K=5$				
j	$s_j^{(0,0,5)}$			$H_j^{(0,0,5)}$		
1	−2.2759	82380	171D+00	7.3769	57186	913E−05
2	−1.4511	12962	779D+00	5.9979	17087	837E−03
3	−8.6476	82555	423D−01	7.3653	91646	751E−02
4	−4.0786	16961	050D−01	3.1203	66356	465E−01
5	−2.7470	54029	303D−04	6.0823	77612	263E−01

		$n=K=10$				
j	$s_j^{(0,0,10)}$			$H_j^{(0,0,10)}$		
1	−4.6806	57541	541D+00	2.2600	47208	314E−12
2	−3.7511	07460	246D+00	2.0991	39260	948E−09
3	−3.0529	50420	540D+00	2.5660	81928	860E−07
4	−2.4909	50734	953D+00	9.3144	07001	406E−06
5	−2.0213	95035	443D+00	1.5170	68553	625E−04
6	−1.6027	52859	312D+00	1.5813	38706	539E−03
7	−1.2001	79848	900D+00	1.2642	38412	945E−02
8	−8.0000	60005	210D−01	7.5818	74944	471E−02
9	−4.0000	00979	270D−01	3.0326	55827	571E−01
10	−6.1635	54859	957D−10	6.0653	06649	903E−01

Table 11.2.3. Values of $p_{r,n}(t)$; $m = r = 0$

The numbers in this table are in floating-point form. The first 13-digit number should be multiplied by 10 to the exponent given by the last two digits and sign. The letter E indicates that the computation was made with single-precision arithmetic.

n	$p_{0,n}(1)$			$p_{0,n}(5)$		
3	8.4802	72499	769E−01	6.4561	50254	396E−01
4	8.4802	93873	406E−01	6.4879	64340	011E−01
5	8.4802	93974	170E−01	6.4898	47620	425E−01
6	8.4802	93974	538E−01	6.4899	33166	252E−01
7	8.4802	93974	540E−01	6.4899	36322	766E−01
8	8.4802	93974	540E−01	6.4899	36420	898E−01
9	8.4802	93974	540E−01	6.4899	36423	535E−01
10	8.4802	93974	540E−01	6.4899	36423	597E−01

n	$p_{0,n}(10)$			$p_{0,n}(20)$		
3	5.8692	58566	381E−01	5.1819	64296	872E−01
4	6.0925	78646	361E−01	5.9253	38150	775E−01
5	6.1186	51741	178E−01	6.0499	46492	322E−01
6	6.1209	35880	399E−01	6.0648	10978	578E−01
7	6.1210	96704	292E−01	6.0662	07480	054E−01
8	6.1211	06189	278E−01	6.0663	16313	169E−01
9	6.1211	06671	055E−01	6.0663	23576	762E−01
10	6.1211	06692	558E−01	6.0663	24001	153E−01

Table 11.2.4. Values of $p_{r,10}(t)$; $m = 0$

t	$p_{0,10}(t)$	$p_{1,10}(t)$	$p_{2,10}(t)$	$p_{3,10}(t)$	$p_{4,10}(t)$	$p_{5,10}(t)$
1	.848029	.139789	.011521	.000633	.000026	.000001
2	.759317	.209067	.028782	.002642	.000182	.000010
3	.705109	.246367	.043041	.005013	.000438	.000031
4	.670956	.267746	.053422	.007106	.000709	.000057
5	.648994	.280581	.060652	.008741	.000945	.000082
6	.634676	.288550	.065593	.009941	.001130	.000103
7	.265255	.293617	.068940	.010791	.001267	.000119
8	.619019	.296893	.071198	.011383	.001365	.000131
9	.614874	.299037	.072716	.011788	.001433	.000139
10	.612111	.300450	.073737	.012064	.001480	.000145
⋮						
15	.607283	.302889	.075534	.012558	.001566	.000156
20	.606632	.303214	.075778	.012625	.001578	.000158

values of t. Analogous results are given in Table 11.2.5 for $m=1$. From these two tables one can see the rate at which $p_{r,10}(t)$ approaches its limiting value as t increases. One also sees the variation of $p_{r,10}(t)$ for fixed t as r increases. These and other similar examples have been discussed by Murphy and O'Donohoe [1975].

Table 11.2.5. Values of $p_{r,10}(t)$; $m=1$

t	$p_{0,10}(t)$	$p_{1,10}(t)$	$p_{2,10}(t)$	$p_{3,10}(t)$	$p_{4,10}(t)$	$p_{5,10}(t)$
1	.279578	.614537	.097502	.007932	.000433	.000018
2	.418134	.456310	.109789	.014387	.001287	.000087
3	.492734	.384538	.104281	.016467	.001816	.000153
4	.535492	.349153	.096694	.016457	.002000	.000188
5	.561162	.330440	.090416	.015766	.002000	.000198
6	.577099	.319949	.085819	.014989	.001929	.000196
7	.587234	.313784	.082603	.014327	.001846	.000189
8	.593787	.310024	.080398	.013821	.001773	.000181
9	.598073	.307666	.078900	.013453	.001716	.000175
10	.600899	.306158	.077889	.013194	.001674	.000170
⋮						
15	.605778	.303643	.076098	.012714	.001593	.000160
20	.606429	.303331	.075854	.012647	.001581	.000158

CHAPTER 12

Miscellaneous Results

We give here a number of recent discoveries which have not as yet been treated in any monograph and which we feel deserve mentioning.

12.1 T-Fraction Expansions for Families of Bounded Functions

The following result was proved by Waadeland [1964].

THEOREM 12.1. *There exists an $R_0 > 1$, and for every $R > R_0$ there exists a $K(R) > 0$, such that if $f(z)$ is holomorphic for $|z| < R$, $f(0) = 1$, and*

$$|f(z) - 1| < K(R) \qquad \text{for} \quad |z| < R, \qquad (12.1.1)$$

then $f(z)$ has a T-fraction expansion

$$1 + d_0 z + \mathop{\mathrm{K}}_{n=1}^{\infty}\left(\frac{z}{1 + d_n z}\right) \qquad (12.1.2)$$

which corresponds to the Taylor series expansion of $f(z)$ at $z = 0$. Moreover, the T-fraction expansion (12.1.2) converges uniformly to $f(z)$ on every compact subset of the disk $|z| < 1$. Finally,

$$\lim_{n \to \infty} d_n = -1. \qquad (12.1.3)$$

Waadeland also observed that, for $R > 2$, one could choose $K(R) = R/2 - 1$. For $R > 2\sqrt{2}$, the choice

$$K(R) = \frac{R-1}{2} - \frac{1}{2(R+1)}$$

ENCYCLOPEDIA OF MATHEMATICS and Its Applications, Gian-Carlo Rota (ed.). Vol. 11: William B. Jones and W. J. Thron, Continued Fractions.   ISBN 0-201-13510-8

is permissible. One cannot improve the 1 in $|z| < 1$, since the periodic T-fraction

$$1 - z + \mathop{\mathbf{K}}_{n=1}^{\infty}\left(\frac{z}{1-z}\right)$$

is known to converge to 1 for $|z| < 1$ and to $-z$ for $|z| > 1$. Following ideas developed in Section 8.4, Waadeland [1966] also investigated the convergence behavior of the sequence $\{T_n(z)\}$ defined by

$$T_n(z) = \frac{A_n(z) + zA_{n-1}(z)}{B_n(z) + zB_{n-1}(z)}, \qquad (12.1.4)$$

where $A_n(z)$ and $B_n(z)$ denote the nth numerator and denominator, respectively, of the T-fraction (12.1.2). The result, as improved by Hovstad [1975], is

THEOREM 12.2. *Let R_0, R and $K(R)$ be as defined in Theorem 12.1, and let, as before, $f(z)$ be a function holomorphic in $|z| < R$, with $f(0) = 1$ and*

$$|f(z) - 1| < K(R) \qquad for \quad |z| < R.$$

Then the sequence $\{T_n(z)\}$ defined by (12.1.4), where the T-fraction (12.1.2) corresponds to $f(z)$ at $z = 0$, converges to $f(z)$ uniformly on every compact subset of $|z| < R$.

The sequence $\{T_n(z)\}$ has been called the *modified T-fraction expansion of $f(z)$*. From our discussion in Section 8.4, it follows that $\{T_n(z)\}$ converges more rapidly to $f(z)$, for $|z| < 1$, than the T-fraction in Theorem 12.1.

In two papers, Waadeland [1967, 1979a] has also investigated general T-fraction expansions of functions holomorphic in a neighborhood of $z = \infty$. In the second of the two articles he proved:

THEOREM 12.3. *For every fixed $c_1 \neq 0$, there exist two ordered pairs of positive numbers (α, R) and (β, ρ), with $\rho < 1/|c_1| < R$, such that if $f(z)$ is a function holomorphic in $|z| < R$ satisfying $f(0) = 1$, $f'(0) = c_1$ and*

$$|f(z) - 1 - c_1 z| \leqslant \alpha \qquad for \quad |z| < R, \qquad (12.1.5)$$

and if $g(z)$ is a function holomorphic for $|z| > \rho$ satisfying

$$|g(z)| \leqslant \beta \qquad for \quad |z| > \rho, \qquad (12.1.6)$$

then there exists a non-terminating general T-fraction

$$1 + \overset{\infty}{\underset{n=1}{\mathrm{K}}} \left(\frac{F_n z}{1 + G_n z} \right) \tag{12.1.7}$$

which corresponds at $z=0$ to the Taylor series for $f(z)$ at $z=0$, and corresponds at $z=\infty$ to the Laurent series of $g(z)$ for $|z|>\rho$ at $z=\infty$. Moreover,

$$\lim_{n\to\infty} F_n = - \lim_{n\to\infty} G_n = F \neq 0. \tag{12.1.8}$$

Finally, the general T-fraction (12.1.7) converges to $f(z)$ uniformly on compact subsets of the disk $|z|<1/|F|$, and converges to $g(z)$ uniformly on compact subsets of $|z|>1/|F|$.

Let $\{T_n(z)\}$ denote a modified general T-fraction defined by (12.1.4), where $A_n(z)$, $B_n(z)$ are nth numerator and denominator, respectively, of the general T-fraction (12.1.7). Then it appears likely that $\{T_n(Fz)\}$ at $z=0$ and $\{T_n(-1)\}$ at $z=\infty$ will lead to larger regions of convergence, as well as faster convergence, than was obtained by the general T-fraction in Theorem 12.3.

Hovstad [1975] was able to show that, under the assumptions of Theorem 12.1, for every $R>1$, there exists an R', $1<R'<R$, and a positive number C such that

$$|f_n(z) - 1| < \frac{C}{(R')^n}.$$

Using Schwarz's lemma, one can deduce that

$$|d_n + 1| < \frac{C}{R(R')^n}.$$

This result motivated Waadeland [1979b] to look for a converse to the theorems discussed above. He was able to prove:

THEOREM 12.4. *Let $\mathfrak{T}_{R,C}$ be the family of all limit periodic general T-fractions*

$$1 + \overset{\infty}{\underset{n=1}{\mathrm{K}}} \left(\frac{F_n z}{1 + G_n z} \right) \tag{12.1.9}$$

with

$$\lim_{n\to\infty} F_n = - \lim_{n\to\infty} G_n = 1$$

and

$$|F_n - 1| < \frac{C}{R^n}, \qquad |G_n + 1| < \frac{C}{R^n}.$$

Let $R > 1$ and $0 < \varepsilon < 1 - 1/R$ be given. Then there exists a C such that every general T-fraction in $\mathfrak{T}_{R,C}$ corresponds at $z = 0$ to the Taylor series at $z = 0$ of a function holomorphic in $|z| < R - \varepsilon$ and corresponds at $z = \infty$ to the Laurent expansion at $z = \infty$ of a function holomorphic for $|z| > 1/R + \varepsilon$.

12.2   *T*-Fractions Corresponding to Rational Functions

For C-fractions and associated continued fractions one has the simple rule that the continued fraction corresponds to a rational function iff it is terminating. Since ordinary T-fractions

$$1 + d_0 z + \overset{\infty}{\underset{n=1}{\mathbf{K}}}\left(\frac{z}{1 + d_n z}\right) \tag{12.2.1}$$

are by definition always non-terminating, new criteria for correspondence to rational functions must be looked for. Since

$$1 - z + \mathbf{K}\left(\frac{z}{1 - z}\right) = -z \qquad \text{for} \quad |z| > 1, \tag{12.2.2}$$

one might try for periodic and/or limit periodic T-fractions. Jefferson [1969a] was able to show that the only periodic T-fractions of period $k = 1, 2, 3$ which correspond to rational functions are of the form (12.2.2). Surprisingly, however, there exists a rational function of the form

$$\frac{1 + a_1 z + a_2 z^2 + a_3 z^3}{1 + b_1 z + b_2 z^2 + b_3 z^3},$$

where a_3 and b_3 are either both different from zero or both zero, to which there corresponds a periodic T-fraction for period 4. Jefferson also proved that if a periodic T-fraction converges to a rational function $R(z)$, then $R(z) = P(z)/Q(z)$, where $P(z)$ and $Q(z)$ are polynomials of the same degree. For non-periodic T-fractions representing rational functions, only a finite number of d_n can vanish, and after that the tail corresponds to a rational function of the form described above.

Influenced by the results of Section 12.1, Hag [1970, 1972] raised the question of which limit periodic T-fractions could be rational. She arrived at the following result:

THEOREM 12.5. *A limit periodic T-fraction corresponds to a rational function only if*

$$\lim_{n\to\infty} d_n = -1. \tag{12.2.3}$$

In that case the T-fraction converges to that function for $|z| < 1$. (There may be poles in this disk.)

For general T-fractions, terminating continued fractions are included, and one has the possibility that the general T-fraction corresponds to a rational function at either $z = 0$, ∞, at both or at neither. Very little is as yet known in this situation. A partial result by Waadeland [1978] is the following:

THEOREM 12.6. *Let $R(z)$ be a rational function of the form*

$$R(z) = \frac{a_1 z + a_2 z^2 + \cdots + a_n z^n}{1 + b_1 z + b_2 z^2 + \cdots + b_n z^n}, \qquad a_n \neq 0, \quad b_n \neq 0; \tag{12.2.4}$$

let $L(z)$ and $L^*(z)$ be the Laurent expansions of $R(z)$ at $z = 0$ and $z = \infty$, respectively. Furthermore, let $R(z)$ be equal to a terminating general T-fraction

$$\mathop{\mathrm{K}}_{k=1}^{n} \left(\frac{F_k z}{1 + G_k z} \right). \tag{12.2.5}$$

Then (12.2.5) corresponds to $L(z)$ at $z = 0$ and to $L^*(z)$ at $z = \infty$ iff

$$G_k \neq 0 \qquad \text{for} \quad k = 1, 2, \ldots, n.$$

12.3 Location of Singular Points of Analytic Functions Represented by Continued Fractions

This is an area in which surprisingly few results are known. While Taylor series, which are so far the best means of locating singularities, do not easily distinguish between poles and other singular points, continued-fraction expansions in general converge at poles. Thus they promise to be a better device for obtaining information about singularities which are not poles. It appears that the convergence of continued fractions can be disturbed by other phenomena, such as value behavior of the function, but almost nothing is known about that.

We have encountered many functions whose continued-fraction expansions converge in the region $R = [z : |\arg z| < \pi]$. From this it does not follow that $f(z)$ has a singularity at $z = 0$ or $z = \infty$ or anywhere on the negative real axis. One way of testing for the existence of branch points would be to study the values taken on by the function $f(z)$ at points $-s+i\varepsilon$, $s>0$, $\varepsilon>0$, to see whether

$$\lim_{\varepsilon\to 0} f(-s+i\varepsilon) \neq \lim_{\varepsilon\to 0} f(-s-i\varepsilon). \qquad (12.3.1)$$

If this turns out to be the case, then the function has singularities (most likely branch points) between 0 and $-s$ and between $-s$ and $-\infty$. To the best of our knowledge, Wall [1942] is the only one who has carried out such an investigation. He showed that (12.3.1) is valid for S-fractions

$$\mathop{\mathbf{K}}_{n=1}^{\infty}\left(\frac{(1-g_{n-1})g_n z}{1}\right), \qquad 0<g_n<1, \qquad (12.3.2)$$

provided

$$\sum_{n=1}^{\infty} |g_n - \tfrac{1}{2}| < \infty \quad \text{and} \quad s>1. \qquad (12.3.3)$$

(See Corollary 4.60 for the convergence of these continued fractions.)
    When $d_n>0$, the T-fraction

$$\mathop{\mathbf{K}}_{n=1}^{\infty}\left(\frac{z}{1+d_n z}\right)$$

is known to converge to a function $f(z)$ holomorphic in $R=[z:|\arg z|<\pi]$. So again the question of singularities of $f(z)$ arises. Jones and Thron [1966] showed that there is a singularity of $f(z)$ at $z=0$ if, in addition to $d_n>0$, $n \geqslant 1$, the sequence $\{d_n\}$ is unbounded.
    Around 1940 Leighton conjectured that the C-fraction

$$\mathop{\mathbf{K}}_{n=1}^{\infty}\left(\frac{a_n z^{\alpha_n}}{1}\right), \qquad \alpha_n \text{ a positive integer,} \quad a_n \in \mathbb{C}, \quad a_n \neq 0,$$

where $\alpha_n \to \infty$ and the a_n are suitably restricted, represents a function $f(z)$ having the unit circle $|z|=1$ as a natural boundary. Scott and Wall [1940b] were able to prove the conjecture in the very special case with $a_n=a$, $\alpha_n=m^{n-1}$, where either a is real and m is an odd integer or a is negative and m is an arbitrary integer. A substantial improvement is the following result of Thron [1949]:

THEOREM 12.7. *Let* $\{\alpha_n\}$ *be a sequence of positive integers satisfying:* (a)

$$\lim \alpha_n = \infty, \qquad \alpha_n > \alpha_{n-1} \quad \text{for } n \geqslant 2, \qquad (12.3.4a)$$

and (b) *there exists a sequence of positive integers* $\{\mu_k\}$, *with* $\lim \mu_k = \infty$, *which has the property that for each k there is an* $n(k)$ *such that* μ_k *divides* α_k *for all* $n > n(k)$. *Let* $\{a_n\}$ *be a sequence of non-zero complex numbers satisfying*

$$\lim_{n \to \infty} |a_n|^{1/\alpha_n} = 1. \tag{12.3.4b}$$

Then the C-fraction $\mathbf{K}(a_n z^{\alpha_n}/1)$ *converges to a function* $f(z)$, *meromorphic in* $|z| < 1$, *which has the unit circle* $|z| = 1$ *as a natural boundary.*

In both of these theorems one requires that $\{\alpha_n\}$ tends monotonically to ∞ and that $\{\alpha_n\}$ satisfies some very strong arithmetical requirements. The conditions in Theorem 12.7 do not force $\{\alpha_n\}$ to tend to ∞ very fast. The choice

$$\alpha_n = 2^{[\log n]}$$

clearly satisfies the requirements.

In a series of papers [Thron, 1953; Singh and Thron, 1956b; Callas and Thron, 1968, 1972], the arithmetical requirements were dropped and were replaced by certain conditions on

$$k(\{n_m\}) = \liminf_{m \to \infty} \frac{\rho_{n_m}}{h_{n_m} - \rho_{n_m}}. \tag{12.3.5}$$

Here

$$h_n = \sum_{m=1}^{n} \alpha_m \tag{12.3.6}$$

and

$$s_n = \deg A_n(z), \qquad t_n = \deg B_n(z), \qquad \rho_n = \max[s_n, t_n]. \tag{12.3.7}$$

$A_n(z)$ and $B_n(z)$ denote the nth numerator and denominator, respectively, of the C-fraction $\mathbf{K}(a_n z^{\alpha_n}/1)$. The conditions were successively improved, and the best result is given by:

THEOREM 12.8. *For the C-fraction* $\mathbf{K}(a_n z^{\alpha_n}/1)$, *let*

$$\lim_{n \to \infty} (4|a_n|)^{1/\alpha_n} = 1. \tag{12.3.8}$$

Further, let $\{n_m\}$ *be a sequence of positive integers such that*

$$\lim_{m \to \infty} \frac{n_m + 1}{h_{n_m} - \rho_{n_m}} = 0, \tag{12.3.9}$$

and assume that $k(\{n_m\}) > 0$, where $k(\{n_m\})$ is defined by (12.3.5). Then the meromorphic function $f(z)$ to which the C-fraction converges, for $|z| < 1$, has at least

$$\beta(\{n_m\}) = \left[\left[\frac{1 + k(\{n_m\})}{k(\{n_m\})}\right]\right] + 1 \qquad (12.3.10)$$

singular points which are not poles on the unit circle $|z| = 1$. The function $f(z)$ cannot be meromorphic on any arc of the unit circle of angular measure greater than

$$\frac{2\pi k(\{n_m\})}{k(\{n_m\}) + 1} \qquad (12.3.11)$$

radians. If $k(\{n_m\}) = 0$, then $f(z)$ has the unit circle as a natural boundary.

Clearly one would like a sequence $\{n_m\}$ to satisfy (12.3.9) and yield a $k(\{n_m\})$ as small as possible. To get some feeling for the relation of $k(\{n_m\})$ to the sequence $\{\alpha_n\}$, let us assume that $\alpha_n < \alpha_{n+1}$, $n \geqslant 1$. For then

$$s_{2n} = \sum_{k=1}^{n} \alpha_{2k}, \qquad s_{2n+1} = \sum_{k=0}^{n} \alpha_{2k+1},$$

$$t_{2n} = \sum_{k=1}^{n} \alpha_{2k}, \qquad t_{2n+1} = \sum_{k=1}^{n} \alpha_{2k+1}$$

[see (12.3.7)]. It follows that

$$\frac{\rho_{2n}}{h_{2n} - \rho_{2n}} = \frac{\displaystyle\sum_{k=1}^{n} \alpha_{2k}}{\displaystyle\sum_{k=0}^{2n+1} \alpha_{2k+1}}.$$

Thus, for example, if $\lim_{n\to\infty} \alpha_n/\alpha_{n-1} = \infty$, we have $k(\{n_m\}) = 0$, and hence $f(z)$ has the unit circle as a natural boundary without requiring $\{\alpha_n\}$ to satisfy any arithmetical conditions.

We conclude this section by describing results of Callas and Thron [1967] for regular C-fractions.

THEOREM 12.9. *Let the regular C-fraction $K(a_n z/1)$ have a bounded sequence of coefficients $\{a_n\}$. Let $f(z)$ be the meromorphic function to which the continued fraction converges in a neighborhood of the origin. Then the* *modulus r of the singularity of f(z), other than poles, nearest to z=0 satisfies*

$$r \leqslant \inf_{m>0} \frac{1+q}{mq^2}, \qquad q = \limsup_{\mu \to \infty} \left[\prod_{\substack{\nu=1 \\ |a_\nu| \leqslant m}}^{2\mu+1} \frac{|a_\nu|}{m} \right]^{1/2\mu}.$$

Here m is an arbitrary positive number.

THEOREM 12.10. *Let the regular C-fraction* $K(a_n z/1)$ *satisfy*

$$\lim_{n \to \infty} |a_n| = d \neq 0.$$

Then $r \leqslant 2/d$. *Here r is defined in Theorem 12.9.*

12.4 Univalence of Functions Represented by Continued Fractions

Since a good deal is known about the values taken on by continued fractions, one might expect that a great deal is known about univalence of continued fractions. This, however, is not so. The first and main contribution to the subject was by Thale [1956]. He has a number of results, of which the following is typical and is probably the most interesting:

THEOREM 12.11. *In the continued fraction*

$$\frac{1}{1} + \frac{a_1 z}{1} + \frac{a_2 z}{1} + \frac{a_3 z}{1} + \cdots, \qquad (12.4.1)$$

let $|a_n| \leqslant \frac{1}{4}$, $n \geqslant 1$. *Then the function* $f(z)$ *to which* (12.4.1) *converges, for* $|z| < 1$, *is univalent for*

$$|z| < 4(3\sqrt{2} - 4) = 0.968\ldots. \qquad (12.4.2)$$

Thale thought that this constant probably could be improved, but Perron [1956] was able to show that it is best possible. More recently Thale's results have been extended by Merkes [1959], Merkes and Scott [1960] and Hayden and Merkes [1964, 1965].

Classification of Special Types of Continued Fractions

In the analytic theory of continued fractions a number of different types of continued fractions

$$b_0(z) + \frac{a_1(z)}{b_1(z)} + \frac{a_2(z)}{b_2(z)} + \frac{a_3(z)}{b_3(z)} + \cdots$$

are studied where the $a_n(z)$ and $b_n(z)$ are functions of a complex variable z involving some restricted parameters. To provide a convenient reference, we summarize in this appendix some of the most widely used types of these continued fractions. In some cases we also point out connections with Padé tables. We adopt the convention that if an arbitrary function $b_0(z)$ is added to a continued fraction of a certain type, the new continued fraction is of the same type. Also the type of a continued fraction is unchanged under an equivalence transformation.

1. *C-fractions* (*corresponding continued fractions*) are continued fractions of the form

$$\frac{a_1 z^{\alpha_1}}{1} + \frac{a_2 z^{\alpha_2}}{1} + \frac{a_3 z^{\alpha_3}}{1} + \cdots, \qquad a_n \neq 0 \qquad (A.1)$$

where each α_n is a positive integer and each a_n is a non-zero complex number. If $\alpha_n = 1$ for all n, we obtain a *regular C-fraction*

$$\frac{a_1 z}{1} + \frac{a_2 z}{1} + \frac{a_3 z}{1} + \cdots, \qquad a_n \neq 0. \qquad (A.2)$$

If $a_n > 0$ for all n, then (A.2) is called an *S-fraction* (or *Stieltjes fraction*).

ENCYCLOPEDIA OF MATHEMATICS and Its Applications, Gian-Carlo Rota (ed.). Vol. 11: William B. Jones and W. J. Thron, Continued Fractions.   ISBN 0-201-13510-8

The following three continued fractions are equivalent to the regular C-fraction (A.2):

$$\frac{a_1}{\zeta} + \frac{a_2}{1} + \frac{a_3}{\zeta} + \frac{a_4}{1} + \cdots, \qquad a_n \neq 0, \quad z = 1/\zeta, \qquad \text{(A.3a)}$$

$$\frac{a_1/\zeta}{1} + \frac{a_2}{\zeta} + \frac{a_3}{1} + \frac{a_4}{\zeta} + \frac{a_5}{1} + \cdots, \qquad a_n \neq 0, \quad z = 1/\zeta, \qquad \text{(A.3b)}$$

$$\frac{a_1/w}{w} + \frac{a_2}{w} + \frac{a_3}{w} + \frac{a_4}{w} + \cdots, \qquad a_n \neq 0, \quad z = 1/w^2. \qquad \text{(A.3c)}$$

Closely related to (A.3b) is the *modified regular C-fraction*

$$\frac{a_1}{1} + \frac{a_2}{z} + \frac{a_3}{1} + \frac{a_4}{z} + \cdots, \qquad a_n \neq 0, \qquad \text{(A.4)}$$

which becomes a *modified S-fraction* if $a_n > 0$ for all n. Henrici [1977] calls a continued fraction of the form

$$\frac{a_1}{1} + \frac{a_2 z}{1} + \frac{a_3 z}{1} + \frac{a_4 z}{1} + \cdots, \qquad a_n \neq 0, \qquad \text{(A.5)}$$

a RITZ *fraction*; it is called a SITZ *fraction* if $a_n > 0$ for all n. A RITZ (or SITZ) fraction becomes a RITZ^{-1} (or SITZ^{-1}) *fraction* if in (A.5) z is replaced by z^{-1}. Thus a RITZ^{-1} (or SITZ^{-1}) fraction

$$\frac{a_1}{1} + \frac{a_2(1/z)}{1} + \frac{a_3(1/z)}{1} + \frac{a_4(1/z)}{1} + \cdots \qquad \text{(A.6)}$$

is equivalent to a modified regular C- (or S-) fraction (A.4). By an equivalence transformation, (A.4) can be written in the form

$$\frac{1}{b_1} + \frac{1}{b_2 z} + \frac{1}{b_3} + \frac{1}{b_4 z} + \cdots, \qquad b_n \neq 0. \qquad \text{(A.7)}$$

If all b_n are real, $b_{2n-1} \neq 0$ and $b_{2n} > 0$, then (A.7) is called an *H-fraction*. A *g-fraction* is a special type of SITZ fraction of the form

$$\frac{s_0}{1} + \frac{g_1 z}{1} + \frac{(1-g_1)g_2 z}{1} + \frac{(1-g_2)g_3 z}{1} + \frac{(1-g_3)g_4 z}{1} + \cdots, \qquad \text{(A.8a)}$$

where

$$s_0 > 0 \quad \text{and} \quad 0 < g_n < 1, \qquad n = 1, 2, 3, \ldots. \qquad \text{(A.8b)}$$

We summarize the following connections between regular C-fractions and

Padé tables. If f_n denotes the nth approximant of the regular C-fraction

$$c_0 + \frac{a_1 z}{1} + \frac{a_2 z}{1} + \frac{a_3 z}{1} + \cdots, \qquad a_n \neq 0, \qquad (A.9)$$

which corresponds (at $z=0$) to the fps

$$L = c_0 + c_1 z + c_2 z^2 + \cdots, \qquad (A.10)$$

and if $R_{m,n}$ denotes the (m, n) Padé approximant of L, then

$$f_{2m} = R_{m,m} \quad \text{and} \quad f_{2m+1} = R_{m+1, m}, \qquad m = 0, 1, 2, \ldots \qquad (A.11)$$

(see Theorem 5.19). Somewhat more generally, suppose that a fps (A.10) is normal (Section 5.5.1). Then for each $k = 1, 2, 3, \ldots$ there exists a regular C-fraction

$$\frac{a_1^{(k)} z}{1} + \frac{a_2^{(k)} z}{1} + \frac{a_3^{(k)} z}{1} + \cdots, \qquad a_n^{(k)} \neq 0, \qquad (A.12)$$

corresponding (at $z=0$) to the fps

$$c_k z + c_{k+1} z^2 + c_{k+2} z^3 + \cdots. \qquad (A.13)$$

If $f_n^{(k)}$ denotes the nth approximant of the continued fraction

$$c_0 + c_1 z + \cdots + c_{k-1} z^{k-1} + \frac{a_1^{(k)} z^k}{1} + \frac{a_2 z}{1} + \frac{a_3 z}{1} + \frac{a_4 z}{1} + \cdots, (A.14)$$

then

$$f_{2m}^{(k)} = R_{k+m-1, m} \quad \text{and} \quad f_{2m+1} = R_{k+m, m}, \qquad m = 0, 1, 2, \ldots, \quad (A.15)$$

where $R_{m,n}$ is the (m, n) Padé approximant of L (see [Wall, 1948, Theorem 96.1] for a proof of this and for a similar result for the reciprocal series $1/L$).

2. *Associated continued fractions* are of the form

$$\frac{k_1 z}{1 + l_1 z} - \frac{k_2 z^2}{1 + l_2 z} - \frac{k_3 z^2}{1 + l_3 z} - \frac{k_4 z^2}{1 + l_4 z} - \cdots, \qquad k_n \neq 0, \quad (A.16)$$

where the k_n and l_n are complex constants. The even part of a regular C-fraction (A.2) is the associated continued fraction

$$\frac{a_1 z}{1 + a_2 z} - \frac{a_2 a_3 z^2}{1 + (a_2 + a_3) z} - \frac{a_4 a_5 z^2}{1 + (a_5 + a_6) z} - \cdots. \qquad (A.17)$$

However, there exist associated continued fractions which are not even parts of regular C-fractions. Let

$$L = c_1 z + c_2 z^2 + c_3 z^3 + \cdots \tag{A.18}$$

denote the fps to which (A.16) corresponds at $z=0$. Then the nth approximant f_n of (A.16) is the (n, n) Padé approximant of L. By making a simple equivalence transformation in (A.16) we obtain

$$\frac{k_1}{l_1 + \dfrac{1}{z}} - \frac{k_2}{l_2 + \dfrac{1}{z}} - \frac{k_3}{l_3 + \dfrac{1}{z}} \cdots, \qquad k_n \neq 0, \tag{A.19}$$

which is closely related to a J-fraction, as is seen below.

3. J-fractions are continued fractions of the form

$$\frac{1}{d_1 + z} - \frac{c_1^2}{d_2 + z} - \frac{c_2^2}{d_3 + z} - \frac{c_3^2}{d_4 + z} - \cdots, \qquad c_n^2 \neq 0, \tag{A.20}$$

where the c_n^2 and d_n are complex constants. A J-fraction (A.20) is said to be positive definite if there exists a sequence of positive numbers $\{g_n\}$ such that

$$|c_n^2| - \operatorname{Re}(c_n^2) \leqslant 2\delta_n \delta_{n+1}(1 - g_{n-1})g_n, \qquad n = 1, 2, 3, \ldots, \tag{A.21a}$$

where

$$\delta_n = \operatorname{Im}(d_n) \geqslant 0 \text{ and } 0 < g_{n-1} < 1, \qquad n = 1, 2, 3, \ldots. \tag{A.21b}$$

It is easily shown that $|c_n^2| - \operatorname{Re}(c_n^2) = 2[\operatorname{Im}(c_n)]^2$, so that (A.21a) can be written as

$$\operatorname{Im}(c_n^2) \leqslant \delta_n \delta_{n+1}(1 - g_{n-1})g_n, \qquad n = 1, 2, 3, \ldots. \tag{A.22}$$

Wall [1948, Definition 16.1] calls the J-fraction (A.20) positive definite if the quadratic form

$$\sum_{j=1}^{p} \delta_j \xi_j^2 - 2 \sum_{j=1}^{p-1} \operatorname{Im}(c_j) \xi_j \xi_{j+1} > 0 \tag{A.23}$$

for all $p = 1, 2, 3, \ldots$ and for all $\xi_1, \xi_2, \xi_3, \ldots$. He then shows that this definition is equivalent to (A.21) [Wall, 1948, Corollary 16.2] and that the inequality in (A.20a) can be replaced by equality [Wall, 1948, Theorem 16.2]. The J-fraction (A.20) is called a *real J-fraction* if all c_n and d_n are real numbers. It is easily seen that every real J-fraction is positive definite.

Perron [1957a] refers to a real J-fraction as a G-fraction.

4. *P-fractions* are continued fractions of the form

$$b_0(z) + \frac{1}{b_1(z)} + \frac{1}{b_2(z)} + \frac{1}{b_3(z)} + \cdots, \qquad \text{(A.24a)}$$

where each $b_n(z)$ is a polynomial in $1/z$:

$$b_n(z) = \sum_{k=-N_n}^{0} a_{-k}^{(n)} z^k, \qquad \text{(A.24b)}$$

where

$$N_0 \geqslant 0; \qquad N_n \geqslant 1 \text{ and } a_{-N_n}^{(n)} \neq 0, \quad n = 1, 2, 3, \ldots. \qquad \text{(A.24c)}$$

The transformed associated continued fraction (A.19) can be seen to be an example of a P-fraction. Let

$$L = c_0 + c_1 z + c_2 z^2 + \cdots \qquad (c_0 \neq 0) \qquad \text{(A.25)}$$

be a given fps, and let (A.24) denote the P-fraction corresponding to L at $z = 0$ (see Corollary 5.4). In this case $N_0 = 0$. Then the sequence of approximants $\{f_n\}$ of the P-fraction is the sequence of consecutive distinct approximants along the main diagonal of the Padé table of L. More generally, let s be a fixed integer, and let

$$b_0^{(s)} + \frac{1}{b_1^{(s)}} + \frac{1}{b_2^{(s)}} + \frac{1}{b_3^{(s)}} + \cdots, \qquad \text{(A.26)}$$

with nth approximant $f_n^{(s)}$, be the P-fraction corresponding to the fLs

$$L_s = z^s L = \sum_{n=0}^{\infty} c_n z^{n+s}. \qquad \text{(A.27)}$$

Then $\{z^{-s} f_n^{(s)}\}$ is the sequence of consecutive distinct approximants in the diagonal number s ($\{R_{m, m-s}\}$ if $s < 0$, $\{R_{m+s, m}\}$ if $s \geqslant 0$) of the Padé table of L [Magnus, 1974, Theorem 6].

5. *General T-fractions* are continued fractions of the form

$$\frac{z}{e_1 + d_1 z} + \frac{z}{e_2 + d_2 z} + \frac{z}{e_3 + d_3 z} + \cdots, \qquad e_n \neq 0, \qquad \text{(A.28)}$$

where the e_n and d_n are complex constants. If $e_n = 1$ for all n, then (A.28) is called a *T-fraction*. General T-fractions are also expressed in the equivalent form

$$\frac{F_1 z}{1+G_1 z} + \frac{F_2 z}{1+G_2 z} + \frac{F_3 z}{1+G_3 z} + \cdots, \qquad F_n \neq 0, \qquad (A.29)$$

where the F_n and G_n are complex constants. The continued fractions (A.28) and (A.29) are equivalent if

$$F_n = \frac{1}{e_n e_{n-1}}, \quad G_n = \frac{d_n}{e_n} \ (e_0 = 1), \qquad n = 1, 2, 3, \ldots, \qquad (A.30)$$

or equivalently, if

$$e_1 = \frac{1}{F_1}, \qquad d_1 = \frac{G_1}{F_1}, \qquad (A.31a)$$

$$e_{2n-1} = \frac{\displaystyle\prod_{k=1}^{n-1} F_{2k}}{\displaystyle\prod_{k=1}^{n} F_{2k-1}}, \quad d_{2n-1} = \frac{G_{2n-1} \displaystyle\prod_{k=1}^{n-1} F_{2k}}{\displaystyle\prod_{k=1}^{n} F_{2k-1}}, \qquad n = 2, 3, 4, \ldots,$$

$$(A.31b)$$

and

$$e_{2n} = \frac{\displaystyle\prod_{k=1}^{n} F_{2k-1}}{\displaystyle\prod_{k=1}^{n} F_{2k}}, \quad d_{2n} = \frac{G_{2n} \displaystyle\prod_{k=1}^{n} F_{2k-1}}{\displaystyle\prod_{k=1}^{n} F_{2k}}, \qquad n = 1, 2, 3, \ldots. \ (A.31c)$$

The general T-fraction (A.28) (or A.29) is called a *positive T-fraction* if $e_n > 0$ and $d_n > 0$ (or $F_n > 0$ and $G_n > 0$) for all n. Closely related to general T-fractions are *M-fractions*

$$\frac{F_1}{1+G_1 z} + \frac{F_2 z}{1+G_2 z} + \frac{F_3 z}{1+G_3 z} + \cdots, \qquad F_n, G_n \neq 0, \qquad (A.32)$$

where the F_n and G_n are non-zero complex constants. M-fractions and general T-fractions are related to two-point Padé tables as follows. Let

$$L = \sum_{m=0}^{\infty} c_m z^m \quad \text{and} \quad L^* = \sum_{m=0}^{\infty} c^*_{-m} z^{-m} \qquad (A.33)$$

be a given pair of fLs. A rational function of type $[m, n]$,

$$R_{m,n}(z) = \frac{P_{m,n}(z)}{Q_{m,n}(z)}, \qquad (A.34)$$

where $P_{m,n}$ and $Q_{m,n}$ are polynomials in z of degrees not exceeding m and n, respectively, is called the (m, n) two-point Padé approximant of (L, L^*) if

$$Q_{m,n}L - P_{m,n} = O(z^r) \quad \text{and} \quad Q_{m,n}L^* - P_{m,n} = O\left(\left(\frac{1}{z}\right)^{s-\max(m,n)}\right),$$

(A.35)

where

$$r = \left[\!\left[\frac{m+n+2}{2}\right]\!\right], \quad s = \left[\!\left[\frac{m+n+1}{2}\right]\!\right].$$

Here $O(z^r)$ means that the fps in increasing powers of z starts with a power greater than or equal to r. Then the nth approximant of an M-fraction is the $(n, n-1)$ two-point Padé approximant of a pair of series (A.33) with $c_0^* = 0$ and $c_0 \neq 0$. The nth approximant of a general T-fraction (A.29), with $F_n \neq 0$ and $G_n \neq 0$ for all n, is the (n, n) two-point Padé approximant of a pair of series (A.33) with $c_0 = 0$ and $c_0^* \neq 0$ (see Theorem 7.18).

6. Perron [1957a, p. 176–178] considers continued fractions of the form

$$b_0 + \frac{a_1 z}{b_1 z} + \frac{a_2}{b_2} + \frac{a_3 z}{b_3 z} + \frac{a_4}{b_4} + \cdots, \quad a_n \neq 0, \; b_{2n-1} \neq 0, \quad (A.36)$$

where the a_n and b_n are complex constants. We note that if $b_{2n} \neq 0$ for $n \geq 1$, then the even part of (A.36) is given by

$$b_0 + \frac{a_1 b_2 z}{a_2 + b_1 b_2 z} - \frac{a_2 a_3 b_4 z}{a_4 b_2 + (a_3 b_4 + b_2 b_3 b_4)z} - \frac{a_4 a_5 b_2 b_6 z}{a_6 b_4 + (a_5 b_6 + b_4 b_5 b_6)z}$$

$$- \frac{a_6 a_7 b_4 b_8 z}{a_8 b_6 + (a_7 b_8 + b_6 b_7 b_8)z} - \cdots \quad (A.37)$$

[see (2.4.24)], which can be seen to be equivalent to a general T-fraction. A special case of (A.36) closely related to an infinite process studied by Schur [1917, 1918] has the form

$$\gamma_0 + \frac{(1-\gamma_0\bar{\gamma}_0)z}{\bar{\gamma}_0 z} - \frac{1}{\gamma_1} + \frac{(1-\gamma_1\bar{\gamma}_1)z}{\bar{\gamma}_1 z} - \frac{1}{\gamma_2} + \cdots, \quad \gamma_n \neq 0, \; |\gamma_n| \neq 1.$$

(A.38)

Another special case of (A.36) studied by Frank [1952] has the form

$$k_0\gamma_0 + \frac{k_0(1-\gamma_0\bar{\gamma}_0)z}{\bar{\gamma}_0 z} - \frac{1}{k_1\gamma_1} + \frac{k_1(1-\gamma_1\bar{\gamma}_1)z}{\bar{\gamma}_1 z} - \frac{1}{k_2\gamma_2} + \cdots, \quad (A.39a)$$

where the k_n and γ_n are complex coefficients satisfying

$$k_n \neq 0, \quad \gamma_n \neq 0, \quad |\gamma_n| \neq 1, \quad n = 0, 1, 2, \ldots . \tag{A.39b}$$

7. *Thiele continued fractions* are of the form

$$b_0 + \frac{z - z_0}{b_1} + \frac{z - z_1}{b_2} + \frac{z - z_2}{b_3} + \cdots, \tag{A.40}$$

where the b_n are complex constants and $\{z_n\}$ is a sequence of distinct complex constants. Thiele continued fractions arise in the following manner. Let $f(z)$ be a given function, and let $\{v_n(z)\}$ be defined (if possible) by

$$v_0(z) = f(z); \quad v_{k+1}(z) = \frac{z - z_k}{v_k(z) - v_k(z_k)}, \quad k = 0, 1, 2, \ldots . \tag{A.41}$$

The function $v_k(z)$ is sometimes called the kth *inverted difference* of $f(z)$. From (A.41) we obtain

$$v_k(z) = v_k(z_k) + \frac{z - z_k}{v_{k+1}(z)}, \quad k = 0, 1, 2, \ldots \tag{A.42}$$

and hence, for $n = 0, 1, 2, \ldots,$

$$f(z) = v_0(z) + \frac{z - z_0}{v_1(z_1)} + \frac{z - z_1}{v_2(z_2)} + \cdots + \frac{z - z_{n-2}}{v_{n-1}(z_{n-1})} + \frac{z - z_{n-1}}{v_n(z)}. \tag{A.43}$$

We are thus led to consider the Thiele continued fraction

$$v_0(z_0) + \frac{z - z_0}{v_1(z_1)} + \frac{z - z_1}{v_2(z_2)} + \frac{(z - z_2)}{v_3(z_3)} + \cdots. \tag{A.44}$$

If $f_n(z)$ denotes the nth approximant of (A.44), then it can be seen that for $k = 0, 1, \ldots, n$,

$$f_n(z_k) = v_0(z_0) + \frac{z_k - z_0}{v_1(z_1)} + \cdots + \frac{z_k - z_{k-1}}{v_k(z_k)} \tag{A.45}$$

provided

$$v_{k+1}(z_{k+1}) + \frac{z_k - z_{k+1}}{v_{k+2}(z_{k+2})} + \cdots + \frac{z_k - z_{n-1}}{v_n(z_n)} \neq 0. \tag{A.46}$$

It follows from (A.43) and (A.45) that

$$f_n(z_k) = f(z_k), \quad k = 0, 1, \ldots, n. \tag{A.47}$$

It follows from the difference equations (2.1.6) that for each $m=0,1,2,\ldots,$ $f_{2m}(z)$ is a rational function of type $[m, m]$ and $f_{2m+1}(z)$ is a rational function of type $[m+1, m]$. Thus if $R_{m,n}$ denotes the (m, n) Newton-Padé approximant of $f(z)$, then

$$f_{2m} = R_{m,m} \text{ and } f_{2m+1} = R_{m+1,m}, \qquad m=0,1,2,\ldots. \qquad (A.48)$$

Further details on Thiele continued fractions and Newton-Padé tables can be found in Section 5.5.2 and references given there.

APPENDIX B

Additional Results on Minimal Solutions of Three-Term Recurrence Relations

The contents of this appendix are largely due to P. Henrici, who very kindly allowed us to make use of them.

Let $\{a_n\}$ and $\{b_n\}$ be two sequences of complex numbers with $a_n \neq 0$, $n \geq 1$. As was discussed in Section 5.3, the computation of a sequence $\{y_n\}$ which satisfies the three-term recurrence relations

$$y_{n+1} = b_n y_n + a_n y_{n-1}, \qquad n \geq 1, \tag{B.1}$$

may present difficulties. Such relations are of interest because they are satisfied by sequences of orthogonal polynomials (see Section 7.2.2) as well as by certain non-polynomial sequences, as for example Bessel functions (Section 5.3) and other functions discussed in Chapter 6.

The next two examples illustrate what can happen if one attempts to compute the y_n directly from (B.1).

Example 1. It is well known that $y_n^{(1)} = \cos n\varphi$ and $y_n^{(2)} = \sin n\varphi$ are solutions of the system of three-term recurrence relations

$$y_{n+1} = 2(\cos \varphi) y_n - y_{n-1}, \qquad n = 1, 2, 3, \dots. \tag{B.2}$$

If $\cos \varphi$ and $\sin \varphi$ are known, then (B.2) provides a stable method for computing $\cos n\varphi$ and $\sin n\varphi$. For $\varphi = \pi/10$, starting with $y_0^{(1)} = 1$, $y_1^{(1)} = \cos(\pi/10) = 0.951056516$ and $y_0^{(2)} = 0$, $y_1^{(2)} = \sin(\pi/10) = 0.309016994$, one obtains the sample of results shown in Table B.1, based on calculations with 10-decimal-digit floating-point arithmetic. The cumulative roundoff error in each $y_n^{(1)}$ and $y_n^{(2)}$ is less than two units in the eighth decimal place.

ENCYCLOPEDIA OF MATHEMATICS and Its Applications, Gian-Carlo Rota (ed.). Vol. 11: William B. Jones and W. J. Thron, Continued Fractions. ISBN 0-201-13510-8

Table B.1

n	$y_n^{(1)} = \cos n\varphi$		$y_n^{(2)} = \sin n\varphi$	
45	0.0000	00015	1.0000	00004
46	-0.3090	16981	0.9510	56524
47	-0.5877	85242	0.8090	17006
48	-0.8090	16989	0.5877	85267
49	-0.9510	56516	0.3090	17010
50	-1.0000	00004	0.0000	00015

Example 2.   The situation changes completely if one considers the recurrence relations

$$y_{n+1} = 3y_n - y_{n-1}, \qquad n = 1, 2, 3, \cdots. \tag{B.3}$$

Two independent solutions of this system are

$$y_n^{(1)} = \left(\frac{3 - \sqrt{5}}{2} \right)^n \quad \text{and} \quad y_n^{(2)} = \left(\frac{3 + \sqrt{5}}{2} \right)^n.$$

Starting with $y_0^{(1)} = 1$ and $y_1^{(1)} = (3 - \sqrt{5})/2 = 0.381966012$, one finds that the computed values for $y_n^{(1)}$ using (B.3) are as given in Table B.2. Since these values are increasing from $n = 13$ on, instead of decreasing exponentially to zero, it is clear that the method of computing $y_n^{(1)}$ recursively from (B.3) fails completely in this case.

Example 3.   Let ν be real, $x > 0$, and let $J_n(x)$ denote the Bessel function of the first kind of order n. It can be seen from (5.2.13) that the numbers

$$y_n = J_{\nu + n}(x)$$

Table B.2

n	$y_n^{(1)}$	
2	0.1458	98035
3	0.0557	28094
4	0.0212	86246
$\vdots$	$\vdots$	
10	0.0000	69090
11	0.0000	33059
12	0.0000	30088
13	0.0000	57204
14	0.0001	41523
15	0.0003	67366

satisfy the recurrence relations

$$y_{n+1} = \frac{2(v+n)}{x} y_n - y_{n-1}, \qquad n > 1, \tag{B.4}$$

which are of the form (B.1). The difficulty encountered in generating the sequence $\{J_n(x)\}$ by (B.4), if values of $J_0(x)$ and $J_1(x)$ are known, has been discussed in Section 5.3 (see Table 5.3.1).

Recalling that the solutions $\{y_n\}$ of the recurrence relations (B.1) form a vector space of dimension 2 over C, we can give an explanation for the unfortunate results of the computation in Examples 2 and 3. The reason lies in the fact that in each case the difference equation has in addition to the solution $\{y_n^{(1)}\}$ or $\{y_n\}$ a second solution that grows rapidly in comparison to $\{y_n\}$. In Example 2 the second solution is $\{y_n^{(2)}\}$. In Example 3, assuming $v=0$ for simplicity, the system (B.4), in addition to $y_n=J_n(x)$, has the solution $z_n = Y_n(x)$, where $Y_n(x)$ denotes the Bessel function of the second kind of order n [see (5.3.11)]. From (5.3.14) it can be seen that $z_n = Y_n(x)$ grows rapidly, especially in comparison with $y_n = J_n(x)$, which tends to zero. The explanation that follows is an elaboration of ideas mentioned in Section 5.3.

If in the numerical evaluation of a recurrence system such as (B.3) or (B.4) the value y is in error by $\varepsilon > 0$, while y_0 is accurate, then the solution w_n actually obtained is a linear combination of $\{y_n\}$ and $\{z_n\}$ given by

$$w_n = \left(1 - \frac{z_0 \varepsilon}{z_1 y_0 - z_0 y_1}\right) y_n + \frac{y_0 \varepsilon}{z_1 y_0 - z_0 y_1} z_n.$$

The relative error thus becomes

$$\frac{w_n - y_n}{y_n} = \frac{y_0 \varepsilon}{z_1 y_0 - z_0 y_1} \frac{z_n}{y_n} - \frac{z_0 \varepsilon}{z_1 y_0 - z_0 y_1}.$$

It is then clear that if $z_n/y_n \to \infty$, the relative error grows without bounds no matter how small $\varepsilon > 0$ is. This is true provided all subsequent arithmetical operations are performed without further error. If, as is likely to be the case, additional errors are introduced at each step, then these errors will further modify the component of the rapidly growing solution $\{z_n\}$.

We recall from Section 5.3 that a non-zero solution $\{y_n\}$ of (B.1) is called *minimal* if there exists another solution $\{z_n\}$ such that

$$\lim_{n \to \infty} \frac{y_n}{z_n} = 0. \tag{B.5}$$

It is easily seen that if $\{y_n\}$ is a minimal solution of (B.1) and if $\{w_n\}$ is any solution that is linearly independent of $\{y_n\}$ and $w_n \neq 0$ for n sufficiently large, then $\lim_{n\to\infty} y_n/w_n = 0$.

According to this definition, $\{y_n^{(1)}\}$ in Example 2 and $\{J_n(x)\}$ in Example 3 are minimal solutions of their respective recurrence relations.

The problem of constructing a minimal solution $\{y_n\}$ of (B.1), provided such a solution exists, will now be considered in greater generality. The following result, which can be deduced readily from the definition, will be useful in this context.

THEOREM B.1. *Any two minimal solution of* (B.1) *are linearly dependent.*

We first consider the case where we want to determine a minimal solution $\{y_n\}$ of (B.1) with $y_0 = c \neq 0$. Instead of trying forward recursion, we shall use backward recursion. Because $a_n \neq 0$, $n \geqslant 1$, the system (B.1) can be inverted to yield

$$y_{n-1} = -\frac{b_n}{a_n}y_n + \frac{1}{a_n}y_{n+1}. \tag{B.6}$$

With starting values of y_N and y_{N-1} this can be used to generate the values y_n for $n = N-2, N-3, \ldots, 0$. The starting values y_N and y_{N-1} may be unknown. So we arbitrarily use values

$$y_N = 0, \qquad y_{N-1} = 1, \tag{B.7}$$

and argue as follows. Let the solution of (B.6) satisfying (B.7) be denoted by $y_n^{(N)}$. Then $y_n^{(N)}$ will, in general, contain a component of a non-minimal solution $\{z_n\}$. Since non-minimal solutions grow rapidly in comparison with $\{y_n\}$ as n increases, they will decrease rapidly as n decreases. Thus, for small enough n the value of $y_n^{(N)}$ will be close to the value of its minimal component. Since our process is not likely to yield $y_0^{(N)} = c$, we form

$$w_n^{(N)} = \frac{y_n^{(N)}}{y_0^{(N)}}c \tag{B.8}$$

and hope that $w_n^{(N)}$ comes close to the desired solution y_n for N sufficiently large. This is indeed what happens.

THEOREM B.2. *If the system* (B.1) *has a minimal solution* $\{y_n\}$ *with* $y_0 = c \neq 0$, *then for* $n = 0, 1, \ldots, N-1$,

$$w_n^{(N)} - y_n = \frac{z_0 y_n - z_n c}{c - \frac{y_N}{z_N}z_0}\frac{y_N}{z_N}, \tag{B.9}$$

where $\{z_n\}$ *is any solution of* (B.1) *which is linearly independent of* $\{y_n\}$ *and* where $w_n^{(N)}$ is defined by (B.7) and (B.8). In particular, for each fixed n

$$\lim_{N\to\infty} w_n^{(N)} = y_n. \tag{B.10}$$

Proof. The solution $\{y_n^{(N)}\}$ defined by (B.6) and (B.7) is a linear combination of $\{y_n\}$ and $\{z_n\}$ and hence is given by

$$y_n^{(N)} = \frac{z_N y_n - y_N z_n}{z_N y_{N-1} - y_N z_{N-1}}. \tag{B.11}$$

It follows from (B.8) and (B.11) that

$$w_n^{(N)} = \frac{z_N y_n - y_N z_n}{z_N y_0 - y_N z_0} c = \frac{y_n - \dfrac{y_N}{z_N} z_n}{y_0 - \dfrac{y_N}{z_N} z_0} c.$$

From this (B.9) can be derived. (B.10) is a consequence of the fact that

$$\lim_{N\to\infty} \frac{y_N}{z_N} = 0,$$

since $\{y_n\}$ is a minimal solution and $\{z_n\}$ is linearly independent of it. ∎

In some applications it may not be convenient to characterize the desired minimal solution $\{y_n\}$ by specifying the value of y_0, since this value may be unknown or hard to compute. It is also possible that y_0 is zero or very small, in which case the algorithm for obtaining $w_n^{(N)}$ will break down. However, it may be known that for some sequence $\{c_k\}$ the desired minimal solution satisfies

$$\sum_{k=0}^{\infty} c_k y_k = s, \tag{B.12}$$

where $s \neq 0$ is also known.

An illustration of Equation (B.12) can be found in the sequence of Bessel functions $\{J_n(x)\}$, for here

$$J_0(x) + 2J_2(x) + 2J_4(x) + \cdots = 1. \tag{B.13}$$

This follows from [Henrici, 1974, (4.5-5)] on letting $t = 1$.

To compute $\{y_n\}$ subject to the condition (B.12) one could use backward recursion as before, but normalize y_n in this case by setting

$$w_n^{(N)} = \frac{y_n^{(N)}}{\displaystyle\sum_{k=0}^{N} c_k y_k^{(N)}}. \tag{B.14}$$

For fixed n one would again expect to have $\lim_{N\to\infty} w_n^{(N)} = y_n$. Note that the condition $y_0 = c \neq 0$ is a special case of (B.12). It is obtained by setting $c_0 = 1$, $c_k = 0$, $k \geq 1$, $s = c$.

There are also applications where one is not interested in the individual values of y_n but rather in the value of an infinite series

$$t = \sum_{k=0}^{\infty} d_k y_k, \tag{B.15}$$

where the y_k are values of a minimal solution. For y_n we may either have $y_0 = c \neq 0$ or, more generally, a condition of type (B.12).

Example 4. The *Fresnel integrals* are, for real τ, defined by

$$C(\tau) = \int_0^\tau \cos\left(\frac{\pi}{2}t^2\right) dt, \qquad S(\tau) = \int_0^\tau \sin\left(\frac{\pi}{2}t^2\right) dt$$

[see (6.1.39)]. For small $|\tau|$ the Fresnel integrals are easily evaluated by power series [see (6.1.40)], and for large $|\tau|$ by means of asymptotic expansions or continued fractions [see, for example, (6.1.41) and (7.3.109)]. For in-between values of $|\tau|$ these methods are slow or inaccurate. For these values of $|\tau|$ the expansions

$$C(\tau) = \sum_{k=0}^{\infty} J_{\frac{1}{2}+2k}\left(\frac{\pi}{2}\tau^2\right), \tag{B.16a}$$

$$S(\tau) = \sum_{k=0}^{\infty} J_{\frac{3}{2}+2k}\left(\frac{\pi}{2}\tau^2\right) \tag{B.16b}$$

can be helpful (see [Abramovitz and Stegun, 1964, p. 301]).

The sequence $\{J_{\frac{1}{2}+n}((\pi/2)\tau^2)\}$ is a minimal solution of the recurrence relation

$$y_{n+1} = \frac{2n+1}{(\pi/2)\tau^2} y_n - y_{n-1}$$

with

$$y_0 = J_{1/2}\left(\frac{\pi}{2}\tau^2\right) = \frac{2}{\tau}\sin\left(\frac{\pi}{2}\tau^2\right),$$

$$y_{-1} = J_{-1/2}\left(\frac{\pi}{2}\tau^2\right) = \frac{2}{\tau}\cos\left(\frac{\pi}{2}\tau^2\right).$$

Thus $C(\tau)$ and $S(\tau)$ can be evaluated using (B.15), with the explicit formulas in this case being given by (B.16a) or (B.16b).

THEOREM B.3. *If the system* (B.1) *has a minimal solution* $\{y_n\}$ *such that*

$$\sum_{k=0}^{\infty} c_k y_k = s \neq 0,$$

if $\sum_{k=0}^{\infty} d_k y_k$ *converges to* t, *and if* $\{z_n\}$ *is a solution of* (B.1) *linearly independent of* $\{y_n\}$, *then*

$$w_N - t = \frac{(t_N - t)s - (s_N - s)t - s\tau_N\dfrac{y_N}{z_N} + t\sigma_N\dfrac{y_N}{z_N}}{s_N - \sigma_N\dfrac{y_N}{z_N}}, \qquad \text{(B.17a)}$$

where

$$w_N = \frac{\displaystyle\sum_{k=0}^{N} d_k y_k^{(N)}}{\displaystyle\sum_{k=0}^{N} c_k y_k^{(N)}} - s \qquad \text{(B.17b)}$$

[$y_k^{(N)}$ *is defined in* (B.11)] *and*

$$s_N = \sum_{k=0}^{N} c_k y_k, \quad \sigma_N = \sum_{k=0}^{N} c_k z_k, \quad t_N = \sum_{k=0}^{N} d_k y_k, \quad \tau_N = \sum_{k=0}^{N} d_k z_k.$$

$$\text{(B.17c)}$$

Moreover

$$\lim_{N\to\infty} w_N = t \qquad \text{(B.18)}$$

holds if

$$\lim_{N\to\infty} \frac{y_N \sigma_N}{z_N} = 0 \quad \text{and} \quad \lim_{N\to\infty} \frac{y_N \tau_N}{z_N} = 0. \qquad \text{(B.19)}$$

Proof. Substituting in (B.17b) the values for $y_n^{(N)}$ obtained in (B.11), one arrives at

$$w_N = \frac{\displaystyle\sum_{k=0}^{N} d_k y_k - \frac{y_N}{z_N} \sum_{k=0}^{N} d_k z_k}{\displaystyle\sum_{k=0}^{N} c_k y_k - \frac{y_N}{z_N} \sum_{k=0}^{N} c_k z_k}\, s$$

and hence

$$w_N = \frac{t_N - \dfrac{y_N}{z_N} \tau_N}{s_N - \dfrac{y_N}{z_N} \sigma_N}\, s,$$

from which we obtain (B.17a). From this (B.18) follows provided (B.19) is satisfied, since $\lim_{N \to \infty} t_N = t$, $\lim_{N \to \infty} s_N = s$. ■

The conditions (B.19) are in general easily verified. For instance, in Example 2 we have

$$y_n^{(1)} = r^n, \qquad y_n^{(2)} = r^{-n}, \qquad r = \frac{3 - \sqrt{5}}{2}.$$

Using the normalizing condition

$$\sum_{k=0}^{\infty} y_k^{(1)} = \sum_{k=0}^{\infty} r^n = \frac{1}{1-r}$$

and choosing $\{z_n\} = \{y_n^{(2)}\}$, one arrives at

$$\frac{y_N^{(1)}}{z_N} \sigma_N = r^{2n} \sum_{k=1}^{N} r^{-k} = r^{2N} \frac{\left(\dfrac{1}{r}\right)^{N+1} - 1}{\left(\dfrac{1}{r}\right) - 1} = r^N \frac{1 - r^{N+1}}{1-r},$$

and thus the quantity tends to zero geometrically.

The $y_N \sigma_N / z_N$ obtained from the normalization condition (B.13) tends to zero at an even faster rate.

In the discussion up to now we have derived stable and efficient methods for computing minimal solutions. However the following questions remain:

1. How can one determine whether a given difference equation has a minimal solution?

2. If there is a minimal solution, is there a way of estimating the ratio y_n / z_n?

Both questions are answered by the classical theorem of Pincherle (a more general version of which was given in Theorem 5.7).

THEOREM B.4 (Pincherle). *Let* $\{a_n\}$, $\{b_n\}$ *be two sequences of complex numbers with* $a_n \neq 0$ *for all* $n \geqslant 1$.

(A) *The system of three-term recurrence relations* (B.1) *has a minimal solution* $\{y_n\}$ *iff the continued fraction*

$$\frac{a_1}{b_1} + \frac{a_2}{b_2} + \frac{a_3}{b_3} + \cdots \tag{B.20}$$

converges (to a finite value or to infinity).

(B) *Suppose that* (B.1) *has a minimal solution* $\{y_n\}$. *Then for* $m = 1, 2, 3, \ldots$,

$$\frac{y_m}{y_{m-1}} = -\frac{a_m}{b_m} + \frac{a_{m+1}}{b_{m+1}} + \frac{a_{m+2}}{b_{m+2}} + \cdots . \tag{B.21}$$

By (B.21) *we mean the following: If* $y_{m-1} = 0$, *then* $y_m \neq 0$ *and the continued fraction* (B.21) *converges to* $\infty = y_m / y_{m-1}$. *If* $y_{m-1} \neq 0$, *then the continued fraction* (B.21) *converges to the finite value* y_m / y_{m-1}.

(C) *Suppose that* (B.20) *converges to a finite value* w *and has* n *th numerator and denominator* A_n *and* B_n, *respectively. Then every minimal solution is proportional to*

$$y_n = A_{n-1} - w B_{n-1}, \tag{B.22}$$

and

$$w = -\frac{y_0}{y_{-1}}$$

holds for any minimal solution $\{y_n\}$

The importance of Theorem B.4 lies in the fact that there exists a considerable body of knowledge concerning the convergence of continued fractions and concerning the speed with which $f_n = A_n / B_n \rightarrow w$. (Some of the known results are given in Chapters 4 and 8.) Let $\alpha_n = A_{n-1}$ and $\beta_n = B_{n-1}$. It follows from the difference equations (2.1.6) that $\{\alpha_n\}$ and $\{\beta_n\}$ are solutions of (B.1). Moreover, they are linearly independent. Since in view of (B.22) we have

$$\frac{y_n}{\beta_n} = \frac{A_{n-1}}{B_{n-1}} - w,$$

it is thus possible to estimate the speed with which y_n / β_n tends to zero, and to obtain more explicit error estimates in Theorems B.2 and B.3.

Bibliography

Abramowitz, M., and Stegun, I. A., *Handbook of Mathematical Functions with Formulas, Graphs and Mathematical Tables*, National Bureau of Standards, Appl. Math. Ser. 55, U.S. Govt. Printing Office, Washington, D.C., 1964.

Akhieser, N. I., *The Classical Moment Problem and Some Related Questions in Analysis*, Hafner, New York, 1965.

Anderson, Chr., "The qd-Algorithm as a Method for Finding the Roots of a Polynomial Equation When All Roots Are Positive," Tech. Rep. CS9, Computer Sci. Div., Stanford Univ., 1964.

Andrews, George E., "On q-Difference Equations for Certain Well-Poised Basic Hypergeometric Series," *Quart. J. Math.* (Oxford Second Series), **19** (1968), 433–447.

Andrews, George E., "An Introduction to Ramanujan's "lost" Notebook," *Amer. Math. Monthly* **86** (1979), 89–108.

Arms, Robert J., and Edrei, Albert, "The Padé Tables and Continued Fractions Generated by Totally Positive Sequences," in *Mathematical Essays dedicated to A. J. Macintyre*, Ohio U. P., Athens, Ohio, 1970, pp. 1–21.

Askey, R. A., and Ismail, M. E. H., "Recurrence Relations, Continued Fractions and Orthogonal Polynomials," to appear.

Auric, A., "Recherches sur les fractions continues algébriques," *J. Math. Pures Appl.* (6), **3** (1907).

Baker, George A., Jr., "The Theory and Application of the Padé Approximant Method," K. A. Brueckner (ed.), *Advances in Theoretical Physics*, Vol. 1, Academic Press, New York, 1965, pp. 1–58.

Baker, George A., Jr., "Best Error Bounds for Padé Approximants to Convergent Series of Stieltjes," *J. Math. Phys.* **10** (1969), 814–820.

Baker, George A., Jr., *Essentials of Padé Approximants*, Academic Press, New York, 1975.

Baker, George A., Jr., and Gammel, John L. (eds.), *The Padé Approximant in Theoretical Physics*, Academic Press, New York, 1970.

Baker, George A., Jr., and Graves-Morris, P. R., *Padé Approximants*, to appear.

Baker, George A., Jr., Rushbrooke, G. S., and Gilbert, H. E., "High Temperature Series Expansions for the Spin-$\frac{1}{2}$ Heisenberg Model by the Method of Irreducible Representations of the Symmetry Group," *Phys. Rev.* **135**, No. 5A (August 31, 1964), A1272–A1277.

Bandemer, Hans, "Über die Konvergenz des Quotienten-Differenzen-Algorithmus von Rutishauser für Eigenwertprobleme gewisser linearer Operatoren," *Math. Nachr.* **27** (1964), 353–375.

Bankier, J. D., and Leighton, Walter, "Numerical Continued Fractions," *Amer. J. Math.* **64** (1942), 653–668.

Barnsley, M., "The Bounding Properties of the Multiple-Point Padé Approximant to a Series of Stieltjes," *Rocky Mountain J. Math.* **4** (1974), 331–333.

Bauer, F. L., "Beiträge zur Entwicklung numerischer Verfahren für programmgesteuerte Rechenanlagen. I: Quadratisch konvergente Durchführung der Bernoulli-Jacobischen Methode zur Nullstellenbestimmung von Polynomen. II: Direkte Faktorisierung eines Polynoms," *S. B. Bayr. Akad. Wiss. Math. Nat. Kl.* (1954), 275–303; (1956), 163–203.

Bauer, F. L., "The Quotient-Difference and Epsilon Algorithms," in R. E. Langer, (ed.), *On Numerical Approximations*, Univ. of Wisconsin Press, Madison, 1959, pp. 361–370.

Bauer, F. L., "The g-Algorithm," *J. Soc. Indust. Appl. Math.* **8** (1960), 1–17.

Bauer, F. L., "Nonlinear sequence transformations," in H. L. Garabedian, (ed.), *Approximation of Functions*, Elsevier, Amsterdam, 1965, pp. 134–151.

Bauer, F. L., "QD-method with Newton Shift," Tech. Rep. 56, Computer Sci. Dept., Stanford Univ., 1967, 6 pp.

Bernoulli, D., "Adversaria Analytica Miscellanea de Fractionibus Continuis," *Novi Comm. Acad. Sci. Imp. Petropolitanae* **20**, pro anno 1775.

Bernstein, F., "Über eine Anwendung der Mengenlehre," *Math. Ann.* **71** (1912), 417–439.

Blanch, G., "Numerical Evaluation of Continued Fractions," *SIAM Rev.* **7** (1964), 383–421.

Bombelli, Rafaele, *L'Algebra*, Venezia, 1572.

Borel, E., "Contributions a l'analyse arithmetique du continu," *J. Math. Pures Appl.* (5) **9** (1903), 329–375.

Borel, E., "Les probabilités denombrables et leurs applications arithmetiques," *Rend. Circ. Mat. di Palermo* **27** (1909), 247–271.

Brezinski, C., "Padé Approximants and Orthogonal Polynomials," in E. B. Saff and R. S. Varga (eds.), *Padé and Rational Approximation*, Academic Press, New York, 1977a, pp. 3–14.

Brezinski, C., "A Bibliography on Padé Approximation and Related Subjects," Publ. No. 96 du Lab. de Calcul, Univ. de Lille, 1977b.

Brezinski, C., *Padé-Type Approximation and General Orthogonal Polynomials*, Birkhäuser Boston Inc. (1980).

de Bruin, M. G., "Some Classes of Padé Tables Whose Upper Halves are Normal," *Nieuw Archief Voor Wiskuunde* (3), **XXV** (1977), 148–160.

Cabannes, H., *Padé Approximant Method and Its Applications in Mechanics*, Lecture Notes in Physics 47, Springer-Verlag, New York, 1976.

Cajori, Florian, *A History of Mathematical Notations*, Vol. 2, Open Court, Chicago, 1929.

Callas, N. P., and Thron, W. J., "Singularities of Meromorphic Functions Represented by Regular C-Fractions," *Kgl. Norske Vid. Selsk. Skr.* (Trondheim) No. 6 (1967), 11 pp.

Callas, N. P., and Thron, W. J., "Singular Points of Certain Functions Represented by C-Fractions," *J. Indian Math. Soc.* **32**, Supplement 1 (1968), 325–353.

Callas, N. P., and Thron, W. J., "Singularities of a Class of Meromorphic Functions," *Proc. Amer. Math. Soc.* **33** (1972), 445–454.

Caratheodory, C., *Conformal Representation*, Cambridge Tract in Mathematics and Mathematical Physics, No. 28, Cambridge, 1932.

Caratheodory, C., *Theory of Functions of a Complex Variable, I*, English edition, Chelsea, New York, 1950.

Cataldi, Pietro, *Trattato del modo brevissimo di trovare la radice quadra delli numeri*, Bologna, 1613.

Cazacu, Cabiria Andreian, *Theorie der Funktionen mehrerer komplexer Veränderlicher*, Birkhäuser Verlag, Basel, 1976.

Chihara, T. S., *Introducion to Orthogonal Polynomials*, Mathematics and Its Applications Ser., Gordon, 1978.

Chisholm, J. S. R., *N*-variable rational approximants, in E. B. Saff and R. S. Varga (eds.) *Padé and Rational Approximation*, Academic Press, New York, 1977, pp. 23–42.

Christoffel, E. B., *Gesammelte Mathematische Abhandlungen*, Vols. 1, 2, B. G. Teubner, Leipzig, Berlin, 1910.

Chui, C. K., "Recent Results on Padé Approximants and Related Problems," in Lorentz, G. G., Schumaker, L. L. (eds.), *Proceedings, Conference on Approximation Theory, 1976*, Academic Press.

Claessens, G., "A New Algorithm for Oscullatory Rational Interpolation," *Numerische Mathematik* **27** (1976/77), 77–83.

Claessens, G., "Some Aspects of the Rational Hermite Interpolation Table and Its Applications," Ph.D. Thesis, Universitaire Instelling Antwerpen, Wilrijk, 1976.

Claessens, G., "A Useful Identity for the Rational Hermite Interpolation Table," *Numer. Math.* **29** (1977/78), 227–231.

Clausen, Th., "Die Function $\dfrac{1}{a\ +}\dfrac{1}{a\ +}\dfrac{1}{a\ +}\cdots$ durch die Anzahl der a ausgedrückt," *Reine Angew. Math.* **3** (1828).

Cody, W. J., Paciorek, Kathleen A., and Thacher, Henry C., Jr., "Chebychev Approximation for Dawson's Integral," *Math. Comp.* **24** (1970), 171–178.

Common, A. K., "Padé Approximants and Bounds to Series of Stieltjes," *J. Math. Phys.* **9**, No. 1 (1968), 32–38.

Copp, George, "Some Convergence Regions for a Continued Fraction," Ph.D. Thesis, Univ. of Texas, Austin, 1950.

Copson, E. T., *An Introduction to the Theory of Functions of a Complex Variable*, Oxford at the Clarendon Press, 1935.

Cowling, V. F., Leighton, W., and Thron, W. J., "Twin Convergence Regions for Continued Fractions," *Bull. Amer. Math. Soc.* **50** (1944), 351–357.

Davis, P. J., *Interpolation and Approximation*, Blaisdell, New York, 1963.

Dawson, D. F., "Convergence of Continued Fractions of Stieltjes Type," *Proc. Amer. Math. Soc.* **10** (1959), 12–17.

Dawson, D. F., "Concerning Convergence of Continued Fractions," *Proc. Amer. Math. Soc.* **11** (1960), 640–647.

Dawson, D. F., "A Theorem on Continued Fractions and the Fundamental Inequality," *Proc. Amer. Math. Soc.* **13** (1962), 698–701.

Dawson, D. F., "Remarks on Some Convergence Conditions for Continued Fractions," *Proc. Amer. Math. Soc.* **18** (1967), 803–805.

Dennis, J. J. and Wall, H. S., "The Limit Circle Case for a Positive Definite *J*-Fraction," *Duke Math. J.* **12** (1945), 255–273.

De Pree, J. D., and Thron, W. J., "On Sequences of Moebius Transformations," *Math. Z.* **80** (1962), 184–193.

Drew, D. M., and Murphy, J. A., "Branch Points, *M*-Fractions and Rational Approximants Generated by Linear Equations," *J. Inst. Maths. Applics.* **19** (1977), 169–185.

Edrei, Albert, "Sur des suites de nombres liées a la théorie des fractions continues," *Bull. Sci. Math.* (2) **72** (1948), 45–64.

Edrei, Albert, "Proof of a Conjecture of Schoenberg on the Generating Function of a Totally Positive Sequence," *Canad. J. Math.* **5** (1953a), 86–94.

Edrei, Albert, "On the Generating Function of a Doubly Infinite Totally Positive Sequence," *Trans. Amer. Math. Soc.* **74** (1953b), 367–383.

Edrei, Albert, "The Padé Table of Meromorphic Functions of Small Order with Negative Zeros and Positive Poles," *Rocky Mtn. J. Math.* **4** (1974), 175–180.

Elliott, D., "Truncation Errors in Padé Approximations to Certain Functions: An Alternative Approach," *Math. Comp.* **21** (1967), 398–406.

Erdelyi, A., Magnus, W., Oberhettinger, F., and Tricomi, F. G., *Higher Transcendental Functions*, Vols. 1, 2, 3, McGraw-Hill, New York, 1953.

Erdös, P., and Piranian, G., "Sequences of Linear Fractional Transformations," *Michigan Math. J.* **6** (1959), 205–209.

Euler, L., *Introductio in Analysin Infinitorum*, Vol. I, 1748, Chapter 18.

Euler, L., *Opera Omnia*, Teubner, Leipzig, Berlin, 1911–.

Fair, Wyman, "A Convergence Theorem for Noncommutative Continued Fractions," *J. Approx. Theory* **5** (1972), 74–76.

Farinha, João, "On a Case of Continued Fractions with Complex Elements," *Gaz. Mat. Lisboa* **12** (1951).

Farinha, João, "Une condition de convergence uniforme," *Rev. Fac. Ci. Univ. Coimbra* **23** (1954a), 17–20.

Farinha, João, Sur la convergence de $\phi(a_i/1)$, *Portugal. Math.* **13** (1954b), 145–148.

Favard, J., "Sur les polynomes de Tchebicheff," *C. R. Acad. Sci. Paris* **200** (1935), 2052–2053.

Field, David A., "Series of Stieltjes, Padé Approximants and Continued Fractions," *J. Math. Phys.* **17**, (1976), 843–844.

Field, David A., "Estimates of the Speed of Convergence of Continued Fraction Expansions of Functions," *Math. Comp.* **31** (1977), 495–502.

Field, David A., "Error Bounds for Elliptic Convergence Regions for Continued Fractions," *SIAM J. Numer. Anal.* **15** (1978a), 444–449.

Field, David A., "Error Bounds for Continued Fractions $\mathrm{K}(1/b_n)$," *Numer. Math.* **29** (1978b), 261–267.

Field, David A., and Jones, William B., "A Priori Estimates for Truncation Error of Continued Fractions $\mathrm{K}(1/b_n)$," *Numer. Math.* **19** (1972), 283–302.

Ford, L. R., *Automorphic Functions*, second edition, Chelsea, New York, 1929.

Frank, Evelyn, "Corresponding Type Continued Fractions," *Amer. J. Math.* **68** (1946a), 89–108.

Frank, Evelyn, "On the Zeros of Polynomials with Complex Coefficients," *Bull. Amer. Math. Soc.*, **52** (1946b), 144–157.

Frank, Evelyn, "On the Properties of Certain Continued Fractions," *Proc. Amer. Math. Soc.* **53** (1952), 921–936.

Frank, Evelyn, "A New Class of Continued Fraction Expansions for the Ratios of Heine Functions," *Trans. Amer. Math. Soc.* **88** (1958), 288–300.

Frank, Evelyn, "A New Class of Continued Fraction Expansions for the Ratios of Heine Functions II," *Trans. Amer. Math. Soc.* **95** (1960a), 17–26.

Frank, Evelyn, "A New Class of Continued Fraction Expansions for the Ratios of Heine Functions III," *Trans. Amer. Math. Soc.* **96** (1960b), 312–321.

Frank, Evelyn, and Perron, Oskar, "Remark on a Certain Class of Continued Fractions,"
Proc. Amer. Math. Soc. 5 (1954), 270–283.

Franzen, N. R., "Some Convergence Results for the Padé Approximants," *J. Approximation Theory* 6 (1972), 254–263.

Fried, Burton D., and Conte, Samuel D., *The Plasma Dispersion Function, The Hilbert Transform of the Gaussian*, Academic Press, New York, 1961.

Frobenius, G., "Über Relationen zwischen den Näherungsbrüchen von Potenzreihen," *J. Reine Angew. Math.* 90 (1881), 1–17.

Gallucci, Michael A., and Jones, William B., "Rational Approximation Corresponding to Newton Series (Newton-Padé Approximants)," *J. Approx. Theory* 17 (1976), 366–392.

Galois, E., "Théorème sur les fractions continues périodiques," *Annales de Mathematiques* (Gergonne) 19 (1828/9), 294–; *Oeuvres mathematiques*, Gauthier Villars, Paris, 1951.

Gargantini, I., and Henrici, P., "A Continued Fraction Algorithm for the Computation of Higher Transcendental Functions in the Complex Plane," *Math. Comp.* 21 (1967), 18–29.

Gauss, C. F., "Disquisitiones Generales circa Seriem Infinitum...," in *Commentationes Societatis Regiae Scientiarum Goettingensis Recentiores*, Vol. 2, 1813; *Werke*, Vol. 3, pp. 134–138.

Gauss, C. F., "Methodus Nova Integralium Valores per Approximationem Inveniendi," (1814); *Werke*, Vol. 3, Göttingen, 1876, pp. 165–196.

Gautschi, Walter, "Computational Aspects of Three-Term Recurrence Relations," *SIAM Review* 9 (1967), 24–82.

Gautschi, Walter, "On the Condition of a Matrix Arising in the Numerical Inversion of the Laplace Transform," *Math. Comp.* 23 (1969a), 109–118.

Gautschi, Walter, "An Application of Three-Term Recurrences to Coulomb Wave Functions," *Aequationes Mathematicae* 2 (1969b), 171–176.

Gautschi, Walter, "Efficient Computation of the Complex Error Function," *SIAM J. Numer. Anal.* 7 (1970), 187–198.

Gautschi, Walter, Anomalous Convergence of a Continued Fraction for Ratios of Kummer Functions, *Math. Comp.* 31, No. 140 (October 1977), 994–999.

Gautschi, Walter, and Slavik, Josef, "On the Computation of Modified Bessel Function Ratios," *Math. Comp.* 32, No. 143 (July 1978), 865–875.

Gilewicz, Jacek, *Approximants de Padé*, Lecture Notes in Mathematics No. 667, Springer-Verlag, New York, 1978.

Gill, John, "Infinite Compositions of Möbius Transformations," *Trans. Amer. Math. Soc.* 176 (1973), 479–487.

Gill, John, "The Use of Attractive Fixed Points in Accelerating the Convergence of Limit-Periodic Continued Fractions," *Proc. Amer. Math. Soc.* 47 (1975), 119–126.

Gill, John, "A Generalization of Certain Corresponding Continued Fractions," *Bull. Calcutta Math. Soc.* 69, (1977), 331–340.

Gill, John, "Modifying Factors for Sequences of Linear Fractional Transformations," *Kgl. Norske Vid. Selsk. Skr.* (Trondheim) (1978a), No. 3.

Gill, John, "Enhancing the Convergence Region of a Sequence of Bilinear Transformations," *Math. Scand.* 43, (1978b), 74–80.

Glaisher, J. W. L., "On the Transformation of Continued Products into Continued Fractions," *Proc. Lond. Math. Soc.* 5 (1873/74).

Gragg, W. B., "Truncation Error Bounds for g-Fractions," *Numer. Math.* 11 (1968), 370–379.

Gragg, W. B., "Truncation Error Bounds for π-Fractions," *Bull. Amer. Math. Soc.* **76** (1970), 1091–1094.

Gragg, W. B., "The Padé Table and its Relation to Certain Algorithms of Numerical Analysis," *SIAM Review* **14** (1972), 1–62.

Gragg, W. B., "Matrix Interpretations and Applications of the Continued Fraction Algorithm," *Rocky Mountain J. Math.* **4** (1974), 213–225.

Gragg, W. B., "Laurent, Fourier, and Chebychev-Padé Tables," in E. B. Saff and R. S. Varga (eds.), *Padé and Rational Approximation*, Academic Press, New York, 1977, pp. 61–72.

Gragg, W. B., and Johnson, G. D., "The Laurent-Padé Table," in *Proc. IFIP Congress*, Vol. 3, Information Processing 74, North-Holland, Amsterdam, 1974, pp. 632–637.

Graves-Morris, P. R. (ed.), *Padé Approximants and Their Applications*, Proceedings of a Conference held at the University of Kent, Canterbury, England (July, 1972), Academic Press, New York, 1973.

Graves-Morris, P. R. (ed.), *Padé Approximants*, Institute of Physics, London, 1973.

Graves-Morris, P. R., "Generalizations of the theorem of de Montessus using Canterbury approximants," in E. B. Saff and R. S. Varga (eds.), *Padé and Rational Approximation*, Academic Press, New York, 1977, pp. 73–82.

Grommer, J., "Ganze transcendente Functionen mit lauter reellen Nullstellen," *J. Reine Angew. Math.* **144** (1914), 212–238.

Grundy, R. E., "Laplace Transform Inversion Using Two-Point Rational Approximants," *J. Inst. Maths. Applics.* **20** (1977), 299–306.

Grundy, R. E., "The Solution of Volterra Integral Equations of the Convolution Type Using Two-Point Rational Approximants," *J. Inst. Maths. Applics.* **22** (1978a), 147–158.

Grundy, R. E., "On the Solution of Non-linear Volterra Integral Equations Using Two-Point Padé Approximants," *J. Inst. Maths. Applics.* **22** (1978b), 317–320.

Günther, S., *Darstellung der Näherungswerte von Kettenbrüchen in Independenter Form*, Habilschrift Erlangen, 1872.

Haddad, Hadi M., "Chain Functions, a Generalization of Chain Sequences," *Bull. College Sci.* (Baghdad) **9** (1966), 191–196.

Hag, K., "A Theorem on T-Fractions Corresponding to a Rational Function," *Proc. Amer. Math. Soc.* **25** (1970), 247–253.

Hag, K., "A Convergence Theorem for Limitärperiodisch T-Fractions of Rational Functions," *Proc. Amer. Math. Soc.* **32** (1972), 491–496.

Hamburger, H., "Über eine Erweiterung des Stieltjesschen Momentenproblems," Parts I, II, III, *Math. Ann.* **81** (1920), 235–319; **82** (1921), 120–164, 168–187.

Hamel, G., "Eine charakteristische Eigenschaft beschränkter analytischer Funktionen," *Math. Ann.* **78** (1918a), 257–269.

Hamel, G., "Über einen limitärperiodischen Kettenbruch," *Arch. d. Math. u. Phys.* **27** (1918b), 37–43.

Hayden, T. L., "Some Convergence Regions for Continued Fractions," *Math. Zeitschr.* **79** (1962), 376–380.

Hayden, T. L., "A Convergence Problem for Continued Fractions," *Proc. Amer. Math. Soc.* **14** (1963), 546–552.

Hayden, T. L., "Continued Fraction Approximation to Functions," *Numer. Math.* **7** (1965), 292–309.

Hayden, T. L., "Continued Fractions in Banach Spaces," *Rocky Mountain J. Math.* **4** (1974), 367–370.

Hayden, T. L., and Merkes, E. P., "Chain Sequences and Univalence," *Illinois J. Math.* **8** (1964), 523–528.

Hayden, T. L., and Merkes, E. P., "On Classes of Univalent Continued Fractions," *Proc. Amer. Math. Soc.* **16** (1965), 252–257.

Heilermann, J. B. H., "Über die Verwandlung der Reihen in Kettenbrüche," *J. Reine Angew. Math.* **33** (1846).

Hellinger, E., and Wall, H. S., "Contributions to the Analytic Theory of Continued Fractions and Infinite Matrices," *Ann. of Math.* (2) **44** (1943), 103–127.

Henrici, P., "The Quotient-Difference Algorithm," *Nat. Bur. Standards Appl. Math. Ser.* **49** (1958), 23–46.

Henrici, P., "Bounds for Eigenvalues of Certain Tridiagonal Matrices," *J. Soc. Indust. Appl. Math.* **11** (1963a), 281–290; Errata: **12** (1964), 497.

Henrici, P., "Some Applications of the Quotient-Difference Algorithm," in *Proc. Symposium Appl. Math.*, Vol. 15, Amer. Math. Soc., Providence, R.I., 1963b, 159–183.

Henrici, P., "The Quotient-Difference Algorithm," in A. Ralston and H. S. Wilf (eds.) *Mathematical Methods for Digital Computers, II*, Wiley, 1967, pp. 35–62.

Henrici, P., *Applied and Computational Complex Analysis, Vol. 1, Power Series, Integration Conformal Mapping and Location of Zeros*, Wiley, New York, 1974.

Henrici, P., *Applied and Computational Complex Analysis, Vol. 2, Special Functions, Integral Transforms, Asymptotics and Continued Fractions*, Wiley, New York, 1977.

Henrici, P., and Pfluger, Pia, "Truncation Error Estimates for Stieltjes Fractions," *Numer. Math.* **9** (1966), 120–138.

Henrici, P., and Watkins, Bruce O., "Finding Zeros of a Polynomial by the QD-Algorithm," *Comm. ACM* **8** (1965), 570–574.

Herschel, J. F. W., *Collections of Examples of $\cdots$ Finite Differences*, Cambridge, 1820.

Hildebrand, F. B., *Introduction to Numerical Analysis*, McGraw-Hill, New York, 1956.

Hillam, K. L., and Thron, W. J., "A General Convergence Criterion for Continued Fractions $K(a_n/b_n)$," *Proc. Amer. Math. Soc.* **16** (1965), 1256–1262.

Hille, E., *Analytic Function Theory*, Vol. I, Blaisdell, Boston, 1959.

Hille, E., *Analytic Function Theory*, Vol. II, Ginn, Boston, 1962.

Householder, A. S., *The Numerical Treatment of a Single Nonlinear Equation*, McGraw-Hill, New York, 1970.

Householder, A. S., "The Padé Table, the Frobenius Identities, and the qd Algorithm," *Linear Algebra Appl.* **4** (1971), 161–174.

Hovstad, R. M., "Solution of a Convergence Problem in the Theory of *T*-Fractions," *Proc. Amer. Math. Soc.* **48** (1975), 337–343.

Hurwitz, A., "Über die angenäherte Darstellung der Irrationalzahlen durch rationale Brüche," *Math. Ann.* **39** (1891), 279–284.

Hurwitz, A., "Über die Bedingungen unter welchen eine Gleichung nur Wurzeln mit negativen reellen Teilen besitzt," *Math. Ann.* **46** (1895), 273–284.

Huygens, Christian, *Descriptio Automati Planetarii*, 1698.

Ince, E. Lindsay, "On the Continued Fractions Connected with the Hypergeometric Equation," *Proc. London Math. Soc.* (2) **18** (1919), 236–248.

Isihara, A., and Montroll, E. W., "A Note on the Ground State Energy of an Assembly of Interacting Electrons," *Proc. Nat. Acad. Sci. U.S.A.* **68** (1971), 3111–3115.

Jacobi, C. G. J., "Über Gausz' neue Methode die Werthe der Integrale näherungsweise zu finden," *J. Reine Angew. Math.* 1 (1826), 301–308.

Jacobi, C. G. J., "Über eine besondere Gattung algebraischer Funktionen die aus der Entwicklung der Funktion $(1 - 2xz + z^2)^{-1/2}$ entstehen," *J. Reine Angew. Math.* 2 (1827).

Jefferson, Thomas H., Jr., "Some Additional Properties of T-Fractions," Ph.D. thesis, Univ. of Colorado, Boulder, Colorado, 1969a.

Jefferson, Thomas H., "Truncation Error Estimates for T-Fractions," *SIAM J. Numer. Anal.* 6 (1969b), 359–364.

Jensen, J. L. W. V., "Bitrag til Kaedebrøkernes Teori," *Festskrift til H. G. Zeuthen*, (1909).

Jones, William B., "Multiple point Padé tables," in E. B. Saff and R. S. Varga (eds.), *Padé and Rational Approximation* Academic Press, New York, 1977, pp. 163–171.

Jones, William B., and Magnus, Arne, "Application of Stieltjes Fractions to Birth-Death Processes," in E. B. Saff and R. S. Varga (eds.) *Padé and Rational Approximation*, Academic Press, New York, 1977, pp. 173–179.

Jones, William B., and Magnus, Arne, "Computation of Poles of Two-Point Padé Approximants and their Limits," Journal CAM (1980), 105–119.

Jones, William B., and Snell, R. I., "Truncation Error Bounds for Continued Fractions," *SIAM J. Numer. Anal.* 6 (1969), 210–221.

Jones, William B., and Snell, R. I., "Sequences of Convergence Regions for Continued Fractions $K(a_n/1)$," *Trans. Amer. Math. Soc.* 170 (1972), 483–497.

Jones, William B., and Thron, W. J., "Further Properties of T-Fractions," *Math. Annalen* 166 (1966), 106–118.

Jones, William B., and Thron, W. J., "Convergence of Continued Fractions," *Canad. J. Math.* 20 (1968), 1037–1055.

Jones, William B., and Thron, W. J., "Twin-Convergence Regions for Continued Fractions $K(a_n/1)$," *Trans. Amer. Math. Soc.* 150 (1970), 93–119.

Jones, William B., and Thron, W. J., "A Posteriori Bounds for the Truncation Error of Continued Fractions," *SIAM J. Numer. Anal.* 8 (1971), 693–705.

Jones, William B., and Thron, W. J., "Numerical Stability in Evaluating Continued Fractions," *Math. Comp.* 28 (1974a), 795–810.

Jones, William B., and Thron, W. J., Rounding error in evaluating continued fractions, *Proceedings of the ACM*, San Diego (1974b), 11–19.

Jones, William B., and Thron, W. J. (ed.), "Proceedings of the International Conference on Padé Approximants, Continued Fractions and Related Topics," *Rocky Mountain J. Math.* 4 (1974c), 135–397.

Jones, William B., and Thron, W. J., "On Convergence of Padé Approximants," *SIAM J. Math. Anal.* 6 (1975), 9–16.

Jones, William B., and Thron, W. J., "Truncation Error Analysis by Means of Approximant Systems and Inclusion Regions," *Numer. Math.* 26 (1976), 117–154.

Jones, William B., and Thron, W. J., "Two-Point Padé Tables and T-Fractions," *Bull. Amer. Math. Soc.* 83 (1977), 388–390.

Jones, William B., and Thron, W. J., "Sequences of Meromorphic Functions Corresponding to a Formal Laurent Series," *SIAM J. Math. Anal.* 10 (1979), 1–17.

Jones, William B., Thron, W. J., and Waadeland, H., "A Strong Stieltjes Moment Problem," *Trans. Amer. Math. Soc.* (1980).

Jordan, J. Q., and Leighton, W., "On the Permutation of the Convergents of a Continued Fraction and Related Convergence Criteria," *Ann. of Math.* **39** (1938), 872–882.

Kahl, E., "Über einen Kettenbruch von zweigliedriger Periode," *Archiv für Math. und Phys.* **19** (1852).

Karlin, S., *A First Course in Stochastic Processes*, Academic Press, New York, 1969.

Karlin, S., and McGregor, J. L., The differential equations of birth-and-death processes, and the Stieltjes moment problem, *Trans. Amer. Math. Soc.* **85** (1957), 489–546.

Karlsson, J., Rational interpolation and best rational approximation, *J. Math. Anal. Appl.* **53** (1976), 38–51.

Karlsson, J., and Wallin, H., "Rational Approximation by an Interpolation Procedure in Several Variables," in E. B. Saff and R. S. Varga (eds.), *Padé and Rational Approximation*, Academic Press, New York, 1977, pp. 83–100.

Khintchine, A., "Einige Sätze über Kettenbrüche, mit Anwendungen auf die Theorie der Diophantischen Approximationen," *Math. Ann.* **92** (1924), 115–125.

Khintchine, A., "Zur metrischen Kettenbruchtheorie," *Comp. Math.* **3** (1936), 276–285.

Khintchine, A. Ya., Continued Fractions (translated by Peter Wynn), P. Noordhoff, Groningen, The Netherlands, 1963.

Khovanskii, A. N., *The Application of Continued Fractions and Their Generalizations to Problems in Approximation Theory* (translated by Peter Wynn), P. Noordhoff, Groningen, The Netherlands, 1963.

Lagrange, J. L., *Oeuvres* (J. A. Serret, ed.) Gauthier Villars, Paris, 1867.

Laguerre, E., *Oeuvres* (C. Hermite, H. Poincaré, E. Rouché, eds.) Vol. 1, Paris, 1898.

Lambert, J. H., "Memoire sur quelques propriétés remarquables des quantités transcendantes circulaires et logarithmiques," *Memoires de l'Acad. de Berlin*, Année 1761 (1768), 265–322.

Landsberg, G., "Zur Theorie der periodischen Kettenbrüche," *J. Reine Angew. Math.* **109** (1892), 231–237.

Lane, R. E. "The Convergence and Values of Periodic Continued Fractions," *Bull. Amer. Math. Soc.* **51** (1945), 246–250.

Lane, R. E., and Wall, H. S., "Continued Fractions with Absolutely Convergent Even and Odd parts," *Trans. Amer. Math. Soc.* **67** (1949), 368–380.

Lange, L. J., "Divergence, Convergence, and Speed of Convergence of Continued Fractions $1 + \mathrm{K}(a_n/1)$," Doctoral Thesis, Univ. of Colorado, Boulder, 1960.

Lange, L. J., "On a Family of Twin Convergence Regions for Continued Fractions," *Illinois J. Math.* **10** (1966), 97–108.

Lange, L. J., and Thron, W. J., "A Two Parameter Family of Best Twin Convergence Regions for Continued Fractions," *Math. Zeitschr.* **73** (1960), 295–311.

Larkin, F. W., "Some Techniques for Rational Interpolation," *Computer J.* **10** (1967), 178–187.

Legendre, I. M., *Traites des fonctions elliptiques et des intégrales Euleriennes*, Vol. 2, Paris, 1826.

Lehner, J., *A Short Course in Automorphic Functions*, Holt, Rinehart, and Winston, New York, 1966.

Leighton. W., "Sufficient Conditions for the Convergence of a Continued Fraction," *Duke Math. J.* **4** (1938), 775–778.

Leighton, W., "Convergence Theorems for Continued Fractions," *Duke Math. J.* **5** (1939), 298–308.

Leighton, W., and Scott, W. T., "A General Continued Fraction Expansion," *Bull. Amer. Math. Soc.* **45** (1939), 596–605.

Leighton, W., and Thron, W. J., "Continued Fractions with Complex Elements," *Duke Math. J.* **9** (1942), 763–772.

Leighton, W., and Wall, H. S., "On the Transformation and Convergence of Continued Fractions," *Amer. J. Math.* **58** (1936), 267–281.

Levinson, N., and Redheffer, R. M., *Complex Variables*, Holden-Day, San Francisco, 1970.

Liouville, J., "Sur des classes très-étendues de quantités dont la valeur n'est ni algébrique, ni même reductible à des irrationnelles algébriques," *J. Math. Pures Appl.* (1) **16** (1851), 133–142.

Luke, Yudell L., "The Padé Table and the τ-Method," *J. Math. Phys.* **37** (1958), 110–127.

Luke, Yudell L., *The Special Functions and Their Approximations*, Vols. I, II, Academic Press, New York, 1969.

Macon, N., and Baskervill, M., "On the Generation of Errors in the Digital Evaluation of Continued Fractions," *J. Assoc. Comp. Mach.* **3** (1956), 199–202.

Magnus, Arne, "Certain Continued Fractions Associated with the Padé Table," *Math. Z.* **78** (1962a), 361–374.

Magnus, Arne, "Expansion of Power Series into P-Fractions," *Math. Z.* **80** (1962b), 209–216.

Magnus, Arne, "On P-Expansions of Power Series," *Norske Vid. Selsk, Skr.* (Trondheim) (1964), No. 3, 1–14.

Magnus, Arne, "P-Fractions and the Padé Table," *Rocky Mountain J. Math.* **4** (1974), 257–259.

Mall, J., "Ein Satz über die Konvergenz von Kettenbrüchen," *Math. Z.* **45** (1939).

Mandell, M., and Magnus, Arne, "On Convergence of Sequences of Linear Fractional Transformations," *Math. Z.* **115** (1970), 11–17.

Markoff, A., "Deux demonstrations de la convergence de certaines fractions continues," *Acta Math.* **19** (1895).

Maurer, G. V., "On the Expansion into Continued Fractions of Some Limiting Cases of Heine Functions," *Volzskii Matematicheskii Sbornik* **5** (1966), 211–221.

McCabe, J. H., "A Continued Fraction Expansion, with a Truncation Error Estimate for Dawson's Integral," *Math. Comp.* **28**, No. 127 (July 1974), 811–816.

McCabe, J. H., "A Formal Extension of the Padé Table to Include Two Point Padé Quotients," *J. Inst. Maths. Applics.* **15** (1975), 363–372.

McCabe, J. H., and Murphy, J. A., "Continued Fractions Which Correspond to Power Series Expansions at Two Points," *J. Inst. Maths. Applics.* **17** (1976), 233–247.

Merkes, E. P., "Bounded J-Fractions and Univalence," *Michigan Math. J.* **6** (1959), 395–400.

Merkes, E. P., "On Truncation Errors for Continued Fraction Computations," *SIAM J. Numer. Anal.* **3**, No. 3 (1966), 486–496.

Merkes, E. P., and Scott, W. T., "Periodic and Reverse Periodic Continued Fractions," *Michigan Math. J.* **7** (1960), 23–29.

Miller, S. C., "Continued Fraction Solutions of the One-Dimensional Schrödinger Equation," *Phys. Rev. D* **12**, No. 12 (15 December 1975), 3838–3842.

de Montessus de Ballore, R., "Sur les fractions continues algébriques," *Bull. Soc. Math. France* **30** (1902), 28–36.

Murphy, J. A., "Certain Rational Function Approximations to $(1+x^2)^{-1/2}$," *J. Inst. Maths. Applics.* **7** (1971), 138–150.

Murphy, J. A., and O'Donohoe, M. R., "Some Properties of Continued Fractions with Applications in Markov Processes," *J. Inst. Math. Applics.* **16** (1975), 57–71.

Murphy, J. A., and O'Donohoe, M. R., "A Continued Fraction Method for Obtaining Approximations to Hypergeometric Functions," Numerical Analysis Report No. 15, April 1976, Univ. of Manchester, Manchester M13 9P1, England.

Murphy, J. A., and O'Donohoe, M. R., "A Class of Algorithms for Obtaining Rational Approximants to Functions Which are Defined by Power Series," *J. Appl. Math. and Phys.* (*ZAMP*) **28** (1977), 1121–1131.

Nörlund, Niels E., *Vorlesungen über Differenzenrechnung*, Springer-Verlag, OHG, Berlin, 1924.

Padé, H., "Sur la représentation approchée d'une fonction par des fractions rationelles," Thesis, *Ann. Ecole Normal* (3), **9** (1892), 1–93, supplement.

Paydon, J. F., and Wall, H. S., "The Continued Fraction as a Sequence of Linear Transformations," *Duke Math. J.* **9** (1942), 360–372.

Perron, O., "Über die Konvergenz periodischer Kettenbrüche," *S.-B. Bayer Akad. Wiss. Math. -Nat. Kl.* **35** (1905).

Perron, O., "Erweiterung eines Markoffschen Satzes über die Konvergenz gewisser Kettenbrüche," *Math. Ann.* **74** (1913).

Perron, O., *Die Lehre von den Kettenbrüchen*, Chelsea, New York, 1929.

Perron, O., "Über eine Formel von Ramanujan," *S.-B. Bayer. Akad. Wiss. Math.-Nat. Kl.* (1952), 197–213.

Perron, O., "Über die Preece'schen Kettenbrüche," *S.-B. Bayer. Akad. Wiss. Math.-Nat. Kl.* (1953), 21–56.

Perron, O., *Die Lehre von den Kettenbrüchen*, Band I, Teubner, Stuttgart, 1954.

Perron, O., "Über eine Schlichtheitsschranke von James S. Thale," *S.-B. Bayer. Akad. Wiss. Math.-Nat. kl.* (1956), 233–236.

Perron, O., *Die Lehre von den Kettenbrüchen*, Band II, Teubner, Stuttgart, 1957a.

Perron, O., "Über zwei Kettenbrüche von H. S. Wall," *S.-B. Bayer Akad Wiss. Math.-Nat. Kl.* (1957b), No. 1, 1–13.

Perron, O., "Über einen Kettenbruch von Ramanujan," *S.-B. Bayer. Akad. Wiss. Math-Nat. Kl.* (1958a), 19–23.

Perron, O., "Über zwei ausgeartete Heinesche Reihen und einen Kettenbruch von Ramanujan," *Math. Z.* **70** (1958b), 245–249.

Phipps, Thomas E., Jr., "A Continued Fraction Representation of Eigenvalues," *SIAM Rev.* **13** (1971), 390–395.

Pincherle, S., "Delle funzioni ipergeometriche e di varie questioni ad esse attinenti," *Giorn. Mat. Battaglini* **32** (1894), 209–291. (Also in *Opere Selecte*, Vol. 1, pp. 273–357.)

Pincherle, S., *Rend. Acad. dei Lincei* **4** (1889), 640.

Piranian, G., and Thron, W. J., "Convergence Properties of Sequences of Linear Fractional Transformations," *Michigan Math. J.* **4** (1957), 129–135.

Poincaré, H., "Sur les intégrales irrégulières des équations linéares," *Acta Math.* **8** (1886), 295–344.

Possé, M. C., *Sur quelques applications des fractions continues*, St. Petersbourg, 1886.

Preece, C. T., "Theorems Stated by Ramanujan (VI): Theorems on Continued Fractions," *J. Lond. Math. Soc.* **4** (1929), 34–39.

Preece, C. T., "Theorems Stated by Ramanujan (X)," *J. Lond. Math. Soc.* **6** (1930).

Pringsheim, A., "Über die Konvergenz unendlicher Kettenbrüche," *S.-B. Bayer. Akad. Wiss. Math.-Nat. Kl.* **28** (1899), 295–324.

Pringsheim, A., "Über die Konvergenz periodischer Kettenbrüche," *S.-B. Bayer. Akad. Wiss. Math.-Nat. Kl.* (1900), 463–488.

Pringsheim, A., "Über Konvergenz und funktionentheoretischen Charakter gewisser limitärperiodischer Kettenbrüche," *S.-B. Bayer. Akad. Wiss. Math.-Nat. Kl.* (1910).

Ramanujan, S., *Collected Papers*, Cambridge, 1927.

Reid, Walter M., "Uniform Convergence and Truncation Error Estimates of Continued Fractions $K(a_n/1)$," Ph.D. Thesis, Univ. of Colorado, Boulder, Colorado 80309, 1978.

Riemann, B., *Gesammelte Mathematische Werke*, 2nd ed., (H. Weber, ed.), 1892.

Ritt, J. F., "On the Derivatives of a Function at a Point," *Ann. Math.* **18** (1916), 18–23.

Roach, F. A., "Diophantine Approximation in a Vector Space," *Rocky Mountain J. Math.* **4** (1974), 379–382.

Roach, F. A., "Boundedness of Value Regions and Convergence of Continued Fractions," *Proc. Amer. Math. Soc.* **62** (1977), 299–304.

Rouché, E., "Mémoire sur le dévelopement des fonctions en séries ordonnéés suivant les denominateurs des réduites d'une fraction continue," *J. École Polytechnique Paris* **37** (1858).

Rutishauser, H., "Der Quotienten-Differenzen-Algorithmus," *Z. Angew. Math. Phys.* **5** (1954a), 233–251.

Rutishauser, H., "Ein infinitesimales Analogon zum Quotienten-Differenzen-Algorithmus," *Arch. Math.* **5** (1954b), 132–137.

Rutishauser, H., "Anwendungen des Quotienten-Differenzen-Algorithmus," *Z. Angew. Math. Phys.* **5** (1954c), 496–508.

Rutishauser, H., "Bestimmung der Eigenwerte und Eigenvektoren einer Matrix mit Hilfe des Quotienten-Differenzen-Algorithmus," *Z. Angew. Math. Phys.* **6** (1955), 387–401.

Rutishauser, H., "Eine Formel von Wronski und ihre Bedeutung für den Quotienten-Differenzen-Algorithmus," *Z. Angew. Math. Phys.* **7** (1956), 164–169.

Rutishauser, H., "Der Quotienten-Differenzen-Algorithmus," *Mitt. Inst. Angew. Math. Zürich* **7** (1957), 74 pp.

Rutishauser, H., "On a modification of the QD-algorithm with Graeffe-type convergence," *Z. Angew. Math. Phys.* **13** (1962), 493–496.

Rutishauser, H., "Stabile Sonderfälle des Quotienten-Differenzen Algorithmus," *Numer. Math.* **5** (1963), 95–112.

Sack, R. A., and Donovan, A. F., "An Algorithm for Gaussian Quadrature Given Modified Moments," *Numer. Math* **18** (1972), 465–478.

Saff, E. B., "An Extension of Montessus de Ballore's Theorem on the Convergence of Interpolating Rational Functions," *J. Approx. Theory* **6** (1972), 63–67.

Saff, E. B., and Varga, R. S. (eds.), *Padé and Rational Approximation Theory and Application*, Academic Press, New York, 1977.

Schlömilch, O., "Über eine Kettenbruchentwicklung für unvollständige Gammafunktionen," *Zeitschr. Math. Phys.* **16** (1871).

Schoenberg, I. J., "Some Analytic Aspects of the Problem of Smoothing," in *Studies and Essays presented to R. Courant on his 60th Birthday, Jan. 8, 1948*, Interscience, New York, 1948, pp. 351–370.

Schur, I., "Über Potenzreihen, die im Innern des Einheitskreises beschränkt sind," *J. Reine Angew. Math.* **147** (1917), **148** (1918).

Schwenter, Daniel, *Deliciae Physico-Mathematicae*, Nürnberg, 1636.

Schwerdtfeger, H., "Möbius Transformations and Continued Fractions," *Bull. Amer. Math. Soc.* **52** (1946), 307–309.

Scott, W. T., and Wall, H. S., "A Convergence Theorem for Continued Fractions, *Trans. Amer. Math. Soc.* **47** (1940a), 155–172.

Scott, W. T., and Wall, H. S., "Continued Fraction Expansions for Arbitrary Power Series," *Ann. Math.* (2) **41** (1940b), 328–349.

Scott, W. T., and Wall, H. S., "On the Convergence and Divergence of Continued Fractions," *Amer. J. Math.* **69** (1947), 551–561.

Seidel, L., *Untersuchungen über die Konvergenz and Divergenz der Kettenbrüche*, Habilschrift München, 1846.

Shanks, D., "Nonlinear Transformations of Divergent and Slowly Convergent Sequences," *J. Mathematical Phys.* **34** (1955), 1–42.

Sheng, P., "Application of Two-Point Padé Approximants to Some Solid State Problems," *Rocky Mountain J. Math.* **4** (1974), 385–386.

Sheng, P., and Dow, J. D., "Intermediate Coupling Theory: Padé Approximants for Polarons," *Phys. Rev. B* **4** (1971), 1343–1359.

Shohat, J. A., and Tamarkin, J. D., *The Problem of Moments*, Mathematical Surveys No. 1, Amer. Math. Soc., Providence, R.I., 1943.

Singh, D., and Thron, W. J., "A Family of Best Twin Convergence Regions for Continued Fractions," *Proc. Amer. Math. Soc.* **7** (1956a), 277–282.

Singh, D., and Thron, W. J., "On the Number of Singular Points, Located on the Unit Circle of Certain Functions Represented by *C*-Fractions," *Pacific J. Math.* **6** (1956b), 135–143.

Sobhy, M. I., "Applications of Padé Approximants in Electrical Network Problems," in P. R. Graves-Morris (ed.), *Padé Approximants and their Applications* Academic Press, New York, 1973, pp. 321–336.

Stegun, Irene A., and Zucker, Ruth, "Automatic Computing Methods for Special Functions," *J. Research, National Bureau of Standards—B. Mathematical Sciences* **74B**, No. 3 (July–Sept. 1970), 211–224.

Stern, M. A., "Theorie der Kettenbrüche und ihre Anwendung," *J. Reine Angew. Math.* **10, 11** (1832).

Stern, M. A., "Über die Kennzeichen der Konvergenz eines Kettenbruchs," *J. Reine Angew. Math.* **37** (1848).

Stern, M. A., *Lehrbuch der Algebraischen Analysis*, Leipzig, 1860.

Stieltjes, T. J., "Quelques recherches sur la théorie des quadratures dites mecaniques," *Ann. Sci. Ec. Norm. Paris* (3) **1** (1884), 409–426.

Stieltjes, T. J., "Recherches sur quelques séries semi-convergentes," *Ann. Sci. Éc. Norm. Paris* (3) **3** (1886), 201–258.

Stieltjes, T. J., "Recherches sur les fractions continues," *Ann. Fac. Sci. Toulouse* **8** (1894), J, 1–122; **9** (1894), A, 1–47; *Oeuvres* **2**, 402–566. Also published in *Memoires Présentés par divers savants à l'Académie de sciences de l'Institut National de France* **33**, 1–196.

Stoer, J., "Über zwei Algorithmen zur Interpolation mit rationalen Funktionen," *Numer. Math.* **3** (1961), 285–304.

Stolz, O., "Vorlesungen über allgemeine Arithmetik," Teubner, Leipzig, 1886.

Sweezy, W. B., and Thron, W. J., "Estimates of the Speed of Convergence of Certain Continued Fractions," *SIAM J. Numer. Anal.* **4**, No. 2 (1967), 254–270.

Szasz, O., *Collected Mathematical Papers*, (H. D. Lipsich, ed.), Cincinnati, 1955.

Szegö, G., *Orthogonal Polynomials*, Colloquium Publications, Vol. 23, Amer. Math. Soc., New York, 1959.

Szegö, G., "An Outline of the History of Orthogonal Polynomials," in Deborah Tepper Haimo (ed.), *Orthogonal Expansions and Their Continuous Analogues* Southern Illinois U. P., Feffer and Simon, London, 1968, pp. 3–11.

Tannery, Jules, "Sur les integrales euleriennes," *C. R. Acad. Sci. Paris* **94** (1882), 1698–1701.

Tchebycheff, P. L., "Sur les fractions continues," *J. Math. Pures Appl. Ser. II* **3** (1858), 289–323.

Tchebycheff, P. L., *Oeuvres*, (A. Markoff and N. Sonin, eds.), St. Petersbourg, Vol. 1, 1899; Vol. 2, 1907.

Thacher, Henry C., Jr., "Computation of the Complex Error Function by Continued Fractions," in *Blanche Anniversary Volume*, Aerospace Research Laboratory, Wright-Patterson Air Force Base, Ohio, AD 657 323, February 1967, pp. 315–337.

Thacher, Henry C., Jr., "New Backward Recurrences for Bessel Functions," Technical Report No. 38-77, Univ. of Kentucky, Lexington, Kentucky 40506.

Thacher, Henry C., Jr., and Tukey, J., "Recursive Algorithm for Interpolation by Rational Functions," unpublished manuscript, 1960.

Thale, James S., "Univalence of Continued Fractions and Stieltjes Transforms," *Proc. Amer. Math. Soc.* **7** (1956), 232–244.

Thiele, T. N., "Bemaerkninger om periodiske Kjaedebrøkers Konvergens," *Tidsskrift for Mathematik* (4) **3** (1879).

Thiele, T. N., *Interpolationsrechnung*, Teubner, Leipzig, 1909.

Thomé, L. W., "Über die Kettenbruch Entwicklung der Gauszschen Quotienten $F(\alpha, \beta+1, \gamma+1; x)/F(\alpha, \beta, \gamma; x)$," *J. Reine Angew. Math.* **67** (1867), 299–309.

Thron, W. J., "Two Families of Twin Convergence Regions for Continued Fractions," *Duke Math. J.* **10** (1943a), 677–685.

Thron, W. J., "Convergence Regions for the General Continued Fraction," *Bull. Amer. Math. Soc.* **49** (1943b), 913–916.

Thron, W. J., "Convergence Regions for Continued Fractions," Ph.D. Thesis, Rice Institute, Houston, Texas, 1943c.

Thron, W. J., "Twin Convergence Regions for Continued Fractions $b_0 + \mathrm{K}(1/b_n)$," *Amer. J. Math.* **66** (1944a), 428–438.

Thron, W. J., "A Family of Simple Convergence Regions for Continued Fractions," *Duke Math. J.* **11** (1944b), 779–791.

Thron, W. J., "Some Properties of Continued Fraction $1 + d_0 z + \mathrm{K}\left(\dfrac{z}{1+d_n z}\right)$," *Bull. Amer. Math. Soc.* **54** (1948), 206–218.

Thron, W. J., "Twin Convergence Regions for Continued Fractions $b_0 + \mathrm{K}(1/b_n)$, II," *Amer. J. Math.* **71** (1949), 112–120.

Thron, W. J., "Singular Points of Functions Defined by C Fractions," *Proc. Nat. Acad. Sci. U.S.A.* **36** (1950), 51–54.

Thron, W. J., *Introduction to the Theory of Functions of a Complex Variable*, Wiley, New York, 1953.

Thron, W. J., "On Parabolic Convergence Regions for Continued Fractions," *Math. Zeitschr.* **69** (1958), 173–182.

Thron, W. J., "Zwillingskonvergenzgebiete für Kettenbrüche $1 + \mathrm{K}(a_n/1)$, deren eines die Kreisscheibe $|a_{2n-1}| < \rho^2$ ist," *Math. Zeitschr.* **70** (1959), 310–344.

Thron, W. J., "Convergence of Sequences of Linear Fractional Transformations and of Continued Fractions," *J. Indian. Math. Soc.* **27** (1963), 103–127.

Thron, W. J., "On the Convergence of the Even Part of Certain Continued Fractions," *Math. Zeitschr.* **85** (1964), 268–273.

Thron, W. J., "Two-point Padé Tables, *T*-Fractions and Sequences of Schur," in E. B. Saff and R. S. Varga (eds.), *Padé and Rational Approximation*, Academic Press, New York, 1977, pp. 215–226.

Thron, W. J., and Waadeland, H., "Accelerating Convergence of Limit Periodic Continued Fractions $K(a_n/1)$," *Numer. Math.* **34** (1980), 155–170.

Thron, W. J., and Waadeland, H., "Analytic Continuation of Functions Defined by Means of Continued Fractions," *Math. Scand.* (to appear).

Van Vleck, E. B., "On the Convergence of Continued Fractions with Complex Elements," *Trans. Amer. Math. Soc.* **2** (1901a), 215–233.

Van Vleck, E. B., "On the Convergence and Character of the Continued Fraction...," *Trans. Amer. Math. Soc.* **2** (1901b), 476–483.

Van Vleck, E. B., "On the Convergence of Algebraic Continued Fractions Whose Coefficients Have Limiting Values," *Trans. Amer. Math. Soc.* **5** (1904), 253–262.

Van Vleck, E. B., "Selected Topics in the Theory of Divergent Series and of Continued Fractions," in *The Boston Colloquium*, Macmillan, New York, 1905, pp. 75–187.

Waadeland, H., "On *T*-Fractions of Functions Holomorphic and Bounded in a Circular Disk," *Norske Vid. Selsk. Skr.* (Trondheim) (1964), No. 8, 1–19.

Waadeland, H., "A Convergence Property of Certain *T*-Fraction Expansions," *Norske Vid. Selsk. Skr.* (Trondheim) (1966), No. 9, 1–22.

Waadeland, Haakon, "On *T*-Fractions of Certain Functions with a First Order Pole at the Point of Infinity," *Kgl. Norske Vid. Selsk. Forh. Bind* **40** (1967), No. 1, 1–6.

Waadeland, H., *Some Properties of General T-Fractions*, Universitetet i Trondheim, Matematisk Institutt, NLHT, 7000 Trondheim, 1978.

Waadeland, H., "On General *T*-Fractions Corresponding to Functions Satisfying Certain Boundedness Properties," *J. Approx. Theory* **26** (1979a), 317–328.

Waadeland, H., "On Limit-Periodic General *T*-Fractions and Holomorphic Functions," *J. Approx. Theory* **27** (1979b), 329–345.

Wall, H. S., "On Continued Fractions and Cross-Ratio Groups of Cremona Transformations," *Bull. Amer. Math. Soc.* **40** (1934), 587–592.

Wall, H. S., "Continued Fractions and Totally Monotone Sequences," *Trans. Amer. Math. Soc.* **48** (1940), 165–184.

Wall, H. S., "The Behavior of Certain Stieltjes Continued Fractions near the Singular Line," *Bull. Amer. Math. Soc.* **48** (1942), 427–431.

Wall, H. S., "Polynomials Whose Zeros Have Negative Real Parts," *Amer. Math. Monthly* **52** (1945), 308–322.

Wall, H. S., *Analytic Theory of Continued Fractions*, Van Nostrand, New York, 1948.

Wall, H. S., "Partially Bounded Continued Fractions," *Proc. Amer. Math. Soc.* **7** (1956), 1090–93.

Wall, H. S., "Some Convergence Problems for Continued Fractions," *Amer. Math. Monthly* **54** (1957), 95–103.

Wall, H. S., and Wetzel, Marion, "Quadratic Forms and Convergence Regions for Continued Fractions," *Duke Math. J.* **11** (1944a), 89–102.

Wall, H. S., and Wetzel, Marion, "Contributions to the Analytic Theory of J-Fractions," *Trans. Amer. Math. Soc.* **55** (1944b), 373–397.

Wallis, J., *Tractatus de algebra*, 1685.

Warner, D. D., "Hermite Interpolation with Rational Functions," Ph.D. Thesis, Univ. of California, San Diego, 1974.

Warner, D. D., "An Extension of Saff's Theorem on the Convergence of Interpolating Rational Functions," *J. Approximation Theory* **18** (1976), 108–118.

Watson, G. N., "Theorems Stated by Ramanujan VII, IX," *J. Lond. Math. Soc.* **4** (1929), 39–48, 231–237.

Watson, G. N., *A Treatise on the Theory of Bessel Functions*, second edition, Cambridge at the University Press, 1952.

Wheeler, J. C., "Gaussian Quadrature and Modified Moments," *Rocky Mtn. J. Math.* **4**, No. 2 (Spring 1974), 287–296.

Widder, David Vernon, *The Laplace Transform*, Princeton U. P., Princeton, 1946.

Widder, D. V., "Homogeneous Solutions of the Heat Equation," in R. P. Gilbert and R. G. Newton (eds.), *Analytic Methods in Mathematical Physics*, Gordon and Breach, 1968.

Worpitsky, J., "Untersuchungen über die Entwickelung der monodromen und monogenen Funktionen durch Kettenbrüche," in *Friedrichs-Gymnasium und Realschule Jahresbericht*, Berlin, 1865, pp. 3–39.

Wuytack, L., "An Algorithm for Rational Interpolation Similar to the qd-Algorithm," *Numer. Math.* **20** (1973), 418–424.

Wuytack, L., "On the Osculatory Rational Interpolation Problem," *Math. Comp.* **29** (1975), 837–843.

Wuytack, L., (ed.) *Padé Approximation and its Applications*, Lecture Notes in Mathematics **765**, Springer-Verlag, New York (1979).

Wynn, P., "On a Device for Computing the $e_m(S_n)$ Transformation," *Math. Tables Aids Comput.* **10**, (1956), 91–96.

Wynn, P., "Converging Factors for Continued Fractions," *Numer. Math.* **1** (1959), 272–320.

Wynn, P., "The Rational Approximation of Functions Which are Formally Defined by a Power Series Expansion," *Math. Comp.* **14** (1960), 147–186.

Wynn, P., "Continued Fractions Whose Coefficients Obey a Non-commutative Law of Multiplication," *Arch. Rational Mech. Anal.* **12** (1963), 273–312.

Wynn, P., "On Some Recent Developments in the Theory and Application of Continued Fractions," *J. SIAM Numer. Anal.*, Ser. B, **1** (1964), 177–197.

Author Index

Numbers set in *italics* indicate pages on which complete literature citations are given.

Subject Index